Methods in Enzymology

Volume 61
ENZYME STRUCTURE
Part H

METHODS IN ENZYMOLOGY

EDITORS-IN-CHIEF

Sidney P. Colowick   Nathan O. Kaplan

Methods in Enzymology

Volume 61

Enzyme Structure

Part H

EDITED BY

C. H. W. Hirs

DEPARTMENTS OF BIOCHEMISTRY,
BIOPHYSICS, AND GENETICS
UNIVERSITY OF COLORADO MEDICAL SCHOOL
DENVER, COLORADO

Serge N. Timasheff

GRADUATE DEPARTMENT OF BIOCHEMISTRY
BRANDEIS UNIVERSITY
WALTHAM, MASSACHUSETTS

ACADEMIC PRESS   New York   San Francisco   London   1979
A Subsidiary of Harcourt Brace Jovanovich, Publishers

ACADEMIC PRESS, INC.
111 Fifth Avenue, New York, New York 10003

United Kingdom Edition published by
ACADEMIC PRESS, INC. (LONDON) LTD.
24/28 Oval Road, London NW1 7DX

Library of Congress Cataloging in Publication Data

Main entry under title:

Enzyme structure.

    (Methods in enzymology, v. 11, 25–27, 47–   )
    Part H–   edited by C. H. W. Hirs and S. N. Timasheff.
    Includes bibliographical references.
    1. Enzymes–Analysis. I. Hirs, Christophe Henri
Werner,   ed. II. Timasheff, Serge N.
III. Series: Methods in enzymology, v. 11 [etc.]
QP601.M49 vol. 11, etc. 574.1'925'08s [547'.758]
ISBN 0–12–181961–2 (v. 61)   79–26910

PRINTED IN THE UNITED STATES OF AMERICA

79 80 81 82   9 8 7 6 5 4 3 2 1

Table of Contents

Section I. Molecular Weight Determinations and Related Procedures

Section II. Interactions

Section III. Conformation and Transitions

Contributors to Volume 61, Part H

Article numbers are in parentheses following the names of contributors.
Affiliations listed are current.

GARY K. ACKERS (9), *Department of Biology and McCollum-Pratt Institute, The Johns Hopkins University, Baltimore, Maryland 21218*

ADAM ALLERHAND (19), *Department of Chemistry, Indiana University, Bloomington, Indiana 47401*

D. S. AULD (15), *Biophysics Research Laboratory, Department of Biological Chemistry, Harvard Medical School, Boston, Massachusetts 02115*

MUGUREL G. BADEA (17), *Mergenthaler Laboratory for Biology, The Johns Hopkins University, Baltimore, Maryland 21218*

RODNEY L. BILTONEN (13, 14), *Departments of Biochemistry and Pharmacology, University of Virginia School of Medicine, Charlottesville, Virginia 22903*

LUDWIG BRAND (17), *Mergenthaler Laboratory for Biology, The Johns Hopkins University, Baltimore, Maryland 21218*

JOHN R. CANN (10), *Departments of Biophysics and Genetics, University of Colorado Medical Center, Denver, Colorado 80262*

CHRISTOPH DE HAËN (8), *Division of Metabolism, Endocrinology, and Gerontology, Department of Medicine, University of Washington, Seattle, Washington 98195*

JOHN P. ELDER (2), *Mettler Instrument Corporation, Princeton-Hightstown Road, Hightstown, New Jersey 08520*

KUNIHIKO GEKKO (3), *Department of Food Science and Technology, Faculty of Agriculture, Nagoya University, Furo-cho, Chikusa-ku, Nagova, Japan*

OTTO GLATTER (11), *Institut für Rontgenfeinstrukturforschung der Osterreichischen Akademie der Wissenschaften, und des Forschungszentrums Graz, A-8010 Graz, Steyregasse 17/V1 Austria*

PETER C. KAHN (16), *Departments of Biochemistry and Microbiology, 327 Lipman Hall, Rutgers University, New Brunswick, New Jersey 08903*

OTTO KRATKY (11), *Institut für Rontgenfeinstrukturforschung der Osterreichischen Akademie der Wissenschaften, und des Forschungezentrums Graz A-8010 Graz, Steyregasse 17/V1 Austria*

JAMES A. LAKE (12), *Institute of Molecular Biology and Department of Biology, University of California, Los Angeles, Los Angeles, California 90024*

NEAL LANGERMAN (13, 14), *Departments of Chemistry and Biochemistry, Utah State University, Logan, Utah 84321*

JAMES C. LEE (3, 4), *Department of Biochemistry, St. Louis University School of Medicine, St. Louis, Missouri 63104*

RUFUS LUMRY (6), *Department of Chemistry, University of Minnesota Minneapolis, Minnesota 55455*

WARNER L. PETICOLAS (18), *Department of Chemistry, University of Oregon, Eugene, Oregon 97403*

INGRID PILZ (11), *Institut für physikalische Chemie der Universitut A-8010 Graz, Austria*

JACQUELINE A. REYNOLDS (5), *Whitehead Medical Research Institute, Department of Biochemistry, Duke University Medical Center, Durham, North Carolina 27710*

ERIC SWANSON (8), *Department of Biochemistry, University of Washington, Seattle, Washington 98195*

DAVID C. TELLER (8), *Department of Biochemistry, University of Washington, Seattle, Washington 98195*

SERGE N. TIMASHEFF (3,4), *Graduate Department of Biochemistry, Brandeis University, Waltham, Massachusetts 02154*

ROLAND VALDES, JR. (9), *Department of Pathology, Clinical Chemistry, and Toxicology, University of Virginia Medical Center, Box 168, Charlottesville, Virginia 22908*

ROBLEY C. WILLIAMS, JR. (7), *Department*

of Molecular Biology, Vanderbilt University, Nashville, Tennessee 37235

ZACHARY YIM (6), *Department of Chemistry University of Wisconsin, Madison, Wisconsin 53706*

DAVID A. YPHANTIS (1), *Biochemistry and Biophysics Section, Biological Sciences Group, University of Connecticut, Storrs, Connecticut 06268*

Preface

This is the third of three volumes of "Enzyme Structure" devoted to updating the treatment of physical methods (parts F and G appeared last year). Although coverage of the various techniques is not exhaustive, it is hoped that the intent of presenting a broad coverage of currently available methods has been reasonably fulfilled. The organization of the material is the same as in the previous volumes.

As in the past, these volumes present not only techniques that are currently widely available but some which are only beginning to make an impact and some for which no commercial standard equipment is as yet available. In the latter cases, an attempt has been made to guide the reader in assembling his own equipment from individual components and to help him find the necessary information in the research literature.

In the coverage of physical techniques, we have departed somewhat in scope from the traditional format of the series. Since, at the termination of an experiment, physical techniques frequently require much more interpretation than do organic ones, we consider that brief sections on the theoretical principles involved are highly desirable as are sections on theoretical and mathematical approaches to data evaluation and on assumptions and, consequently, limitations in the applications of the various methods.

We wish to acknowledge with pleasure and gratitude the generous cooperation of the contributors to this volume. Their suggestions during its planning and preparation have been particularly valuable. Academic Press has provided inestimable help in the assembly of this volume. We thank them for their many courtesies.

C. H. W. Hirs
Serge N. Timasheff

METHODS IN ENZYMOLOGY

EDITED BY

Sidney P. Colowick and Nathan O. Kaplan

VANDERBILT UNIVERSITY
SCHOOL OF MEDICINE
NASHVILLE, TENNESSEE

DEPARTMENT OF CHEMISTRY
UNIVERSITY OF CALIFORNIA
AT SAN DIEGO
LA JOLLA, CALIFORNIA

METHODS IN ENZYMOLOGY

EDITORS-IN-CHIEF

Sidney P. Colowick Nathan O. Kaplan

VOLUME VIII. Complex Carbohydrates
Edited by ELIZABETH F. NEUFELD AND VICTOR GINSBURG

VOLUME IX. Carbohydrate Metabolism
Edited by WILLIS A. WOOD

VOLUME X. Oxidation and Phosphorylation
Edited by RONALD W. ESTABROOK AND MAYNARD E. PULLMAN

VOLUME XI. Enzyme Structure
Edited by C. H. W. HIRS

VOLUME XII. Nucleic Acids (Parts A and B)
Edited by LAWRENCE GROSSMAN AND KIVIE MOLDAVE

VOLUME XIII. Citric Acid Cycle
Edited by J. M. LOWENSTEIN

VOLUME XIV. Lipids
Edited by J. M. LOWENSTEIN

VOLUME XV. Steroids and Terpenoids
Edited by RAYMOND B. CLAYTON

VOLUME XVI. Fast Reactions
Edited by KENNETH KUSTIN

VOLUME XVII. Metabolism of Amino Acids and Amines (Parts A and B)
Edited by HERBERT TABOR AND CELIA WHITE TABOR

VOLUME XVIII. Vitamins and Coenzymes (Parts A, B, and C)
Edited by DONALD B. MCCORMICK AND LEMUEL D. WRIGHT

VOLUME XIX. Proteolytic Enzymes
Edited by GERTRUDE E. PERLMANN AND LASZLO LORAND

Section I

Molecular Weight Determinations and Related Procedures

[1] Pulsed Laser Interferometry in the Ultracentrifuge

By DAVID A. YPHANTIS

The principal advantages of continuous laser light sources for Rayleigh interferometry in the ultracentrifuge lie in the increased convenience and lower operating costs associated with relatively long-lived lasers that require no water cooling and in the greatly increased monochromaticity and coherence length of the laser light compared to the usual high pressure mercury illumination. These advantages are discussed in detail in Volume 48 of this series.[1] The additional benefits of pulsed laser interferometry stem largely from the precise control of illumination timing and the high light intensities that are readily available.

Two experimental arrangements have been described for utilizing pulsed laser light in the ultracentrifuge: One arrangement employs a pulsed argon ion laser of moderately high intensity[2] while the other modulates a low power helium–neon gas laser either externally with a Pockel cell[3] or by means of internal switching.[1] Both arrangements make possible the illumination of (tangentially) selected regions of the rotor and cells. The Pockel cell modulation system makes possible finer control of the region to be illuminated because of its shorter pulse length, down to 0.8 μsec compared to the ~6 μsec for the unmodified pulsed argon ion laser. On the other hand, the argon ion laser emits significantly higher light intensity than the He–Ne lasers used and, in addition, can provide a number of laser lines in the green and blue region of the spectrum. The pulsed argon ion laser system used in this laboratory for several years is described here.

Apparatus

The argon ion laser used has been manufactured intitially by TRW as the 83-A (and 83-AR, when a manual refill system has been included) and later by Quantrad as the Model 83-AR (83-AB with a modified refill system) and 93 laser. The maximum repetition rate of the ~6 μsec long pulses is limited by the line frequency to 60 pulses/sec. The lasers used provide 0.8 W (peak power) of light in the TEM 00 Mode at 5145 A. Nearly as much power is available at 4880 A and considerably less at

[1] R. C. Williams, Jr., Vol. 48, p. 185.
[2] C. H. Paul and D. A. Yphantis, *Anal. Biochem.* **48,** 588 (1972).
[3] J. A. Lewis and J. W. Lyttleton, *Anal. Biochem.* **56,** 52 (1973).

4579, 4765 and 4965 A. The configuration of the laser optics has been described *in extenso*.[2] It differs only in minor detail from the continuous laser installation described in Volume 48 of this series,[1] most notably in the inclusion of a wavelength selector or, more recently, interference filters immediately in front of the laser so as to isolate the desired laser line.

The pulsed argon ion laser emits polarized light and the plane of polarization is adjusted (by rotating the laser about its long axis) to lie in the radial direction of the ultracentrifuge in order to minimize distortion effects with the usual sapphire windows of the ultracentrifuge cells.[4] This polarization is also of use when a single laser flash provides too much light for proper exposures: a sheet of properly oriented Polaroid in front of the laser then serves to attenuate the light over a wide range.

Several timing arrangements of varying degrees of complexity have been used to trigger the laser in phase with the rotating rotor and within the constraint of the line frequency rate of pulsing. The simplest of these[2] utilizes a monostable multivibrator that is continuously triggered by the Schmitt trigger of the (commercial) absorption scanner of the Spinco model E ultracentrifuge. The output of this monostable is enabled at line frequency so as to trigger the laser. The timing of this simple system is limited by the stability of the analog components of the monostable. Unfortunately, long-term stability and reproducibility is marginal, making for cumbersome overall operation. Considerable improvement can be obtained by replacing the monostable with a precise crystal-controlled oscillator whose output can be counted to provide a reproducibility of $\pm 0.5 \ \mu$sec. The complete circuit[5] makes possible more convenient and more flexible operation of the laser. Both of these control circuits pulse the laser a fixed time interval after receiving a synchronizing pulse from the scanner optical system. Accordingly both suffer from two common problems: (1) The scanner pulses are derived from the absorption optical system and this is located 180° across the rotor from the refractometric optics. One of the components of the (ever present) rotor precession induces timing variations that would be minimized if the synchronizing pulses originated from the same optical system as the laser. (2) With a fixed *time* delay, speed fluctuations of the rotor are directly reflected as variations in synchronization. Replacement of a fixed time delay by a delay of a fixed fraction of a rotation should minimize these variations.

These considerations led to the development of a computer-interfaced laser controller.[6] This controller obtains its synchronization pulses from

[4] A. T. Ansevin, D. E. Roark, and D. A. Yphantis, *Anal. Biochem.* **34**, 237 (1970).

[5] C. H. Paul and D. A. Yphantis, *Anal. Biochem.* **48**, 605 (1972).

[6] T. M. Laue, R. Domanik, D. Rhodes, and D. A. Yphantis, *Biophys. J.* **17**, 101a (1977).

an avalanche photodiode mounted between the stationary interference slits that, in turn, are located directly over the lower collimating lens under the rotor.[7] This photodiode views, through the rotor reference hole, an infrared light emitting diode mounted on the top chamber lens. The pulses indicating the arrival of the rotor reference hole are amplified and used to generate a frequency (4096 times the rotor frequency) that is precisely locked to the rotation of the ultracentrifuge rotor. This frequency is then counted, instead of the crystal clock of the previous design, to provide laser timing that is relatively independent of minor speed fluctuations. The synchronization achieved appears to be considerably better than 1/1000 of a revolution of the rotor from a speed of 2000 rpm on up. The use of a computer to control the number of "rotor frequency" pulses to be counted, to count the number of laser flashes, activate the camera, and take care of other such "bookkeeping" chores makes possible simple execution of experimental protocol.

The alignment of the optical system differs only in minor ways from the usual procedure. The laser and the spatial filter must be arranged so that the chamber is evenly illuminated with collimated light, parallel to the axis of rotation. This can be accomplished in several ways. The current procedure[8] used in this laboratory is as follows.

The convergence point of the spatial filter is positioned in the focal plane of the lower collimating lens by installing a corner-cube reflector in the evacuated rotor chamber and a cube beam-splitter between the spatial filter and the interference filter (or wavelength selector) in front of the laser. First the converging lens of the spatial filter is adjusted to pass the laser beam through the pinhole and the pinhole is then removed from the spatial filter. The components (laser, spatial filter, and prism) are arranged so that the light returned by the corner-cube reflector passes again through the spatial filter and is partially reflected, at right angles to the laser, by the beam-splitter. The spatial filter is then translated longitudinally so that the cross section of this returned and reflected beam is minimized. With a long optical lever arm the spatial filter can be positioned to within 30 μm. The beam splitter and corner cube are then removed.

Lateral and vertical positioning of the spatial filter is required to set the axis of the collimated chamber light parallel to the axis of rotation: A cell with a fully metalized lower window is spun at a few thousand revolutions per minute (not in synchrony with the line frequency), and the laser is triggered at the line frequency. The light reflected back from the cell window retraces its optical path and can be identified easily by its irregular timing. The spatial filter is then positioned laterally and vertically so

[7] D. A. Yphantis, *Ann. N.Y. Acad. Sci.* **88,** 586 (1960).

[8] D. A. Yphantis, C. H. Paul, and D. Rhodes, in preparation.

that this reflection is coincident with the (diverging) laser beam just after the convergence plane of the spatial filter. Adjustment to a fraction of a millimeter is easily obtained if the rotor is spun rapidly enough to avoid significant precession. This is basically the reflection procedure described by Gropper[9] and is subject to the same limitations from imperfect rotors and cell components.

The laser is positioned vertically and laterally to illuminate the bottom collimating lens evenly. The angle of the laser determines, for a given position of the spatial filter lens, the exact position, perpendicular to the optic axis, of the convergence point of the spatial filter (where the pinhole will be located), and this is of secondary importance. Last, the laser is rotated about its long axis to orient the electric vector of the emitted polarized light in the radial direction. This completes the alignment of the laser and spatial filter.

One more special procedure is useful for determining the position of the phase-plate along the optic axis: The empty chamber is evacuated and the usual phase-plate is replaced by an optical flat of the same dimensions that has been fully metalized on its front surface. With the pinhole removed, the beam-splitter cube is again placed between the interference filter and the spatial filter in order to sample the light returned from the phase-plate mirror through the spatial filter. Collimated light in the chamber will be focused onto the mirror at the phase-plate position and returned as collimated light in the chamber only when the mirror is located at the focal plane of the top collimating lens. In turn, the returned beam sampled by the beam-splitter will be collimated, and therefore uniformly of minimal diameter, only when the mirror is appropriately located. The sensitivity of this adjustment is better than 300 μm.

The remainder of the alignment procedure is essentially the same as with mercury lamp illumination (see, for example, Richards *et al.*[10-12]) with but one significant difference: The great coherence length of the laser light source introduces troublesome random interference effects (stationary speckle) during the focusing of the camera and cylinder lenses if ordinary ground or opal glass is used as a diffuser. This annoying effect can be eliminated by using a suspension of a strong scatterer as a diffuser. A layer of ~2 cm of milk diluted with water to OD_{5145}^{1cm} ~1 is adequate to convert the stationary speckle to a moving speckle that is averaged out during exposure times of several seconds.

[9] L. Gropper, *Anal. Biochem.* **7,** 401 (1964).
[10] E. G. Richards, D. C. Teller, and H. K. Schachman, *Anal. Biochem.* **41,** 189 (1971).
[11] E. G. Richards, D. C. Teller, V. D. Hoagland, R. H. Haschemeyer, and H. K. Schachman, *Anal. Biochem.* **41,** 215 (1971).
[12] E. G. Richards, T. Bell-Clark, M. Kirschner, A. Rosenthal, and H. K. Schachman, *Anal. Biochem.* **46,** 295 (1971).

Applications

The ability to control the timing of the laser pulses simplifies and improves existing experiments and makes feasible new experimental arrangements. Obviously, it is simple (by changing delays) to multiplex the interferometry with several cells in a rotor without requiring awkward and expensive wedge cells. Illumination of the Rayleigh slits only when both sample and reference channels are simultaneously located directly behind their corresponding interference slits significantly enhances fringe contrast over the usual continuous illumination.

Interferometry without cell slits appears to be feasible. Indeed one can scan across the paired solution–solvent sectors by simply incrementing laser timing. The fringe displacements then observed with ordinary double-sector cells, in a water versus water region, are found to be linearly proportional to timing delay, whereas the displacements of the fringes in the air versus air region are virtually independent of timing delay.[13] This implies that the cell windows used must have been bowed even at the low speeds necessarily used for this experiment because of the spatial resolution limits imposed by the pulse length of the laser. It is this bowing, along with the additional bowing occasioned by high centrifugal fields, that requires cell slits in order to spatially delimit the area contributing to interferometry. At low speeds it is practical to use the pulse length of the laser to delimit this area for interferometry, but fringe quality deteriorates at the higher speeds since the temporally delimited area is essentially proportional to speed. With the current pulse length of the argon ion laser in use this temporally delimited area is about 1.2 mm wide at 30,000 rpm. Shorter laser pulses, available, for example, with different lasers or by modification of the pulse forming network of the laser power supply, should make possible good interferometry at the higher speeds without using cell slits. This should make feasible simultaneous observations of solutions, at any speed, by both Rayleigh interferometry and the absorption scanner. Such simultaneous observations are generally impractical with continuous illumination, because the narrow cell slits required for interferometry strongly attenuate available light and drastically lower the signal to noise ratio of the scanner.

The polychromatic emission of the argon ion laser has the disadvantage that it requires filtration before use in interferometry. On the other hand, the several laser lines that are present make possible a promising "white light fringe" approach. Preliminary experiments[13] suggest that it is practical to use interferograms taken with different laser wavelengths to calculate the positions of an effective "white light fringe." This calculated position can then be used to label the fringes unambiguously and thus con-

---

[13] C. H. Paul, Ph.D. Dissertation, University of Connecticut, Storrs (1974).

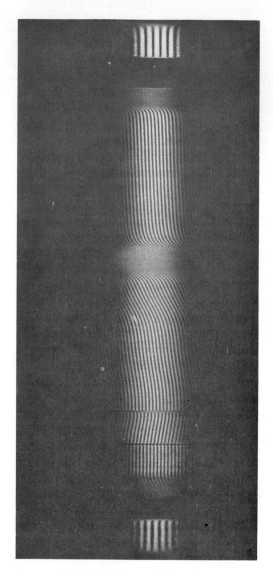

FIGURE 1a

vert the Rayleigh interferometer into an absolute interferometer. This
procedure differs from previous "white light fringe" procedures (see, for
example, Edelstein and Ellis[14]) in that it is possible to compensate directly
for chromatic dispersion both of the optical system and of the solution

[14] S. J. Edelstein and G. H. Ellis, *Anal. Biochem.* **43,** 89 (1971).

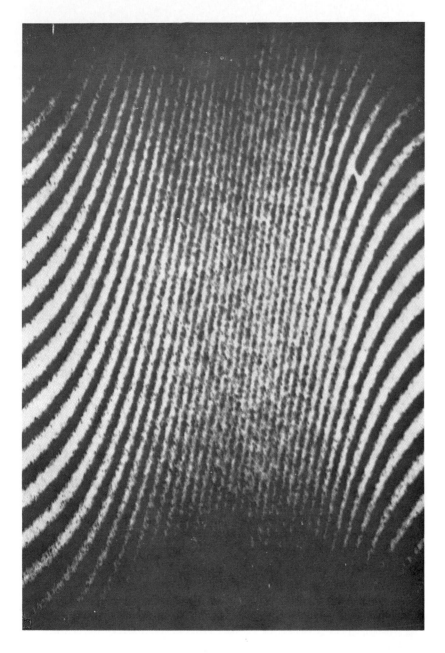

FIGURE 1b

versus solvent, since the position of the white light fringe is not observed directly but rather is calculated. So far, the procedure appears applicable only in regions where refractive index gradients are small, since radial magnification, and thus imaging position, is somewhat wavelength

9

FIGURE 1c

dependent even with the nominally achromatic lenses of the ultracentrifuge.

The combination of precise timing and high light intensities makes a number of experiments possible. For example, it is easy to project, through the usual viewing lens and onto a sheet of paper about 50 cm from the lens, magnified images of the interference fringes that are readily visible in a darkened room. More useful is the ability to obtain fairly high quality interferograms with a single laser flash (see Fig. 1) on the same photographic emulsions that require several minutes exposure,[4] with the AH-6 mercury lamp, to obtain images of somewhat poorer quality and definition. Such short exposures obviate the blurring effects of precession that can be so ubiquitous and make possible good interferograms of rapidly sedimenting and/or diffusing systems.

The much increased light intensity also makes practical the use of slow fine grain emulsions such as the Kodak type V-F glass plates and the more recently available SO-115 and SO-141 films on a stable Estar base. Indeed, it is possible to obtain images of extreme contrast on graphic arts films with usefully small exposures, e.g. with about 300 flashes on Ortho type 3 film. These fine grain, high resolution images appear to be especially useful for rapid, precise determinations of fringe positions using automatic image measurement systems, such as those described by DeRosier et al.,[15] by Carlisle et al.,[16] and by Domanik et al.[17]

[15] D. DeRosier, P. Munk, and D. Cox, *Anal. Biochem.* **50,** 139 (1972).
[16] R. M. Carlisle, J. I. H. Patterson, and D. E. Roark, *Anal. Biochem.* **61,** 248 (1974).
[17] R. Domanik, D. A. Yphantis, and T. M. Laue, *Biophys. J.* **17,** 101a (1977).

FIG. 1. Rayleigh interference patterns from a velocity experiment on ~1% hemocyanin from a spider crab (*Libinia emarginata*) run at 20,000 rpm. (a) Interferogram obtained with a single laser flash (~6 μsec) on II-G spectroscopic emulsion. Note the negatively sloping fringes in the air versus solvent region centripetal to the solution meniscus. This negative slope represents the largely compressive refractive index gradient in the longer solvent column compared to the virtually uniform air space above the solution column. The hemocyanin used here shows three major sedimenting components (~5 S, 16 S, and 25 S). Note also the double fringe spacing in the counterweight reference holes generated by interference slits of half-normal spacing.[4] (b) Magnified central boundary region for the 25 S component from another single flash exposure on a II-G plate. The laser flash for this exposure was attenuated about fourfold by insertion of a piece of Polaroid at an angle to the electric vector of the laser illumination. Note the clear and distinct resolution of the fringes at this maximum gradient of about 530 fringes/cm in the cell. This refractive index gradient corresponds to about 1.5 times the maximum usable gradient usually giving resolvable fringes with the AH-6 lamp. As pointed out by Williams,[1] care must be exercised in using data obtained at such high refractive index gradients because of the potential error from Wiener skewing. Note also the prominent emulsion grains of this relatively sensitive emulsion. (c) The same region observed a little later with a 100 flash exposure on a V-F Spectroscopic emulsion. Note the finer grain structure and the increased contrast compared to the II-G emulsion. The maximum gradient of ~330 fringes/cm in the cell that is seen here roughly corresponds to the maximum useful refractive index gradient with mercury arc illumination. (The vertical spacing of the interference fringe of the solution pattern was 267 μm on the original photographs.) (Reproduced by permission from Paul and Yphantis.[2])

Last, the pulsed argon ion laser seems particularly promising as a light source for real-time data acquisition with the Rayleigh optics. In preliminary experiments, single laser pulses provided much more illumination than needed to form good quality magnified fringe images on a closed-circuit television system equipped with an ordinary vidicon. The fine timing control and the high light intensities available should make feasible a number of approaches using various present-day light sensors.

[2] Density Measurements by the Mechanical Oscillator

By JOHN P. ELDER

There has long been a need for a simple and rapid method for measuring fluid density with extremely high accuracy and precision. This is of particular significance to those in the biophysicochemical fields, interested in investigating molecular interactions of biological macromolecules in solution. An important parameter characterizing such systems is the partial specific volume of the solute. This parameter assists in the interpretation of other physical information regarding the state of the system, resulting from complex conformational changes and molecular associations. It is also an important function relating sedimentation rates, and low-angle X-ray scattering intensity data with the molecular weight of the biological polymer. In order to reduce the error in such molecular weight determinations to the order $\pm 1\%$, the partial specific volume must be measured with an uncertainty of less than $\pm 0.15\%$, which in turn requires accurate density and differential density measurements with a precision of $\pm -4.10^{-6}$ g/cm³.

Recognizing this fact, Stabinger, Kratky, and Leopold[1] developed a simple but extremely elegant technique for such precision density measurements. The overall determination has been reduced to the measurement of a time interval, which can be performed digitally with great accuracy and precision. They have admirably described the original digital density meter, and discussed pertinent applications in the biophysical and related fields.[2]

Such is the ever-increasing demand for a rapid method for determining accurate fluid densities with varying levels of precision, that a range of instruments and accessories are now available. (Such instruments, manu-

---

[1] H. Stabinger, O. Kratky, and H. Leopold, *Monatsh. Chem.* **98**, 436 (1967); O. Kratky, H. Leopold, and H. Stabinger, *Z. Angew. Phys.* **4**, 273 (1969).

[2] O, Kratky, H. Leopold, and H. Stabinger, Vol. 27, Part D, p. 98.

factured by the A. Paar Company, Graz, Austria, are marketed in the United States by the Mettler Instrument Corp.) It is the purpose of this brief review to discuss the advantages of recent design developments, to acquaint the potential user with experimental techniques, and to discuss briefly some recent applications in various fields.

Principle of the Mechanical Oscillator Technique

In any generalized coordinate system, if a point mass m is attracted toward an origin by a force proportional to its distance x from the origin, then in the absence of any external restraint, the Hooke's law force is exactly compensated by the Newton's second law force, and the motion is thus described by Eq. (1)

$$m\ddot{x} + Cx = 0 \qquad (1)$$

where C is the elastic modulus. At any time t the location of the point mass is given by the real part of Eq. (2).

$$x = a \exp (i\omega_0 t) \qquad (2)$$

The motion is said to be simple harmonic, with a complex amplitude a and angular velocity, $\omega_0 = (C/m)^{1/2}$. In a real system, the motion will decay to zero due to the restraining action of a frictional force f proportional to and acting in opposition to the natural inertia of the system. The motion is now described by Eq. (3), and the position of the moving mass is given by the real part of Eq. (4).

$$m\ddot{x} + Cx = -f\dot{x} \qquad (3)$$

$$x = a \exp (i\omega_1 - \lambda)t \qquad (4)$$

One now has damped harmonic motion, which is simple harmonic motion with an exponentially decreasing amplitude. The angular velocity, $\omega_1 = (\omega_0^2 - \lambda^2)^{1/2}$, is reduced by the damping coefficient, $\lambda = f/2m$. If an external force is to be applied to counterbalance the damping effects of friction, then it must have the form $b \exp (i\omega_0 t)$, where b is a complex amplitude. Figure 1a is a simple vector diagram of the balanced forced harmonic motion system. If, for any reason, there is a phase shift in the external force, the system is out of balance (Fig. 1b) and a component of the external force acts in the direction of the dynamic force, thereby altering the effective mass of the oscillating system. It is important, therefore, in a real system to monitor continually the forced motion and thereby control the phase and magnitude of the external force. Furthermore, it is important that this rigorous control be maintained for an oscillator of varying frequency. When balance prevails, one has resonance. It is the measure-

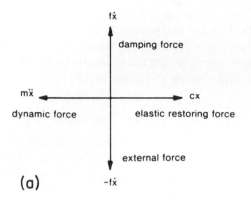

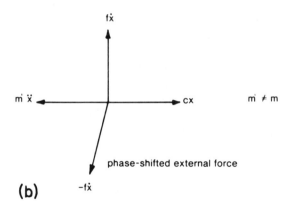

Fig. 1. Vector diagram of forced harmonic oscillation. (a) Balanced system. (b) Unbalanced system.

ment of the change in the resonant frequency of a hollow, vibrating tube, when filled with fluids of varying density, which is the heart of the mechanical oscillator technique.

The relationships discussed apply to any real situation provided certain limits are not exceeded. Thus, a hollow U-shaped tube excited into a bending-type vibration orthogonal to its plane about a nodal axis will exhibit forced harmonic motion, provided that the material remains in the elastic region of deformation. Thus, the amplitude must be limited such that the elastic modulus C remains constant, and Hooke's law applies. In Eq. (1), m now represents the total mass of the system, comprising the tube body M of volume V and its contents of density ρ.

$$m = M + V\rho \qquad (5)$$

The resonance period T is

$$T = \frac{1}{\nu} = \frac{2\pi}{\omega_0} = 2\pi \left(\frac{M + V\rho}{C}\right)^{1/2} \tag{6}$$

From Eq. (6), one has

$$T^2 = A\rho + B \tag{7}$$

where $A = 4\pi^2 V/C$ and $B = T_M^2 = 4\pi^2 M/C$. T_M is the resonance period of the tube alone, which may be calculated from calibration data or measured directly. The density differential between two fluids is therefore

$$\Delta\rho = \rho_1 - \rho_2 = (1/A)(T_1^2 - T_2^2) = K(\Delta T^2) \tag{8}$$

The procedure for measuring density differentials is therefore reduced to developing a method for measuring a time period accurately and precisely.

Mechanical Oscillator Design and Operational Procedure

It is pertinent to consider those factors which ultimately govern the accuracy and precision of a density measurement. It was such a consideration that led Kratky et al.[2] to select and develop the mechanical oscillator technique.

The initial primary requirement was the ability to measure fluid density differentials with a maximum absolute error of $\pm 2.10^{-6}$ g/cm^3. Perhaps the most significant point to be considered in developing any method and procedure is the criticality of temperature control. This can be calculated from a consideration of the fluid thermal expansion coefficient, γ. Essentially Eq. (9) applies

$$\partial\rho/\partial T = -\gamma \tag{9}$$

Since for aqueous solutions, $\gamma \sim 10^{-4}$ cm^3/°C, an absolute density error of 10^{-6} g/cm^3 requires a temperature control to within $\pm 10^{-2}$°. Such precise temperature control immediately indicates that one should use a very small sample volume and that the experimental measurement should be completed within a short time period, of the order, 1–2 min. The sample holder should be small and designed for easy access and removal of the sample. Fluid characteristics, which could result in solid–liquid interfacial phenomena, e.g., viscosity and surface tension, should not affect the measurement. One should be able to use samples in volatile solvents, without loss of solvent. The sample should be capable of being injected into the container very quickly, without the necessity of precise volume dosing.

The U-shaped sample holder, of internal volume about 0.7 cm³, capable of being housed in a thermostatted jacket with easily accessible inlet and exit ports met these requirements admirably.

To achieve a precision of $\pm 10^{-n}$ in determining density, it is necessary to measure a time lapse, at least 10^{n+1} times a chosen unit. If the resonance period of the oscillator, which is of the order 2–5 msec, is selected as the time unit, then a 5–6 hr measurement time would be required. Since this is completely out of the question, Kratky et al.[1] measured by independent means the total time lapse in small but precisely defined units for a preset number of oscillations of the sample holder. For reasons defined by the electronic detection circuitry, (vide infra), the smallest time unit which can be measured is 10 μsec. Precision density measurements in the range 10^{-5}–10^{-6} g/cm³ are thereby attainable by measuring time lapses in the range 10–100 sec. Provided the number of vibrations for which the time lapse is measured remains constant, the basic differential density relationship [Eq. (8)] applies, irrespective of the number.

Digital Precision Density Meter Instrumentation

Instrumentation covering the absolute measurement precision range, $\pm 1.10^{-4}$ to $\pm 1.5.10^{-6}$ g/cm³ is available. Figure 2 is a general block diagram for these instruments, describing the oscillator excitation, amplitude control, and time lapse measuring, and data reception circuitry which Leopold[3] has discussed in detail. The motion of a transducer, attached to the vibrating U-tube, induces an alternating voltage in the sensor, proportional to its time-dependent velocity. The signal is amplified and used in two circuits.

Excitation Control. The amplified signal is voltage-limited and fed back to control the voltage amplitude of the excitation source. In this manner, a fixed amplitude, about 0.01 mm forced harmonic oscillation is maintained. Phase shifts are made negligible by setting a large bandwidth (3–100,000 Hz) for the excitation and amplitude-control loop relative to the resonance frequency of the tube when filled with fluids covering the density range 0–3 g/cm³. However, this chosen bandwidth results in an uncertainty of 10 μsec in detecting the zero voltage crossing of the velocity-proportional amplified sensor voltage. Thus, the sensitivity of the electronic detection circuitry defines the small time unit for the time lapse measurement.

Time Lapse Measurement. The amplified sensor signal is concurrently impressed on the Schmidt trigger. A start command, which may be effected by manual depression of a push-button automatically following a

[3] H. Leopold, *Electronik* **19**, 297 (1970).

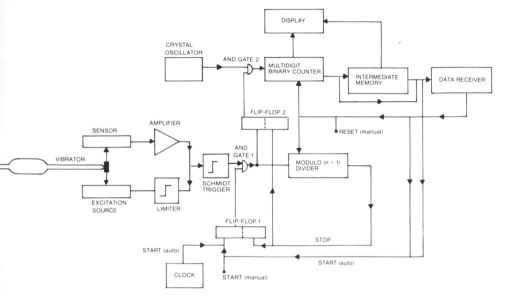

FIG. 2. General block diagram of a digital density meter.

measurement or at a specified time by remote command from an external pulse generator, sets flip-flop 1, thereby allowing the square wave output of the trigger to pass through and-gate 1. Therefore, the first positive zero crossing of the amplified sensor signal will set flip-flop 2, thereby opening and-gate 2 to the 10 μsec pulses emanating from the 100 kHz quartz crystal oscillator. These pulses are summed continually in the binary counter and displayed. In certain instruments, this count is stored in an intermediate memory. Following $n + 1$ Schmidt trigger output pulses, the modulo divider resets flip-flop 1 which closes and-gate 1. This automatically resets flip-flop 2, thereby closing and-gate 2. The total time lapse for n pulses is now displayed. The value may be transferred to a suitable receiver, printer, programmable calculator, or computer, as desired. The modulo divider and multidigit counter may then be reset either manually or by remote command. In the most precise instrument, n may be varied manually from 10 to 40,000, depending upon the desired accuracy.

In the original instruments,[1] the oscillator was a simple U-tube cemented into a metallic collar housed in an air-filled thermostatted jacket. The transducer was a small magnet glued to the apex of the tube. This was perturbed into oscillatory motion by means of an excitation coil in its vicinity. The moving magnet in turn induced in a pick-up coil the velocity-proportional voltage signal. It has since been realized that this method for sustaining controlled amplitude oscillation is not sufficiently accurate for the most precise differential density measurements. Furthermore, in order to obtain the necessary temperature control ($\pm 0.01°$) the sample had to be thermally equilibrated for at least 15 min prior to initiating the time-lapse measurement.

Recent Developments in Instrumental Design

Figure 3 shows the coimplete instrument for the most precise differential density measurements, model DMA02D, with the newly designed measuring cell on the right-hand side. A side view of this all-glass sample holder is shown in Fig. 4. The U-tube, is completely enclosed in an envelope, containing a high thermal conductivity gas. This envelope is surrounded by a glass jacket, through which a therrmostatted liquid is circulated. By this means, the temperature equilibration time has been reduced to less than 4 min. In this model, forced harmonic oscillation of fixed amplitude is obtained by use of an electrooptical sensor circuit. A phototransistor is irradiated by infrared light from a light-emitting diode. A small vane, coated with magnetic material, is attached to the apex of the vibrating U-tube. In its motion, the vane chops the light beam, thereby modulating the phototransistor output. This signal, when amplified and limited, is fed back to control the magnetization of a ferrite material, which in turn controls the motion of the oscillator. The absolute precision of this model is $1.5 . 10^{-6}$ g/cm^3, when the time-lapse is measured for a preset count of 10^4 or greater.

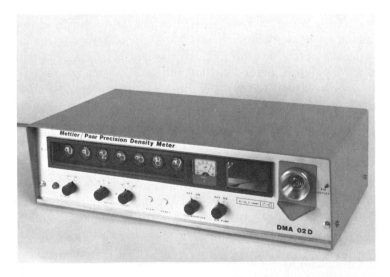

FIG. 3. Complete instrument for most precise differential density measurements.

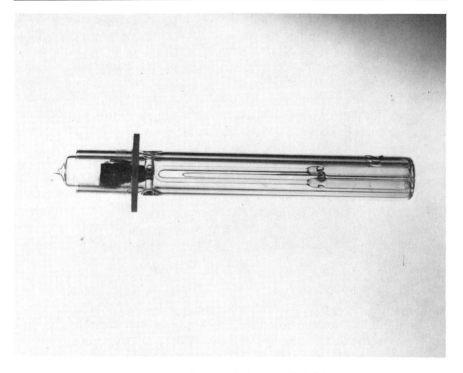

Fig. 4. Side view of all-glass sample holder.

Experimental Procedure

Since one is performing a differential measurement, one ensures that equal volumes of both the sample and reference fluid are used by completely filling the measuring cell. This is achieved by continuously injecting the fluid into the lower port of the measuring cell until the excess flows out of the upper port. Disposable, polyethylene syringes about 2 ml in volume are used for this purpose. The ports are then plugged with small Teflon stoppers, which prevent any loss of fluid even if highly volatile liquids are used. One ensures that the inner surface of the cell is completely wetted, first by starting with a clean dry tube and, second, by injecting the sample at such a rate that the advancing meniscus is always concave to the direction of the flow. Following the measurement, the sample may be removed by reverse use of the filling syringe with minimal loss of sample. The tube should then be well flushed with an appropriate solvent followed by air-drying. For this purpose, a small pump with a sintered glass filter is built into each instrumental model. When the relative humidity is high or when the measurement temperature is below the dew point, it is wise to attach a tube containing a suitable desiccant to the pump outlet port. A clean dry tube results if the time-lapse measurement for the empty cell returns to its original value prior to sample introduction.

The requirements of precise temperature control cannot be overemphasized. Single thermostatting systems are being employed by a number of users. Cascade systems are also used. Here, one thermostat provides the coolant for the coils of an ultrathermostat, which, in turn, provides the measuring cell temperature control. Even good ultrathermostats, with a control specification of $\pm 0.01°$, can only maintain such control for short periods of time. During extended use, random fluctuations outside the specified band width are to be expected. As previously indicated, [cf. Eq. (9)], this limitation results in a density variation, $\partial \rho_t$. This must be added to the absolute error of the instrument ($\partial \rho_{abs}$) resulting from the 10 μsec uncertainty in the time-lapse measurement. Since all measurements are differential, the practical error is the mean of two compounded errors.

$$\partial \rho_{prac} = \pm 2^{1/2}(\partial \rho_t + \partial \rho_{abs}) \tag{10}$$

For aqueous solutions, $\partial \rho_t \approx 2.10^{-6}$ g/cm^3. For the most precise instrument, $\partial \rho_{abs} = \pm 1.5.10^{-6}$ g/cm^3, hence $\partial \rho_{prac} = \pm 5.10^{-6}$ g/cm^3.

The thermostatting liquid temperature may be monitored at the inlet and outlet of the density meter thermostatting jacket by means of a precise thermometric device, e.g., a differential mercury in glass thermometer (Beckmann type) or a quartz thermometer (Hewlett-Packard type). The actual cell temperature may be monitored *in situ*. The cell is provided with a small single-ended tube which lies close to the vibrating U-tube within the gas-filled envelope (cf. Fig. 4). Special thermometric devices, e.g., thermistor probes, may be inserted here. Heat sink devices, e.g., platinum resistance thermometers, cannot be used for obvious reasons.

Instrument Calibration

The instrument constants, A, B, and $K (= 1/A)$ in Eqs. (7) and (8) are obtained by determining the time lapse, (T value), for the measuring cell containing two fluids of known density. Dry air and air-free, pure water are universally used, since they are readily available, and precise, accurate density data are available.

The density of dry air at a temperature, $t°C$, and atmospheric pressure, p torr, is given by

$$\rho_{t,p} = \frac{0.0004646p}{t + 273.13} \tag{11}$$

The density of air-free, pure water has been given at 0.1° intervals from 0°–40°. The values given in Table I are taken from the complete data of

TABLE I
DENSITY OF AIR-FREE PURE WATER

t (°C)	ρ (g/cm³)	t (°C)	ρ (g/cm³)	t (°C)	ρ (g/cm³)	t (°C)	ρ (g/cm³)
0	0.9998396	10	0.9996987	20	0.9982019	30	0.9956454
1	0.9998985	11	0.9996039	21	0.9979902	31	0.9953391
2	0.9999399	12	0.9994961	22	0.9977683	32	0.9950243
3	0.9999642	13	0.9993756	23	0.9975363	33	0.9947010
4	0.9999720	14	0.9992427	24	0.9972944	34	0.9943694
5	0.9999637	15	0.9990977	25	0.9970429	35	0.9940296
6	0.9999399	16	0.9989410	26	0.9967818	36	0.9936819
7	0.9999011	17	0.9987728	27	0.9965113	37	0.9933263
8	0.9998477	18	0.9985934	28	0.9962316	38	0.9929629
9	0.9997801	19	0.9984030	29	0.9959430	39	0.9925920
						40	0.9922136

Wagenbreth and Blanke.[4] Over the temperature range, 0°–40°, the constant K increases linearly with temperature.

$$K = K_0 + K_1 t \tag{12}$$

Figure 5 shows typical air-water calibration data at 5° intervals from 10°–40°. A linear regression yielded the following values for the temperature functional parameters, $K_0 = 1.2638306 \times 10^{-3}$, $K_1 = 1.3406 \times 10^{-7}$ with a correlation coefficient greater than 99.95%. For 10,000 oscillations of the U-tube, typical T values for air and water at 25° are 28.39144 and 39.89911 sec, respectively. For the most precise differential measurements, the instrument should be calibrated prior to each set of sample density measurements.

Various other density standards are available for confirmatory checks on the accuracy and precision of the instrument, for example, National Bureau Standards standard reference 217b; 2,2,4-trimethyl pentane. A number of American Petroleum Institute standard reference hydrocarbons are also available (Carnegie-Mellon University, Pittsburgh, Pennsylvania.)

Density Measurements at High and Low Temperatures

For the standard instrument models, the temperature range over which time lapse measurements may be made is −10° to 60°. As indicated, water density data are only available for part of this range. The temperature range for secondary nonaqueous standards is even more limited. Density values at the extreme temperatures may be evaluated, using standard calibration data, in the following manner. The volume of the empty U-tube at any temperature, is given by Eq. (13).

[4] H. Wagenbreth and W. Blanke, *PTB—Mitt.* **6,** 412 (1971).

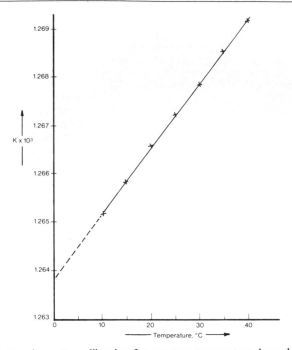

FIG. 5. Density meter calibration for constant temperature dependence.

$$V_t = V_0 \ [1 + 3\alpha(t - t_0)] \tag{13}$$

where V_0 is the tube volume at t_0, at which temperature standard density data are available and α is the coefficient of linear expansion of the tube material.

Generally, glass tubes are used, although certain other materials, e.g., nickel, are available for specific industrial purposes. The α value for the glass, Duran 50, is $3.25.10^{-6} \ °C^{-1}$. Using Eq. (13) together with the definitions of constants, A and B, [cf. Eqs. (6) and (7).], it is easily shown that the required sample density at temperature t is given by

$$\rho_s^t = \frac{\rho_c^0}{(T_c^0/T_m^0)^2 - 1} \ \frac{(T_s^t/T_m^t)^2 - 1}{1 + 3\,\alpha(t - t_0)} \tag{14}$$

where T_c^0 is the time lapse for the density calibration standard ρ_c^0 measured at temperature t_0, T_s^t is the sample time lapse at temperature t, and T_m^t and T_m^0 are the time lapses for the empty tube, i.e., under vacuum, at temperatures t and t_0, respectively.

Instrument Performance

The speed with which the measuring cell equilibrates thermally can be demonstrated by rapidly injecting cold and hot water into a dry cell, main-

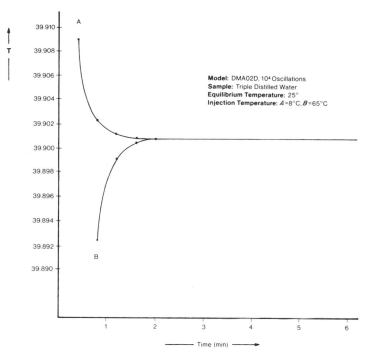

FIG. 6. Temperature equilibrium.

tained at some intermediate temperature, and recording consecutive time-lapse measurements. The results of two such tests are shown in Fig. 6. Even for an initial temperature differential of $+40°C$, the cell equilibrates within 2 min.

The precision and accuracy of the density meters may be evaluated by determining the density of aqueous sodium chloride solutions, for which highly accurate values are known from several sources.[5-7] These solutions are easily made up by weight. Some results over a wide concentration range, together with the pertinent air–water calibration data, are shown in Table II. In all cases, the precision is better than $\pm 5.10^{-6} \, g/cm^3$. Measurements have been made using the National Bureau of Standards (N.B.S.) standard, 2,2,4-trimethyl pentane. Values of 0.691820 at 20°, and 0.687714 at 25°, agree very well with the N.B.S. values of 0.69183 ($\pm 2.10^{-5}$) and 0.68772 ($\pm 2.10^{-5}$), respectively.

[5] D. W. Kaufmann, "Sodium Chloride," p. 611. Van Nostrand-Reinhold, Princeton, New Jersey, 1960.
[6] "International Critical Tables," Vol. 3, p. 79. McGraw-Hill, New York.
[7] Landolt-Börnstein, "Physikalisch-Chemische Tabellen," Suppl. 1, p. 202. Springer-Verlag, Berlin and New York, 1927.

TABLE II

PRECISION DENSITY MEASUREMENTS, AQUEOUS SODIUM CHLORIDE SOLUTIONS[a]

Concentration (w/o)	T Value (sec)	ρ (g/cm^3)	Literature value
0.99	36.26212	1.005287	1.005285 (-2×10^{-6})
4.01	36.45784	1.026970	1.26973 $(+3 \times 10^{-6})$
16.01	37.25241	1.116402	1.116398 (-4×10^{-6})

[a] Density meter, Mettler/Paar Precision Density Meter model DMA02D; temperature $20° \pm 0.01°$; Oscillations, 10^4, Calibrations: T (air, 759 torr): 25.61005 sec; T (water): 36.19795 sec; $K = 1.523493 \times 10^{-3}$.

Examples of Applications

Although the mechanical oscillator technique was developed initially for the express purpose of determining partial specific volumes of proteins in solution with a greater accuracy and precision than had hitherto been possible, the instruments developed in the last four years have found wide use in a variety of academic and industrial fields, research, and quality control.

Persinger et al.[8] have indicated the ease and rapidity with which one can measure departures from ideality in such systems as acetic acid–water, where strong hydrogen bonding markedly affects the density. Excess volume studies in benzene–cyclohexane mixtures have been made by Kiyohara and Benson.[9] Bøje and Hvidt[10] have investigated interactions in aqueous mixtures of nonelectrolytes. In the author's laboratory the density meter has been used with success to measure specific gravities of mixed halogenated hydrocarbon polymers. These liquids, used as gyroscopic fluids, are extremely viscous at room temperature. Their density at high temperature, 50° and above, can be determined using calibration data determined at lower temperatures. [cf. Eq. (14)]. Hydrography will benefit from the sea water density measurements of Kremling.[11] Henning and Hicke[12] recommend the digital density meter as the standard reference method for density measurements in the wine industry.

Tilz Leopold, and their co-workers have discussed the use of the density meter as an aid in clinical diagnostics[13] and in the medical laboratory.[14] The Kratky school continue their low angle X-ray scattering studies

[8] H. E. Persinger, J. Lowen, and W. S. Tamplin, Am. Lab. September (1974).
[9] O. Kiyohara and G. C. Benson, Can. J. Chem. 51, 2489 (1973).
[10] L. Bøje and A. Hvidt, J. Chem. Thermodyn. 3, 663 (1971).
[11] K. Kremling, Deep-Sea Res. 19, 1 (1972).
[12] G. Henning and E. Hicke, Z. Anal. Chem. 265, 97 (1973).
[13] G. P. Tilz and H. Leopold, Wien. Klin. Wochenshr. 84, 697 (1972).
[14] G. P. Tilz, H. Leopold, R. Hesch, and S. Sailer, GIT Fachz. Lab. 16, 1048 (1972).

of conformational interactions of biological systems.[15,16] Pilz and Czerwenka[17] have investigated the concentration and temperature dependence of protein solutions partial specific volume. Behlke and Wandt[18] have directed their efforts toward hemoglobin and myoglobin. Lee and Timasheff[19] have measured the partial specific volumes of a number of proteins in 6 M guanidine hydrochloride and have obtained values of preferential interactions of the proteins with the denaturant, as well as of the volume changes which accompany denaturation.

In conclusion, a recent and provoking inovation in the mechanical oscillator technique should be mentioned, namely, its use in measuring the density of powdered material. By measuring the density of a suspension of a uniformly dispersed powder in an inert fluid phase, one can determine the density of the particulate matter. The particle size distribution should lie in a narrow band, and the liquid phase should be relatively viscous in order to reduce the natural sedimentation rate following dispersion of the solid in the medium. It is easily shown that the density of the solid is given by

$$\rho_{solid} = \frac{m_{solid}\rho_{suspension}}{m_{solid} + m_{liquid}(1 - \rho_{suspension}/\rho_{liquid})} \tag{15}$$

By performing a partial differentiation of Eq. (15) with respect to the four measurable parameters, mass of solid and liquid phase and density of liquid phase and suspension, one can evaluate the total error $\partial\rho_{solid}$. As an example, 3.5988 g of coke dust was dispersed in 10.7131 g of mineral oil, of density 0.9158 g/cm^3. The density of the resulting suspension was 1.0697 g/cm^3. The errors in weighing and measuring density were ± 0.1 mg and $\pm 2.10^{-4}$ g/cm^3, respectively. Thus, the density of the coke dust is 2.1405 $\pm$ 6.5.10^{-3} g/cm^3.

[15] I. Pilz, O. Kratky, A. Licht, and M. Sela, *Biochemistry* **12**, 4998 (1973).
[16] I. Pilz, O. Kratky, and F. Karush, *Eur. J. Biochem.* **41**, 91 (1974).
[17] I. Pilz and G. Czerwenka, *Makromol. Chem.* **170**, 185 (1973).
[18] J. Behlke and I. Wandt, *Acta Biol. Med. Ger.* **31**, 383 (1973).
[19] J. C. Lee and S. N. Timasheff, *Biochemistry* **13**, 257 (1974).

[3] Measurements of Preferential Solvent Interactions by Densimetric Techniques

By JAMES C. LEE, KUNIHIKO GEKKO, and SERGE N. TIMASHEFF

I. Basic Theory

The preferential interaction, or preferential "binding" parameter

$$\xi_i \equiv \left(\frac{\partial g_i}{\partial g} \right)_{T,P,\mu_i}$$

is a thermodynamic quantity which is ubiquitously found explicitly or implicitly in equations describing processes which involve interaction between macromolecules and components of the solvent system. This parameter is usually expressed as grams of component i per gram of component k (or moles of i per mole of k). Here the concentration, g_i, is expressed in grams of i per gram of principal solvent, usually water in biological systems. Thus, the concentration is on a molal scale, m_i (moles of component i per 1000 grams of water), since $g_i = m_i M_i/1000$, where M_i is the molecular weight of component i. We shall use the notation of Scatchard[1] and Stockmeyer,[2] according to which, in a three-component system, component 1 is principal solvent, usually water in biological systems, component 2 is macromolecule, component 3 is the added solvent component, e.g., guanidine hydrochloride, sucrose, CsCl, etc. This parameter enters explicitly or implicitly into the calculations in sedimentation velocity or equilibrium, as well as small-angle X-ray scattering measurements. It is also found to enter into a variety of relations which describe biological processes involving interactions with solvent components. The preferential interaction parameter may be measured by a variety of techniques, such as differential refractometry, densimetry, vapor pressure equilibrium and equilibrium dialysis, or gel permeation chromatography, in their many ramifications. In this chapter, we shall discuss its measurement by densimetry.

Density is an intensive property. In a three-component system, its change with variation in solvent composition is given by

$$d\rho = \left(\frac{\partial \rho}{\partial g_2} \right)_{T,P,g_3} dg_2 + \left(\frac{\partial \rho}{\partial g_3} \right)_{T,P,g_2} dg_3 \qquad (1)$$

---

[1] G. J. Scatchard, *J. Am. Chem. Soc.* **68**, 2315 (1946).
[2] W. J. Stockmeyer, *J. Chem. Phys.* **18**, 58 (1950).

where ρ is the density of the solution, T is the thermodynamic (Kelvin) temperature and P is pressure. Taking the partial derivative of Eq. (1) with respect to protein concentration at constant chemical potential of component 3 results in the working equation for calculating preferential interactions from density increments:

$$\xi_3 = \left(\frac{\partial g_3}{\partial g_2}\right)_{T,P,\mu_3} = \frac{(\partial \rho / \partial g_2)^0_{T,P,\mu_3} - (\partial \rho / \partial g_2)^0_{T,P,m_3}}{(\partial \rho / \partial g_3)_{T,P,m_2}} \quad (2)$$

where μ_3 is the chemical potential of component 3. Since in laboratory practice concentrations are most frequently measured on the molar, or mass per volume, scale, the preferential interaction parameter can also be expressed as

$$\left(\frac{\partial g_3}{\partial g_2}\right)_{T,P,\mu_3} = \frac{(1 - C_2 \bar{v}_2)}{(1 - C_3 \bar{v}_3)} \frac{(\partial \rho / \partial C_2)^0_{T,P,\mu_3} - (\partial \rho / \partial C_2)^0_{T,P,m_2}}{(\partial \rho / \partial C_3)_{T,P,m_2}} \quad (3)$$

where C_i is concentration in grams per milliliter of solution, $(C_i = g_i / V)$ and $\bar{v}_2$ and $\bar{v}_3$ are the partial specific volumes of components 2 and 3, defined by

$$\left(\frac{\partial \rho}{\partial g_i}\right)_{T,P,m_{j \neq i}} = \frac{1}{V}(1 - \bar{v}_i \rho) \quad (4)$$

where V is the volume of solution containing 1 g of principal solvent. The superscript 0 indicates extrapolation to zero protein concentration. It should be noted that in the density increment measurements of Eq. (3), it is the molal and not molar concentration of component 3 which is kept constant. As we shall see below, should the molar concentration be kept constant, the preferential interaction term measured will assume a different meaning from ξ_i. It can be shown[3] that, within a close approximation, at extrapolation to zero protein concentration and within the assumption that the density of the solution is a linear function of C_2, i.e.,

$$\rho = \rho_0 + (\partial \rho / \partial C_2)_{T,\mu_1,\mu_3} C_2 \quad (5)$$

where ρ_0 is the density of the solvent, Eq. (3) reduces to

$$\left(\frac{\partial g_3}{\partial g_2}\right)_{T,\mu_1,\mu_3} = \xi_3 = \frac{(1 - \phi_2' \rho_0) - (1 - \bar{v}_2^0 \rho_0)}{1 - \bar{v}_3 \rho_0} = \frac{\bar{v}_2^0 - \phi_2'}{1/\rho_0 - \bar{v}_3} \quad (6)$$

Here ϕ_2' is an apparent partial specific volume, defined as[4]

$$\left(\frac{\partial \rho}{\partial C_2}\right)_{T,\mu_1,\mu_3} \equiv 1 - \phi_2' \rho_0 \quad (7)$$

[3] E. Reisler, Y. Haik, and H. Eisenberg, *Biochemistry* **16**, (1977).
[4] E. F. Casassa and H. Eisenberg, *Adv. Protein Chem.* **19**, 287 (1964).

where ϕ_2' is a quantity obtained by measuring macromolecule solution densities after bringing these solutions into dialysis equilibrium with a large excess of solvent. Eq. (6) is more rigorously obtained from Eq. (2). In this approximation, we also set

$$\left(\frac{\partial g_3}{\partial g_2}\right)_{T,P,\mu_3} \approx \left(\frac{\partial g_3}{\partial g_2}\right)_{T,\mu_1,\mu_3}$$

This approximation eliminates the necessity of carrying out the operations under a hydrostatic head equal to the osmotic pressure of the protein solution; it results in an uncertainty which is negligibly small in aqueous solutions of proteins.

What is the meaning of the term "preferential interaction" or "preferential binding", as it is frequently called? This parameter may be considered from two points of view.

First, it is a measure of the excess of a solvent component in the immediate vicinity of the macromolecule, relative to the composition of bulk mixed solvent. In this mode, it may be regarded as the number of grams of component 3 per gram of macromolecule found inside of a dialysis bag, in excess (or deficit) of its concentration outside the bag, and it assumes the characteristics of binding. Yet, this term in no way implies the formation of stoichiometric complexes between macromolecules and solvent components, nor any strong binding. In fact, it is an average quantity over all degrees of interaction of the macromolecule with solvent components, covering the spectrum from strong stoichiometric complexation down to the weakest interaction, which reflects a momentary perturbation by the macromolecule of the rotational or translational freedom of motion of a molecule of a solvent component, and is manifested as a small perturbation of the activity coefficient of that component.

In its second mode, preferential interaction is a pure thermodynamic quantity. It is simply a reflection of the perturbation of the chemical potential or activity coefficient of component 3 by addition of macromolecule, since

$$
\begin{aligned}
\left(\frac{\partial m_3}{\partial m_2}\right)_{T,P,\mu_3} &= \left(\frac{\partial g_3}{\partial g_2}\right)_{T,P,\mu_3} \frac{M_2}{M_3} \\
&= -\frac{(\partial \mu_3/\partial m_2)_{T,P,m_3}}{(\partial \mu_3/\partial m_3)_{T,P,m_2}} \\
&= \frac{(\partial \ln \gamma_3/\partial m_2)_{T,P,m_3}}{1/m_3 + (\partial \ln \gamma_3/\partial m_3)_{T,P,m_2}}
\end{aligned}
\tag{8}
$$

$$
\begin{aligned}
\mu_i &= \mu_i^0 + RT \ln a_i = \mu_i^0 + RT \ln m_i\gamma_i \\
&= \mu_i^0 + RT \ln m_i + \mu_i^{(e)}
\end{aligned}
$$

where a_i is the molal activity of component i, γ_i is its activity coefficient and $\mu_i^{(e)}$ is the excess chemical potential and is equal to $RT \ln \gamma_i$. At chemical equilibrium, the chemical potentials and hence activities of solvent components must be identical throughout the system, namely, in the immediate vicinity of the macromolecule and at infinite distance in the bulk solvent. Since unless $(\partial \mu_3/\partial m_2)_{T,P,m_3}$ is zero, introduction of the macromolecule into the system causes a change in the activity coefficients of solvent components in its domain, the concentrations of the solvent components in this domain must also change. It is this factor which makes it possible to express preferential interaction in terms of net "binding" of solvent molecules to the macromolecule. This situation is depicted schematically in Fig. 1 for the case in which interaction between component 3 and macromolecule is favorable. It is shown in Fig. 1A that as the distance from the surface of the macromolecule increases, the free energy of interaction of component 3 with the macromolecule decreases. The corresponding distribution of solvent components is shown in Fig. 1B. The increasingly weak interaction as the distance from the macromolecule increases results in a situation similar to the distribution of small ions next to a charged macromolecule. The effect measured by ξ_3 is the total of the deviations of the amounts of components 1 and 3 from bulk solvent composition in the area from the surface of the macromolecule to the distance at which the interactions became vanishingly small.

Interaction between solvent components and macromolecules may be favorable or unfavorable. In fact, as shown in Fig. 1B, excess of one component in the domain of the macromolecule must mean by necessity deficiency of the other one. Thus, depending on whether component 3 interacts preferentially with component 2, or is preferentially excluded from its domain, the interaction parameter will take on positive or negative values. A negative value of $(\partial g_3/\partial g_2)_{T,\mu_1,\mu_3}$ indicates a deficiency of component 3, i.e., preferential interaction of the protein with component 1 (when the principal solvent is water, this is preferential hydration)[5]

$$\left(\frac{\partial g_1}{\partial g_2}\right)_{T,\mu_1,\mu_3} = -\frac{1}{g_3}\left(\frac{\partial g_3}{\partial g_2}\right)_{T,\mu_1,\mu_3} \tag{9}$$

The degree of preferential interaction $(\partial g_3/\partial g_2)_{T,\mu_1,\mu_3}$ is defined as zero when the number of grams of third component per gram of water is identical on both sides of the membrane at osmotic equilibrium. On the other hand, it is also possible to define as zero preferential interaction the state in which the amount of third component per milliliter of solution is identical on the two sides of the dialysis membrane, $(\partial C_3/\partial C_2)_{T,\mu_1,\mu_3}$, where C_i is the concentration of component i in gram per milliliter of solution. This

[5] S. N. Timasheff and M. J. Kronman, *Arch. Biochem. Biophys.* **83**, 60 (1959).

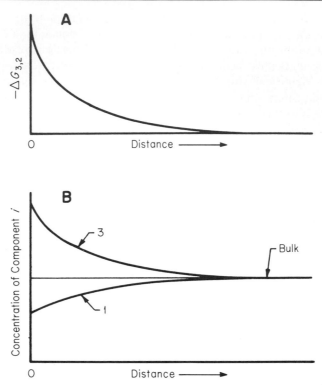

FIG. 1. Preferential interaction of component 3 with a macromolecule. (A) Variation of free energy of attraction, $-\Delta G_{3,2}{}^0$, of component 3 to macromolecule, as a function of its distance from the macromolecule. (B) Distribution of solvent components resulting from the interaction depicted in (A). As the distance to the macromolecule decreases, the amount of component 3 per unit volume increases over its value in the bulk solvent; simultaneously, the amount of component 1 decreases.

definition does not correctly reflect the thermodynamic interaction, since it does not consider the volume occupied by the protein and includes, as part of the interaction, dilution of the solvent by the addition of the protein.[6] At extrapolation to zero protein concentration, the two quantities are related by[7]

$$\left(\frac{\partial C_3}{\partial C_2}\right)^0_{T,P,\mu_3} = \frac{\bar{v}_1 C_3}{g_3}\left(\frac{\partial g_3}{\partial g_2}\right)^0_{T,P,\mu_3} - C_3\bar{v}_2 \tag{10}$$

The last term on the right-hand side of Eq. (10) represents the mass of component 3 displaced by the protein when the total volume of solution is

[6] H. Inoue and S. N. Timasheff, J. Am. Chem. Soc. **90**, 1890 (1968).
[7] M. E. Noelken and S. N. Timasheff, J. Biol. Chem. **242**, 5080 (1967).

kept constant. For a given system, the difference between the numerical values of

$$\left(\frac{\partial g_3}{\partial g_2}\right)_{T,P,\mu_3} \quad \text{and} \quad \left(\frac{\partial C_3}{\partial C_2}\right)_{T,P,\mu_3}$$

may be dramatic. In fact they may be of opposite signs. It is, thus, incorrect to say that the concentration units employed in these measurements are immaterial, since measurements at constant molarity, C_3, give numbers that can be misleading, indicating preferential hydration when, in fact, the opposite is true.

Preferential interaction, expressed as "binding," is related to total interactions between the macromolecule and solvent components by[8]

$$\left(\frac{\partial g_3}{\partial g_2}\right)_{T,\mu_1,\mu_3} = A_3 - g_3 A_1 \tag{11}$$

where A_i is the total amount of compound i "bound" to component 2, expressed in grams of component i per gram of component 2. Examination of Eq. (11) reveals that the sign of the preferential interaction parameter is defined by a balance between the total amounts of solvent components interacting with the protein, i.e., A_i and A_3 at a given solvent composition g_3. In fact, it is possible to have a situation in which the degree of interaction between a solvent component and protein increases monotonely as the bulk solvent composition is enriched with respect to that component, while preferential interaction displays a more complicated pattern, possibly even becoming negative at high contents of component 3.[9] This is depicted schematically in Fig. 2 for a situation in which hydration is taken as being independent of solvent composition, while solvation is proportional to the concentration of component 3. We see that at high contents of component 3, $(\partial g_3/\partial g_2)_{T,P,\mu_3}$ becomes negative, while A_3 is positive and, in fact, increasing in value.

A negative value of ξ_3 may also signify exclusion of component 3 from contact with the macromolecule. In fact, Eq. (11) may be rewritten as[3,10]

$$\left(\frac{\partial g_3}{\partial g_2}\right)_{T,\mu_1,\mu_3} = A_3 - E_3 - g_3 A_1 \tag{12}$$

where E_3 is the number of grams of component 3 excluded per gram of component 2. If A_1, A_3, and E_3 are independent of solvent composition, a plot of $(\partial g_3/\partial g_2)_{T,\mu_1,\mu_3}$ as a function of g_3 should give the hydration as the

[8] H. Inoue and S. N. Timasheff, *Biopolymers* **11**, 737 (1972).

[9] S. N. Timasheff and H. Inoue, *Biochemistry* **7**, 2501 (1968).

[10] D. W. Kupke, *in* "Physical Principles and Techniques of Protein Chemistry" (S. J. Leach, ed.), Part C, p. 1. Academic Press, New York, 1973.

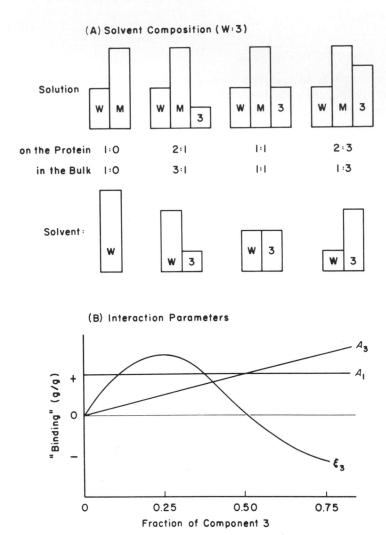

FIG. 2. Comparison of preferential and total interactions of solvent components with a macromolecule as a function of bulk solvent composition. (A) Composition of solvent bound to the macromolecule (upper diagrams) in equilibrium with bulk solvent of various compositions (lower diagrams). It is assumed that hydration is independent of solvent composition, while interaction with the co-solvent is proportional to its contents in the medium. Here M stands for macromolecule, W for water, and 3 for the cosolvent (component 3). (B) Resulting interaction parameters, A_1, A_3, and ξ_3. We note that because of the relation between solvent compositions on the protein and in the bulk, the preferential binding ξ_3 passes through a maximum and becomes negative at high concentrations of component 3, even though its absolute "binding" to the macromolecule continuously increases.

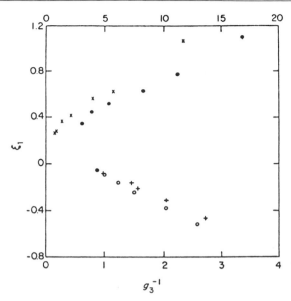

FIG. 3. Preferential interaction (hydration) parameter ξ_1 as a function of reciprocal salt concentration g_3 for (○) bovine serum albumin and (+) aldolase solutions in guanidine hydrochloride (lower scale); and for (●) Na DNA in NaCl and (×) Cs DNA in CsCl solutions (upper scale). Analysis of the data in terms of Eq. (13) gives, for serum albumin, $A_1 = 0.2$ g of water per gram of protein and a minimal value of $A_3 = 0.28$ g of guanidine hydrochloride per gram of protein (assuming that $E_3 = 0$); for aldolase, $A_1 = 0.16$ g/g, A_3 (minimal) $= 0.23$ g/g. For DNA, a similar analysis gives for Na DNA, $A_1 = 0.2$ g/g and a minimum E_3 of 0.054 (setting $A_3 = 0$); for Cs DNA, $A_1 = 0.24$ g/g and $(E_3 - B_3) = 0.07$ g/g. Therefore, while with the proteins both solvent components bind to the protein and there is preferential binding of component 3, in the case of DNA, both salts are preferentially excluded from the domain of the macromolecule due to Donnan exclusion, and there is a large preferential hydration which increases with decreasing salt concentration, even though the total hydration is independent of salt concentration. (Reprinted from Reisler *et al.*[3] with the kind permission of the American Chemical Society.)

slope and the balance between binding and exclusion of cosolvent as the intercept. Similarly, one may write

$$\left(\frac{\partial g_1}{\partial g_2}\right)_{T,\mu_1} = \xi_1 = A_1 - \frac{A_3 - E_3}{g_3} \tag{13}$$

An excellent example of such treatment of preferential interaction data, taken from studies by Reisler *et al.*[3], is given in Fig. 3. Here the authors clearly demonstrate a balance between hydration and solvation for two proteins in aqueous guanidine hydrochloride, and exclusion of salt in the case of DNA in CsCl and NaCl.

II. Significance of Preferential Interactions

At this point, it might seem pertinent to ask under what circumstances are preferential interactions significant? The most general answer is that preferential interactions are significant in all situations in which a chemical process is affected by solvent components, whether they are specific ligands, nonspecific additives, or solvent components, such as glycerol, CsCl, sucrose, urea, or alcohol.

Probably, the most familiar example of the importance of preferential interactions is found in the measurement of molecular weights by sedimentation equilibrium in a concentrated denaturant, such as the measurement of the molecular weight of the subunits of an enzyme in 6 M guanidine hydrochloride. In sedimentation experiments, the preferential interaction term enters through the buoyancy term, $(\partial\rho/\partial C_2)$ or its more familiar form, $(1 - \bar{v}_2\rho)$.

In order to obtain a correct molecular weight in sedimentation equilibrium, the buoyancy term must be expressed for macromolecules in chemical equilibrium with solvent components, namely, through ϕ_2'.

$$M_2 = \frac{2RT}{\omega^2(1 - \phi_2'\rho)} \frac{d \ln C_2}{d(r^2)} \tag{14}$$

If, in a three-component system, the partial specific volume used is that of the noninteracting macromolecule $\bar{v}_2$, then introduction of Eq. (6) into Eq. (14) shows that the molecular weight obtained is an apparent quantity $M_{2,\text{app}}$.

$$M_{2,\text{app}} = M_2\left[1 + \frac{(1 - \bar{v}_3\rho_0)}{(1 - \bar{v}_2{}^0\rho_0)}\left(\frac{\partial g_3}{\partial g_2}\right)_{T,P,\mu_3}\right] \tag{15}$$

$M_{2,\text{app}}$, which may be very different from M_2 (commonly by 20–35%), contains information not only on the molecular weight of the macromolecule but also on its interactions with solvent components. When ϕ_2' cannot be measured directly, a knowledge of the preferential interaction parameter measured by other techniques, such as differential refractometry or vapor phase equilibrium, nevertheless permits the calculation of the correct molecular weight. Neglect of interactions with solvent components may result in molecular weight values which are off (too high or too low) by 20–35% and which make it impossible to establish whether a given subunit enzyme is a trimer or a tetramer.

In similar manner, this parameter enters into the measurement of molecular weights by light scattering[11] and into the analysis of small-angle x-ray scattering data.[12]

[11] E. P. Pittz, J. C. Lee, B. Bablouzian, R. Townend, and S. N. Timasheff, Vol. 27[10], 209.
[12] H. Pessen, T. F. Kumosinski, and S. N. Timasheff, Vol. 27 [9], 151.

The preferential interaction term enters explicitly into expressions for the dependence of equilibrium constants on solvent components, or ligands. This has been rigorously demonstrated by Wyman.[13] In his theory of linked functions, Wyman[13] has shown that if a reaction is affected by a ligand L then the variation of the equilibrium constant for that reaction with ligand concentration is given by, at constant activity of all components other than ligand L,

$$\frac{d \ln K}{d \ln a_L} = \left(\frac{\partial m_L}{\partial m_{prod}}\right)_{T,P,\mu_L} - \left(\frac{\partial m_L}{\partial m_{react}}\right)_{T,P,\mu_L} \tag{16}$$

where a_L is the activity of the ligand and the right-hand side of the equation is the difference between the preferential interactions, expressed in molal units, of the ligand with the products and with the reactants.

Combination of Eq. (16) with Eq. (11) gives

$$\frac{d \log K}{d \log a_L} = \bar{\nu}_L^{prod} - \bar{\nu}_L^{react} - \frac{m_x}{m_{H_2O}}(\bar{\nu}_{H_2O}^{prod} - \bar{\nu}_{H_2O}^{react}) \tag{17}$$

where $\bar{\nu}_L$ and $\bar{\nu}_{H_2O}$ are the number of ligand and water molecules, respectively, bound to the products and reactants and m is molal concentration. Thus, a plot of log K versus log a_L should give as its slope the number of ligand molecules which enter into the reaction, if hydration does not change. Biochemists frequently use this relation to determine the number of protons involved in a pH-dependent enzymic reaction or indeed the number of other ligands, such as specific metal ions or cofactors. When ligand concentration is low, i.e., when interaction between ligand and macromolecule is strong, the change in hydration makes a negligibly small contribution to the results, since $m_{H_2O} = 55.5$, so that Eq. (17) reduces to

$$\frac{d \log K}{d \log a_L} \simeq \bar{\nu}_L^{prod} - \bar{\nu}_L^{react} = \Delta\bar{\nu}_L \tag{18}$$

where $\Delta\bar{\nu}_L$ is the difference between the number of ligand molecules bound to the products and to the reactants. Under such circumstances, the normal procedure used by biochemists gives the correct results. It should be noted, however, that in the plot log K must be plotted as a function of the log of the activity, and not the concentration, of the ligand. On the other hand, when the ligand concentration is high, for example, when an enzyme reaction is affected by salt at high concentration, (such as ribonuclease by Na_2SO_4),[14] i.e., when the interaction between the ligand and the macromolecule is weak, the approximation of Eq. (18) is no longer valid,

[13] J. Wyman, *Adv. Protein Chem.* **19**, 224 (1964).
[14] J. Winstead and F. Wold, *J. Biol. Chem.* **240**, PC3694 (1965).

the quantity measured is a change in preferential interactions, and the change in hydration must be taken into account in the calculation of the number of ligand molecules involved in the reaction.

An interesting consequence of Eq. (16) is that, in order for a ligand to drive a reaction, it need not bind to the macromolecule; in fact, it may be preferentially excluded from contact. This is true because the quantity of importance is the difference between the extents of preferential interaction with the two end products of a reaction. Therefore, if both parameters on the right-hand side of Eq. 16 are negative, and

$$\left| \frac{\partial m_L}{\partial m_{prod}} \right| < \left| \frac{\partial m_L}{\partial m_{react}} \right|, \text{ the } \frac{d \ln K}{d \ln a_L} \text{ is positive}$$

and the reaction will be driven by the nonbinding ligand. It is striking, therefore, that a solvent component may drive a reaction by actually not coming into contact with the molecules on which it is acting. One may refer to such a substance as a thermodynamic booster,[15] since its activity is strictly through a nonspecific indirect thermodynamic effect. An example of such a situation is found in the enhancement by glycerol of the *in vitro* self-assembly of microtubules.[16]

One may visualize this situation by a simple model, depicted schematically on Fig. 4A. Here, we have a hypothetical dimerization reaction. Each molecule, shown as a square, has surrounding it a volume from which a solvent component, say glycerol or CsCl, is preferentially excluded. Therefore, ξ_3 for monomer is negative. When a dimer is formed, however, part of this volume of exclusion is eliminated at the surface of intermolecular contact. As a result, while ξ_3 of the dimer still remains negative, its absolute numerical value is less than twice that of a monomer,

$$|\xi_3 \text{ (dimer)}| < |2\xi_3 \text{ (monomer)}|$$

and $d \ln K / d \ln a_L$ is positive. As a result, the reaction will be driven by an increase in the concentration of the solvent component which is preferentially excluded from contact with the macromolecules. The same situation may be described rigorously in terms of chemical potential changes by using Eq. (8).

The same argument may be extended to the precipitation or crystallization of enzymes by solvent components, such as 2-methyl-2,4-pentanediol (MPD) or indeed ammonium sulfate. MPD has been used in recent years to crystallize several proteins for x-ray structural studies. A notable case is ribonuclease A.[17] It is found that MPD is strongly ex-

[15] J. C. Lee, N. Tweedy, and S. N. Timasheff, *Biochemistry* **17**, 2783 (1978).
[16] J. C. Lee and S. N. Timasheff, *Biochemistry* **16**, 1754 (1977).
[17] M. V. King, B. S. Magdoff, M. B. Adelman, and D. Harker, *Acta Crystallogr.* **9**, 460 (1956).

(A) Self–Association

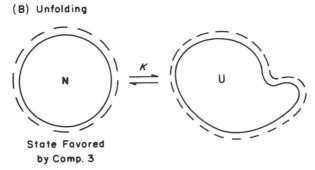

State Favored
by Comp. 3

(B) Unfolding

State Favored
by Comp. 3

FIG. 4. Schematic diagrams of changes in volume from which a cosolvent is excluded during chemical reactions. (A) Boosting of the self-association of two monomer (M) protein molecules into a dimer by exclusion of a solvent component. The zone of exclusion of component 3 is represented by the dashed lines around the macromolecules. It is seen that dimerization reduces the zone of exclusion; therefore, ξ_3 (dimer) is less negative than $2\xi_3$ (monomer) and the reaction is driven to the right. (B) Stabilization of a native protein structure against unfolding (or denaturation) by a solvent component which is excluded from contact with the protein. Change of the native (N) structure to the unfolded (U) structure increases the zone of exclusion of component 3 (shown by the dashed lines); therefore, $\xi_3(U)$ is more negative than $\xi_3(N)$ and the reaction is driven to the left.

cluded from contact with the enzyme.[18] Proper use of the preferential interaction parameter has shown rigorously the conditions at which the protein must come out of solution in this medium. These conditions are exactly those found empirically by the crystallographers in their studies.

In a similar manner, macromolecule structure stabilizing inert additives, such as glycerol and sucrose, which are used extensively by biochemists to stabilize enzymic activity, act through preferential exclusion rather than binding.[19] Again, if we wish, we may visualize this through simple models, such as that shown in Fig. 4B.

[18] E. P. Pittz and S. N. Timasheff, *Biochemistry* **17**, 615 (1978).
[19] S. N. Timasheff, J. C. Lee, E. P. Pittz, and N. Tweedy, *J. Colloid Surf. Chem.* **55**, 658 (1976).

Finally, the effect of solvent components, such as salts or alcohols, which are known to affect enzymic reactions without themselves entering into the pathway, may be explained by the effect of preferential interactions on the binding of substrate to the enzyme. The requirements for this are that $(\partial m_3/\partial m_{substrate})$ be negative and that the binding site on the enzyme have a lower concentration of component 3 (e.g., the salt) than the bulk solvent. Then, considerations such as those of Eqs. (8) and the discussion that follows show that the binding constant of the substrate to the enzyme will be increased, resulting, in effect, in an increase of the concentration of substrate seen by the enzyme and a corresponding effect on the rate of the enzymic reaction. This, of course, is not the only manner in which a solvent component may affect an enzymic reaction. There are, obviously, other modes, for example *via* a conformational change, or a change in charge distribution on the surface of the enzyme. Nevertheless, the presently described mechanism, which has been mostly neglected, should be seriously considered along with others when an attempt is made to interpret such solvent effects.

III. Methodology

A. What Must be Measured?

As shown in Eqs. (2) and (3), a determination of preferential interactions by densimetry requires the measurement of the density increment under two sets of circumstances namely, at constant m_3 and at constant μ_3.

The first $(\partial\rho/\partial g_2)_{T,P,m_3}$ is a density increment measured under conditions at which the molal concentrations of diffusible species are kept identical in the solvent and the macromolecule solution. This results in a partial specific volume $\bar{v}_2$ of the macromolecule in the given solvent without any consideration of interactions between it and components of the solvent. This will be referred to as the partial specific volume at constant molality. Operationally this is done by measuring the densities of pure solvent and of a protein solution dissolved in it.

The second $(\partial\rho/\partial g_2)_{T,P,\mu_3}$ is a density increment measured under conditions at which the chemical potential of component 3 is kept identical in the solvent and the macromolecule solution. This results in an apparent partial specific volume ϕ_2' of the macromolecule in chemical equilibrium with solvent components (i.e., in ϕ_2' interactions with solvent components are taken into account). This will be referred to as the partial specific volume at constant chemical potential. Operationally this is obtained by bringing the macromolecule solution to dialysis equilibrium with solvent and measuring the densities of the dialyzed protein and the dialysate. It should be noted that in this measurement the compositions of the solvent in the macromolecule solution and the reference solvent are no longer identical.

A third way of measuring the density increment has been used at times. This involves measurements at conditions at which it is the molar concentration of cosolvent which is kept identical in the macromolecule solution and the reference solvent. As pointed out above, the resulting density increment $(\partial\rho/\partial C_2)_{T,P,C_3}$ contains a contribution from interaction with solvent simply through dilution due to the volume occupied by the protein, and it must be corrected for this effect prior to use in any calculations.

B. Experimental Procedures

1. Preparation of Solvents. A water–nonaqueous solvent mixture of a given molality is prepared by mixing known weights of the two components. In protein chemistry, a dilute aqueous buffer is usually taken as the water component. However, even in the presence of buffers, mixing with the nonaqueous solvent component can lead to large shifts in pH. This requires adjustment of the pH of the buffer prior to mixing, to a value such that the final mixture will have the desired pH. Adjustment of pH after mixing is excluded, since this changes the solvent molality by an unknown amount. The pH of the buffer can be determined best in preliminary trial experiments. For example, the preparation of a pH 5.80 solution, containing 30% glycerol by volume from an 0.01 M sodium acetate, 0.02 M NaCl buffer requires that the pH of the initial buffer be 5.67. A similar solution of pH 5.80 containing 40% 2-chloroethanol by volume requires that the pH of the initial buffer be 5.20.

All solvent systems should be devoid of dust particles and should be made up with high purity distilled water. Since some dust particles are usually present, as can be detected easily by visual inspection under a strong beam of light, all solvents should be filtered through a sintered-glass filter before using them either for dissolving protein or for dialysis.

2. Preparation of Sample Solutions. In preferential interaction studies, the partial specific volumes of protein are usually measured under conditions at which the chemical potential and the molality of solvent components are, in turn, kept identical in the reference solvent and in the protein solution. The former condition can be attained operationally to a close approximation by dialyzing the protein solution against the solvent.

A most important requirement in measurements of preferential interactions is that the partial specific volume of the protein should be identical before and after dialysis against the dilute buffer solution in the absence of the third component, e.g., organic solvent or guanidine hydrochloride. A difference between the partial specific volumes of the protein solutions in dilute buffer prepared by procedures (a) and (b) (see below) is symptomatic of the presence of extraneous material in the protein sample. This could be buffer ions preferentially bound to the protein, some impurities (e.g., salts present in the protein sample), degradation products of the protein, or water molecules present because of incomplete drying of the protein; it is also possible that the protein may have partly denatured during the drying procedure in the oven or during dialysis against the solvent. Therefore, prior to starting density measurements in a three component system, the sample should be carefully checked for freedom from these undesirable factors.

a. PREPARATION OF SAMPLES AT CONSTANT MOLALITY OF COMPONENT 3, m_3. The protein is completely deionized either by exhaustive dialysis against distilled water (containing 10^{-3} M HCl in some cases, when the protein does not dissolve in or is not stable in water) at 4° or by passage through a mixed-bed ion-exchange column (Amberlite MB-1). The salt-free aqueous solution of protein obtained is filtered through a clean sintered-glass filter to eliminate dust, and then freeze-dried. Five or more protein aliquots, containing 5 to 20 mg of protein, are weighed into dry test tubes which had been tared before the addition of protein. These protein samples contained in their respective test tubes are then further dried under vacuum at 40° in the presence of phosphorus pentoxide for 24–48 hr. After the oven cools down to room temperature, air is admitted into it through concentrated sulfuric acid. As fast as possible after that 1–1.5 ml of solvent is added to each tube, and the tubes are sealed quickly with Parafilm. The protein is helped to dissolve completely by gentle rocking. Such motion not only helps to obtain a uniform protein solution but also helps to rid the sample of bubbles which are a source of errors in density measurements. If some bubbles persist, the solution may be kept still for some time until they disappear before starting the density measurements. The solution should also be free from precipitates and dust particles.

In an alternate procedure, described by Eisenberg and co-workers,[20,21] a stock solution of macromolecules is dialyzed against distilled water until salt-free. Densities and concentrations of aliquots of these solutions are measured, which permits the calculation of the weight of water and macromolecule in the salt-free solutions. In the next step, dried samples of component 3 are weighed into tightly closing plastic bottles. Next, appropriate weighed amounts of the desalted concentrated stock solution are added to component 3 in the plastic bottles. In this way, the molalities of the two solute components are known exactly. Reference solutions are

[20] G. Cohen and H. Eisenberg, *Biopolymers* **6,** 1077 (1968).
[21] E. Reisler and H. Eisenberg, *Biochemistry* **11,** 4572 (1969).

prepared by adding water to identical amounts of component 3. The densities of solvent and solutions are then measured. A small density correction then should be made to the solvent density using calibration curves, since it is difficult to add water to the same molality as in the macromolecule solutions.

b. PREPARATION OF SAMPLES AT CONSTANT CHEMICAL POTENTIAL OF COMPONENT 3, μ_3. From the protein solution remaining from the constant molality experiment or freshly prepared protein solutions within the concentration range of 5–20 mg/ml, 1.3 to 1.5 ml are introduced with a Pasteur pipette into thoroughly cleaned No Jax dialysis tubing (Union Carbide Corp., New York, New York). This tubing is precleaned by boiling several times in sodium bicarbonate and then in distilled water. If the solution shows the presence of some dust particles, it may be filtered through a Millipore filter before being placed into the dialysis bag. This material is then dialyzed against a large excess of clean solvent, after taking good care to rinse the sealed bag with a small amount of the same solvent in order to eliminate dust or small amounts of solution that may adhere to the top knot of the dialysis bag. Dialysis is usually carried out at 4° for 15–20 hr to avoid denaturation of the protein and then at 22° ± 1° for 3–4 hr before taking out any sample for density measurements.

3. *Protein Concentration Determination.* Protein concentrations are always determined *after* the density measurements. Most frequently, this is done spectrophotometrically by measuring the absorbance of the solution at the wavelength of maximal absorption. Since the protein concentration normally employed in density measurements are high, these solutions must be diluted with the same solvent prior to the absorbance measurements. This dilution is done gravimetrically on a balance sensitive to 10^{-5} g and should always be done immediately after the density measurements in order to minimize any concentration changes caused by the evaporative loss of water from the solutions. The gravimetrically obtained dilution factor may be converted to a volumetric one by using the measured density of the solvent and the solutions with the assumption that their volumes are additive. For those proteins whose extinction coefficients are not known, the concentrations should be determined by other methods, for example dry weight.[22] The extinction coefficient of a protein in a mixed solvent system may be different from that in water or low salt solvents. In such cases, the extinction coefficient in the mixed solvent system can be determined by uv spectroscopy. For this, absolutely identical aliquots of native protein stock solutions, 10–20 mg/ml in concentration, are diluted volumetrically to identical extents with the low salt buffer and with the mixed solvent. The uv spectra of these dilute solutions are

[22] D. W. Kupke and T. E. Dorrier, (1978) Vol. 48 [6], 155–162.

recorded on a spectrophotometer. If there is a time lag before a stable spectrum is obtained in the mixed solvent, repeated scans are taken until the spectrum no longer changes. Knowledge of the extinction coefficients of the protein in its native state and of the absorbance ratio measured at a given wavelength, ($OD_{native}/OD_{mixed\ solvent}$) makes it possible to calculate the extinction coefficient of the same protein in the mixed solvent.

When the concentration is determined by uv absorbance, the measured absorbance must be properly corrected for light scattering. For this, the absorption spectrum from 240 to 400 nm is recorded on a spectrophotometer. The light-scattering contribution to the absorbance is corrected for by the method of Leach and Scheraga.[23] In this procedure, extreme care must be exercised in the extrapolation of the log absorbance versus log wavelength plots. Any sample for which the point at 310 nm falls on the extrapolated line from 380 to 320 nm should be omitted, since this is an indication of a light scattering contribution sufficiently large to render the measured concentration value too uncertain to use in density calculations.

4. Density Measurements. The densities of the solvents and the protein solutions, a series of concentrations for each set of conditions, are measured with a precision density meter. The description here will be limited to the Anton Paar DMA-02C precision density meter, available from the Mettler Co.,[24] since this is the only instrument for which the present authors have first-hand knowledge. The principle of the technique has been described elsewhere in this series.[25] In summary, the density of an unknown liquid is measured by reference to a known standard. The difference between the densities of two samples is given by

$$\rho_1 - \rho_2 = (1/A)(T_1^2 - T_2^2) \tag{19}$$

where T is the time lapse during a preset number of periods, A is an instrument constant, and ρ is density. The instrument constant is obtained from calibration measurements with standard solutions of known density. The standards used are aqueous solutions of sucrose or NaCl of known concentrations. These materials are dried in a vacuum oven overnight before use. The density values for these standard solutions are listed in Table I. Experience has shown that the instrument constant A does not change within a testing period of 6 months. Usually measurements are made at 20° with the cell compartment maintained at this temperature ($\pm 0.02°$) with a refrigerated and heated circulating bath. It is found that a bath with a large tank capacity, e.g., 40 liters, will maintain the temperature suffi-

[23] S. J. Leach and H. A. Scheraga, *J. Am. Chem. Soc.* **82,** 4790 (1960).
[24] J. P. Elder, this volume [2].
[25] O. Kratky, H. Leopold, and H. Stabinger, Vol. 27 [5], p. 98.

TABLE I

DENSITIES OF NaCl OR SUCROSE SOLUTIONS IN
DISTILLED WATER AT 20°

Concentration (g/100 g solution)	Density (g/ml)	
	NaCl	Sucrose
1	1.00534	1.002120
2	1.01246	1.006015
4	1.02680	1.013881
6	1.04127	1.021855
8	1.05589	1.029942
10	1.07068	1.038143
12	1.08566	1.046462
14	1.10085	1.054900
16	1.11621	1.063460
18	1.13190	1.072147
20	1.14779	1.080959

ciently stable so that no highly sophisticated temperature regulating system is required. It is desirable, although not necessary, to keep the instrument in a constant temperature room of low relative humidity. This serves not only as a constant sink for the heat generated by the machine but also facilitates the drying of the cell. Experience has shown that the top of the instrument should be free of any material, e.g., notebook or solutions, which seem to hinder the radiation of heat.

The general procedures used for density measurements are as follows. The instrument is warmed up and the process of stabilization is monitored by using deionized distilled water as standard. It is not ready for further measurements until at least three consecutive samples of water give reading within $\pm 2 \times 10^{-5}$ sec at a preset count of 1×10^4. Once the instrument is stabilized, the cell is thoroughly washed with several 5-ml aliquots of deionized distilled water, followed by three aliquots, 2 ml each, of absolute ethanol. The cell is then air-dried with the pump on the instrument, a dry cell being indicated by a constant instrument reading. The densities of solvent and solutions can then be measured, the usual sequence of measurements being solvent, protein solutions, and again solvent. At the beginning, the densities of at least three solvent samples are measured, and any reading which varies by more than $\pm 5 \times 10^{-5}$ sec is discarded. For each individual sample, the precision is generally $\pm 2 \times 10^{-5}$ sec. Each experiment usually consists of measurements on four to five protein solutions within the concentration range of 3–20 mg/ml. Since all solution measurements are either at constant molality or constant chemical potential, extreme care should be taken to avoid evaporation and excessive handling. Generally, all systems are well sealed with Parafilm until the samples are ready for measurement. For measurements at constant molality, the protein solutions, which are well sealed in test tubes, are transferred to the density meter immediately before the measurement. This operation is carried out with a sterile disposable 1-ml syringe or a tuberculin syringe with a female Luer adapter, which permits easy injection of the solution into the cell. For measurements at constant chemical potential, the protein solutions are kept in the dialysis system with continuous stirring until just before the measurement. The protein solutions in their dialysis bags are then retrieved individually from the dialysis system with a stainless steel tricep or forceps. The solution is transferred from the dialysis bag to the density meter with a clean syringe with a needle. The needle facilitates the transfer by minimizing the exposure of the solution to air. The dialysis bag is pierced at the top with the needle and the solution in the bag is slowly drawn into the syringe, taking care that more than 0.8 ml is transferred to the syringe. A too rapid drawing of the solution may result in the formation of bubbles. Following the direct transfer of the solution from the bag to the syringe, the needle is replaced by a female Luer adapter for injection of the solution into the cell. The injection should be done carefully to avoid the formation of any bubbles, and the process can be followed by inspection through the window on the instrument. After addition of sample to the cell, 15–20 min are required for thermal equilibration. During the duration of the experiment, the syringe is covered tightly with a small plastic cap to prevent any concentration change by the evaporative loss of solvent water. After reading the time lapse T, the sample is withdrawn from the cell, using the same syringe, and it is transferred to a new tube which is immediately covered with Parafilm. This solution is used to determine the concentration as described above. After the measurement of each protein sample, the cell is washed and dried as described above. It is of particular importance that the readings for air before and after each sample be similar. A deviation of $>20 \times 10^{-5}$ sec indicates the presence of impurities in the cell, and the washing and drying procedures must be repeated. Upon completion of the experiments on the protein samples, the instrument constant is rechecked using the solvent and then deionized distilled water as standard. The time lapse measured for the solvent before and after the sample measurements should be the same within experimental error. If there is a big change in the time lapse, the data must be discarded. After all measurements, generally, it may be advantageous to fill the cell with a protein denaturant, e.g., 6 M guanidine hydrochloride or dilute soap solution, to ensure that no film of protein adheres to the walls of the cell. In this laboratory, it is found that such a precaution is necessary for successful experiments.

C. Sample Calculations

A sample calculation of density and partial specific volume measurements is given in Table II. The value of the density of water at 20° is 0.998232 gm/ml, taken from the International Critical Tables, 1928. Since the average values of T for water and solvent are 25.12847 and 25.48161 sec, respectively, and A has been determined to be 302.435 ml/g, then according to Eq. (19)

$$\Delta\rho = \rho_0 - \rho_{H_2O} = (1/302.435)(25.48161^2 - 25.12847^2)$$
$$= 0.0059095 \text{ g/ml}$$

where the subscripts denote solvent and water. Knowing $\rho_{H_2O} = 0.998232$ g/ml and $\rho_0 = \rho_{H_2O} + \Delta\rho$ gives $\rho_0 = 1.057327$ g/ml. Densities of the protein solutions at the various concentrations can be calculated in a similar manner (e.g., for solution 2 in Table II).

$$\Delta\rho = \rho - \rho_0 = (1/302.435)(25.49007^2 - 25.48161^2)$$
$$= 0.0014258 \text{ g/ml}$$

The apparent partial specific volume ϕ is calculated from the density data with the following equation.

$$\phi = (1/\rho_0)(1 - \Delta\rho/c)$$

where c is the concentration of protein in grams per millileter. For solution 2 in Table II, the protein concentration c is 6.277×10^{-3} g/ml. Hence,

TABLE II

DENSITY AND PARTIAL SPECIFIC VOLUME MEASUREMENTS
OF CHYMOTRYPSINOGEN A[a]

Samples	T (sec) 1	2	3	ρ (g/ml)	Protein concentration (g/ml $\times 10^3$)	$(\rho - \rho_0)/c$	ϕ (ml/g)
H_2O	25.12847	25.12847	25.12847	0.998232			
Solvent	25.48161	25.48160	25.48161	1.057327			
Solution 1	25.48617	25.48618	25.48616	1.058096	3.371	0.2282	0.7300
Solution 2	25.49007	25.49008	25.49007	1.058753	6.277	0.2271	0.7309
Solution 3	25.49333	25.49333	25.49333	1.059303	8.792	0.2247	0.7332
Solution 4	25.49696	25.49697	25.49696	1.059915	11.43	0.2264	0.7317
Solution 5	25.49959	25.49959	25.49959	1.060358	13.51	0.2244	0.7335
Solvent	25.48162	25.48163	25.48162				
H_2O	25.12846	25.12846	25.12848				

[a] Solvent is 30% glycerol, 0.01 M HCl, pH2, at 20°. Constant molality experiments; $A = 302.435$.

$$\frac{\Delta\rho}{c} - \frac{1.4258 \times 10^{-3}}{6.277 \times 10^{-3}} = 0.2271$$

and

$$\phi = \frac{(1 - 0.2271)}{1.057327} = 0.7309 \text{ ml/g}$$

The calculated values of ϕ are then plotted as a function of protein concentration and the extrapolated value to infinite dilution is taken as the partial specific volume $\bar{v}_2$. As an example, Fig. 5 shows plots of ϕ against protein concentration for chymotrypsinogen A in aqueous glycerol solutions under conditions of constant chemical potential and constant molality of glycerol. In most cases, we have obtained a good linear relationship between the apparent partial specific volume and protein concentration, which allows extrapolation of ϕ to infinite dilution of protein. This extrapolation is usually made by using the least-squares method. At constant molality of diffusible component, the extrapolated value is denoted by $\bar{v}_2$, whereas that at constant chemical potential is ϕ_2'.

From density measurements carried out at conditions at which the molality and the chemical potential of the diffusible components is kept identical, in turn, in the protein solution and the reference solvent, it becomes possible to determine the extent of preferential interaction $(\partial g_3/\partial g_2)_{T,\mu_1,\mu_3} = \xi_3$ of the solvent components with the macromolecules

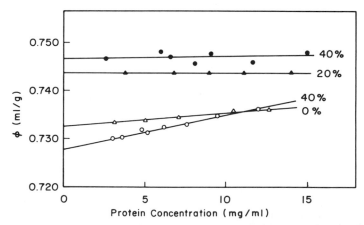

FIG. 5. Apparent partial specific volumes as a function of protein concentration for chymotrypsinogen A in aqueous glycerol solutions (0.01 M HCl, pH 2) at 20°. Open symbols, experiments at constant molality of glycerol ($\bar{v}_2$); closed symbols, experiments at constant chemical potential of glycerol (ϕ_2'). The numbers in the figure indicate the glycerol concentration in volume percent (from K. Gekko and S. N. Timasheff, unpublished data).

TABLE III

PARTIAL SPECIFIC VOLUMES AND PREFERENTIAL INTERACTION PARAMETERS OF CHYMOTRYPSINOGEN A IN GLYCEROL–WATER MIXTURES AT $20°$[a,b]

Glycerol (vol %)	g_3 (g/g H$_2$O)	ϕ_{2,m_3}^* (ml/g)	ϕ_{2,μ_3}' (ml/g)	$-\left(\dfrac{\partial g_3}{\partial g_2}\right)_{T,\mu_1,\mu_3}$ (g/g)	$\left(\dfrac{\partial g_1}{\partial g_2}\right)_{T,\mu_1,\mu_3}$ (g/g)	$-\left(\dfrac{\partial m_3}{\partial m_2}\right)_{T,P,m_2}$ (mole/mole)
0	0	$0.732_7 \pm 0.001$	$0.732_3 \pm 0.001$			
10	0.1398	$0.730_1 \pm 0.001$	$0.738_4 \pm 0.002$	0.040 ± 0.012	0.285	11.3 ± 3.4
20	0.3134	$0.729_4 \pm 0.001$	$0.743_8 \pm 0.001$	0.081 ± 0.008	0.258	22.6 ± 2.4
30	0.5348	$0.726_9 \pm 0.001$	$0.745_4 \pm 0.001$	0.123 ± 0.010	0.229	34.2 ± 2.8
40	0.8273	$0.727_1 \pm 0.001$	$0.747_4 \pm 0.002$	0.161 ± 0.032	0.195	45.0 ± 8.9

[a] All solutions contain 0.01 M HCl (pH 2).
[b] K. Gekko and S. N. Timasheff, unpublished date.

TABLE IV

PARTIAL SPECIFIC VOLUMES AND PREFERENTIAL INTERACTION PARAMETERS OF PROTEINS[a]

Protein	$\bar{v}_2^0$ (native) (ml/g)	$\bar{v}_2^0$ (GuHCl) (ml/g)	ϕ_2' (ml/g)	ξ_3 (g/g)	$(\partial m_3/\partial m_2)_{\mu_1,\mu_3}$ (mole/mole)
RNase A	0.696 ± 0.001	0.694 ± 0.001	0.694 ± 0.001	0.0 ± 0.01	0
Lysozyme	0.702 ± 0.001	0.704 ± 0.002	0.694 ± 0.001	0.09 ± 0.02	14 ± 3
Tubulin[b]	0.736 ± 0.001	0.736 ± 0.002	0.725 ± 0.002	0.10 ± 0.02	55 ± 11
Chymotrypsinogen A	0.733 ± 0.001	0.729 ± 0.001	0.712 ± 0.002	0.15 ± 0.02	41 ± 5
α-Chymotrypsin	0.738 ± 0.001	0.732 ± 0.001	0.713 ± 0.002	0.17 ± 0.03	44 ± 7
Bovine serum albumin	0.735 ± 0.002	0.724 ± 0.001	0.717 ± 0.001	0.06 ± 0.01	44 ± 7
Carboxypeptidase A[b]	0.748 ± 0.001	0.741 ± 0.002	0.735 ± 0.001	0.05 ± 0.01	20 ± 4
Lactate dehydrogenase (BH)[b]	0.741 ± 0.001	0.739 ± 0.001	0.736 ± 0.002	0.03 ± 0.01	10 ± 3
Catalase[b]	0.730 ± 0.001	0.726 ± 0.001	0.725 ± 0.002	0.01 ± 0.01	6 ± 6
β-Lactoglobulin	0.750 ± 0.002	0.728 ± 0.002	0.719 ± 0.001	0.08 ± 0.02	16 ± 4
Lima bean trypsin inhibitor	0.699 ± 0.001	0.699 ± 0.001	0.698 ± 0.003	0.01 ± 0.04	1 ± 4
α-Lactalbumin	0.704 ± 0.001	0.701 ± 0.001	0.698 ± 0.002	0.03 ± 0.02	5 ± 3

[a] From Lee and Timasheff.[27]

[b] Proteins which are reduced and S-carboxymethylated for $\bar{v}$ measurements in 6 M guanidine hydrochloride.

with the use of Eq. (2) or in terms of the partial specific volumes at infinite dilution, since

$$(1 - \phi_2'\rho_0) = (1 - \bar{v}_2{}^0\rho_0) + \xi_3(1 - \bar{v}_3\rho_0)$$

The partial specific volume of the third component $\bar{v}_3$ is measured with the density meter following the same procedure as those described above. Taking $\bar{v}_3 = 0.7680$ ml/g, $\bar{v}_2 = 0.729_4$ ml/g, and $\phi_2' = 0.743_8$ ml/g, gives

$$\begin{aligned}\xi_3 &= \rho_0(\bar{v}_2{}^0 - \phi_2')/(1 - \bar{v}_3\rho_0) \\ &= 1.057327(0.7294 - 0.7438)/(1 - 0.7680 \times 1.057327) = -0.081\end{aligned}$$

Positive values of ξ_3 signify preferential binding of component 3 to protein; negative values indicate preferential hydration.

Tables III and IV show data for chymotrypsinogen A in aqueous glycerol solutions[26] as an example of preferential hydration, and for a variety of proteins in 6 M guanidine hydrochloride[27] as an example of preferential binding, respectively.

[26] K. Gekko and S. N. Timasheff, to be published.
[27] J. C. Lee and S. N. Timasheff, *Biochemistry* 13, 257 (1974).

[4] The Calculation of Partial Specific Volumes of Proteins in 6 M Guanidine Hydrochloride

By JAMES C. LEE and SERGE N. TIMASHEFF

In recent years, sedimentation equilibrium has been increasingly used for the determination of the number of subunits in associated systems. The particular advantage of this method is that it is based on firm thermodynamic theory, whereas some of the more popular techniques, such as column chromatography and gel electrophoresis in the presence of denaturants are based on empirical correlations. When a protein does not assume the proper conformation in the presence of the denaturant, the values of molecular weights obtained by these techniques may be subjected to major errors. Sedimentation equilibrium, on the other hand, yields data which are independent of the structure of the protein, be it random coil, rod-like, or globular.

A commonly used procedure to determine the number of subunits in a protein is to measure the molecular weight of the native associated protein in the presence of low ionic strength buffers, followed by a measurement of the molecular weight of the dissociated subunits in the presence of high concentrations of a denaturant, such as guanidine hydrochloride.

Copyright © 1979 by Academic Press, Inc.

The determination of accurate molecular weights from ultracentrifuge data requires exact knowledge of the partial specific volume $\bar{v}$ of the protein in the given medium. Lack of knowledge of the correct value of $\bar{v}$ in the presence of the denaturant can lead to erroneous conclusions on the stoichiometry of subunit composition in the assembly of macromolecules, the uncertainty at times being sufficient to render a distinction between three and four subunits impossible.[1,2] Unfortunately, the frequent unavailability of proper means to measure the partial specific volume of a protein, especially when it is present only in small quantity, has led to the practice of estimating this quantity in the presence of the denaturant.

The partial specific volumes of proteins can be calculated from their amino acid compositions,[3] correct values for $\bar{v}$ usually being obtained in dilute buffers.[4] Although values of the partial specific volume have been calculated for glycoproteins from the amino acid and carbohydrate compositions,[5] it has not yet been shown systematically that the calculated value correctly reflects this quantity for the glycoprotein in dilute buffers. These calculations, however, cannot be applied directly to proteins in the presence of high concentrations of denaturants, since in that case protein–solvent interaction effects must be taken into account. In a solution of macromolecules and two low molecular weight components, such as guanidine hydrochloride (GuHCl) and water, the macromolecules generally interact with one or both of the solvent components. Preference for one or the other component may result not only from the existence of specific binding sites for that component but also from nonspecific attractive forces, as well as the exclusion of the other solvent component from the domain of the macromolecule. Such preferential interactions are reflected in changes in the measured value of the apparent partial specific volume, and, as a consequence, they must be taken into account when molecular weights are determined in three-component systems. Furthermore, there may be a volume change involved when the protein is transferred from dilute buffer to denaturant and such a change may also be reflected in the value of $\bar{v}$. In the past, the partial specific volumes of proteins in GuHCl have been rarely measured, the practice being to decrease the value of the partial specific volume of the native protein by 0 to 0.02

[1] E. Reisler and H. Eisenberg, *Biochemistry* **8**, 4572 (1969).
[2] D. L. Barker and W. P. Jencks, *Biochemistry* **8**, 3879 (1969).
[3] E. J. Cohn, and J. T. Edsall, "Proteins, Amino Acids, and Peptides," p. 372. Van Nostrand-Reinhold, Princeton, New Jersey, 1943.
[4] T. L. McMeekin and L. Marshall, *Science* **116**, 142 (1952).
[5] R. A. Gibbons, *in* "Glycoproteins" (A. Gottschalk, ed.), p. 29. Elsevier, Amsterdam, 1966.

ml/g, following the few observations reported in the literature.[6-9] Our recent study on the interactions of proteins with GuHCl by density measurements, however, have shown that while in about 50% of the cases this approximation is quite valid, there are proteins in which this change may be as great as 0.03 ml/g.[10] An error of such magnitude in the partial specific volume may lead to very serious errors in the calculated molecular weight, when this value of $\bar{v}$ is used in conjunction with a sedimentation equilibrium experiment in 6 M GuHCl.

Based on the results of our study,[10] a simple method is proposed to calculate the apparent partial specific volumes of proteins in 6 M GuHCl, which greatly reduces the uncertainties inherent in the previously used procedures. The only information required is the amino acid composition of the protein.

Theory

The buoyancy of a sedimenting protein may be affected by GuHCl denaturation. It has been found that, in most cases, the major change in the buoyancy is due to interactions of the denatured protein with solvent components and that the change in volume of protein upon denaturation in 6 M GuHCl is generally small, its effect on the partial specific volume of the protein being negligible. In practice, therefore, one may assume that ϕ_2, the partial specific volume at infinite dilution of the protein, measured in 6 M GuHCl in such manner that the molalities of solvent components in the protein solution and in the reference solvent are identical is equal to $\bar{v}_2$, the partial specific volume of the native protein in dilute buffer.

In a three-component system, such as water = component 1, protein = component 2, GuHCl = component 3, the apparent molecular weight $M_{2,\text{app}}$ determined in a standard sedimentation equilibrium experiment is related to the true molecular weight M_2 after extrapolation to zero protein concentration, by:

$$M_{2,\text{app}} = \frac{2RT}{(1 - \bar{v}_2\rho)_{m_3}\omega^2} \frac{d \ln c_2}{d(r^2)}$$
$$= M_2 \left[1 + \frac{(1 - \bar{v}_3\rho)_{m_2}}{(1 - \bar{v}_2\rho)_{m_3}} \left(\frac{\partial g_3}{\partial g_2}\right)_{T,p,\mu_3} \right] \tag{1}$$

[6] W. W. Kielley and W. F. Harrington, *Biochim. Biophys. Acta* **41**, 401 (1960).
[7] E. P. K. Hade and C. Tanford, *J. Am. Chem. Soc.* **89**, 5034 (1967).
[8] E. Marler, C. A. Nelson, and C. Tanford, *Biochemistry* **3**, 279 (1964).
[9] P. A. Small and M. E. Lamm, *Biochemistry* **5**, 259 (1966).
[10] J. C. Lee and S. N. Timasheff, *Biochemistry* **13**, 257 (1974).

where ρ is the density of the solvent, c_2 is the protein concentration in any units, R is the universal gas constant, T is the absolute temperature, r is the distance from the center of rotation, p is pressure, ω is the angular acceleration, g_i is the concentration of component i in grams per gram of water in the system, $\bar{v}_i$ is its partial specific volume, and μ_3 is the chemical potential of component 3. The subscripts m_j indicate that the partial specific volume of component i is measured at conditions at which the molality m_j of component $j(j \neq i)$ is identical in the bulk solvent and in the solution of component i. Thus, the experimental value of the apparent molecular weight, after extrapolation of the protein concentration to zero, contains a contribution from the preferential interaction between the protein and solvent components,[11] $(\partial g_3/\partial g_2)_{T,p,\mu_3}$. The true molecular weight M_2 may be obtained from sedimentation equilibrium, if the partial specific volume of the protein is measured at conditions at which the chemical potential of component 3 in the protein solution is identical with that in the bulk solvent. Thus, at extrapolation to zero protein concentration

$$M_2 = \frac{2RT}{(1 - \phi_2'\rho)} \frac{d \ln c_2}{d(r^2)} \tag{2}$$

where ϕ_2' is the apparent partial specific volume of the protein in chemical equilibrium with the solvent. At infinite dilution of the macromolecular species, this quantity is related to the preferential interaction parameter

$$\left(\frac{\partial g_3}{\partial g_2}\right)_{T,p,\mu_3} \simeq \left(\frac{\partial g_3}{\partial g_2}\right)_{T,\mu_1,\mu_3} \equiv \xi_3$$

by

$$\phi_2' = \phi_2 - \xi_3(1/\rho - \bar{v}_3) \tag{3}$$

since,[12,13] at constant temperature and pressure,

$$\xi_3 = \left[\left(\frac{\partial \rho}{\partial g_2}\right)_{\mu_3} - \left(\frac{\partial \rho}{\partial g_2}\right)_{m_3} \right] \Big/ \left(\frac{\partial \rho}{\partial g_3}\right)_{m_2} \tag{4}$$

The preferential interaction parameter is related to the actual amount of solvent components bound to the protein by[14]

$$\xi_3 = A_3 - g_3 A_1 \tag{5}$$

where all interactions are expressed in units of grams ligand bound per gram of protein; A_3 is absolute solvation, i.e., the actual amount of dena-

[11] S. N. Timasheff, Adv. Chem. Ser. **125**, 327 (1973).
[12] E. F. Casassa and H. Eisenberg, Adv. Protein Chem. **19**, 287 (1964).
[13] G. Cohen and H. Eisenberg, Biopolymers **6**, 1077 (1968).
[14] H. Inoue and S. N. Timasheff, Biopolymers **11**, 737 (1972).

turant bound to the protein; A_1 is the total hydration, and g_3 is the solvent composition, expressed as grams of denaturant per gram of water. Combining Eqs. (3) and (5) gives

$$\phi_2' = \phi_2 - (1/\rho - \bar{v}_3)(A_3 - g_3 A_1) \tag{6}$$

Therefore, the apparent partial specific volume of the protein in chemical equilibrium with solvent ϕ_2' can be calculated if ϕ_2, A_3, A_1, g_3, ρ, and $\bar{v}_3$ are known. The parameter g_3 is defined by solvent composition; for 6 M GuHCl, its value is 1.007 g GuHCl per gram of water.[10] Vales of ρ and $\bar{v}_3$ are available in the literature for a number of solvent systems; for 6 M GuHCl, $\rho = 1.1418$ g/ml at 20° and $\bar{v}_3 = 0.763$ ml/g. The problem reduces itself, therefore, to the calculation of the parameters A_1, A_3, and ϕ_2. We propose simple internally consistent rules for calculating these parameters from the amino acid compositions of the proteins. A_1 can be calculated[10] for each protein from the hydration of its constituent amino acids according to the method of Kuntz.[15] A_3 is calculated by assuming that one GuHCl molecule is bound to each pair of peptide bonds and to each aromatic amino acid side chain.[10] As stated above, ϕ_2 may be assumed to have a value identical to $\bar{v}_2{}^0$, which, in turn, can be calculated from the amino acid composition by the method of Cohn and Edsall.[3]

Method of Calculation

The calculation for α-chymotrypsin is presented to illustrate the proposed method and the amino acid composition of the protein is expressed as residues per 10^5 grams of protein.

Calculation of $\bar{v}_2{}^0$. The value of the partial specific volume of a protein is related to the specific volumes $\bar{v}_i$ of its constituent amino acids by[3]

$$\bar{v}_2{}^0 = \frac{\Sigma N_i(W_i \bar{v}_i)}{\Sigma N_i W_i} \tag{7}$$

where N_i is the number of residue of amino acid of type i and W_i is the residue weight, i.e., molecular weight minus 18.0. For convenience, the values of W_i and $\bar{v}$ are listed in Table I. Using the amino acid composition of α-chymotrypsin, in Table II, a value of $\bar{v}_2{}^0 = 0.732$ is calculated. This agrees reasonably well with the value of 0.738 ± 0.001 measured in dilute buffer. The calculated value for $\bar{v}_2{}^0$ is then set equal to ϕ_2.

Calculation of A_1. The total hydration of a protein, A_1, is given by

$$A_1 = \Sigma H_i N_i \tag{8}$$

where H_i is the hydration of amino acid species i. The values of H_i, determined by Kuntz,[15] are listed in Table I. As summarized in Table II, for

[15] I. D. Kuntz, *J. Am. Chem. Soc.* **93**, 514 (1971).

TABLE I
AMINO ACID DATA FOR ESTIMATION OF $\bar{v}_2^0$ AND
HYDRATION OF PROTEINS[a]

Amino acid[a]	W_i (g/mole)	$\bar{v}_i$ (ml/g)	$W_i\bar{v}_i$ (ml/mole)	Hydration (mole H_2O/mole residue)
Ala	71.1	0.74	52.6	1.5
Arg	156.2	0.70	109.3	3.0
Asn				2.0
Asp	115.1	0.60	69.1	6.0
½ Cys	103.2	0.61	63.0	1
Gln				2
Glu	129.1	0.66	85.2	7.5
Gly	57.1	0.64	36.5	1
His	137.2	0.67	91.9	4
Ile	113.2	0.90	101.9	1
Leu	113.2	0.90	101.9	1
Lys	128.2	0.82	105.1	4.5
Met	131.2	0.75	98.4	1
Phe	147.2	0.77	113.3	0
Pro	97.1	0.76	73.8	3.0
Ser	87.1	0.63	54.9	2
Thr	101.1	0.70	70.8	2
Trp	186.2	0.74	137.8	2
Tyr	163.2	0.71	115.9	3
Val	99.1	0.86	85.2	1

[a] M_i = molecular weight in neutral form; $W_i = M_i - 18.0$; $\bar{v}_i$ = specific volume of amino acid residue.

α-chymotrypsin, a total hydration of 1948.5 moles of H_2O per 10^5 g of protein is obtained. This corresponds to $(1948.5 \times 18/10^5) = 0.351$ g of H_2O per gram of protein.

Calculation of A_3. The extent of GuHCl binding to protein in 6 M GuHCl is given by[10]

$$\frac{\text{moles GuHCl}}{\text{mole protein}} = \frac{\Sigma N_i - 1}{2} + \Sigma N_{\text{aromatic}} \qquad (9)$$

Setting arbitrarily, as above, the protein molecular weight equal to 10^5 grams per mole,

$$A_3 = \frac{\text{mole GuHCl}}{\text{mole protein}} \times \frac{\text{MW (GuHCl)}}{\text{MW (protein)}}$$

$$= \frac{95.54}{100,000} \left(\frac{\Sigma N_i - 1}{2} + \Sigma N_{\text{aromatic}} \right) \qquad (9a)$$

For α-chymotrypsin, 954 residues per 10^5 grams corresponds to 953 peptide bonds. Assuming an interaction of one GuHCl molecule per pair of

TABLE II

CALCULATION OF PARTIAL SPECIFIC VOLUME AND TOTAL
HYDRATION FOR α-CHYMOTRYPSIN[a]

Amino acid	N_i (residues/10^5 g protein)	N_iW_i	$N_iW_{i'i}$	H_iN_i
Ala	86	6114.6	4523.6	129
Arg	15	2343	1639.5	45
Asn	54			108
Asp	35	10243.9	6149.9	210
½ Cys	39	4024.8	2457	39
Gln	39			78
Glu	20	7616.9	5026.8	150
Gly	90	5139	3285	90
His	8	1097.6	735.2	32
Ile	39	4414.8	3974.1	39
Leu	74	8367.8	7540.6	74
Lys	55	7051	5780.5	247.5
Met	8	1049.6	787.2	8
Phe	23	3385.6	2605.9	0
Pro	35	3398.5	2583	105
Ser	108	9406.8	5929.2	216
Thr	89	8997.9	6301.2	178
Trp	31	5772.2	4271.8	62
Tyr	16	2611.2	1854.4	48
Val	90	8919	7668	90
Total	954	99954.2	73112.9	1948.5

[a] $\bar{v}_2^0 = 0.732$ ml/g; $A_1 = 0.351$ g/g.

peptide bonds, there are 477 molecules of GuHCl interacting with the protein backbone. Furthermore, assuming an interaction of one GuHCl molecule per aromatic side chain, another 78 molecules of GuHCl are involved in the complex, giving a total of 555 moles of GuHCl per 10^5 g of α-chymotrypsin, or $A_3 = 0.530$ g of GuHCl/g of protein.

Calculation of ϕ_2'. Knowing $\rho = 1.1418$ g/ml, $g_3 = 1.007$ g/g, $\bar{v}_3 = 0.763$ ml/g, substitution of the calculated values of $\bar{v}_2^0$, A_1, and A_3 into Eq. (6) yields directly ϕ_2'. For α-chymotrypsin, this results in a calculated value of $\phi_2' = 0.712$ml/g, which is in good agreement with the experimentally determined value of 0.713 ml/g.[10]

The applicability of this procedure has been tested on a few cases taken from the literature.[16] A general improvement in the agreement between the molecular weights of the native and denatured proteins is obtained when this method of estimating ϕ_2' is used. More recently, there have appeared in the literature several reports in which our method to es-

[16] J. C. Lee and S. N. Timasheff, *Arch. Biochem. Biophys.* **165**, 268 (1974).

TABLE III
SUMMARY OF MOLECULAR WEIGHT AND $\bar{v}$ OF PROTEINS
AND THE CALCULATED ϕ_2' IN GuHCl

Protein	$M_w(N)^a$	$M_w(D)$	$\bar{v}_2{}^b$	ϕ_2' calc.
Pseudomonas 7A	140,000		0.735	
Glutaminase–Asparaginase[c]		36,000		0.735
Pig Heart Coenzyme A	92,000		0.634	
Transferase[d]		45,600		0.730[e]
		45,300		0.729[f]
Horse Serum Butyryl—	320,000		0.723	
cholinsterase[g]		86,700		0.718[e]
		81,000		0.707[f]

[a] Molecular weight of native protein (N) and of protein denatured in 6 M GuHCl (D).

[b] Partial specific volume of protein either calculated from amino acid composition or measured in dilute buffer.

[c] Holcenberg *et al.*[17]

[d] White and Jencks.[18]

[e] Calculated values of ϕ_2' corrected for preferential binding of solvent components to the protein only.

[f] Calculated values of ϕ_2' with additional correction for preferential binding of GuHCl to the carbohydrates.

[g] Teng *et al.*[19]

timate ϕ_2' in GuHCl has been used successfully.[17–19] Two of these systems involve glycoproteins.[18,19] These data are summarized in Table III. For the glycoproteins, additional GuHCl binding to the protein is calculated by assuming that each residue of carbohydrate binds one molecule of GuHCl. In the case of coenzyme A transferase, the total carbohydrate contents are only about 1.6% by weight. Therefore, the final results are essentially identical whether or not corrections for additional GuHCl binding to carbohydrates are made. When the amount of carbohydrate increases, such as in butyrylcholinesterase which contains about 20% carbohydrate by weight, the values of ϕ_2' and, therefore, the apparent molecular weight are substantially different from those obtained with neglect of carbohydrate. The resultant molecular weight of the denatured protein is much closer to the expected value when the additional correction of GuHCl binding to carbohydrate is made. In all the calculations with glycoproteins, A_1 is calculated without any contribution of carbohydrate hydration, since such data are not available. Furthermore, the assumption

[17] J. S. Holcenberg, D. C. Teller, and J. Roberts, *J. Biol. Chem.* **251,** 5375 (1976).

[18] H. White and W. P. Jencks, *J. Biol. Chem.* **251,** 1708 (1976).

[19] T. L. Teng, J. A. Harpst, J. C. Lee, A. Zinn, and D. M. Carlson, *Arch. Biochem. Biophys.* **176,** 71 (1976).

that each residue of carbohydrate binds one molecule of GuHCl is not based on any experimental observations. Even with such drawbacks, the initial attempts to estimate ϕ_2' for glycoproteins seem to yield satisfactory results, although great caution must be exercised in drawing conclusions from calculations based on untested hypotheses. The simplicity of this method should permit protein chemists to arrive at better estimates of ϕ_2' than was possible up to the present and to achieve better accuracy in molecular weight determinations in 6 M GuHCl by the method of sedimentation equilibrium.

A cautionary note, however, seems in order. As has been stated above, a volume change which occurs when a protein system is transferred from water to 6 M GuHCl should also be reflected in the change in the partial specific volume. In such a case ϕ_2' is not equal to $\bar{v}_2^0$. Generally, the magnitude of such volume changes is small and can be neglected. There are cases, however, where this change is significant. For these proteins, the decrease in apparent partial specific volume calculated from binding alone is insufficient. At present, there is no way of predicting the volume change which accompanies the unfolding of any given protein. There are indications, however, that the magnitude of this change is similar in various denaturants. For example, β-lactoglobulin denaturation, whether in 6 M GuHCl,[10] 6.4 M urea,[20] or 40% 2-chloroethanol,[11] results in a volume decrease close to 600 ml/mole. Thus, the knowledge of a volume change when a protein is unfolded in any given system might be used as an indication of whether this will affect the calculation of ϕ_2' in 6 M GuHCl by the method of this chapter. Furthermore, since the presence of lipids may interfere with the protein–GuHCl interactions, as well as with the calculation of $\bar{v}_2$, the presently described method of estimating ϕ_2' is not applicable to lipoproteins. The best values of molecular weights will be obtained, quite evidently, if ϕ_2' is measured directly either by density measurements[10,21,22] or by the D_2O–H_2O technique.[23-25]

[20] L. K. Christensen, *C. R. Trav. Lab. Carlsberg* **28,** 37 (1952).

[21] O. Kratky, H. Leopold, and H. Stabinger, Vol. 27, p. 98.

[22] D. V. Ulrich, D. W. Kupke, and J. W. Beams, *Proc. Natl. Acad. Sci. U.S.A.* **52,** 349 (1964).

[23] S. J. Edelstein and H. K. Schachman, Vol. 27 p. 82.

[24] J. O. Thomas, and S. J. Edelstein, *Biochemistry* **10,** 477 (1971).

[25] W. L. Gagen, *Biochemistry* **5,** 2553 (1966).

[5] The Role of Micelles in Protein–Detergent Interactions[1]

By JACQUELINE A. REYNOLDS

I. Introduction

The interaction of proteins with amphiphilic ligands has received increasing attention in recent years. The practical as well as theoretical importance of these interactions are illustrated by the following examples.

1. Investigations of the molecular properties of membrane proteins and serum lipoporoteins have for the most part required the use of detergents as solubilizing agents and as probes for hydrophobic binding sites.[2]

2. The popular technique of identifying and cataloging polypeptides on the basis of their mobilities in sodium dodecyl sulfate–polyacrylamide gel electrophoresis is based on a specific type of detergent–protein interaction.[3]

3. Two-dimensional polyacrylamide gel electrophoresis using sodium dodecyl sulfate in one direction and the nonionic detergent, Triton X-100, in the other has been used to identify polypeptides containing long hydrophobic sequences or regions.[4] This technique relies on differences in binding characteristics between water-soluble and intrinsic membrane proteins in that the former do not in general bind nonionic detergents.

It is apparent from these few examples that an understanding of the thermodynamics of detergent–protein and detergent–detergent interactions is of central importance in many areas of research. It is the purpose of this chapter to outline the theoretical and practical aspects of these interactions with particular emphasis on the competitive effects of micelle formation and protein–detergent binding.

II. Thermodynamic Equilibria

A. Micelle Formation. In aqueous solution amphiphilic molecules self-associate at a specific concentration (critical micelle concentration) to form well-defined interaction products. The theoretical aspects of this

[1] This work was supported in part by National Institutes of Health Grants HL 14882 and NS 12213.
[2] C. Tanford and J. A. Reynolds, *Biochim. Biophys. Acta* **457,** 133 (1976).
[3] T. B. Nielsen and J. A. Reynolds, see Vol. 48, p. 3.
[4] A. Helenius and K. Simons, *Proc. Natl. Acad. Sci. U.S.A.* **74,** 529 (1977).

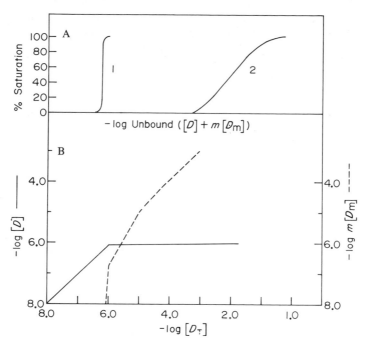

FIG. 1. (A) The binding of detergent to protein as a function of total unbound detergent concentration where the amphiphile has the self-association properties shown in (B). Protein concentration $= 10^{-5}\ M$, $n = 100$. (1) $\Delta G^0 = -8.4$ kcal/mole; (2) $\Delta G^0 = -8.21$ kcal/mole. (B) The increase in free monomer concentration (—) and micelle concentration (----) as a function of total detergent concentration. Critical micelle concentration $= 0.906 \times 10^{-6}\ M$; $m = 100$, $\Delta G_{mic}^0 = -8.15$ kcal/mole monomer.

phenomenon are discussed in detail in a number of recent publications.[5–7] We can describe this process by means of the following equation.

$$mD \rightleftharpoons D_m \qquad (1)$$

where m is the average association number, D is the concentration of monomeric amphiphile and D_m is the concentration of micelles. The association constant is

$$K = \frac{[D_m]}{[D]^m} \qquad (2)$$

Figure 1A shows the increase in concentration of D and D_m as a function of total amphiphilic concentration; where $m = 100$ and the free energy of micellization ΔG_{mic}^0 is -8.15 kcal/mole monomer. It is of particular im-

[5] C. Tanford, "The Hydrophobic Effect." Wiley (Interscience), New York, 1973.
[6] C. Tanford, *J. Mol. Biol.* **67**, 59 (1972).
[7] C. Tanford, *J. Phys. Chem.* **78**, 2469 (1974).

portance to note that above the critical micelle concentration of 0.91×10^{-6} M the monomer concentration increases very little with increasing total amphiphile. As m increases, the increase in D above the critical micelle concentration becomes even less. For an infinitely large aggregate, such as a phospholipid bilayer, the monomer concentration is effectively constant above the critical micelle concentration.

B. *Protein–Amphiphile Interaction.* The simplest cases of protein–detergent interactions are illustrated here for the purpose of demonstrating the interdependence of micelle formation and binding to proteins. It will be obvious that more complicated situations can ensue and can be treated thermodynamically using the principles discussed here and in a more extensive treatise by Steinhardt and Reynolds.[8]

If a protein contains n specific binding sites which interact only with D in the monomeric form by means of a two-state process, we can represent the reaction by

$$P + nD \rightleftharpoons PD_n \tag{3}$$

The association constant for this process is

$$J = \frac{[PD_n]}{[P][D]^n} \tag{4}$$

If we assume $n = 100$ (the same value as m in the previous example of micelle formation) and an association constant corresponding to $\Delta G^0 = -8.4$ kcal/mole monomer, we obtain the binding isotherm labelled "1" in Fig. 1B. If we now lower the free energy of binding to $\Delta G^0 = -8.21$ kcal/mole, we observe the isotherm labeled "2" in Fig. 1B. Since the protein must compete with the micellar aggregate for monomeric detergent, it is apparent that if n, ΔG^0, or both are equal to or lower than m and $\Delta G_{\mathrm{mic}}^0$, the binding isotherm is shifted to higher total unbound detergent concentrations and is significantly less cooperative. From a practical standpoint, if one wishes to saturate a protein with detergent, it is essential to maintain an appropriate concentration of monomeric ligand if the mode of interaction is between monomer and protein. The critical micelle concentrations of ionic amphiphiles can be manipulated by altering the ionic strength.[5] High salt concentrations lower the critical micelle concentrations, and low salt concentrations increase it. However, nonionic detergents are unaffected within experimental error by the ionic strength of the solution. In this latter case, if the association constant or number of binding sites is significantly smaller than the average aggregation number or free energy of micellization of pure detergent micelles, one may not be able to fill all binding sites on the protein.

[8] J. Steinhardt and J. A. Reynolds, "Multiple Equilibria in Proteins." Academic Press, New York, 1969.

A second type of protein–detergent interaction can involve direct binding of the protein to preformed micelles.

$$P + D_m \rightleftharpoons P\text{–}D_m \tag{5}$$

In this case, little or no interaction will be observed until the unbound detergent concentration in solution is close to the critical micelle concentration. The binding will appear highly cooperative, and if the detergent is ionic, the apparent association constant will be ionic strength dependent. Examples of this type of binding have been found with cytochrome b_5 and the major glycoprotein from human red cell membranes.[9,10]

III. Experimental Procedures

The determination of interaction between proteins and amphiphiles has been described in detail by Steinhardt and Reynolds.[8] Here we will summarize the techniques applicable to the investigation of monomeric and micellar detergent binding.

A. *Equilibrium Dialysis and Ultrafiltration.* These techniques are feasible when the total unbound detergent concentration is below the critical micelle concentration. Both methods require the use of a semipermeable membrane which allows rapid passage of unbound ligand and retains the protein–ligand complex. The amphiphile concentration is determined on both sides of the membrane, the difference being the concentration of bound ligand. If micelles are present, equilibration times are excessively long and the techniques consequently unreliable.

B. *Gel Filtration Chromatography.* This method is applicable either above or below the critical micelle concentration but requires large amounts of amphiphile. A column buffer is prepared containing a specific concentration of ligand, and the sample is applied in an excess of ligand. The eluted protein peak is analyzed for protein and *excess* amphiphile. Each experimental binding point at a specific unbound ligand concentration requires a separate elution experiment. Additionally, it is necessary to show that the protein–detergent complex has a different elution position than that of pure micelles, since overlapping of these two fractions will give erroneously high binding values.

C. *Analytical Ultracentrifugation.* Both the binding of ligand and the molecular weight of the protein are obtained by this procedure. The quantity determined directly at equilibrium in the analytical ultracentrifuge is $M_p(1 - \phi'\rho)$ where M_p is the molecular weight of the protein, ϕ' is the effective partial specific volume of the particle, and ρ is the solvent density. The factor $(1 - \phi'\rho)$ can be set equal to $(1 - \bar{v}\rho) + \delta_L(1 - \bar{v}_L\rho)$. In this

[9] N. C. Robinson and C. Tanford, *Biochemistry* **14**, 369 (1975).
[10] S. P. Grefrath and J. A. Reynolds, *Proc. Natl. Acad. Sci. U.S.A.* **71**, 3913 (1974).

latter relation, $\bar{v}$ is the partial specific volume of the protein, $\bar{v}_L$ is the partial specific volume of the ligand, and δ_L is the grams of bound ligand per gram protein. Measurements are made at several densities, obtained by replacement of H_2O with D_2O, and δ_L is obtained from the slope of a plot of $M_p(1 - \phi'\rho)$ versus ρ, i.e.,

$$\frac{-d[M_p(1 - \phi'\rho)]}{d\rho} = M_p(\bar{v} + \delta_L\bar{v}_L)$$

This procedure has the significant advantage that very little material is required. It is important to note that it is assumed that D_2O freely exchanges with bound H_2O, thus eliminating the buoyant density term due to bound water. The method cannot be used when the density is altered by adding sucrose or salts, since these solutes do not exchange with bound water, thus requiring an *a priori* knowledge of the amount of hydration in order to calculate this portion of the buoyant density term. The exact method of determining molecular weight and extent of binding by this procedure is published elsewhere in some detail.[11]

D. Indirect Methods. Alteration in optical properties is often used to determine the extent of interaction between proteins and amphiphilic ligands. Any such procedure must be calibrated against a rigorous thermodynamic method to ensure that optical properties are indeed a unique linear function of extent of binding.

E. Sources of Information Regarding Critical Micelle Concentrations. The most complete current listing of critical micelle concentrations is found in a United States Department of Commerce Publication by P. Mukerjee and K. J. Mysels entitled "Critical Micelle Concentrations of Aqueous Surfactant Systems," NSRDS-NBS 36, 1971. Additional and somewhat newer data on a more limited number of detergents is available in Tanford.[5]

---

[11] J. A. Reynolds and C. Tanford, *Proc. Natl. Acad. Sci. U.S.A.* **73**, 4467 (1976).

Section II

Interactions

[6] The Dilution Method and Concentration Difference Spectrophotometry: New Designs for an Old Method[1]

By ZACHARY YIM *and* RUFUS LUMRY

I. Introduction

The dilution method for determining dissociation equilibrium or rate constants depends on the cratic entropy increase on producing two or more fragments from one. The formal cratic entropy increase is 8 gibbs for each mole of new fragment with 1 M standard states, so the method is not particularly powerful. However, given sufficient changes in one or more observables upon dissociation, and newer measurement systems of high precision, the method can be very useful. More extensive and more detailed information about the dissociation process, e.g., the precise changes in electronic spectrum produced by dissociation, correct mechanisms for dissociation, and solvent effects upon dissociation can be obtained rapidly, often with very good reliability using late-model instruments, especially spectrophotometers of increasingly high photometric precision. The power available when such instruments are coupled with modern digital techniques for rapid data acquisition and noise rejection, suggest a reexamination of the dilution method seeking present improvement and estimates of future promise. In addition to the fact that the method can usually be used in such a way that concentrations rather than activities are determined with higher virial coefficient effects relegated to the group of small errors, simple modifications of procedure eliminated by maintaining constant products of parameters such as cuvette pathlength times total amount of dissociating species greatly reduce errors and provide sensitive tests for systematic errors. These lumped parameters make calculations easier as shown by Fisher and Cross,[1a] who have brought the method back to general attention, but its application requires instrumental developments, the first of which is described herein. To experienced instrument designers, our development is only a beginning, but it has been successful in demonstrating design problems and often the path to their solution. So little application of the method has thus far been made that our coverage in this chapter is no more than first-order, but the

[1] The development of the method and the equipment was supported by National Institutes of Health Grant HL-16833 and by aid on several forms from the Cary Instruments Co. and the Varian Corp. Publication LBC70 from the Laboratory for Biophysical Chemistry.

[1a] H. F. Fisher and D. G. Cross, *Arch. Biochem. Biophys.* **110**, 217 (1965).

low cost of the method, ("everyone has a good new spectrophotometer"), its efficiency and versatility, and the wide variety of kinds of important information extractable, should in time rejuvenate this dusty corpse for macromolecular chemistry. Of course, it has continued to be a major tool in inorganic chemistry through the ages.

Basically, the method involves the measurement of the concentrations of the associated or dissociated species or both, after each dilution step of solvent addition. The measurements can be made by light absorption, fluorescence intensity, circular dichroism, conductance, sedimentation techniques, and others, with the limits set only by the properties of the solutes and the sensitivities of the measuring equipment available. Spectrographic light absorption measurements of the solute concentrations have been the most used in experiments. Fisher and Cross[1a] in their excellent article "Spectrophotometric Studies of the Quarternary Structure of Proteins. I. Method of Concentration-Difference Spectra" have shown how data acquisition and workup can be facilitated with a double-beam spectrophotometer employing reference and sample cuvettes. However, computers and highly stable spectrophotometers now eliminate the need for using the reference cuvette for difference spectra.

We have applied the method to study the dissociation of mammalian hemoglobins. Deoxyhuman A is known to dissociate detectably only near $10^{-10} M$ and lower, in the absence of solvent additives favoring dissociation. With the stability of commercial photometric systems, this is at least two orders of magnitude outside the range of the method. Of course, thus far no static method of concentration measurement has yet provided the necessary precision. A few, such as sedimentation, can give data in this range. Fully oxygenated Hb A presents a simpler problem since dissociation begins on dilution to $10^{-5} M$ protein. The tetramer to dimer process effects a maximum absorbance change of about 10%. The accuracy can depend on such matters as protein loss to walls so the accuracy may be no better than perhaps 30% until these complications are removed or accommodated in the calculations. For Hb_4O_8 A, the dimer to monomer dissociation constant has been estimated by the multiple cuvette method to be about $10^{-5} M$ also. This establishes a rough lower total concentration limit of $10^{-8} M$ for heme proteins. Because of the richness of its spectrum, hemoglobin is particularly easy to study, but the method has also been applied to glucagon,[2] glutamic dehydrogenase,[2a] and insulin.[3]

Application data are not yet available in the number and variety necessary to assess the full range of practical advantages and disadvantages of

[2] M. H. Blanchard and M. V. King, *Biochem. Biophys. Res. Commun.* **25**, 298 (1966).

J. C. Swann and G. G. Hammes, *Biochemistry* **8**, 1 (1969).

[2a] D. G. Cross and H. F. Fisher, *Arch. Biochem. Biophys.* **110**, 222 (1965).

[3] R. S. Lord, F. Gubensek, and J. A. Rupley, *Biochemistry* **12**, 4385 (1973).

the method. With small complex proteins having only a single kind of subunit or subunits of very similar nature, e.g., hemoglobin, determinations of the stoichiometry of dissociation processes can be determined by trial and error fitting to the possible dissociation schemes. Even for such systems, this procedure becomes dangerous if there is more than one kind of dissociation process. For oxyhemoglobin

$$\alpha_1\beta_1\alpha_2\beta_2 \leftrightarrow \alpha_1\beta_1 + \alpha_2\beta_2$$
$$\alpha_1\beta_1\alpha_2\beta_2 \leftrightarrow \alpha_1\beta_2 + \alpha_2\beta_1$$
$$\alpha_1\beta_1\alpha_2\beta_2 \leftrightarrow \alpha_1\alpha_2 + \beta_1\beta_2$$

processes have been distinguished by the tryptophan $C3\beta$ absorbance behavior. The tryptophan residue, fortuitously present only in β chains, lies in the $\alpha_1\beta_2$ interface and proves a good indicator of dissociation occurring between the two $\alpha_1\beta_1$ dimer units. Equally fortuitously, the spectrum of this tryptophan is not altered on dimer to monomer dissociation. In less favorable cases and particularly when several dissociation alternatives are possible, recourse to some additional methods at a few protein concentrations, e.g. sedimentation, osmotic pressure or light scattering, may be utilized if these methods are adequate at the given low protein concentrations. In complicated cases, recourse to combinations of methods is likely to be required. With high enough precision, the dilution method can often become the method of choice despite the trial and error approach required in analysis of complicated cases.

When the spectra of the intermediates can be resolved from the aggregate spectra, solvent-perturbation information and other sophisticated resources of spectroscopy become available to provide important molecular structure information not usually available.

The opportunities to obtain such a variety of information in single experiments makes spectrophotometry attractive despite some limitation on detection sensitivity. Nevertheless, in spite of these virtues, the method has rarely been applied to macromolecular association–dissociation reactions. Continuing improvements in sedimentation techniques maintain the popularity of that powerful but expensive method for determining dissociation constants. Lack of familiarity with high performance spectrophotometers and a lack of suitable cuvettes have prevented wider use.

Insufficient change of spectrum on dissociation provides the major limitation with absorbance detection and makes necessary development of cuvette systems with precision equal to that of the best commercial spectrophotometers, roughly 10^{-4} absorbance units. We have developed a continuous dilution cuvette system to match this precision limit. Additional improvements should make it possible to match improved precision in future photometric instruments.

II. Basis of the Concentration-Difference Spectral Method

In part, the method as applied to protein association using spectrophotometry can be considered to be an outgrowth of the studies of the solvent perturbation of exposed protein chromophores. The spectrum of a protein changes as its chromophores move to an environment with different polarizability. Usually the largest changes are produced when aromatic groups are moved from an internal position with moderate or high polarizability to bulk water with a low polarizability. Yanari and Bovey[4] appear to be the first to explain this behavior for aromatic protein chromophores, but there has been extensive work on the subject since then. Excellent review articles are available, e.g., Wetlaufer,[5] Herskovitz,[6] and Donovan.[7] These articles on solvent perturbation difference spectroscopy have direct relevance because the spectral changes produced in this way are those most often available to measure the degree of dissociation. Electronic and vibrational spectral changes intrinsic to the chromophore and the state of the protein, such as appear in heme spectra, provide useful often unique information. In addition to tryptophan, tyrosine, histidine, cysteine, and cystine residues can demonstrate similar useful electronic and vibrational shifts. The chromophore may be in an inhibitor or substrate as well as in the protein. Any perturbation which enchances the differences between the microenvironments of "buried" and "exposed" chromophores can be employed to amplify the concentration difference spectrum, but spectral perturbations produced by drastic changes in solvent and solvent composition may complicate interpretation because of artifactual and unwanted effects of solvent on protein conformation. On the other hand, Galley and Edelman,[8] Steiner and Edelhoch,[9] Bello,[10] Leach,[11] and others have found that the spectra of buried and exposed protein chromophores often have large differences in temperature dependencies. This behavior can be very useful for spectral interpretation but it may complicate the determination of the temperature dependence of the dissociation equilibrium constants.

Chromophoric molecules of any size which bind in different ways or to different degrees with associated and dissociated species can give infor-

[4] F. A. Bovey and S. Yanari, *Nature (London)* **186,** 1042 (1960).

[5] D. B. Wetlaufer, *Adv. Protein Chem.* **17,** 303 (1962).

[6] T. T. Herskovits, Vol. 11, p. 748.

[7] J. W. Donovan, Vol. 27, p. 497.

[8] J. A. Gally and G. M. Edelman, *Biochim. Biophys. Acta* **60,** 499 (1962), *Biopolym. Symp.* **1,** 367 (1964).

[9] R. F. Steiner, R. F. Lippolot, H. Edelhoch, and V. Frattal, *Biopolym. Symp.* **1,** 355 (1964).

[10] J. Bello, *Biochemistry* **11,** 4542 (1969).

[11] N. A. Nicola and S. J. Leach, *Int. J. Pept. Protein Res.* **8,** 393 (1976).

mation about the dissociation through their own absorbance changes. Obviously, they must be bound differently by each macromolecule species involved. Such added "reporters" inevitably perturb the dissociation equilibrium so they can be used only with considerable work. In situations such as that provided by ferrous hemoglobins, the allosteric effector diphosphoglycerate (DPG) alters the protein spectrum upon binding so its binding equilibrium can be employed to measure the hemoglobin dissociation processes if due consideration is given to the shifting of the equilibrium constant by the binding of the DPG. In this experiment two linked processes are simultaneously described quantitatively.

In effect, it is inevitable with such "third-party molecules" that the equilibrium between the latter and the macromolecules of interest must be characterized quantitatively. Of course, the method can be exploited just as effectively to determine small molecule association with large molecule as large molecule with large molecule.

A potentially useful type of spectral change is that produced by nicotinamide, which forms charge-transfer complexes with exposed aromatic side chains of proteins. Such charge-transfer bands, as reported by Deranleau,[12] have some generality, but as with the nicotinamide reported used by Deranleau *et al.*, dangerously high concentrations of the complexing third-party molecule may be required.

III. Conventional Experimental Procedures

Concentration-difference spectra can be obtained simply but expensively with a serial dilution method which depends on the availability of a set of conventional cuvettes with different path lengths. The solution of interest is diluted so that the product of solute concentration and path length is constant. In principle, difference spectra against some solution chosen as standard reveal spectral changes only due to the effects of dilution. The workup of data is similar to that for continuous dilution experiments, but the procedure is laborious and inherently less precise. Cuvettes with different path lengths can be purchased, and the procedure of mixing and solvent addition is straightforward. However, this type of sample handling is difficult at the 0.01% level no matter how good the spectrophotometer. Furthermore, it is very difficult to be certain that the difference spectra observed are not due to concentration mismatch, that is, due to $c_1l_1 \neq c_2l_2$. The calibration method and the testing procedure for concentration mismatch have been thoroughly treated by Fisher and Cross.[1a]

The continuous dilution method uses a single cuvette of variable path

[12] D. A. Deranleau, L. M. Hinman, and C. R. Coan, *J. Mol. Biol.* **94**, 567 (1975).

length. Difference spectra as a function of dilution may be obtained with a
fixed reference sample in a double-beam spectrophotometer, but it is generally preferable to collect the data for each path length in digital form for
subtraction by a computer from the initial spectrum or some reference
spectral data. The stepwise differential curve provides a rough derivative
spectrum.

Both methods require corrections for any pathlength and optical mismatch between pairs of cuvettes by individual measurements of cuvette
path lengths, as well as base line measurements using solvent blanks.

IV. General Equations Related to Concentration-Difference Spectroscopy

The method involves measurements of deviations from the Beer–
Lambert law of the absorption of solutions (Eq. 1). The validity of the
interpretations of such measurements in terms of association and dissociation of oligomeric proteins depends on subsequent extraction of equilibrium constants with the assumption of strict adherence of the Beer–
Lambert law by individual equilibrium species themselves, regardless of
degree of dilution[12a] (Eq. 2). The accuracy of the additivity relation for
multi-species absorption (Eq. 3) (noninterference of component species)
depends on adherence to Beer's Law.
For all λ's

$$A \neq \epsilon cl - \text{(Beer-Lambert law)} \tag{1}$$

Instead

$$A = \epsilon(c)cl$$

and

$$A_i = \epsilon_i c_i l \neq \epsilon_i(c_i)c_i l \tag{2}$$

so

$$A = \Sigma A_i \tag{3}$$
$$A = (\Sigma \epsilon_i c_i)l$$

where A is total absorbance, ϵ is extinction coefficient, c is concentration,
l is path length, λ is wavelength, i is species i. Consider as an example the
following oligomeric formation equilibrium, undergoing successive dilutions by solvent in continuous dilution cuvette.

$$P_i \underset{K_i}{\rightleftharpoons} iP \qquad i = 1, 2, \ldots, N$$

[12a] A different derivation is required when studying continuous solvent perturbation and,
of course, different final equations result.

For all λ's (except for isosbestic points)

$$A(c) = cl\epsilon(c)$$

$$= l \sum_{i=1}^{N} \epsilon_i c_i$$

$$= cl \sum_{i=1}^{N} \epsilon_i \psi_i(c) \quad \psi_i = \text{fraction of } i\text{-mers}$$

Let A_0 be the absorbance of the reference solution and A_k be the absorbance of the sample solution.

$$A_0 = c_0 l_0 \sum_{i=1}^{N} \psi_{0i}(c_0)\epsilon_i$$

$$A_k = c_k l_k \sum_{i=1}^{N} \psi_{ki}(c_k)\epsilon_i \qquad c_k \neq c_0$$

$$\Delta A_k = A_k - A_0$$
$$= c_k l_k \Sigma \psi_{ki} \epsilon_i - c_0 l_0 \Sigma \psi_{0i} \epsilon_i$$

where ΔA_k is the concentration-difference spectrum. The Fisher-Cross condition is that $c_k l_k = c_0 l_0$.

$$\Delta A_k = c_0 l_0 (\Sigma \psi_{ki} \epsilon_i - \Sigma \psi_{0i} \epsilon_i)$$

$$= c_0 l_0 \sum_{i=1}^{N} (\psi_{ki} - \psi_{0i})\epsilon_i$$

$$\Delta \epsilon_k = \frac{\Delta A_k}{c_0 l_0} = \sum_{i=1}^{N} (\psi_{ki} - \psi_{0i})\epsilon_i$$

$$\psi_{ki} = c_{ki} / \sum_{i=1}^{N} c_{ki} = f(c_k, - K_i\text{'s})$$

Therefore,

$$\Delta \epsilon_k = f(c_k, c_0, K_i\text{'s}, \epsilon_i\text{'s}, i = 1 \ldots N)$$

By standard curve-fitting procedure, we can obtain K_i's and ϵ_i's, but only if nucleation steps are not numerous, and if i is small. For large i better methods are available but they may not provide information about nucleation processes so readily.

For the simple case of dimerzation:

$$P_2 \underset{K_2}{\rightleftharpoons} 2P$$

$$\Delta A_k = c_k l_k \sum_{i=1}^{2} \psi_{ki} \epsilon_i - c_0 l_0 \sum_{i=1}^{2} \psi_{0i} \epsilon_i$$

$$= (c_k l_k - c_0 l_0)\epsilon_2 + (c_k l_k \psi_{k1} - c_0 l_0 \psi_{01})(\epsilon_1 - \epsilon_2)$$

If $c_k l_k = c_0 l_0$, then $\Delta A_k = c_0 l_0 (\psi_{k1} - \psi_{01}) \Delta\epsilon$.

$$
\begin{aligned}
K_2 &= P^2/P_2 &\text{dissociation constant} \\
\psi_k &= P/c_k &\text{fraction of monomer} \\
P &= c_k \psi_k &\text{concentration of monomer} \\
2P_2 &= (1 - \psi_k)c_k &\text{concentration of dimer}
\end{aligned}
$$

Therefore

$$K_2 = \frac{(c_k \psi_k)^2}{(1 - \psi_k)c_k/2}$$

$$2c_k \psi_k^2 + \psi_k K_2 - K_2 = 0$$

$$\psi_k = \frac{-K_2 + (K_2^2 + 8c_k K_2)^{1/2}}{4 c_k} \qquad \text{fraction of monomer at } c_k$$

$$\psi_0 = \frac{-K_2 + (K_2^2 + 8c_0 K_2)^{1/2}}{4 c_0} \qquad \text{fraction of monomer at initial}$$

concentration

By assuming a value for K_2, we can generate a series of values of $(\psi_k - \psi_0)$ by substituting the value of c_k's. The correct guessed value of K_2 would give a straight line when ΔA_k is plotted against $(\psi_k - \psi_0)$ to give $c_0 l_0 \Delta\epsilon$ as the slope. This is Kagarise's method.[13] Although this graphical method is simple to use, and gives a rough and ready estimate of the values of K's, the least-square fitting method is always preferrable, since with properly weighted data, it gives best values for the estimates of the K's and their random errors.

It is desirable to be able to go to either extremes in concentration to get the absolute spectrum of the fully associated and fully dissociated species. This removes one fitting parameter from the curve fitting procedure, thus simplifying computation and rendering the estimated values of K's more reliable.

When (molar) absorbance spectra of all species are identical, differ only by a constant or have differences lying below the precision limit of the measuring instrument, the method is useless. However, indicator molecules which do demonstrate spectral changes when bound to different species can be used with care to obtain estimates of the behavior of the indicator-free macromolecular system.

For a system with three equilibrium species, one would normally use

[13] R. E. Kagarise, *Spectrochim. Acta* **19**, 629 (1963).

nonlinear least-square fitting to obtain the K's. However, if one has the good fortune of having bands attributable to single dissociating steps, there is much simplification, and simple straight line plots can be used to obtain K's.

The use of plotting procedures is exemplified by the study of tetramer to dimer to monomer dissociation of horse oxyhemoglobin, an early application of the stepwise dilution procedure by Mizukami and Lumry.[14]

They assume the following two-step equilibrium process for oxyhemoglobin:

$$Hb_4 \overset{K_1}{\rightleftharpoons} 2\ Hb_2$$

$$Hb_2 \overset{K_2}{\rightleftharpoons} 2\ Hb$$

The conservation equation is

$$c_s = c_1 + c_2 + c_4$$

where c_s is the concentration of the sample in moles of heme group. If $r_1 = c_1/c_s$, etc., then $1 = r_1 + r_2 + r_4$

$$K_1 = \frac{r_2^2 c_s}{r_4} \qquad K_2 = \frac{r_1^2 c_s}{r_2}$$

$$\Delta A = (c_1\epsilon_1 + c_2\epsilon_2 + c_4\epsilon_4)l_s - c_r\epsilon_4 l_r$$

where c_r and l_r are concentration, and path length of the reference cuvette.

$$\Delta A = (r_1\epsilon_1 + r_2\epsilon_2 + r_4\epsilon_4)c_s l_s - c_r\epsilon_4 l_r$$

Assuming the reference solution to be all in tetramers.

$$\Delta\epsilon = \frac{\Delta A}{c_r l_r} = [r_1\epsilon_1 + r_2\epsilon_2 - (1 - r_4)\epsilon_4]$$

$$= r_1\epsilon_1 + r_2\epsilon_2 - (r_1 + r_2)\epsilon_4$$

They found that at

$$\lambda = 293\ nm \qquad \epsilon_1 = \epsilon_2$$

and at

$$\lambda = 270\ nm \qquad \epsilon_2 = \epsilon_4$$

Therefore,

$$\Delta\epsilon_{293} = \frac{\Delta A_{293}}{c_r l_r} = (r_1 + r_2)(\epsilon_1 - \epsilon_4)_{293} \qquad (4)$$

$$\Delta\epsilon_{270} = \Delta A_{270}/(c_r l_r) = r_1(\epsilon_1 - \epsilon_4)_{270} \qquad (5)$$

[14] H. Mizukami and R. Lumry, *Arch. Biochem. Biophys.* **118**, 434 (1967).

$(\epsilon_1 - \epsilon_4)_{270}$ and $(\epsilon_1 - \epsilon_4)_{293}$ can be found by extrapolating ΔA_{270} and ΔA_{293} to the value when $r_1 = 1$. Knowing $(\epsilon_1 - \epsilon_4)_{270}$, r_1 can be determined at any concentration, since $r_1 = \Delta\epsilon_{270}/(\epsilon_1 - \epsilon_4)_{270}$. Using r_1, [Eq. (4)] and the expression for K_2 in terms of r_2, they obtain

$$\frac{\Delta\epsilon_{293}}{r_1} = \left(1 + \frac{r_1 c_s}{K_2}\right)(\epsilon_1 - \epsilon_4)_{293}$$

A plot of $\Delta\epsilon_{293}/r_1$ versus $r_1 c_s$ yields K_2 from the slope.

In order to obtain K_1, the following analysis is required. The equilibrium constants are related through the equation $K_1 K_2^2 = r_1^4 c_s^3/r_4$, and since for large c_s, r_1 is small, Eq. (4) may be rewritten as

$$r_2 = \Delta\epsilon_{293}/(\epsilon_1 - \epsilon_4)_{293}$$

and $r_4 \approx 1 - r_2$, so that

$$K_1 K_2^2 = (K_2 r_2/c_s)^2 [c_s^3/(1 - r_2)]$$

or

$$K_1 = r_2^2 c_s/(1 - r_2)$$

Since r_2 is measured, a plot of $r_2^2/(1 - r_2)$ versus c_s^{-1} for sufficiently large c_s should be a straight line with slope K_1. However, as the authors pointed out, "The functional relationships between abscissa and ordinate for the plots represent a rather awkward use of the data from the point of view of statistical fitting and a straightforward computer analysis of the original equilibrium relationships will be more significant in future work in which error control is better."

V. Comments on Data Analysis

The existence of isosbestic points almost always indicates an interacting system with only two species. Under rare solvent conditions isosbestic points can be found in one-species systems and three-species systems. Furthermore, a two-species system may not yield isosbestic points. Fortunately these complications are seldom encountered.

For multispecies systems of a single component, the number of species can be determined from the rank of the absorbance matrix[15–17] and/or the various curve fitting procedures currently available,[18,19] using

[15] J. J. Kankare, *Anal. Chem.* **42**, 1322 (1970).

[16] J. T. Bulmer and H. F. Shurvell, *J. Phys. Chem.* **77**, 256 and 2085 (1973).

[17] D. Katakis, *Ana. Chem.* **37**, 877 (1965).

[18] D. J. Leggett, *Anal. Chem.* **49**, 276 (1977).

[19] D. E. Metzler, C. Harris, I. Y. Yang, D. Siano, and J. A. Thomson, *Biochem. Biophys. Res. Commun.* **46**, 1588 (1972).

either absorbances at selected wavelengths as is conventionally done, or using band shape analysis techniques, fitting the whole absolute spectrum or difference spectrum with "component" bands empirically derived or experimentally determined from studies of suitable model compounds. It is a good practice to make use of several wavelength regions of the difference spectrum rather than a single or a very few wavelengths. Even if the absorbance differences at additional wavelengths are significantly less than the largest differences, the improvement in reliability usually outweighs the loss in precision. Optimum experimental design can be worked out for each kind of experiment. Where spectroscopic information is needed or more commonly when it is desired to determine the spectra of the intermediate species in complex dissociation equilibria, gaussian distributions of absorbance wavelength bands are usually used, but log-normal distributions against wave numbers may be preferable. The latter has the advantage in revealing vibronic fine structure and thus facilitate identification of individual contributions to the total spectrum of the chromophores. The specific method of application in either case depends on whether one uses the absolute spectrum at a given concentration or difference spectra, but many treatments of all these problems exist. Temptations to use polynomials are ill-advised since the characteristics to be expected of spectral bands are well known and provide better bases for analysis. In any event, the statistical problems associated with polynomial fitting must be kept in mind (cf. Bevington[20]).

Interference from substances perturbing the process of interest must be avoided. Presumably no one would add 2,3-diphosphoglycerate in less than saturating concentrations in a hemoglobin experiment since DPG, upon binding, perturbs hemoglobin bands throughout the ultraviolet and visible spectra of hemoglobin.

VI. Special Merits of the Method for Study of the Quarternary Structure of "Oligomeric" Proteins

The method involves the direct measurement of the concentrations of the equilibrium species; most other methods measure quantities which are approximate functions of their concentrations as a result of activity coefficient effects and higher-order virial coefficient effects.

A way to probe the nature and measure the thermodynamics of subunit–subunit interactions is via measurements of the equilibrium constants as a function of the perturbations by solution additives, which change the relative magnitudes of interaction forces and energies[21]—

[20] P. R. Bevington, "Data Reduction and Error Analysis for the Physical Sciences," McGraw-Hill, New York, 1969.
[21] D. Elbaum and T. T. Herskovits, *Biochemistry* **13,** 1268 (1974).

hydrophobic effects, solvation energies, charge effects, hydrogen-bonding, dipolar interactions, etc. The continuous dilution method is particularly useful for such studies and rather unique, since the method of concentration difference spectroscopy does not involve any separately measured solvent composition dependent quantities nor does it depend on corrections for solution nonideality.

It is necessary to study the temperature dependence of the equilibrium constant, in order to get the enthalpy and entropy of intermediate equilibrium process by applying the van't Hoff relation. Calorimetry is often impossible, even for single-step dissociation processes because of low protein concentration. Studies at elevated temperatures may be difficult for methods which require long measuring times, due to protein denaturation, autolysis, etc. Our method requires 2–3 hr for a complete run. In some applications this may be too long, but the speed can be considerably improved. Some other methods do not provide good enough temperature controls for van't Hoff studies. Our method allows for straightforward and highly stable temperature control. The air thermostat, to be discussed, produces a stability of 0.02° from 6° to 60°.

The method not only provides concentration information about the species, but also information about the location and nature of the molecular and solvation changes concomitant with the association–dissociation process. By an analysis of the magnitude of the spectral changes corresponding to the various equilibrium steps, considerable information is potentially available, e.g., the subunit interfaces involved in the various steps may be distinguishable. Such characterization would facilitate identification of the interfaces themselves using X-ray-diffraction information. Detailed insight about structure–function relationships, allosterism, and linkage in general, for example, can be obtained if concentration difference spectra can also be analyzed in terms of chromophoric prosthetic groups, coenzymes, bound transition metal ions, etc.

VII. Critique of the Stepwise-Dilution Procedure

Errors Due to Mismatches. The conventional method of employing many cuvettes of different pathlengths works simply if there is perfect match and reproducibility in optical properties and in pathlengths between cuvette pairs (i.e., identical spectra for identical samples). Otherwise tedious calibration procedures are necessary if one wants to use more than one wavelength. Even then, the difficulty of exact repositioning of cuvettes from run to run leads to random errors in the effective pathlength of the cuvette, and these are compounded by multiple internal reflections.[22] Furthermore, the method depends on a rather imprecise

[22] C. T. Chen and J. D. Winefordner, *Can. J. Spectrosc.* **19**, 120 (1974).

although logically sound method of testing and correcting for concentration mismatch, the major source of error in concentration difference spectroscopy.

We know from Section IV that for the simple case of dimerization

$$\Delta A_k = (c_k l_k - c_0 l_0)\epsilon_2 - (c_k l_k \psi_k - c_0 l_0 \psi_0)\Delta\epsilon \tag{6}$$

i.e., $\Delta A_k{}^{obs}$ (difference observed) = δA (error due to mismatch) plus $\Delta A_k{}^{real}$ (real difference). Let's assume a 1% mismatch, i.e., $\delta c l / c_0 l_0 = 0.01$. Taking the better than average case when $\psi_k = 1$, $\psi_0 = 0.5$, and knowing normally $\Delta\epsilon = 0.1\epsilon_2$, we have

$$\frac{\delta A_k}{\Delta A_k{}^{real}} = \frac{\delta c l \epsilon_2}{(\psi_k - \psi_0)(c_0 l_0)\Delta\epsilon_2} = 20\%$$

Thus, a mismatch of 1% introduces an error of 20% in the difference spectra.

The checking for concentration mismatch in the concentration difference spectra as suggested by Fisher and Cross[1a] is carried out in the following manner. First, from Eq. (6) we note that changes due to concentration mismatch are an order of magnitude larger than those due to dissociation. By the very slight perturbation of the concentration of the sample, and then the reference, those bands which undergo a change of sign would be due to concentration mismatch alone, while the bona fide dissociation induced difference spectra would maintain their sign.

Proteins stick to glass and to the various grades of quartz in both reversible and irreversible ways depending on the surface, the solvent, the protein, and its concentration. The facts are few and often limp, but we have found no adsorption at the 1% level in $10^{-4}\,M$ hemoglobin A solutions in the continuous dilution cuvette. Sakura[23] reports considerably more trouble with hemoglobin loss through adsorption. However, Sakura used the stepwise dilution method which involves a large surface to volume ratio and much solution handling. Losses to surface with myoglobin and many chymotrypsin proteins are very low, but become detectable after much manipulation of solutions. Hence, in addition to volumetric errors in solution manipulation, protein loss errors also occur. The continuous dilution method minimizes these errors. Furthermore, the "dead-volume" correction in the latter not only reveals loss of protein to surfaces and denaturation during a continuous dilution experiment but it also provides the information necessary to correct for such losses even though they may vary with time.

Proteins are often rapidly denatured by exposure to an air–water interface. Rapid stirring or bubbling air through a protein solution is to be

---

[23] J. D. Sakura, Ph.D. Dissertation, Department of Biochemistry, University of Arizona, Tucson (1970).

avoided unless one is applying the Sakaguchi process. Air–water interfaces are to be avoided during any mixing process.

The first few dilution steps are usually the most important, and in complex dissociation processes many dilution steps must be used. All that means lots of cuvettes, lots of solutions, and lots of manipulation using the multiple-cuvette method.

VIII. Continuous Dilution Cuvette—(For detail see Figs. 2, 3, and 4)
A Brief Description

The continuous dilution cuvette is designed as an accessory to spectrophotometers, but with few modifications it can be adopted for fluorescence, light scattering, optical rotation, and circular dichroism. Modifications retaining the all-important stirring method and several other special features make conductimetric and other nonoptical measurements possible. The cuvette, originally suggested by Hawes, consists of a precision-bore glass barrel fitted with two fused silica end windows and a movable piston carrying a fused silica window. The sample is introduced on one side of the piston. The initial path length is usually only a few tenths of a millimeter. Solvent is introduced via a stepping-motor driven pump, which has no slippage (backflow) at pressures at least as high as 20 psi. On the sample side through the inlet port, the piston is moved down the barrel displacing the solvent on the other side of the piston through the outlet port. The mixing is effected by rotating the cuvette several times in one direction and then the other. The mixing cycle is repeated several times to ensure complete mixing. After mixing, the cuvette is positioned using a locking rod for absorbance measurements. The whole procedure is carried out within a sealed, thermostated cuvette compartment. Except for a one-time-only dead volume correction which takes 30 min to determine (see Section X,B), no further calibration procedures are necessary.

IX. Rationale Behind the Design

By making the pathlength of the sample chamber of the cuvette continuously variable, one can have as many experimental points at as many specific lengths as is required by a wide variety of dissociation processes.

By making the pathlength of the sample chamber vary in direct proportion to the actual volume added, the constancy of Cl is assured independent of the measurement of the volume added. This constancy is of major importance for data analysis as well as a test for systematic error. Initially

$$C_0 L_0 = \frac{g}{V_0} \frac{V_0}{A} = \frac{g}{A}$$

where g is the mass of sample, and A is the cross-sectional area of the cuvette. On adding ΔV_k in the kth dilution

$$c_k l_k = \frac{g}{V_0 + \Delta V_k} \frac{V_0 + \Delta V_k}{A} = \frac{g}{A}$$

If there is a mistake δ_V in the volume measurement,

$$c_k' l_k' = \frac{g}{V_0 + \Delta V_k + \delta_V} \frac{V_0 + \Delta V_k + \delta_V}{A} = \frac{g}{A}$$

Hence, CL remains constant even though an error in volume measurement is made and mismatch error is eliminated.

By carrying out the dilution directly in the cuvette, and by employing a unique method (*vide infra*) of mixing which operates in the complete absence of air–liquid interfaces, mechanical processes which often give rise to protein adsorption and surface denaturation are minimized. Errors from cuvette repositioning are reduced to values which are undetectable relative to current spectrophotometric errors.

The auxilliary chamber filled with solvent maintains the optical path constant in effective length so that beam position from the cuvette to detector is independent of sample path length. Only in this way can the photometer respond directly and unambiguously to absorption changes alone.

Dilution in the cuvette avoids the compounding of errors in volumetric measurements, as indicated by the following comparison of the propagation of errors in the sample concentration with each dilution.

For continuous dilution cuvette,

$$c_0 v_0 = c_1 v_1 \qquad c_1 = c_0 v_0 / v_1$$

For the first dilution variance estimator is propagated in the following way if covariance estimates are negligible.

$$\sigma_{c_1}^2 = \left(\frac{\partial(c_0 v_0 / v_1)}{\partial c_0}\right)^2 \sigma_{c_0}^2 + \left(\frac{\partial(c_0 v_0 / v_1)}{\partial v_0}\right)^2 \sigma_{v_0}^2 + \left(\frac{\partial(c_0 v_0 / v_1)}{\partial v_1}\right)^2 \sigma_{v_1}^2$$

$$= \left(\frac{v_0}{v_1}\right)^2 \sigma_{c_0}^2 + \left(\frac{c_0}{v_1}\right)^2 \sigma_{v_0}^2 + \left(\frac{c_0 v_0}{v_1^2}\right)^2 \sigma_{v_1}^2$$

For the Nth dilution

$$\sigma_{c_N}^2 = \left(\frac{v_0}{v_1}\right)^2 \sigma_{c_0}^2 + \left(\frac{c_0}{v_1}\right)^2 \sigma_{v_0}^2 + \left(\frac{c_0 v_0}{v_N^2}\right)^2 \sigma_v^2$$

Hence, the errors are not compounded.

For the conventional sequential dilution method using pipettes and volumetric flasks:

$$c_0 v_0 = c_1 v_1$$

$$(\sigma_{c_1})^2 = \left(\frac{v_0}{v_1}\right)^2 \sigma_{c_0}{}^2 + \left(\frac{c_0}{v_1}\right)^2 \sigma_{v_0}{}^2 + \left(\frac{c_0 v_0}{v_1{}^2}\right)^2 \sigma_{v_1}{}^2$$

$$c_1 v_1' = c_2 v_2$$

Taking an aliquot (v_1') from solution (c_1), and dilute to v_2.

$$c_2 = c_1 v_1'/v_2$$

$$\sigma_{c_2}{}^2 = \left(\frac{v_1'}{v_2}\right)^2 \sigma_{c_1}{}^2 + \left(\frac{c_1}{c_2}\right)^2 (\sigma_{v_1'})^2 + \left(\frac{c_1 v_1}{v_2{}^2}\right)^2 \sigma_{v_2}$$

$$= \left(\frac{v_1'}{v_2}\right)^2 \left(\left(\frac{v_0}{v_1}\right)^2 \sigma_{c_0}{}^2 + \left(\frac{c_0}{v_1}\right)^2 \sigma_{v_0}{}^2 + \left(\frac{c_0 v_0}{v_1{}^2}\right)^2 \sigma_{v_1}{}^2 \right)$$

$$+ \left(\frac{c_1}{v_2}\right)^2 (\sigma_{v_1'})^2 + \left(\frac{c_1 v_1'}{v_2{}^2}\right)^2 \sigma_{v_2}{}^2$$

In this case the errors are compounded.

X. Construction of the Continuous Dilution Cuvette (see Figs. 2, 3, and 4)

A. The Mixing Method

The mixing technique utilizes the principle of boundary layer flow. It operates like a washing-machine without fins, and is best demonstrated by the way tea leaves spiral down to the bottom of the cup as tea is being poured. As the glass barrel is rotated along its axis, it carries along with it the boundary layers of water, which would spiral radially toward the axis of the glass barrel. If the rotation of the glass barrel is then reversed, the same spiraling motion would be started in a direction opposing the motion already in progress. This creates instabilities everywhere and thorough mixing results. To be sure, small eddy currents due to centrifugal and end effects contribute partly toward the longitudinal mixing. The mixing is efficient from a few tenths of a millimeter to at least 10 cm in length of the sample chamber. The lower and upper length limits for efficient stirring have not been explored for more extreme lengths.

B. Dead Volume Corrections

Since the dilution is exactly compensated by the increase in path-length, for any sample which obeys Beer's law, the absorbance should

stay constant regardless of the position of the piston window. We observe
no deviation greater than the noise level of the spectrophotometer (0.0002
absorbance units) as the piston traverses the whole length of the barrel
with blank runs (both sides filled with the same solution); we use water
and glycerol and see no dependence on refractive index, as well as potassium ferricyanide solutions with absorbances ranging from 0.01 to 1 and
see no dependence on absorbance. However, in actual dilution experiments such as with potassium ferricyanide solution, there is a systematic
deviation which is due to end window surface imperfections (vide infra),
and deviations of cuvette body from a perfect cylinder. When taken into
consideration by a calibration procedure, these yield a single correction
factor, called the dead volume correction. After the correction, no deviation greater than 0.0004 absorbance units remains in the data.

The surface imperfections are readily understood from the way the
end windows and piston are constructed. The latter are constructed by
gluing a molded silicon rubber gland onto the optical windows. Owing to
the rubber nature of the gland, it is impossible to glue the pieces together
to form a perfectly flat surface, with no indentations and/or lopsidedness.
Also, for the same reason it is unrealistic to expect the two window surfaces to be perfectly parallel to each other. However, by all indications
they retain their relative orientation as the piston slides along the whole
length of the barrel. Both imperfections would give rise to deviations which
can be lumped together and treated in the following manner (Fig. 1).

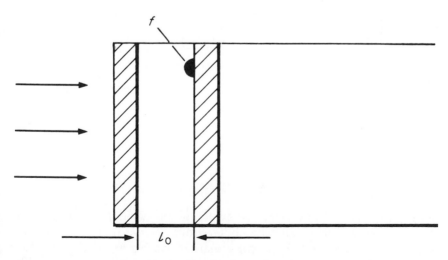

FIG. 1. See text for explanation.

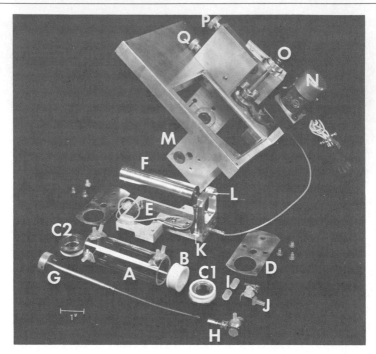

FIG. 2. A, Precision-bore glass barrel (10 cm long, cut from a 50 cc Becton-Dickinson multifit syringe body) with 5 pinholes (diameter = 1/10,000 in.). Epoxied onto the holes are five Teflon nipples machined from Hamilton male-male connectors (Part No. 86506) with no holes, since in our case we need pin-holes which we drilled ourselves. B, Piston with quartz window, glass collar and silicon rubber gland. C1, End window with the rubber gland facing upward. C2, End window with the metal retainer ring facing upward. D, Copper spring. E, Base plate with hollow shaft. F, Aluminum rod for counter weight. G, Flexible shaft. H, Hamilton valve with key hole for the insertion of the flexible shaft. I, Teflon plugs. J, Hamilton valve. K, Guiding tube for the flexible shaft. L, Groove for the locking rod. M, Base block with bearing for the rotating shaft of the baseplate. N, Motor to drive the flexible shaft (G). O, Linkage mechanism to rotate the flexible shaft (G) when inserted. P, Knob for the locking rod. Q, Knob for the alignment of the locking rod.

$$Ab_0 = \epsilon c_0 l_0$$

$$= \epsilon \left(\frac{b_0}{v_0 - f}\right) \frac{v_0}{A}$$

$$Ab_k = \epsilon \frac{b_0 v_0 + \Delta v_k}{v_0 - f + \Delta v_k A}$$

where Ab_0 is initial absorbance, b_0 is total amount of sample, and f is dead volume correction due to surface imperfection, $v_0 = l_0 A$ is initial volume if the cuvette is perfect, A is the cross-sectional area of the cuvettes $v_{in} = v_0 - f$ = actual initial volume, Ab_k is the absorbance after Δv_k has been added.

FIG. 3. A, Hamilton valve with its central rotating plug hooked up with the retrievable flexible shaft [represented as I (2) in Fig. 5]. B, Hamilton valve [represented as I (1) in Fig. 5.] C and E, Teflon tubings threaded through hollow shaft of the baseplate and flexible guiding tubings to prevent entanglement as the cuvette rotates. The cuvette is rotated by a motor via chain F. The motor is controlled by a circuit board so that it rotates an equal number of times in either direction, being stopped after each cycle by dynamic braking, so that the slight twist developed along the Teflon tubing upon rotation in one direction would be completely reversed in the other. G, Knob to switch on the PMT of the spectrophotometer. D, Guiding tube for the flexible shaft.

$$\Delta Ab_k = Ab_k - Ab_0 = \epsilon \frac{b_0}{A} \left(\frac{v_0 + \Delta v_k}{v_0 - f + \Delta v_k} - \frac{v_0}{v_0 - f} \right)$$

$$= \frac{\epsilon b_0}{A} \left(\frac{v_0^2 + \Delta v_k v_0 - v_0 f - \Delta v_k f - v_0^2 + v_0 f - v_0 \Delta v_k}{(v_0 - f + \Delta v_k)(v_0 - f)} \right)$$

$$= - \frac{\epsilon b_0}{A} \left(\frac{\Delta v_k}{(v_0 - f + \Delta v_k)(v_0 - f)} \right) f$$

$$= - \frac{\epsilon b_0}{A} \left(\frac{f}{v_0 - f} \right) \left(\frac{\Delta v_k}{v_0 - f + \Delta v_k} \right)$$

$$\frac{1}{\Delta Ab_k} = - \frac{A}{\epsilon b_0} \left(\frac{v_0 - f}{f} \right) \left(\frac{v_0 - f + \Delta v_k}{\Delta v_k} \right)$$

$$= - \frac{A}{\epsilon b_0} \left(\frac{v_0 - f}{f} \right) \left[\left(\frac{v_0 - f}{\Delta v_k} \right) + 1 \right]$$

$$= - \frac{A}{\epsilon b_0} \left(\frac{v_0 - f}{f} \right) - \frac{A}{\epsilon b_0} \frac{(v_0 - f)^2}{f} \frac{1}{\Delta v_k}$$

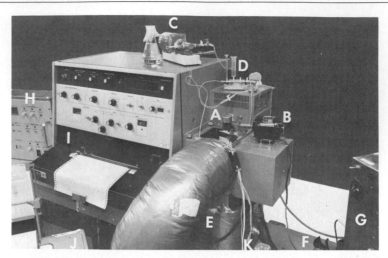

FIG. 4. A, Topview of the continuous dilution cuvette when placed in the cell compartment of the spectrophotometer. B, Motor to rotate the cell inside the cell compartment. C, Reciprocating and rotating pump to meter in solvent. D, Syringes to measure the volume of solvent displaced by solvent inflow. E, Air-thermostat, consisting of blower to keep thermostated air circulating in the system; radiator thermally regulated by the circulation of water from a constant temperature bath. It, in turn, keeps the air which passes through it thermostated; an insulated duct to form a close loop connecting the blower, the radiator, and the cell compartment. F, Constant temperature water bath to supply water circulating through the radiator in the air-thermostat. G, Control circuit board for the motor. H, Data interface accessory for data acquisition and control. I, Cary 118. J, Teletype. K, Time-delay relay for actuating the Hamilton valve via the electromechanical linkage system.

If $1/\Delta A b_k$ is plotted against $1/\Delta v_k$, we should get a straight line with

$$\text{Intercept} = -\frac{A}{\epsilon b_0}\left(\frac{v_0 - f}{f}\right)$$

and

$$\text{Slope} = -\frac{A}{\epsilon b_0}\frac{(v_0 - f)^2}{f}$$

Therefore,

$$\frac{\text{slope}}{\text{intercept}} = v_0 - f$$

Also, since

$$\lim \Delta A b_k = -\epsilon \frac{b_0}{A}\left(\frac{f}{v_0 - f}\right) \qquad \Delta v_k \to \infty$$

Therefore

$$|\Delta Ab_{max}| = \epsilon \frac{b_0}{A} \cdot \left(\frac{f}{v_0 - f}\right)$$

$$\frac{|\Delta Ab_{max}|}{Ab_0} = \frac{\epsilon(b_0/A)[f/(v_0 - f)]}{\epsilon(b_0/A)[v_0/(v_0 - f)]} = \frac{f}{v_0}$$

Knowing both $(v_0 - f)$ and (f/v_0) from the graph, we can find f.

The maximum error in the determination of the dead volume correction is 3%. The effect of this error on the concentration difference spectra is negligible, as shown in the following analysis:

$$\Delta Ab = \epsilon \frac{b_0}{A} \left(\frac{f}{v_0 - f}\right) \frac{\Delta v}{v_0 - f + \Delta v}$$

$$\sigma_{\Delta Ab} = \left(\frac{\partial \Delta Ab}{\partial f}\right) \sigma_f$$

$$= \sigma_f \left[\epsilon \frac{b_0}{A} \Delta v \left(\frac{(v_0 - f)(v_0 - f + \Delta v) - f \, \partial(v_0 - f)(v_0 - f + \Delta v)/\partial f}{(v_0 - f)^2(v_0 - f + \Delta v)^2}\right)\right]$$

$$= \sigma_f \epsilon \frac{b_0}{A} \Delta v \left(\frac{v_0^2 - 2v_0 f + v_0 \Delta v - f\Delta v + f^2 - f(-2v_0 + 2f - \Delta v)}{(v_0 - f)^2(v_0 - f + \Delta v)^2}\right)$$

$$= \sigma_f \epsilon \frac{b_0}{A} \Delta v \left(\frac{v_0^2 + v_0\Delta v - f^2}{(v_0 - f)^2(v_0 - f + \Delta v)^2}\right)$$

$$\frac{\sigma_{\Delta Ab}}{\Delta Ab} = \frac{v_0^2 + v_0\Delta v - f^2}{(v_0 - f)(v_0 - f + \Delta v)} \frac{\sigma_f}{f}$$

$$\sigma_{\Delta Ab} = \frac{v_0^2 + v_0\Delta v - f^2}{(v_0 - f)(v_0 - f + \Delta v)} \frac{\sigma_f}{f} \Delta Ab$$

Taking the two extreme cases of first dilution and final dilution:

For the first dilution

$$\Delta v = 0.2 \text{ ml} \qquad v_0 = 0.3 \text{ ml} \qquad f = 0.03 \text{ ml} \qquad \Delta Ab = 0.01$$

$$\sigma_{\Delta Ab} = \frac{(0.3)^2 + (0.3)(0.2) - (0.03)^2}{(0.3 - 0.03)(0.3 - 0.03 + 0.2)}(0.03)(0.01)$$

$$= 0.00035$$

$$\frac{\sigma_{\Delta Ab}}{\Delta Ab} = 3.5\%$$

For the final dilution

$$\Delta Ab = 0.05$$

$$\sigma_{\Delta Ab} \sim (\sigma_f/f) \, \Delta Ab \sim 0.03 \times 0.05$$
$$\sim 0.001$$

$$\frac{\sigma \Delta Ab}{\Delta Ab} = \frac{0.001}{0.05} = 2\%$$

(as compared with ~20% for the multiple-cuvette case).

XI. Performance Characteristics of the Continuous Dilution Cuvette

For potassium ferricyanide, throughout the possible concentration range studied, the deviation from Beer's law linearity is no greater than 0.0004 absorbance units, the present precision and accuracy limit of the spectrophotometer itself.

Four orders of magnitude in concentration can be spanned by making use of the dilution factor (~200) of the continuous dilution cuvette and the allowable range in initial absorbance (0.01–1), with the necessary precision.

A rough estimate of the smallest absorbance change from which one can obtain a reliable estimate of the equilibrium constant goes as follows: In order to get a "minimum" good fit of data to a dissociation curve, about 10 points evenly spaced along the ΔA (difference absorbance) axis are needed. Let's require that the distance between points to be at least twice the standard deviation of the spectrophotometer reading (2σ), then the total absorbance change needed would be 20σ, which is 0.01 since $\sigma = 0.0005$.

We can then guess at the feasibility of applying the method to any particular system in the following manner:

$$\Delta A = c_0 l_0 \Delta \epsilon > 0.01$$

To get two orders of magnitude in dilution $l_0 < 0.04$ cm, therefore

$$c_0 \Delta \epsilon > 0.01/0.04 = 0.25$$

XII. Studies Done so Far with the Continuous Dilution Cuvette

With potassium ferricyanide, metmyoglobin, and both met- and oxy-hemoglobin at nondissociating concentrations, no systematic deviation from Beer's law linearity has been found, and the random deviation observed is never greater than 0.0005 absorbance units, fluctuations expected from propagation of errors due to photometric error (0.0002), volume measurement error (0.4%), and uncertainty in the value of f (dead volume correction factor) (3%).

Although our past practice has been to use the double reciprocal plots to get our dead volume correction factor, and to test for deviations from Beer's law linearity, it has been found that, in our case, the uncertainty in the value of f, as a result of the mutual dependence of the slope and intercept of the plots, unnecesarily lowers the precision of the method. We have since expressed f explicitly in the following form:

$$f = \frac{\Delta Ab(v_0 + \Delta v)}{Abs_0(\Delta v/v_0) + \Delta Ab}$$

where v_0 is the initial volume, Δv is the volume added, Abs_0 is the initial absorbance, and ΔAb is the absorbance change. With each dilution step, a value of f can be calculated. A complete dilution run would generally consist of 20 dilution steps, each with a calculated f. The value of f used for all subsequent dilution experiments is an average of all the f's, and is found to be independent of wavelengths and solution standards used, insofar as they obey Beer's law. The standard error of the dead volume thus calculated is around 3%.

With oxyhemoglobin at pH 7, in 1 M NaCl and 0.05 M Bis-Tris buffer, we have obtained the lower and middle portion of the dissociation curve by monitoring spectral changes at around 290 nm, and the middle and upper portion of the dissociation curve by monitoring changes at the Soret maximum. The total absorbance change observed was 0.01 absorbance units, constituting a change in molar absorbance of 8%. The tetramer–dimer dissociation constant obtained in both cases are the same. The standard deviation in the dissociation constant for a single experiment is 30%. Data obtained at wavelengths lower than 280 nm have not thus far been used, because of interference by Bis–Tris absorbance. Our data indicate that Bis–Tris binds to hemoglobin with concomitant spectral changes.

With methemoglobin at pH 6 in 0.1 M phosphate buffer, the dissociation constant is found to be near 10^{-7} M, which is considerably lower than published results obtained using laser light scattering[24] and the analytical ultracentrifuge.[25,26] Substantiation of this rather preliminary finding requires closer approximation to the experimental conditions of the experiments reported, and we intend to confirm our results by using the hydrogen-exchange method of Rosenberg, Ide, and Barksdale.[27,28]

[24] S. L. White, *J. Biol. Chem.* **250**, 1263 (1975).
[25] P. Hensley, S. J. Edelstein, D. C. Wharton, and Q. H. Gibson, *J. Biol. Chem.* **259**, 952 (1975).
[26] G. B. Ogunmola, private communication.
[27] G. J. Ide, A. D. Barksdale, and A. Rosenberg, *J. Am. Chem. Soc.* **98**, 1595 (1976).
[28] A. D. Barksdale and A. Rosenberg, Vol. 48, p. 321.

XIII. Possible Improvements in System Design

Because of the movable piston the cuvette outer body must be drawn with precision of 0.001 in. or better. This requirement is not due to nonlinearities in the pathlength–dilution relationship which can be corrected by not particularly laborious computational procedures, but by the need to prevent liquid and gas leakage between the two compartments of the full cuvette formed by the moving window. Glass tubing with 0.001 in. precision is available with internal dimensions varying by a few ten thousands. Many alternative procedures have been examined but the silicone rubber gasket for the piston to wall contact is the only workable solution we have found.

Molds for these gaskets are expensive so the tubing inner diameter must be rather closely matched to the final diameter of the gasket. Differences in operating temperature greater than about 10° impair piston performance (too high driving pressure, on the one hand, and leakage, on the other) so a better gasket material is needed. This problem can be somewhat alleviated by the use of a quartz outer body to match temperature coefficients of windows to those of cuvette body. This improves thermal expansion mismatch between quartz parts but does not eliminate the large residual temperature coefficient of the gasket material. Depending on the future availability of precision bore quartz tubing, the main cuvette body should be a fused silica tube with one end window fused or soldered with a "glass solder" to it and the necessary inlet and outlet holes and connections fabricated by the tubing maker. The second window must be removable for cleaning and to remove and install the piston.

Complications of this kind indicate that additional development by people already skilled in the design of the several parts of the total system is necessary. The cuvette system discussed here can be duplicated in a university or industrial laboratory with the aid of additional technical information available from Yim[29] and subsequent publications, but the ultimate success of the method appears to depend on arousing commercial interest.

The properties of the end-window gaskets and the need to attach nipples of low liquid volume has made it necessary to use epoxy cements which are soft and easily removed. This detail in cuvette assembly is not particularly laborious but should be removed. Hard epoxies can be used for a few seals which can be treated as permanent if there are no chemical implications for the solutes or solvent, but use of soft epoxies such as Extra Fast Setting Epoxy by Harmon Inc., Belleville, New Jersey is on the whole satisfactory. Chemical and adsorption problems with the gasket

[29] Z. Yim, Ph.D. Dissertation, Department of Chemistry, University of Minnesota, Mineapolis (1977).

material can be serious. Some organic solvents, e.g., alkanols, even when present only in a few percent in aqueous solution destroy the gaskets, pose another reason to seek better gasket material. It is probable that a variety of gaskets of different materials may be necessary for general purpose use of the cuvette.

In the limited experiments to date, protein loss to glass and quartz is negligible at concentrations below $10^{-4} M$, but this good fortune will certainly not continue. Sakura[23] had great difficulty with the problems of hemoglobin adsorption, but he used the many-cuvette procedure. Silica appears to provide less serious problems than glass, but the improvements vary with the grade of silica and the many methods now available to treat glass and silica surfaces. Problems of protein adsorption to these and other solid substances are far from eliminated by studies during the last few years, but recently progress in this area has been rapid. Coatings to lubricate piston translation and to keep macromolecules, such as proteins, off the walls are promising but the coatings must be chemically bonded to the solid tubes to prevent their removal by piston friction. Much could be done in any given case to solve these problems, but we have found them minimal so long as there is no gas phase present so that it has been unnecessary to go into a detailed study of the problem. Silica parts are significantly superior to glass parts.

Longer cuvettes would increase dilution factors and thus increase the dynamic range of the method, but they would have to be considerably longer to be worth the trouble. Doubling cuvette length is not only of little value but produces cuvette systems which do not fit the cell compartments of most spectrophotometers. All such matters considered, the length we have chosen is about optimal, unless much longer pathlengths become possible. It is important to note that at least with water solutions light scattering is negligible at our current 0.0002 absorbance unit precision level. Longer cuvettes are, of course, possible, but multiple-pass systems are much to be preferred and certainly can be made available by commercial instrument designers, who can optimize length against scattering and photometric precision loss. Laser light sources do not have the versatility of spectrophotometers for absorbance measurements, but when lasers with suitable wavelengths are available, they may aid greatly in the design and use of multiple-pass systems. While remarkable gains in sensitivity are not to be expected with improvements in spectrophotometer design following conventional lines, improvements in the design of multiple-pass cuvette systems and *ad hoc* instruments for laser illumination should make it possible to increase sensitivity by at least one to two orders of magnitude, particularly since electronic noise supression procedures already can add a factor of ten.

The cuvette system can be used with many types of detection. Spectroscopic detection has been emphasized because of the wide variety of information available with the new generations of high-performance spectrophotometers, but the method is intrinsically weak. Fluorescence detection with the various well-known modifications based on polarization may provide increased sensitivity as well as information useful in characterizing the dissociating species themselves. Straight-through fluorescence measurements are necessary with the cuvette discussed here, and light collection is poor. Better illumination and light collection systems are desirable for large-scale use of fluorescence and phosphorescence detectors. Nonoptical types of detection reduce requirements on cuvette construction and can increase sensitivity. A maximum gain in sensitivity of about 10^3 over spectrophotometry appears possible. The dilution and mixing features of the cuvette system are all that need to be retained, but it is likely that in the future simultaneous nonoptical and optical detection will provide considerable advantages in many cases.

Utilization of up-to-date electronics provide the major source of gain in sensitivity for dissociation experiments. Thus far the utilization has lagged far behind the sophistication and power available. These matters are discussed in Section XV. It is not yet generally realized that interactive coupling of spectrophotometers, etc., through microprocessor units to computers eliminates continuous operator participation in many types of experiments, complex or bulky. Not only can long-standing needs to remove human decisions in data acquisition and statistical problems be eliminated by such systems, but experiments involving so many samples and measurements that were totally impractical for the research laboratory are now possible. The cost is slight and the systems run night and day.

XIV. Detailed Instructions on the Use of the
        Continuous Dilution Cuvette (see Fig. 5)

The cuvette can be washed either fully disassembled, piece by piece, or completely assembled, as described below. With I_2, I_4 and I_5 closed, I_1 connected to I_4 and SY_1 filled with air, SY_2 filled with detergent solution, the sample chamber is first purged of sample solution with air through I_3, then partly filled with detergent solution, which is then churned within the cell either by shaking the whole assemblage (bartender fashion) or by rotating the cell manually back and forth (washing machine fashion). On completion of the washing cycle, SY_2 is then filled with pure water. The cuvette is then alternately filled with water and purged with air several times. Finally, the sample chamber is once more emptied and dried with air. It can now be filled with solvent for the next experiment.

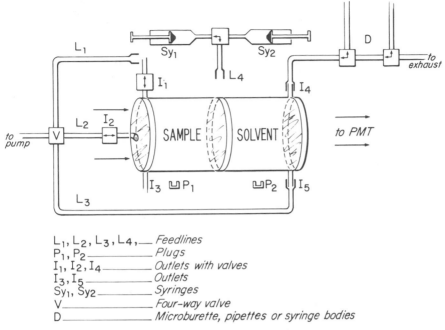

L₁, L₂, L₃, L₄, ____ Feedlines
P₁, P₂ _____ Plugs
I₁, I₂, I₄ _____ Outlets with valves
I₃, I₅ _____ Outlets
Sy₁, Sy₂ _____ Syringes
V _____ Four-way valve
D _____ Microburette, pipettes or syringe bodies

FIG. 5. See text for explanation

Before each dilution experiment, a base line should be obtained with both chambers filled with solvent. It is immaterial where the piston is situated in this case, as the pathlength remains constant irrespective of its position. However, in filling the cell with solvent, one should make sure that there are no bubbles either in the feedline or inside the cuvette. Bubbles can produce errors by working their way into the light path; also, if they appear in the sample chamber, they effectively increase the dead volume correction and, of course, reveal their presence in this correction.

Sample Introduction. Sample can be introduced into the sample chamber in the following manner (see Fig. 5). With V opened to L_3, and with I_4, I_1 and I_2 closed, move the piston toward the left end window (sample side), until it is about 0.5 in. from the end, by pumping solvent into the cuvette through I_5 and allowing the displaced solvent out through I_3. This 0.5 in. space is found to be the minimal space required to remove all the air bubbles when filling the sample chamber with protein solutions.

With the piston situated 0.5 in. from the end window, and I_2, I_4 and I_5 closed, I_1 connected to L_4, and SY_1 filled with air and SY_2 filled with sample, the sample chamber is first purged of solvent with air through I_3, then a small quantity of the sample is introduced to rinse the sample chamber with a few twirls of the cuvette. The sample chamber is subsequently purged with air, and then filled completely with sample. I_3 is then capped with plug, P_1, which has been first filled with sample for the elimination of air bubbles. The piston is moved toward the left end window (sample side) to the desired path length, by pumping in solvent along L_3 through I_5, and with I_2, I_3, and I_4 closed, and I_1 disconnected from L_1 and opened for exhaust. When the desired pathlength is reached, I_1 is closed. Of course, the initial pathlength should be as small as possible in order to maximize the dilution factor; but, the absolute absorbance should not be smaller than 0.1, since the predicted 5–10% change in absorbance upon dissociation would amount to 0.005–0.01, which is close to the lower limit of the applicability of the method for obtaining equilibrium constants.

Initial Volume. In order to know the concentration of the solution at various stages of dilution, we need to know the amount of diluent added at each dilution step, as well as the initial volume. At present, we find it best to measure the volume of inflow by monitoring the volume of outflow with a set of syringe bodies, microburettes or pipettes of different sizes, as shown in the diagram. This system dispenses with the high cost of precision threaded syringe pumps, which would give no better precision. The initial volume can be determined by knowing the initial pathlength l_0 since

$$V_{in} = Al_0 - f$$

where f is the dead volume, A is the cross-sectional area of cuvette, and l_0 is the initial pathlength. The initial pathlength of the sample can be determined by comparing the absorbance of the sample as it is in the cuvette with the absorbance of the sample in a standard cuvette, the pathlength of which has been accurately determined.

Solvent Addition for Dilution. In principle, any fluid metering device can be used. We have tried hand-held syringes, peristaltic pumps, syringe pumps, and various other pumps. A FMI reciprocating and rotating pump (RP-SY-CS, Fluid Metering, Inc., Oyster Bay, New York) was selected for the following considerations:

1. The cost is low.
2. The rate of flow is continuously variable.
3. The flow has very little ripple.
4. There is very little fluid slippage at the operating back-pressure (20 psi).
5. Its use of stepping motor operating at line frequency eliminates inertial effects.
6. It allows easy and inexpensive measurement of solvent volume added, by counting the number of pulses sent to the stepping motor with each dilution.

Sample Backflow Into Feedline. Since very small changes in absorbance are measured, any sample backflow into the feedline destroys the experiment. Owing to frictional forces there needs to be a certain amount of internal pressure developed against the piston before it can be moved along the barrel (about 20 psi in our system); hence, the sample chamber is slightly pressurized each time the inlet valve is closed after diluent addition. It is, therefore, necessary to develop a sufficient pressure head in the feedline before the valve is opened for another diluent addition. This prevents the sample from moving into the feedline, otherwise there is no guarantee that all the sample has been swept back into the sample chamber. The situation is worse if the valve has channels of convoluted geometry. We found serious backflow problems with a number of solenoid and check valves. The Hamilton valve (No. 1) was found to have essentially zero displacement and straightforward channel geometry, so a simple electro-mechanical linkage system was designed to drive this kind of valve in such a way that there is a slight time delay for the on position, and no time delay for the off position.

XV. Interfacing Accessories for Digital Data Acquisition and Instrumental Control (see Fig. 6)

The digital systems discussed in this section were designed by Mr. Blakeley LaCroix, Instrumentation Laboratory, Chemistry Department, University of Minnesota.

The rapid advances in data logging and instrumentation control have been matched by computer development and software techniques for recovering signal from noise and all have advanced in precision and accuracy faster than photometric systems. It is, nevertheless, necessary to use advanced digital instrumentation to obtain maximum data quality and experimental versatility with available spectrophotometers, even though the latter are the limiting factor in concentration difference spectrophotometry. Decreasing costs makes previously impossible experiments possible, and introduces to chemists and biochemists really operator-free experimentation for the first time. Data logging and simple instrument operation are well established, but the major gains lie in the area of interaction between these capabilities and minicomputers. The interactive mode introduces operator-independent logic decisions, operational-research procedures to avoid unnecessary experiments and duplication. Operator supervision can be nearly eliminated to allow 24-hr utilization of instruments. Many of these new improvements are common in large physics experiments, X-ray and other diffraction work, and well-engineered testing programs, but, although now generally available at low cost, it still remains to be utilized to a significant degree by chemists and biochemists.

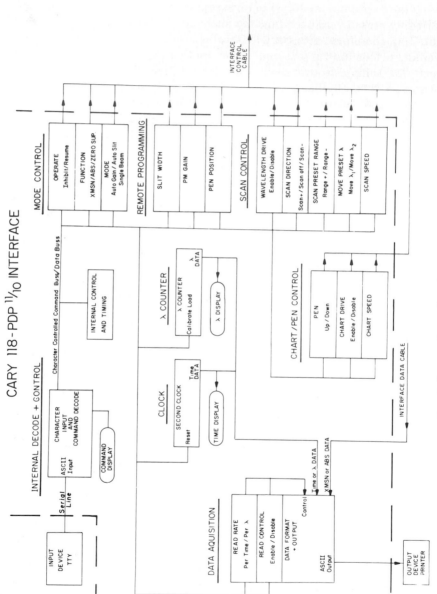

Fig. 6. See text for explanation

The most recent electronics developments in microprocessors are not yet generally used in commercial instruments. These cheap and versatile devices are ideal for data logging and control of instrument operation when only very simple logical decisions are required, and are incorporated in most new instruments of any sophistication, but not for interactive-mode operation. One can afford to dedicate a microprocessor to a single application, but not a computer. Computers are necessary for interaction and must be used on-line, albeit on a time-sharing basis. When necessary, it is possible in the interactive mode, to allow easy random participation of an operator for *ad hoc* commands, program changes, calculation check, etc. The interface discussed in the remainder of this section is of a microprocessor type, but was designed before a suitable microprocessor was available. Descriptions of design modifications which incorporate recent microprocessors for somewhat lower cost and some additional power, are available from Mr. Blakeley LaCroix. The buffer storage and ROM items are not included in Fig. 6 since they might vary with the memory capabilities of the other peripherals and the computer used. The block diagram described an interface for PDP 11-10 computer to Varian-Cary 118 spectrophotometers. It is a minimum system for sophisticated interaction and can be updated for many purposes by a variety of microprocessor or multiprocessor incorporations. The first-generation unit diagrammed in Fig. 6 has several new features. It is applicable to many types of detection instrumentation.

The unit illustrated in Fig. 6 is a serial-line device tailored to operate in an interactive mode between measuring instrument and computer without operator supervision using ASCII2 language, but with the added feature that an operator can enter the experiment at any time using a simplified language which is adjustable to different kinds of experiment. It is controllable on-line so that the program can be changed in real time without the use of a minicomputer. In the interactive mode there appears to be little limit on the sophistication of operation. In routine use without computer interaction, using easily identifiable character sets, the operation of the spectrophotometer can be directed throughout an experiment either by a program of buffer-stored commands or through a real-time input device. One can type in a series of command operations under the inhibit mode and these operations will not be executed until the "resume" command is given. This allows testing of the several parts of a program independently. The command structure is hierarchical so that operations can be stopped instantly, but always in a defined state to assure reproducibility of operations. The system differs from commercially available data interfacing units in allowing more flexibility in the design of the strategies of spectral measurement.

The following is an abbreviated list of the capabilities of the interface unit used in conjunction with a modern, high-performance spectrophotometer.

1. Store command program
2. Scan specified spectrum at specified scan speed and chart speed
3. Programmable slit control
4. Programmable gain control
5. Formatted read wavelength and absorbance mode
6. Formatted read time and absorbance mode
7. Take specified number of absorbance readings at any set of wavelengths
8. Programmed (buffer memory or tape) and real time interactive control. Complete second-level, operator-independent experiment control involving one or many measuring instruments (flexible logic provided by minicomputer)

Acknowledgments

David Anderson worked on the early development phase of the concentration-difference cuvette. Mr. Roland Hawes and Mr. Kenyon George provided advice and ideas, some of which have been patented with assignment of patent to the Cary Instruments, Inc. (now the Varian Associates Corp.) Digital control interface was conceived and designed by Blakeley LaCroix. Major machining problems were solved by Mr. Richard Weizel. The movable-window design for the continuous dilution cuvette has been patented by Mr. George and Mr. Hawes, and a device eliminating the free-volume error has been patented by D. Anderson, R. Lumry, and K. George. Both patents are assigned to the Varian Associates Corp. We are especially indebted to Mr. George, recently deceased, whose support and assistance made the undertaking possible.

[7] Advances in Sedimentation Equilibrium[1]

By ROBLEY C. WILLIAMS, JR.

Sedimentation equilibrium, while being supplanted as a method for the simple measurement of molecular weights, is becoming steadily more useful as a means of studying macromolecular interactions. It has many advantages over other methods, chief among which are the facts that sedimentation equilibrium can be observed in free solution, that relatively great precision is available in modern ultracentrifuges, and that a sound

[1] This work was supported by Grants HL 12901 and HL 20873 of the National Institutes of Health.

INSTRUMENTAL MODIFICATIONS USEFUL IN SEDIMENTATION
EQUILIBRIUM APPLICATIONS

Modification	Optical System	References
Precise rotation and translation devices for the double slits	Interference	4,5
Convenient and precise optical alignment methods	Interference	6–8
Teflon or polyvinyl chloride window liners	Interference	9,10
Polarizing filter	Interference	10
Laser light source and films	Interference	11,12
Improved plate reading equipment	Interference	13,14
Computer-based absorption scanners	Absorption	15–19
Rotor precession detector	Absorption and interference	16
Auxilliary shutter	Interference	—

thermodynamic basis exists for the interpretation of results. The study of macromolecular interactions by sedimentation equilibrium was thoroughly dealt with in this series in 1973 by Teller,[2] and most of that chapter remains current. This article describes some advances in instrumentation and in numerical analysis of data that have been made in the four years since that article was written. For general background, the reader is referred to Teller's chapter, to a review by Van Holde,[3] and to the many references in those articles.

Instrumental Advances

The study of macromolecular interactions by equilibrium ultracentrifugation requires the best instrumental precision currently attainable. The Beckman Model E ultracentrifuge is being improved only slowly by its manufacturer. The burden of modification of the instrument thus has fallen increasingly upon its users in recent years. Small and large improvements have been developed in a number of laboratories which, cumulatively, substantially increase the precision and usefulness of the Model E. A number of such changes are listed in the table and are described below. An investigator could make one or several of the modifications, depending on the requirements of the research at hand and on the available funds.

Interference Optical System. The precision and sensitivity obtainable with the Rayleigh interference optical system makes its use preferable to the use of the Schlieren optical system for sedimentation equilibrium

---

[2] D. C. Teller, see Vol. 27, Part D, p. 346.
[3] K. E. Van Holde, *in* "The Proteins" (H. Neurath, ed.), 3rd ed., Vol. 2, p. 225. Academic Press, New York, 1975.

experiments. The interference system requires careful alignment, however. Rotational and translational alignment of the double-slit mask, difficult with the standard instrument, can be rendered simpler by installation of either a micrometer-driven translator[4] or of a precise mask-rotation device[5] adjustable from outside the vacuum chamber of the ultracentrifuge. The particular design of the rotator makes it possible to adjust the mask without stopping the rotor, a distinct advantage if alignment is carried out by means of observation of a concentration gradient within the centrifuge cell. Alignment of the optical system as a whole has been rendered somewhat simpler and more precise by the excellent protocol of Rees and co-workers,[6] and has been carefully analyzed by Richards et al.[7,8]

Distortion of the windows of the ultracentrifuge cell is an important source of imprecision. The use of either Teflon[9] or polyvinyl chloride window liners[10] has been shown to reduce the blurring of fringes from this source, as has the insertion of a polarizing filter into the incident light beam.[10] In the author's laboratory, we are routinely able to carry out experiments with the interference optical system at 52,000 rpm when these inexpensive and simple measures are employed.

Lasers are superior to the standard mercury arc as light sources for the interference optical system. The construction of a simple laser installation[11] and of a more complex one[12] are described in these volumes. In conjunction with a laser, films can be used in place of glass plates in most applications, with a consequent increase in fringe resolution and saving of money.

With the advent of inexpensive computing gear, the use of automatic[13,14] or semiautomatic apparatus for measurement of the interference fringes on the photographic record has become widespread. The author's laboratory uses a pair of digital micrometers, installed on the Nikon comparator in place of the micrometers supplied with it. The counters of these micrometers are connected directly to the PDP-8/e computer which is used for data reduction. The fringe is positioned manually, but the numbers are entered directly into the computer by pushing a foot

[4] D. J. Cox and A. T. Ansevin, Anal. Biochem. 58, 161 (1974).
[5] W. F. Bowers and R. H. Haschemeyer, Anal. Biochem. 28, 257 (1969).
[6] A. W. Rees, E. A. Lewis, and M. S. DeBuysere, Anal. Biochem. 62, 19 (1974).
[7] E. G. Richards, D. C. Teller, V. D. Hoagland, Jr., R. H. Haschemeyer, and H. K. Schachman, Anal. Biochem. 41, 215 (1971).
[8] E. G. Richards, D. C. Teller, and H. K. Schachman, Anal. Biochem. 41, 189 (1971).
[9] T. A. Horbett and D. C. Teller, Anal. Biochem. 45, 86 (1972).
[10] A. T. Ansevin, D. E. Roark, and D. A. Yphantis, Anal. Biochem. 34, 237 (1970).
[11] R. C. Williams, Jr., see Vol. 48, Part F, 185.
[12] D. A. Yphantis, this volume [1].
[13] D. J. DeRosier, P. Munk, and D. J. Cox, Anal. Biochem. 50, 139 (1972).
[14] R. M. Carlisle, J. I. H. Patterson, and D. E. Roark, Anal. Biochem. 61, 248 (1974).

switch. This method works well at small cost, and is representative of a variety of similar systems which obviate the necessity of writing numbers as part of the fringe-measuring process. Where a local minicomputer is not available, digital micrometers can be used to drive a papertape punch or magnetic tape data logger, and the tape can then be carried to a large computer.

In the author's experience, the opening of the camera shutter of the Model E ultracentrifuge causes the instrument to vibrate, with consequent blurring of the photograph. The vibrations can be observed to die out in about 10 sec. The installation of an auxilliary shutter which is opened, either manually or by means of a time-delay relay, about 20 sec after the main camera shutter yields distinctly better fringe contrast and is extremely simple to build. The solenoid-driven blade shutter ("view shutter") normally installed over the mercury arc lamp works well for this purpose.

Absorption Scanners. Computer-based absorption scanners are becoming more common,[15-19] and their cost is steadily decreasing. The precision provided by these devices makes practical their use for measurement of association constants by sedimentation equilibrium methods.[20,21] There is, as yet, no standard design for such a scanner, but the experience reported in the literature is extensive enough to enable a person who is moderately skilled in electronics to build one with confidence that the outcome will be successful.

The present limitation in attainable precision in sedimentation equilibrium experiments has been ascribed by Teller[2] to rotor movements which can be described as precession or nutation of the rotor, or both. Rapid movements of the rotor (at frequencies of several per second), are small and difficult to detect. The slower nutational swinging of the rotor (at frequencies between 0.5 and 0.05 sec^{-1} and amplitudes of 0 to 40 μm) is responsible for blurring of fringes when the interference optical system is employed and for systematic distortion of the concentration distribution measured by the absorption scanner. This slow movement can be measured by means of an electromagnetic detector mounted near the rotor support ring. In the author's laboratory we have used this "precession de-

[15] R. H. Crepeau, C. P. Hensley, Jr., and S. J. Edelstein, *Biochemistry* **13**, 4860 (1974).
[16] R. C. Williams, Jr., *Biophys. Chem.* **5**, 19 (1976).
[17] S. P. Spragg, W. A. Barnett, J. K. Wilcox, and J. Roche, *Biophys. Chem.* **5**, 43 (1976).
[18] D. L. Rockholt, C. R. Royce, and E. G. Richards, *Biophys. Chem.* **5**, 55 (1976).
[19] R. Cohen, J. Cluzel, H. Cohen, P. Male, M. Moignier, and C. Soulie, *Biophys. Chem.* **5**, 77 (1976).
[20] P. Hensley, S. J. Edelstein, D. C. Wharton, and Q. H. Gibson, *J. Biol. Chem.* **250**, 952 (1975).
[21] R. C. Williams, Jr. and H. Kim, *Biochemistry* **15**, 2207 (1976).

tector'' to correct the apparent radius measured with the scanner.[16] Dr. M. S. Lewis (personal communication) has devised a modification of this circuit to modulate the intensity of a laser light source to expose the film only when the rotor is near the extremum of its swing. This strategy reduces blurring of the interference fringes.

Analysis of Sedimentation Equilibrium Experiments

Teller's article[2] explains clearly the methods generally employed to proceed from the initial data of concentration versus radius to an interpretation in terms of thermodynamic quantities. It has been commonplace to draw a distinction between two routes to the determination of stoichiometries and association constants: through the calculation of molecular weights at numerous points in the concentration distribution, or through direct fitting of the concentration distribution by an appropriate mathematical function representative of the intermolecular interactions thought to be taking place within the solution. This distinction is not really an important one. Regardless of the approach taken, the experimenter must eventually attempt to match observation to an interpretive scheme. The use of molecular weight averages as an intermediate step in analysis provides a useful means for the experimenter to orient himself to the interacting system. Their use enables assessment of the concentration range to see whether or not it is broad enough to cover an expected range of behavior (e.g., if a monomer–tetramer system is hypothesized, does the weight average molecular weight approach monomer and tetramer at the low and high ends of the measured concentration span?). They enable simple classification of the interactions as rapidly or slowly equilibrating, and they allow simple detection of contamination. However, no information is gained (and some may be lost) by first transforming concentration versus radius data to molecular weight data and then fitting the latter to a model. Only computational convenience is gained.

Direct Fitting of Data. Direct fitting of concentration versus radius data to a model has distinct advantages. Foremost among them is the fact that the effects of random experimental error are easier to deal with if their propagation through a molecular weight average does not have to be assessed as part of the curve-fitting process. The use of direct fitting techniques is not subject to easy codification as a ''method,'' rather it is a straightforward means of testing agreement between hypothesis and experiment. In principle, the problem is a simple one, but in practice it is frequently computationally difficult. Access to a large computer is usually required. The feasibility of direct fitting has been demonstrated, and a strategy of approach to the problem has emerged from several recent analyses.[22–27]

In general, experiments yield several sets of data, c_{jk} as a function of r_{jk}, where c_{jk} is the concentration at radial point r_{jk}, and the subscripts j and k signify the jth set of experimental conditions (rotor speed and initial concentration, for instance) and the kth experimental point. One then postulates an explanatory functional relationship to fit the data and adjusts its variable elements to obtain the best fit. Examination of the quality of the fit then provides a measure of the success of the hypothetical functional relationship as an explanation of the observed concentration distributions. For a number N, of ideal macromolecular species present in solution simultaneously, one can write a functional dependence of concentration on radius of the form

$$c_{ijk} = \sum_{i=1}^{N} c_{ij}{}^0 \exp \left[\sigma_{ij}(\xi_{jk} - \xi_j{}^0) \right]$$

Here, the subscript i refers to the ith molecular species, $c_{ij}{}^0$ signifies the concentration of species i at the meniscus of the solution column under the jth set of experimental conditions, σ_{ij} is the reduced molecular weight of species i under the jth set of conditions, and ξ_{jk} and $\xi_j{}^0$ are the values of $r^2/2$ corresponding to the kth point and to the jth meniscus, respectively. The variables to be determined are the $c_{ij}{}^0$ (one for each species at each initial concentration or rotor speed), and the σ_{ij}, one for each species at each speed. For instance, for a hypothetical mixture of two macromolecular species examined at three initial concentrations at each of three rotor speeds, the maximum number of unknown variables would be 24: one $c_{ij}{}^0$ for each species at each of the three initial concentrations at each of the three rotor speeds plus one σ_{ij} for each species at each rotor speed. This clearly intractable number of variables can be reduced by interrelating the unknowns. The interrelationships constitute, together with the number N, the "model" of the macromolecular interactions.

For instance, if the two macromolecular species are a dimer in equilibrium with a monomer, then $c_{2j} = Kc_{1j}{}^2$, and $\sigma_{2j} = 2\sigma_{1j}$, where the subscripts 1 and 2 refer to monomer and dimer, respectively. These relations reduce the number of unknown variables to 13. The rotor speeds will be

[22] S. Szuchet and D. A. Yphantis, *Arch. Biochem. Biophys.* **173**, 495 (1976).
[23] M. L. Johnson and T. M. Schuster, *Biophys. Chem.* **2**, 32 (1974).
[24] A. Rosenthal, *Macromolecules* **4**, 35 (1971).
[25] R. H. Haschemeyer and W. F. Bowers, *Biochemistry* **9**, 435 (1970).
[26] M. L. Johnson, Ph. D. Thesis, University of Connecticut, Storrs (1973).
[27] K. C. Aune and M. F. Rohde, *Anal. Biochem.* **79**, 110 (1977).

known, thus reducing the number of σ_{ij}'s to one, and the number of unknowns to 11. If, in addition, the value of σ_1 is known from chemical determination, the number of variables can be reduced in this example to 10. As Roark has shown in an elegant application to a problem of mixed association in histones,[28] if the positions of the bottoms of the three solution columns are known, then conservation of mass can be applied separately to each solution column for the three speeds, reducing the number of unknown variables in this example to 4, a very tractable number.

After thus introducing a model to particularize the problem as far as knowledge of the system will allow, one searches for the best set of the unknown variables. One pools the available values of c_{ijk}, makes an initial guess at values of the unknowns, and computes a value of χ^2, where

$$\chi^2 = \sum_i \sum_j \sum_k w_{ijk} (c_{ijk}{}^{\text{OBS}} - c_{ijk}{}^{\text{CALC}})^2$$

Here $c_{ijk}{}^{\text{OBS}}$ and $c_{ijk}{}^{\text{CALC}}$ are, respectively, the observed and calculated values of concentration at a given radial point. The w_{ijk} are appropriate statistical weights. An appropriate nonlinear minimization routine is then used to find that set of parameters which produces a minimum in χ^2. Minimization routines often fail, either wandering aimlessly or returning physically unreasonable answers. In our experience with both model data and real data we have found that the "cure" for such failure is to change the model or to supply *more data*. Careful modeling of real data with synthetic data contrived on the basis of canny guesses at the answers will often reveal the necessity for a greater range of concentrations or rotor speeds.

The chapter by Aune[29] in Volume 48 of this series provides sufficient detail to enable the reader to carry out direct fitting. Further information on computational methods for nonlinear fitting can be found in the useful books of Bevington[30] and Bard.[31] Anyone with a large body of experimental sedimentation equilibrium data to analyze should investigate the use of direct fitting.

Difference Sedimentation Equilibrium. A method for measuring small differences in molecular weight has been devised and tested.[32,33] It is essentially a variant of the meniscus depletion method in which the reference side of the ultracentrifuge cell is filled with one of the two solutions

[28] D. E. Roark, *Biophys. Chem.* **5**, 185 (1976).
[29] K. C. Aune, see Vol. 48, Part F, p. 163.
[30] P. R. Bevington, "Data Reduction and Error Analysis for the Physical Sciences." McGraw-Hill, New York, 1969.
[31] Y. Bard, "Nonlinear Parameter Estimation." Academic Press, New York, 1974.
[32] M. S. Springer, M. W. Kirschner, and H. K. Schachman, *Biochemistry* **13**, 3718 (1974).
[33] M. S. Springer and H. K. Schachman, *Biochemistry* **13**, 3726 (1974).

to be compared and advantage is taken of the fact that the Rayleigh interference optical system measures the difference in refractive index
between the two sectors. In this method, a rotor speed is chosen that is
high enough to render the concentration near the meniscus practically
zero. Both the mean of the concentrations in the two channels $\bar{c}$ and the
difference in concentration between them Δc are computed with respect
to the assumed zero concentration at the meniscus. The Schlieren optical
system is employed to measure $\bar{c}$ and the interference system to record
Δc. The difference $\Delta \sigma$ in the reduced molecular weight σ is given as

$$\Delta \sigma = d(\Delta c / \bar{c}) / d(r^2/2)$$

where

$$\sigma = mw^2(1 - \bar{v}\rho)/RT$$

The degree of approximation is quite good for small values of $\Delta \sigma / \sigma$, but
the theory is at present restricted to ideal, single-component solutions. Its
use in the study of macromolecular interactions would thus be restricted
to the study of slowly interacting or very tightly linked molecular complexes.

[8] The Translational Friction Coefficient of Proteins

By DAVID C. TELLER, ERIC SWANSON, and CHRISTOPH DE HAËN

In this article we consider the translational friction coefficient of proteins. Section I deals with the historical background. Section II presents a
semiempirical method for calculating the geometrical arrangement of subunits in oligomeric proteins. In Section III a method is presented to calculate friction coefficients of proteins from atomic coordinates. Section IV
presents the theory by which friction properties of objects can be calculated in a rigorous way, and in Section V the mechanics of such calculations are illustrated. In Section VI the method for finding the principle
axes of translation of a particle is given. Finally, Section VII contains
some results, conclusions and problems that remain for future work.

I. Background

The translational friction coefficient of proteins is easily obtained
from diffusion or sedimentation experiments through the Einstein–
Sutherland equation

$$f = kT/D \tag{1}$$

METHODS IN ENZYMOLOGY, VOL. 61

or the Svedberg equation

$$f = M(1 - \bar{v}\rho)/Ns \tag{2}$$

where D is the translational diffusion coefficient, k is Boltzmann's constant, T the absolute temperature, N is Avogadro's number, ρ the density of the solution, $\bar{v}$ the partial specific volume, M the molecular weight, and s the sedimentation coefficient of the protein.

Rather than reporting the absolute magnitude of f, it is customary to report a friction ratio f/f_0, where f_0 is the friction coefficient of a sphere of equal molecular weight M and partial specific volume v. f_0 is obtained from Stokes' law as

$$f_0 = (6\pi\eta_0)(3M\bar{v}/4\pi N)^{1/3} \tag{3}$$

where η_0 is the viscosity of the solvent.

Oncley[1] developed a method that separated the frictional ratio into two factors

$$f/f_0 = (f/f_e)(f_e/f_0), \tag{4}$$

Protein shapes were modeled as rotational ellipsoids and one factor (f_e/f_0) was related to the axial ratio p. For rotational ellipsoids $p = b/a$, where b is the equatorial radius and a is the semiaxis of revolution. The axial ratio can be calculated from Perrin's formulas[2] (see also Oberbeck[3] and Herzog et al.[4]):

For a prolate ellipsoid ($p < 1$)

$$\frac{f_e}{f_0} = \frac{(1 - p^2)^{1/2}}{p^{2/3} \ln\{[1 + (1 - p^2)^{1/2}]/p\}} \tag{5}$$

For an oblate ellipsoid ($p > 1$)

$$\frac{f_e}{f_0} = \frac{(p^2 - 1)^{1/2}}{p^{2/3} \arctan(p^2 - 1)^{1/2}} \tag{6}$$

The other factor (f/f_e) is attributed to hydration through

$$f/f_e = (1 + w/\rho v)^{1/3} \tag{7}$$

where w is a specific hydration in grams of water per gram of protein.

Oncley[1] and several authors after him were careful in pointing out that without independent measurements of either of the factors, the other

[1] J. L. Oncley, Ann. N.Y. Acad. Sci. **41**, 121 (1941).
[2] F. Perrin, J. Phys. Radium (Paris) [7] **7**, 1 (1936).
[3] A. Oberbeck, J. Reine Angew. Math. (Crelle's J.) **81**, 62 (1876).
[4] R. O. Herzog, R. Illig, and H. Kudar, Z. Phys. Chem. Abt. A **167**, 329 (1934).

could not be validly computed. It has become a custom to assume a specific hydration of proteins of 0.25 H_2O per gram of protein and report resulting axial ratios of equivalent rotational ellipsoids. Frequently the axial ratios resulting from such a treatment are absurd in the light of the present knowledge of protein structure. Unfortunately, the situation is often no better when viscosity data are considered at the same time.[5] The reason why these absurd axial ratios are accepted into the modern literature is that they are rarely interpreted further, i.e., they are deadend numbers. It appears, however, that if a more rigorous treatment of the problem would be available, friction coefficients could indeed provide insights into protein structure in solution.

Molecules with molecular weights smaller than about 2000 have frictional ratios smaller than 1; i.e., they diffuse and sediment faster than spheres of equal M and $\bar{v}$. This means that these molecules do not experience the solvent as a continuum.[6-8] Such molecules most likely move through solvent by a "jump and wait" mechanism for a significant fraction of their net movement, taking advantage of cavities in the liquid.[9]

For proteins with molecular weights in excess of 5000, however, motion of molecules in solution follows a continuous flow pattern. These motions correspond to particles moving at very low Reynolds number ($<10^{-4}$) in a continuous medium. An excellent treatise on the properties of particle motion at low Reynolds number is that of Happel and Brenner, "Low Reynolds Number Hydrodynamics."[10]

From a hydrodynamic viewpoint, proteins are rigid, impermeable, and incompressible.[11,12] Thus protein atoms not exposed to solvent, as well as any solvent molecules trapped within the structure, do not contribute to friction. The friction properties of proteins are dictated by their surface exposed to solvent. The surface of proteins can be described as a shell of beads consisting of atomic groups, such as —CH_3, —CH_2—, —NH_2, etc. If these groups are probed by the solvent molecules, then the pearl necklace model of polymers of Kirkwood[13] and the shell model of Bloomfield et al.[14] should be applicable at this atomic resolution.

[5] H. A. Scheraga and L. Mandelkern, *J. Am. Chem. Soc.* **75**, 179 (1953).
[6] S. Nir and W. D. Stein, *J. Chem. Phys.* **55**, 1598 (1971).
[7] A. Polson, *J. Phys. Colloid Chem.* **54**, 649 (1950).
[8] D. C. Teller and C. de Haën, *Pac. Slope Biochem. Conf. Publ.* **17**, 31 (1975).
[9] J. H. Hildebrand and R. H. Lamoreaux, *Proc. Natl. Acad. Sci. U.S.A.* **69**, 3428 (1972).
[10] J. Happel and H. Brenner, "Low Reynolds Number Hydrodynamics," 2nd ed. Noordhoff Int., Gröningen, Leyden, 1973.
[11] M. H. Klapper, *Biochim. Biophys. Acta* **229**, 557 (1971).
[12] F. M. Richards, *J. Mol. Biol.* **82**, 1 (1974).
[13] J. G. Kirkwood, *Recl. Trav. Chim. Pays-Bas* **68**, 649 (1949).
[14] V. Bloomfield, W. O. Dalton, and K. E. Van Holde, *Biopolymers* **5**, 135 (1967).

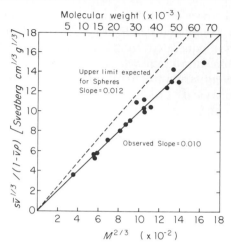

FIG. 1. Relationship of sedimentation coefficient and molecular weight of monomeric proteins according to Eq. (8) and (9) (From D. C. Teller, E. Swanson, and C. de Haën, unpublished results).

In the work of Bloomfield and co-workers, the beads were conceived as subunits of protein oligomers, viruses, and the like. However, in order to test the calculations, Bloomfield *et al.*[14] constructed shell models of objects with known hydrodynamic behavior and extrapolated the results to infinite numbers of beads of infinitesimal radius. Recent theoretical advances[15,16] prompted in part by the work of Bloomfield *et al.*[14] have made it worthwhile to follow this line of research to try to determine the factors which dictate the frictional properties of proteins.

II. Geometry of Oligomeric Proteins

By combination of Stokes' law [Eq. (3)] with the Svedberg equation [Eq. (2)], the following relation is obtained for a series of smooth spherical molecules of differing molecular weight.[17]

$$\frac{s\bar{v}^{1/3}}{1 - \bar{v}\rho} = \left(\frac{4\pi}{3}\right)^{1/3} (6\pi\eta_0)^{-1} N^{-2/3} M^{2/3} \tag{8}$$

In water at 20°, the term preceding $M^{2/3}$ on the right-hand side of Eq. (8) has the value 0.012 Svedberg cm g^{-1} mol$^{2/3}$. This equation is not accu-

[15] J. Rotne and S. Prager, *J. Chem. Phys.* **50**, 4831 (1969).
[16] H. Yamakawa, *J. Chem. Phys.* **53**, 436 (1970).
[17] K. E. Van Holde, *Proteins, 3rd* Ed. **1**, 226 (1975).

rate for proteins because of the rugosity of their surface. For monomeric proteins we found empirically[18] (Fig. 1):

$$s\bar{v}^{1/3}/(1 - \bar{v}\rho) = 0.010M^{2/3} \tag{9}$$

The factor 0.010 is believed to account for the average rugosity of the surface of grossly globular proteins.

In order to compute the sedimentation coefficient of an oligomeric protein one may assume that the subunits behave as if they possess the hydrodynamic behavior of the monomer, but pack together as spheres.[19] Van Holde[17] adapted the bead on a string model of Kirkwood[13,20] to calculate the ratio of the sedimentation coefficient of the n-mer to that of the monomer,

$$s_n/s_1 = n\zeta/f_n \tag{10}$$

where ζ is the friction coefficient of the monomer, f_n that of the n-mer, and n is the number of subunits. Following Teller and de Haën,[18] substitution of Eq. (9) into Eq. (10) yields

$$\frac{s_n\bar{v}^{1/3}}{1 - \bar{v}\rho} = 0.010M_n^{2/3}n^{1/3}(\zeta/f_n) \tag{11}$$

where M_n is the molecular weight of the n-mer. One can define

$$F_n = n^{1/3}(\zeta/f_n) \tag{12}$$

where F_n is purely a geometric factor. The geometric factor F_n can be calculated from either the Kirkwood approximation (Section III) or from the more rigorous theories of hydrodynamic interaction as presented in Sections IV and V. Table I presents the numerical values of F_n for a variety of subunit arrangements calculated from the rigorous theory.[21] With knowledge of s, $\bar{v}$, and M_n, the subunit geometry is determined by calculating F_n and choosing the closest value in Table I. Alternately, a graph such as Fig. 2 can be constructed and the geometry determined from the closest line.

For those oligomeric proteins that can be separated into subunits without denaturation of the polypeptide chains (e.g., in high concentrations of $CaCl_2$), Eqs. (11) and (12) serve as a reasonably accurate means of the determination of subunit number or geometry of the protein. From the

[18] D. C. Teller and C. de Haën, *Fed. Proc., Fed. Am. Soc. Exp. Biol.* **34,** 598 (Abstr. 2143) (1975).
[19] D. C. Teller, *Nature (London)* **260,** 729 (1976).
[20] J. G. Kirkwood, *J. Polym. Sci.* **12,** 1 (1954).
[21] D. C. Teller, E. Swanson, and C. de Haën, *Fed. Proc., Fed. Am. Soc. Exp. Biol.* **36,** 668 (Abstr. 2092) (1977).

TABLE I

GEOMETRIC FACTORS (F_n) FOR VARIOUS
ARRANGEMENTS OF SUBUNITS

n	Arrangement	$F_n{}^a$
Dimer		0.9449
Trimer	Linear	0.8639
Trimer	Equilateral triangle	0.9555
Tetramer	Linear	0.8064
Tetramer	Square	0.9259
Tetramer	Tetrahedron	0.9772

^{a} F_n is defined by Eq. (12) in the text and is computed using the rigorous theory of Section IV.

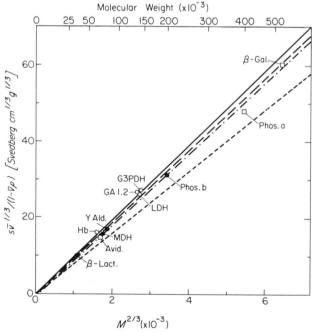

FIG. 2. Relationship of sedimentation coefficient and molecular weight of oligomeric proteins, for which there exists independent evidence for the geometrical arrangement of subunits. ●, dimers; □, square planar tetramers; ○, tetrahedral tetramers; Theoretical lines computed from the values in Table I are ——, tetrahedral tetramers; — — —, dimers; —·—·—, square planar tetramers; - - - -, linear tetramers. β-Lact., β-lactoglobulin; Avid, avidin, hen egg white; MDH, malic dehydrogenase; Hb, hemoglobin; Y Ald., yeast aldolase; LDH, lactic dehydrogenase; GA1.2, asparaginase from *Acinetobacter;* G3PDH, glyceraldehyde-3-phosphate dehydrogenase; Phos. b, phosphorylase b; Phos. a, phosphorylase a; β-Gal., β-galactosidase. (From D. C. Teller, E. Swanson, and C. de Haën, unpublished results.)

sedimentation coefficient and the Svedberg equation [Eq. (2)], the values of ζ and f_n are obtained for the monomer and oligomer. The closest value of F_n of Table I together with Eq. (12) and test integers suffices to determine n. If n is known by other experiments, then the geometry is determined by Eq. (12) and Table I. For geometries not included in Table I, the value of F_n can be calculated approximately by the formula

$$F_n = n^{-2/3} \left(1 + \frac{a}{n} \sum_{\substack{i=1 \\ i \neq j}}^{n} \sum_{j=1}^{n} r_{ij}^{-1} \right) \tag{13}$$

derived from the Kirkwood approximation. Here, r_{ij} is the distance between subunits of equal size and radius a. The radius r may be calculated from the subunit molecular weights and $\bar{v}$ as in Eq. (3). We presently do not know the general accuracy of this procedure but it should be good.

Figure 2 shows the results for some proteins of known subunit arrangement using the values of Table I for the geometric factor F_n. The data in this figure are seen to agree with the geometric arrangements found by independent methods.

III. Friction Properties of Proteins—Approximate Theory

Kirkwood[13,20] derived a very simple formula for calculation of the translational friction coefficient of an array of beads exposed to solvent which was based on the Oseen tensor. Later Bloomfield et al.[14] extended this approach to shells of beads describing solid objects, such as viruses and proteins, that can be modeled as being composed of subunits, e.g., serum albumin at low pH.[22] Kirkwood's result for an assembly of beads of radius a was

$$f_K = 6\pi\eta_0 na / \left(1 + \frac{a}{n} \sum_{\substack{i=1 \\ i \neq j}}^{n} \sum_{j=1}^{n} r_{ij}^{-1} \right) \tag{14}$$

where f_K is the calculated translational friction coefficient, η_0 is the solvent viscosity, n is the number of beads of radius a, and r_{ij} is the distance between beads i and j. This formula was soon recognized to contain an error.[23-25] Later it was realized that it actually constituted an approximation[26] although the expected error could not be estimated. The worst discrepancy between results from Eq. (14) and a more rigorous solution in

[22] V. Bloomfield, *Biochemistry* **5**, 684 (1966).
[23] Y. Ikeda, *Kobayashi Rigaku Kenkyusho Hokoku* **6**, 44 (1956).
[24] J. J. Erpenbeck and J. G. Kirkwood, *J. Chem. Phys.* **29**, 909 (1958).
[25] J. J. Erpenbeck and J. G. Kirkwood, *J. Chem. Phys.* **38**, 1023 (1963).
[26] H. Yamakawa and J. Yamaki, *J. Chem. Phys.* **58**, 2049 (1973).

the case of proteins is 7%.[27] For an infinitely large rigid ring of infinitesimal beads, it is known to be in error by 8.3%.[28,29] The errors always operate to underestimate f when calculated by Eq. (14).

Equation (14) is an approximation to the true value of f and is quite easy to calculate. In contrast, rigorous calculations are usually prohibitively time consuming, even on fast computers. For this reason, it is useful to find empirical ways to correct the discrepancy between values observed and those predicted by Eq. (14) for known objects.

Bloomfield et al.[14] used a shell model of tiny beads to describe the surface of rigid impenetrable objects, such as spheres, ellipsoids, cylinders, etc., for which other and more acurate hydrodynamic calculations were available. The correspondence of the results was sufficiently accurate to encourage further investigation into the limits of applicability of Kirkwood's formula [Eq. (14)], both theoretically[15,16] and empirically.

Indeed, we have found[21,27] that a shell model of proteins together with Eq. (14) can be used to accurately describe the friction properties of proteins. The procedure involves the following calculations:

1. Atomic coordinates of known protein structures are used to calculate the coordinates of nonoverlapping test spheres of radius 1.4 Å, that coat the protein surface.

2. The friction coefficient of this test sphere shell is computed using Eq. (14).

The results of this procedure are in very good agreement with the observed friction coefficients of these globular proteins (Table II).

A. Description of the Test Sphere Shell

For the purpose of this article protein atoms shall be defined as all atoms of proteins except hydrogens. Van der Waals radii of these protein atoms are adjusted so as to include the hydrogens in their van der Waals radius. The shell of test spheres with radius 1.4 Å is created by arranging around each protein atom 12 test spheres in a hexagonal close packing arrangement. The distance from the center of the protein atom to that of the test sphere is 1.4 Å plus the van der Waals radius of the protein atom. Test spheres that overlap with protein atoms are then eliminated. There now remains a set of test spheres some of which may overlap. The rigorous calculations described later can deal with overlapping test spheres, but the Kirkwood formula [Eq. (14)] cannot. Reduction of the set

[27] D. C. Teller, E. Swanson, and C. de Haën, in preparation.
[28] R. Zwanzig, J. Chem. Phys. 45, 1859 (1966).
[29] E. Paul and R. M. Mazo, J. Chem. Phys. 51, 1102 (1969).

TABLE II
FRICTION COEFFICIENTS[a] OF PROTEINS COMPARED TO CALCULATED VALUES

Protein	f_D	f_S	f_K	f_K/f_D	f_K/f_S
Ferredoxin		2.62	2.626		1.002
Ribonuclease S	3.79	3.80	3.623	0.956	0.953
Lysozyme	3.61	3.71	3.592	0.995	0.968
Trypsin	4.18	4.36	4.176	0.999	0.958
Subtilisin BPN'	4.48	4.48	4.382	0.978	0.978
Carboxypeptidase A		4.91	4.679		0.953
Thermolysin		4.57	4.77		1.044
Hemoglobin (deoxy)		6.06	5.913		0.976

[a] f_D, experimental friction coefficient by diffusion; f_S, experimental friction coefficient by sedimentation; f_K, friction coefficient computed by Kirkwood approximation [Eq. (14)]; (From D. C. Teller, E. Swanson, and C. de Haën, unpublished results.)

to nonoverlapping test spheres is done in three passes, in order to retain the maximum number of test spheres. During the first pass, any test sphere in contact with three or more others is eliminated, updating the count of overlap contacts after each elimination. During pass two, test spheres in contact with two others are eliminated, and, in the final pass, test spheres contacting one other are eliminated.

Occasionally test spheres are placed by this procedure into cavities in the interior of the protein where they cannot contribute to the frictional resistance. Such test spheres are identified with the help of a program which displays a slice of the shell of arbitrary thickness on an oscilloscope, and allows the observer to eliminate such test spheres. This yields the final set of test sphere coordinates.

B. Calculations

Calculations of friction coefficients are made on test sphere shells using Eq. (14) and the coordinates of the final test sphere shell. Table II compares such calculated friction coefficients with those observed in sedimentation and diffusion experiments.

Despite the limitations of this approach, it can definitely be concluded that the rugosity of the protein surface plays a major role in determining the friction coefficient. This is a determinant that is overlooked if axial ratios are derived by the procedures of Oncley[1] or later refinements.[5]

The test spheres were chosen to have the radius ascribed to water molecules by Pauling,[30] because water molecules are the smallest units to

[30] L. Pauling, "The Nature of the Chemical Bond." Cornell Univ. Press, Ithaca, New York, 1960.

probe the protein surface. However, the test spheres should not be identified with water of hydration, because this would result in a specific hydration of about 0.5 g water per gram of protein, larger than is reasonable.[31] They are an artifice to expand the frictional surface in a manner that compensates for the low values of friction coefficients obtained with Kirkwood's formula applied to protein surface atoms directly. In early studies, a similar result was obtained by isomorphous expansion of all protein coordinates. A best fit of calculated and measured friction coefficients was obtained, when the surface atoms are moved on the average by 3.6 Å out from the center of mass of the proteins.[18] After considering the rigorous theory, we will return to the question of hydration.

IV. Friction Properties of Proteins-Rigorous Theory

The basis from which Kirkwood's formula [Eq. (14)] was derived is the hydrodynamic interaction tensor for the pairwise interaction between point sources of friction developed by C. W. Oseen[32] in 1927 and applied by Burgers[33] to objects of various geometries. Zwanzig et al.[34] pointed out that even if the error (or approximation) introduced by Kirkwood[13,20] (vide supra) is avoided, for some objects negative friction coefficients will be calculated. The Oseen tensor applied to bead models treats the beads as point sources of friction so the matrix to be inverted is not necessarily positive-definite, leading to nonphysical results. Rotne and Prager[15] and Yamakawa[16] derived a modified Oseen tensor which is always positive-definite provided that Stokes' law is used for the friction coefficient of the beads. Friction coefficients calculated from this modified Oseen tensor are not necessarily exact, but we know from the work of Rotne and Prager[15] that $f^* < f^t$, where f^* is the friction coefficient calculated from the tensor and f^t is the exact (true) value. By true value we mean one that would result from an approximation free solution of the steady state Navier–Stokes equation.[15] We have made calculations of ellipsoids for which an exact theory is known.[34a] Results of the computations extrapolated to an infinite number of beads of infinitesimal size placed on the inside or outside surface of rotational ellipsoids find $f^* = f^t$ within the precision of the calculation ($\pm 0.1\%$). For cylinders, the values of f^* agree with experimental values[35] within experimental error ($\pm 1\%$). Thus, at

[31] I. D. Kuntz, Jr. and W. Kauzmann, Adv. Protein Chem. 28, 239 (1974).

[32] C. W. Oseen, Nevere Methoden und Ergebnisse in der Hydrodynamik, in "Mathematik und ihre Anwendungen in Monographien und Lehrbüchern" (E. Hilb, Ed.). Akad. Verlagsges., Leipzig, 1927.

[33] J. M. Burgers, Verh. K. Ned. Akad. Wet., Afd. Natuurkd., Reeks 1 16 (4), In "Second Report on Viscosity and Plasticity," Chapt. 3, p. 113 (1938).

[34] R. Zwanzig, J. Kiefer, and G. H. Weiss, Proc. Natl. Acad. Sci. U.S.A. 60, 381 (1968).

[34a] E. Swanson, D. C. Teller, and C. de Haën, J. Chem. Phys. 68, 5097 (1978).

[35] J. F. Heiss and J. Coull, Chem. Eng. Prog. 48, 133 (1952).

present we have good confidence in the accuracy of friction coefficients calculated using the modified Oseen tensor.

According to Oseen[32] (as cited, e.g., by McCammon and Deutch[36]) for a rigid object made of an aggregate of n beads moving in a solvent that is at rest at infinite distance from the object, the force, F_i, exerted on the ith bead is given by

$$F_i + \zeta \sum_{j=1}^{n'} \mathbf{T}_{ij} F_i = \zeta(U_i - V_i^0) \tag{15}$$

where ζ is the friction coefficient of a bead of radius a, and $\zeta = 6\pi\eta_0 a$.[37] U_i is the velocity of the ith bead, V_i^0 is the velocity which the solvent would have at the position of the ith bead if the assembly were absent, n is the number of beads, and the prime on the summation indicates that $j = i$ is not included. For all future considerations we take $V_i^0 = 0$. The modified Oseen tensor[15] multiplied by the friction coefficient of a bead is given by

$$\zeta \mathbf{T}_{ij}^{kl} = \frac{6\pi\eta_0 a}{8\pi\eta_0 r_{ij}} \left[\mathbf{I} + \frac{r_{ij}^k r_{ij}^l}{r_{ij}^2} + \frac{2a^2}{r_{ij}^2} \left(\tfrac{1}{3}\mathbf{I} - \frac{r_{ij}^k r_{ij}^l}{r_{ij}^2} \right) \right] \tag{16}$$

r_{ij}^k and r_{ij}^l are the projections of vector r_{ij} onto the cartesian axes k and l, where k and l each in turn are cartesian x, y, and z axes of the coordinate system common to all beads (i.e., x, y, z coordinate system of atomic coordinates of protein). In this equation r_{ij} is the vector between the center of bead i and bead j, r_{ij} is the distance between nonoverlapping beads i and j, and $\mathbf{I}$ is the unit tensor. When the beads overlap, then

$$\zeta \mathbf{T}_{ij}^{kl} = \left(1 - \frac{9r_{ij}}{32a} \right) \mathbf{I} + \frac{3r_{ij}^k r_{ij}^l}{32ar_{ij}} \tag{17}$$

The transition from Eq. (16) to Eq. (17) is continuous and smooth.[34a] Note that the original Oseen tensor[32] only contained the first two terms in the bracket of Eq. (16). The translation of these equations into computational forms will be presented later.

Following McCammon and Deutch,[36] we now write Eq. (15) as an algebraic set of simultaneous linear equations,

$$\mathcal{M}\mathcal{F} = \zeta\mathcal{U} \tag{18}$$

where $\mathcal{M}$ is a $3n \times 3n$ "supermatrix" consisting of 3×3 blocks $\mathbf{M}_{ij}$, where

[36] J. A. McCammon and J. M. Deutch, *Biopolymers* **15**, 1397 (1976).

[37] Note the following conventions or symbols: scalars and scalar elements of matrices: f, ζ, T_{ij}; vectors and 3×3 matrices of vectors: r_{ij}, F_i; tensors and 3×3 matrices of scalars: $\mathbf{T}$, $\mathbf{I}$, $\mathbf{M}_{ij}$, $\mathbf{f}$; "supermatrices" and "supervectors" composed of 3×3 matrices of vectors as scalars: $\mathcal{F}$, $\mathcal{I}$, $\mathcal{M}$, $\mathcal{U}$.

$$\mathbf{M}_{ij} = \left\{ \begin{array}{l} \zeta\,\mathbf{T}_{ij}\,,\, i \neq j \\ \mathbf{I}\quad,\, i = j \end{array} \right. \tag{19}$$

Subscripts i and j indicate the ith and jth bead, respectively.

$\mathscr{F}$ and $\mathscr{U}$ are matrices ("supervectors") with dimensions $3n \times 3$. $\mathscr{F}$ is given by

$$\mathscr{F} = \begin{pmatrix} F_1^{xx} & F_1^{xy} & F_1^{xz} \\ F_1^{yx} & F_1^{yy} & F_1^{yz} \\ F_1^{zx} & F_1^{zy} & F_1^{zz} \\ F_2^{xx} & F_2^{xy} & F_2^{xz} \\ \cdot & \cdot & \cdot \\ \cdot & \cdot & \cdot \\ \cdot & \cdot & \cdot \\ F_n^{zx} & F_n^{zy} & F_n^{zz} \end{pmatrix} \tag{20}$$

where the subscript indicates the bead index, the first superscript the direction of the flow and the second the direction of the force. $\mathscr{U}$ is given by

$$\mathscr{U} = \begin{pmatrix} 1 & 0 & 0 \\ 0 & 1 & 0 \\ 0 & 0 & 1 \\ 1 & 0 & 0 \\ 0 & 1 & 0 \\ 0 & 0 & 1 \\ \cdot & \cdot & \cdot \\ \cdot & \cdot & \cdot \\ \cdot & \cdot & \cdot \\ 0 & 0 & 1 \end{pmatrix} \tag{21}$$

Note that we have used unit velocities. The unknown terms of Eq. (18) are the individual forces on each bead as given in Eq. (20). These forces can be computed either by inversion of $\mathscr{M}$, or by iterative solution of the system of equations. Once $\mathscr{F}$ is obtained, the scalar elements are added to produce a 3×3 total force matrix

$$\mathbf{F}_t = \begin{pmatrix} F^{xx} & F^{xy} & F^{xz} \\ F^{yx} & F^{yy} & F^{yz} \\ F^{zx} & F^{zy} & F^{zz} \end{pmatrix} \tag{22}$$

where

$$\mathbf{F}^{xx} = F_1^{xx} + F_2^{xx} + F_3^{xx} + \cdots + F_n^{xx}$$
$$\mathbf{F}^{xy} = F_1^{xy} + F_2^{xy} + F_3^{xy} + \cdots + F_n^{xy}$$
$$\cdot\qquad\cdot\qquad\cdot\qquad\cdot\qquad\cdot\qquad\cdot$$
$$\cdot\qquad\cdot\qquad\cdot\qquad\cdot\qquad\cdot\qquad\cdot$$
$$\mathbf{F}^{zz} = F_1^{zz} + F_2^{zz} + F_3^{zz} + \cdots + F_n^{zz}$$

In matrix algebra notation this operation is equivalent to the multiplication of $\mathscr{F}$ by a matrix, $\mathscr{I}^t$ of dimension $3 \times 3n$, corresponding to the transpose of Eq. (21). We now have the equality

$$\zeta \mathscr{I}^t \mathscr{M}^{-1} \mathscr{U} = \mathscr{I}^t \mathscr{F} = \mathbf{F}_t \tag{23}$$

where $\mathbf{F}_t$ is a 3×3 total force matrix. In this equation, the first term assumes that we have solved the equations by inversion of $\mathscr{M}$. The middle term assumes that the force matrix $\mathscr{F}$ is known by iteration methods (see below). The total force on a rigid object is given by

$$\mathbf{F}_t = \mathbf{f}\mathbf{U} \tag{24}$$

where $\mathbf{f}$ is the desired friction tensor and $\mathbf{U}$ is the velocity matrix. Consequently,

$$\mathbf{f} = \mathbf{F}_t\mathbf{U}^{-1} \tag{25}$$

Since $\mathbf{U}^{-1}$ is an identity matrix, it only converts the units. Thus numerically

$$\mathbf{f} = \mathbf{F}_t \tag{26}$$

The 3×3 friction matrix $\mathbf{f}$ consists of the nine scalar friction coefficients $f^{xx}, f^{xy}, f^{xz}, \ldots, f^{zz}$. $\mathbf{f}$ is symmetric and positive definite.[10] All friction matrices $\mathbf{f}$ are *similar* regardless of particle orientation. To obtain the average scalar friction coefficient f^*, it is necessary to average the reciprocal values of the coefficients.[2,10] Thus, we obtain the value of f^* from the trace of inverse $\mathbf{f}$ (trace equals sum of diagonal entries).

$$1/f^* = \tfrac{1}{3}\mathrm{tr}\ \mathbf{f}^{-1} \tag{27}$$

It is of interest to note, in passing, that one inverse of the nonsquare matrix $\mathscr{U}$ is $\mathscr{U}^{-1} = (1/n)\mathscr{I}^t$. Conversely one inverse of the nonsquare matrix $\mathscr{I}^t$ is $(\mathbf{I}^t)^{-1} = (1/n)\mathbf{U}$.

By virtue of these equalities and Eq. (26), we may now write

$$\mathbf{f}_{\mathrm{K}}^{-1} = [\zeta \mathscr{I}^t \mathscr{M}^{-1} \mathscr{U}]^{-1} = (1/\zeta)\mathscr{U}^{-1}\mathscr{M}(\mathscr{I}^t)^{-1} = (1/n^2\zeta)\mathscr{I}^t\mathscr{M}\mathscr{U} \tag{28}$$

Equation (28) yields an expression whose identity with Kirkwood's formula [Eq. (14)] can be demonstrated. The error (or approximation) in Kirkwood's formula appears in this derivation in the guise of a lack of uniqueness of the inverses of $\mathscr{U}$ and $\mathscr{I}^t$.

V. Calculation Method for the Rigorous Theory

In this section, we illustrate the mechanics of the calculation of the friction coefficient of a three bead asymmetric object with beads of unit radius a by the theory of Section IV. The coordinates of this object are given in Table III. Bead 3 has been raised out of the x,y plane in order to illustrate some aspects of the calculation. By Eq. (19), the 3×3 block $\mathbf{M}_{ij}$ for $i = 1, j = 1$ is simply

$$
\mathbf{M}_{11} = \begin{pmatrix} 1 & 0 & 0 \\ 0 & 1 & 0 \\ 0 & 0 & 1 \end{pmatrix}
$$

Now consider $i = 1, j = 3, r_{13} = 2\sqrt{2}$. The entries for $\mathbf{M}_{13}$ are obtained through Eq. (16), because beads do not overlap. With $k = x$ and $l = x$,

$$
\zeta T_{13}{}^{xx} = \frac{3a}{4r_{13}} \left[1 + \frac{(x_1 - x_3)^2}{r_{13}{}^2} + \frac{2a}{r_{13}{}^2} \left(\frac{1}{3} - \frac{(x_1 - x_3)^2}{r_{13}{}^2} \right) \right]
$$

$$
= \frac{3}{8\sqrt{2}} \left[1 + \frac{4}{8} + \frac{2}{8} \left(\frac{1}{3} - \frac{4}{8} \right) \right]
$$

$$
= 0.3867
$$

and with $k = x$ and $l = y$,

$$
\zeta T_{13}{}^{xy} = \frac{3a}{4r_{13}} \left[0 + \frac{(x_1 - x_3)(y_1 - y_3)}{r_{13}{}^2} + \frac{2a}{r_{13}{}^2} \left(0 - \frac{(x_1 - x_3)(y_1 - y_3)}{r_{13}{}^2} \right) \right]
$$

$$
= \frac{3}{8\sqrt{2}} \left[\frac{(-2)(-2)}{8} + \frac{2}{8} \left(0 - \frac{(-2)(-2)}{8} \right) \right]
$$

$$
= 0.0703
$$

In analogous manner all six different entries of $\mathbf{M}_{13}$ (not 9, because $\mathbf{M}_{13}$ is symmetric) may be obtained.

$$
\mathbf{M}_{13} = \begin{pmatrix} 0.3867 & 0.0703 & 0.0703 \\ 0.0703 & 0.3370 & 0.0497 \\ 0.0703 & 0.0497 & 0.3370 \end{pmatrix}
$$

TABLE III

COORDINATES OF A TRIANGULAR OBJECT OF THREE NONOVERLAPPING SPHERICAL BEADS OF RADIUS $a = 1$

Bead number	Coordinates (alternative names[a])	x (^1x)	y (^2x)	z (^3x)
1		0	0	0
2		2	0	0
3		2	$\sqrt{2}$	$\sqrt{2}$

[a] The alternative names have been introduced in Eq. (29) for computational ease.

Equation (16) was given in this form, because if only the first two terms in the brackets are considered, one has the original Oseen tensor.[32] However, for computational purposes a more convenient form can be written.[36] Let the x, y, and z directions be denoted ^1x, ^2x, and ^3x, or in general ^kx or ^lx, where k and l assume values 1 to 3. We then get from Eq. (16) and (17)

$$\zeta T_{ij}^{kl} = \frac{3a}{4r_{ij}} \left(B\delta^{kl} + C \frac{(^kx_i - {}^kx_j)(^lx_i - {}^lx_j)}{r_{ij}^2} \right) \tag{29}$$

For $i = j$,

$$\zeta T_{i=j}^{kl} = \delta^{kl} \tag{29a}$$

where $\delta^{kl} = 1$ if $k = l$, and $\delta^{kl} = 0$ otherwise. For $i \neq j$ and nonoverlapping beads,

$$B = 1 + \frac{2a^2}{3r_{ij}^2} \quad \text{and} \quad C = 1 - \frac{2a^2}{r_{ij}^2} \tag{29b}$$

and for $i \neq j$, with overlapping beads,

$$B = 1 - \frac{9r_{ij}}{32a} \quad \text{and} \quad C = \frac{3r_{ij}}{32a} \tag{29c}$$

Finally

$$r_{ij} = \left(\sum_{k=1}^{3} (^kx_i - {}^kx_j)^2 \right)^{1/2} \tag{29d}$$

Obviously it is indicated to compute all in 3×3 blocks of interacting beads i and j.

For small problems such as the 3-bead right triangle, for which the entire matrix $\mathcal{M}$ can be stored in the computer memory and inverted, we use Eqs. (23) and (26) as

$$\mathbf{f} = \mathbf{F}_t = \zeta \mathcal{S}^t \mathcal{M}^{-1} \mathcal{U}$$

As an example, Table IV gives the 9×9 matrix for the 3-bead right triangle described above. Note that the left and right upper corner 3×3 submatrices of the matrix $\mathcal{M}$ are the matrices $\mathbf{M}_{11}$ and $\mathbf{M}_{13}$ given above. In this case, matrix inversion by Gauss elimination is used and $\mathcal{M}^{-1}$ is given in Table IVb. Table IVc gives the forces on each subunit [Eq. (20)] in each direction. Note that the largest force is directed along the velocity direction but there are finite components in other directions. Table IVd is the friction matrix $\mathbf{f}$ [Eq. (26)]. Inversion of this matrix gives the nine entries of $\mathbf{f}^{-1}$ (Table IVe), where $\mathbf{f}^{-1}$ is necessarily symmetric. We will later show

TABLE IV

ILLUSTRATION OF THE MATRICES USED IN THE CALCULATION OF THE FRICTIONAL PROPERTIES OF THE THREE-BEAD OBJECT OF TABLE III

(a) Supermatrix $\mathcal{M}$ [Eq. (19)]

1.0000	0.0000	0.0000	0.6250	0.0000	0.0000	0.3867	0.0703	0.0703
0.0000	1.0000	0.0000	0.0000	0.4375	0.0000	0.0703	0.3370	0.0497
0.0000	0.0000	1.0000	0.0000	0.0000	0.4375	0.0703	0.0497	0.3370
0.6250	0.0000	0.0000	1.0000	0.0000	0.0000	0.4375	0.0000	0.0000
0.0000	0.4375	0.0000	0.0000	1.0000	0.0000	0.0000	0.5313	0.0937
0.0000	0.0000	0.4375	0.0000	0.0000	1.0000	0.0000	0.0937	0.5313
0.3867	0.0703	0.0703	0.4375	0.0000	0.0000	1.0000	0.0000	0.0000
0.0703	0.3370	0.0497	0.0000	0.5313	0.0937	0.0000	1.0000	0.0000
0.0703	0.0497	0.3370	0.0000	0.0937	0.5313	0.0000	0.0000	1.0000

(b) Inverse of $\mathcal{M}$: $\mathcal{M}^{-1}$

1.7363	0.0513	0.0513	−0.9749	0.1087	0.1087	−0.2521	−0.2099	−0.2099
0.0513	1.2736	0.0192	0.0203	−0.4459	0.0378	−0.1196	−0.2004	−0.0517
0.0513	0.0192	1.2736	0.0203	0.0378	−0.4459	−0.1196	−0.0517	−0.2004
−0.9749	0.0203	0.0203	1.7878	−0.0768	−0.0768	−0.4080	0.1087	0.1087
0.1087	−0.4459	0.0378	−0.0768	1.6055	0.1823	0.0203	−0.7293	−0.2456
0.1087	0.0378	−0.4459	−0.0768	0.1823	1.6055	0.0203	−0.2456	−0.7293
−0.2521	−0.1196	−0.1196	−0.4080	0.0203	0.0203	1.2928	0.0513	0.0513
−0.2099	−0.2004	−0.0517	0.1087	−0.7293	−0.2456	0.0513	1.4953	0.2410
−0.2099	−0.0517	−0.2004	0.1087	−0.2456	−0.7293	0.0513	0.2410	1.4953

(c) Force matrix $\mathscr{F}$ [Eq. (20)]

0.096192	-0.009072	-0.009072
-0.009413	0.118479	0.001014
-0.009413	0.001014	0.118479
0.076474	0.009852	0.009852
0.009852	0.081283	-0.004808
0.009852	-0.004808	0.081283
0.119494	-0.009413	-0.009413
-0.009072	0.106828	-0.010636
-0.009072	-0.010636	0.106828

(d) Total force matrix $\mathbf{F}_t$ or friction matrix $\mathbf{f}$ [Eqs. (22) and (26)]

0.292160	-0.008633	-0.008633
-0.008633	0.306590	-0.014430
-0.008633	-0.014430	0.306590

(e) Inverse friction matrix $\mathbf{f}^{-1}$ [Eqs. (26) and (27)]

3.428775	0.101316	0.101316
0.101316	3.271921	0.156853
0.101316	0.156853	3.271921

that the off-diagonal elements can be eliminated by rotation of the object. The average friction coefficient is calculated from Eq. (23) as

$$1/f^* = \tfrac{1}{3}(3.4288 + 3.2719 + 3.2719)$$

or

$$f^* = 0.30082$$

For large problems, inversion of $\mathcal{M}$ may be impractical, and an iterative solution to the system of linear equations corresponding to Eq. (18) is the better method. The Seidel method[38] is well suited since convergence is relatively rapid and computer memory requirements are minimal. For each step of the iteration, the necessary $\mathbf{M}_{ij}$ are recomputed (rather than storing $\mathcal{M}$), and only the progressively refined entries of $\mathcal{F}$ are stored. Required total storage for arrays is $12n$, where n is the number of beads ($9n$ for storage of $\mathcal{F}$, $3n$ for coordinates). The largest problem we have solved is the test sphere set of hemoglobin, which involves about 3300 simultaneous equations. Each iteration by the Seidel method required eight hours of time on a PDP-12 computer with Floating Point Processor. Fifteen iterations were required to estimate f^* to 0.16% uncertainty.

In order to begin the iteration, we use the following procedure: The 3×3 matrices $\mathbf{M}_{ij}$ are summed across the rows of $\mathcal{M}$ to yield a $3n \times 3$ matrix of n blocks of 3×3 submatrices. These 3×3 submatrices are individually inverted to yield an initial estimate of the 3×3 blocks of the force vector $\mathcal{F}$. As an example consider the interaction of bead 1 with all the others:

$$\left(\sum_{j=1}^{n} \mathbf{M}_{1j} \right)^{-1} = \begin{pmatrix} F_1^{xx} & F_1^{xy} & F_1^{xz} \\ F_1^{yx} & F_1^{yy} & F_1^{yx} \\ F_1^{zx} & F_1^{zy} & F_1^{zz} \end{pmatrix} \tag{30}$$

While a Seidel iteration applied to the system of equations represented by Eq. (18) will converge, regardless of the starting point for $\mathcal{F}$, provided the modified Oseen tensor [Eqs. (16) and (17)] is used, it may take an excessively long time. The above procedure [Eq. (30)] gives a much better starting point for the iteration, than choosing (for example) all entries for $\mathcal{F}$ equal to unity.

VI. Location of the Principal Axes of Translation

Because the orientation of particles in an external Cartesian coordinate system is arbitrary, in general, the off-diagonal elements in $\mathbf{f}$ will fre-

[38] J. Westlake, "A Handbook of Numerical Matrix Inversion and Solution of Linear Equations," p. 55ff. Robert E. Krieger, Huntington, 1975.

quently be finite. If an object, such as the 3-bead right triangle, is dropped into a solution parallel to any of the three coordinate axes used, it will move with an angle to the axis because of the off-diagonal resistances.

According to Happel and Brenner[10] (p. 167), because of the symmetry of f, such a particle must possess at least three mutually perpendicular axes, fixed in the particle, such that if it is translating without rotation parallel to one of them, it will experience a force only in this direction; that is, if we can find these axes, there will be no lateral forces and the off-diagonal elements of f should be zero. Happel and Brenner[10] call these axes the "principal axes of translation." If we know the scalar friction coefficients f^{kl} ($k, l = x,y,z$) in our external Cartesian coordinates, then the principal axes of translation can be determined from the eigenvectors. The f^{kl} coefficients corresponding to this special coordinate system are termed the "principal friction coefficients of translation."[10] Here we present the method used when we know the matrix f in the arbitrary coordinate system. First we solve the cubic equation corresponding to

$$\det \begin{pmatrix} f^{xx} - f & f^{xy} & f^{xz} \\ f^{yx} & f^{yy} - f & f^{yz} \\ f^{zx} & f^{zy} & f^{zz} - f \end{pmatrix} = 0 \tag{31}$$

which has three real, positive roots, corresponding to the eigenvalues f^k. For the friction matrix f, of Table IVd, for example, the values are $f^x = 0.27995$, $f^y = 0.30437$, and $f^z = 0.32102$ as presented in Table V. The average of the reciprocals of these values is the same as $1/f^*$ previously obtained.

In order to find the principal translation axes, we must find the eigenvectors. To find these, we must solve three sets of three simultaneous equations for the unknown eigenvector values e^{kx}, e^{ky}, and e^{kz},

$$\begin{aligned}
e^{kx}(f^{xx} - f^k) &+ e^{ky}f^{xy} &+ e^{kz}f^{xz} &= 0 \\
e^{kx}f^{yx} &+ e^{ky}(f^{yy} - f^k) &+ e^{kz}f^{yz} &= 0 \\
e^{kx}f^{zx} &+ e^{ky}f^{zy} &+ e^{kz}(f^{zz} - f^k) &= 0,
\end{aligned} \tag{32}$$

where $k = x,y,z$ indicates the three directions of motion. Table V gives the eigenvectors associated with f of Table IVd, normalized to unit length. Solution of Eq. (32) does not identify which of the eigenvectors correspond to the directions x,y, and z. The correspondence is obtained by inspection of the components of the eigenvectors, the largest positive component indicating the direction of the new axis.

In order to align the principal axes of translation, rigidly fixed in the particle, with the external coordinate system of the laboratory the object with coordinates x, y, and z is rotated through appropriately determined

TABLE V

EIGENVALUES AND EIGENVECTORS OF THE FRICTION MATRIX OF THE
THREE-BEAD OBJECT OF TABLE III[a]

		Eigenvector		
k	Eigenvalue	e^{kx}	e^{ky}	e^{kz}
x	0.27995	0.70711	0.50000	0.50000
y	0.30437	-0.70711	0.50000	0.50000
z	0.32102	0.00000	-0.70711	0.70711

[a] Eulerian angles of the principal axes of translation: $\phi = 135°$, $\theta = 45°$, $\psi = 180°$.

Eulerian angles ϕ, θ, and ψ, yielding new coordinates x', y', and z'.[39] The Eulerian angles represent (1) a rotation by ϕ about z axis, (2) a rotation by θ about the new x axis, and (3) a rotation by ψ about the new z axis created in step 2. The two systems are connected by the product of the three rotation matrices,

$$\begin{pmatrix} x' \\ y' \\ z' \end{pmatrix} = \begin{pmatrix} \cos\phi & -\sin\phi & 0 \\ \sin\phi & \cos\phi & 0 \\ 0 & 0 & 1 \end{pmatrix} \begin{pmatrix} 1 & 0 & 0 \\ 0 & \cos\theta & -\sin\theta \\ 0 & \sin\theta & \cos\theta \end{pmatrix}$$

$$\begin{pmatrix} \cos\psi & -\sin\psi & 0 \\ \sin\psi & \cos\psi & 0 \\ 0 & 0 & 1 \end{pmatrix} \begin{pmatrix} x \\ y \\ z \end{pmatrix} \quad (33)$$

or

$$\begin{pmatrix} x' \\ y' \\ z' \end{pmatrix} = \begin{pmatrix} \cos\psi\cos\phi - \cos\theta\sin\phi\sin\psi & -\sin\psi\cos\phi - \cos\theta\sin\phi\cos\psi & \sin\theta\sin\phi \\ \cos\psi\sin\phi + \cos\theta\cos\phi\sin\psi & -\sin\psi\sin\phi + \cos\theta\cos\phi\cos\psi & -\sin\theta\cos\phi \\ \sin\theta\sin\psi & \sin\theta\cos\psi & \cos\theta \end{pmatrix} \begin{pmatrix} x \\ y \\ z \end{pmatrix} \quad (34)$$

By equating the unit eigenvectors with the elements of the rotation matrix $\mathbf{R}$ of Eq. (34), the values of θ, ϕ, and ψ may be deduced, since $e^{zz} = \cos\theta$, $e^{zy} = \sin\theta\cos\psi$, $e^{yz} = -\sin\theta\cos\phi$, etc., as illustrated in Table V. To rotate the particle, it is not necessary to deduce the angles. The coordinates of the particle rotated such that the former x, y, and z axes are parallel to the principal translation axes are obtained by multiplying the rotation matrix $\mathbf{R}$ times the particle coordinates (x, y, z) as indicated by Eq. (34) and Table VI.

Rounding errors in computing eigenvalues and eigenvectors may become significant, so the eigenvector matrix (Table V) should be checked to ensure that it is a proper rotation matrix. An orthogonal rotation matrix

[39] We have chosen to rotate the particle coordinates in a right-handed manner, rather than to rotate the external coordinate system in a right-handed manner.

TABLE VI

COORDINATES OF THE THREE-BEAD OBJECT OF TABLE III WITH THE AXES OF THE
NEW COORDINATE SYSTEM PARALLEL TO THE PRINCIPAL TRANSLATION
AXES OF THE PARTICLE[a]

Bead number	Coordinates		
	x'	y'	z'
1	0	0	0
2	$\sqrt{2}$	$-\sqrt{2}$	0
3	$2\sqrt{2}$	0	0

[a] These coordinates were obtained by rotating the object according to Eq. (34) through the angles of ϕ, θ, and ψ from Table V.

has a determinant equal to $+1$, and the product of the transpose, R^t, and the matrix R is an identity matrix:

$$R^t R = I \qquad \text{or} \qquad R = (R^t)^{-1} \tag{35}$$

The second equality of Eq. (35) can be used to smooth out any calculation inaccuracies by averaging the elements of R and $(R^t)^{-1}$.

Table VI gives the coordinates of the reoriented three-bead particle that we have been considering. In the initial external orientation that we considered (Table III), it would tend to move off-axis and possibly tumble if dropped into a fluid along the axis x, y, and z since off-diagonal elements appear in the friction matrix f (Table IVd). If dropped into a fluid parallel to x', y', or z' coordinates of Table VI, however, it would follow a stable translational motion along the chosen axis with no tumbling, provided it is not perturbed by Brownian motion. In considering model studies of asymmetric molecules, these considerations may become important.

VII. Results, Conclusions, and Future Directions

Using the rigorous theory on protein atoms alone yields friction coefficients which are less than the observed values. Calculations based on the complete test sphere shell (section III) gives values greater than observed for proteins. Placing test spheres only on charged groups and computing the friction coefficients based on these test spheres and surface protein atoms yields values essentially in agreement with experimental values. Considering that a rigorous theory has been used, and that charged groups are well recognized to be hydrated, it appears legitimate to equate test spheres in this case with water of hydration. Thus, the frictional behavior of proteins is determined by several factors: overall dimensions, rugosity of the surface, and hydration of charged (and perhaps some polar) groups on the surface.

If the rugosity of globular proteins would on the average be similar, one could expect a relationship between the friction coefficient (or sedimentation or diffusion coefficient) and the surface area of proteins accessible to solvent. Indeed, for those monomeric proteins whose accessible surface area has been calculated[40-43] and for which hydrodynamic data are available, the following empirical equation holds quite well.

$$s\bar{v}^{1/3}/(1 - \bar{v}\rho) = kA_s \qquad (36)$$

A_s is the accessible surface area in $\mathring{A}^2$, and $k = 8.9 \pm 0.7 \times 10^{-4}$ Svedberg cm g$^{-1/3}$ $\mathring{A}^{-2}$. Thus, hydrodynamic data can be used to estimate accessible surface area.

One problem area that has not been approached to date is the coupling of translation and rotation of asymmetric objects such as proteins. Consider an object such as a propeller translating through a solvent. It should spin, even if it moves along the principal axes of translation, one of which in this case is the rotational symmetry axis. Translation and rotation are coupled. As required by the properties of principal axis of translation, the object will, however, not tend to leave the direction of translation. We presently do not know the magnitude of translation–rotation coupling for real molecules.

Another area which should be explored is the rotational frictional behavior of proteins. Are the conclusions we have reached for translational friction valid for rotation as well?

Hopefully, the answers to these questions and related ones will become available in the near future.

Acknowledgments

This work was supported by United States Public Health Service Grant GM-13401, and in part by AM-02456.

[40] B. Lee and F. M. Richards, *J. Mol. Biol.* **55,** 379 (1971).
[41] A. Shrake and J. A. Rupley, *J. Mol. Biol.* **79,** 351 (1973).
[42] C. Chothia, *Nature (London)* **254,** 304 (1975).
[43] C. Chothia, *J. Mol. Biol.* **105,** 1 (1976).

[9] Study of Protein Subunit Association Equilibria by Elution Gel Chromatography

By ROLAND VALDES, JR. and GARY K. ACKERS

Introduction

A. Purpose and Scope

In this chapter we describe methods for determining by elution gel chromatography the stoichiometries and equilibrium constants for a self-associating protein. The elution method, although only one of several gel permeation techniques currently utilized for such studies,[1,2] requires only a minimum of equipment which is readily accessible in all laboratories. It proves a simple and yet rigorously exact method for studying equilibria of the type:

$$
\begin{aligned}
M_1 + M_1 &\rightleftharpoons M_2 \\
M_2 + M &\rightleftharpoons M_3 \\
&\text{------------------------} \\
M_{i-1} + M &\rightleftharpoons M_i \\
&\text{------------------------} \\
M_{n-1} + M &\rightleftharpoons M_n
\end{aligned}
\tag{1}
$$

where M_1 represents the fundamental subunit species and M_i are various association complexes. The association reactions may either proceed to a definite stoichiometric complex M_n, as depicted above, or may be of the "indefinite" type in which no final complex is formed; rather the population of higher-level species decreases asymptotically as the probability of breakage increases with degree of association.

In a given experimental system the basic problem consists of determining whether the reaction proceeds to a definite complex, and if so, with what degree of association n. Further, one wants to know how many of the intermediate states are appreciably populated, and what are their free energies of formation. This last piece of information centers around the problem of estimating the equilibrium constants:

$$
k_i = \frac{(m_i)}{(m_{i-1})(m_1)}
\tag{2}
$$

or equivalently:

[1] G. K. Ackers, Adv. Protein Chem. 24, 343 (1970).
[2] G. K. Ackers, Proteins, 3rd Ed. 1, 1 (1975).

METHODS IN ENZYMOLOGY, VOL. 61

$$K_i = \frac{(m_i)}{(m_1)^i} = \prod_1^i k_i \tag{3}$$

where (m_i) represents molar concentration of species i. In cases where the equilibria are established rapidly in relation to the (several hours) duration of a chromatographic experiment, it is quite convenient to obtain experimentally the desired information. This is true provided the number of species M_i present in a given reaction mixture (i.e., at specified solution conditions and concentration range) does not exceed three, or if the reaction is of the indefinite (isodesmic) type. Fortunately, these conditions are met by a great many systems of multisubunit enzymes and other self-associating proteins. Even in complex systems (e.g., tobacco mosaic virus coat protein) the number of appreciably populated intermediates in self-assembly may be small over each of a sequence of overlapping concentration ranges. For a system having many possible geometric arrangements of subunits, as a general rule, only those which form the maximum number of intersubunit bonding domains will be appreciably populated provided the free energy per bond is at least several times kT (cf. Caspar[3]). An additional contributing factor is the appearance of cooperativity which frequently accompanies subunit assembly processes, as in the so-called "nucleated polymerization" systems.[4]

Here we will be primarily concerned with the technique of elution gel chromatography which is based upon the molecular size-dependent penetration of solute molecules into porous networks of gels such as cross-linked dextrans (Sephadex), polyacrylamide, and agaroses. The same considerations to be described here may also be applied to columns of controlled-pore glass.[5] In general, the range of molecular sizes for complexes that can be analyzed by the methods to be described lies between several angstroms and several hundred angstroms in molecular diameter, as a correspondingly wide range of gel porosities is readily available. The basic principles of gel chromatography and applications to single and polydisperse solute systems have been described in several reviews.[1,2] Selected aspects of theory pertaining to elution chromatography are briefly described here in an attempt to present a coherent framework from which the reader can appreciate the theoretical justification for the experimental methods described in the later part of this chapter. Experimental papers dealing with particular applications of the method include those of Ackers et al.[6-10] Additional references which the reader may wish to consult deal with direct optical scanning of gel chromatography columns[11] and with

[3] D. L. D. Caspar, Adv. Protein Chem. 18, 37 (1963).
[4] F. Oosawa and S. Higashi, Prog. Theor. Biol. 1, 79 (1967).
[5] W. Haller, Nature (London) 206, 693 (1965).
[6] D. J. Winzor and H. A. Scheraga, Biochemistry 2, 1263 (1963).

equilibrium gel permeation methods.[12,13] An earlier article in this series[14] deals with methods for study of protein–ligand binding reactions by gel permeation techniques. It should be noted that the method to be described in this article is particularly well-suited to studies in the lower concentration ranges. Using spectrophotometric assay experiments may be conducted down to a few micrograms/milliliter in protein concentration.

B. Theoretical Considerations

1. Solute Partitioning within Porous Gels. The basic phenomenon upon which all gel chromatography techniques are based is the molecular size-dependent partitioning of solute molecules between the solvent spaces within porous particles (the *stationary phase*) and the solvent space exterior to the particles (the *mobile phase*). For any gel partitioning process it is useful to distinguish between three distinct regions within the experimental system: (1) the solvent region exterior to the gel particles which has a volume V_0, termed the *void volume*, (2) solvent within the interior of the gel particles which has an *internal volume* V_i and is involved in diffusional exchange of solute with the void spaces, and (3) the gel-forming or otherwise solid material (the gel matrix) which has a volume V_g. The total system is the sum of these three regions and occupies a volume V_t.

$$V_t = V_0 + V_i + V_g \qquad (4)$$

When solute is introduced into a gel–solvent system it is distributed by diffusion between the solvent regions inside and outside the gel. At equilibrium the distribution is described by a partition isotherm that defines the relationship between the weight of solute Q_i inside the gel, and the solute concentration, C (g/liter), in the void space exterior to the gel.

$$Q_i = f(C) \qquad (5)$$

[7] G. K. Ackers and T. E. Thompson, *Proc. Natl. Acad. Sci. U.S.A.* **53**, 342 (1965).
[8] E. Chincone, L. M. Gilbert, G. A. Gilbert, and G. L. Kellett, *J. Biol. Chem.* **243**, 1212 (1968).
[9] S. W. Henn and G. K. Ackers, *Biochemistry* **8**, 3829 (1969).
[10] R. Valdes and G. K. Ackers, *J. Biol. Chem.* **252**, 74, (1977).
[11] G. K. Ackers, *in* "Methods of Protein Separation" (N. Catsimpoulis, ed.), p. 1. Plenum, New York, 1976.
[12] G. K. Ackers, E. E. Brumbaugh, S. Ip, and H. R. Halvorson, *Biophys. Chem.* **4**, 171 (1976).
[13] L. P. Vickers and G. K. Ackers, *Biophys. Chem.* **6**, 399 (1977).
[14] G. K. Ackers, Vol. 27, Part D, p. 441.

The exact form of the function $f(C)$ depends upon the type of solute studied.

It is convenient for our purposes to formulate this isotherm in terms of equilibrium partition coefficients, and we will express the pertinent relationships in this article in terms of the partition coefficient σ referred to internal solvent volume. It should be noted that a coefficient K_{av} based upon total volume of stationary phase is also in common usage.[15] The coefficient σ is defined as the amount of solute distributed into the gel per unit internal volume V_i and external concentration C. In terms of this coefficient the isotherm Eq. (5) can be written (see Ackers[1]) as

$$Q_i = \sigma V_i C \tag{6}$$

The dimensionless quantity σ is thus a measure of the degree of solute penetration within the interior solvent region of the gel. Equation (6) is applicable to thermodynamically nonideal systems as well as ideal ones. For the case of thermodynamic ideality a simple interpretation of σ can be made. The penetrable volume V_p occupied within the gel by solute molecules at equilibrium is

$$V_p = Q_i/C_p \tag{7}$$

where C_p is the solute concentration within the regions of distribution and under ideal conditions may be equated with C. The partition coefficient then represents the fraction of the internal volume V_i that is penetrable by the solute molecules under consideration.

$$\sigma = V_p/V_i \tag{8}$$

This relationship is a good approximation for many systems of interest, especially at low solute concentration where the partition isotherm is found to be essentially linear in solute concentration. However, under more general conditions the partition coefficient must be written to include effects of nonideality which give rise to concentration dependence of σ represented by

$$\sigma = \sigma^0 (1 - gC) \tag{9}$$

where $\sigma^0 = V_p/V_i$, the limiting value of σ at infinite dilution, and g is the coefficient of concentration dependence. The value of g is generally found to be in the range -0.001 to -0.01 dl/g.

The value of σ for a particular gel and solute may be viewed primarily as a result of molecular steric exclusion effects. A statistical treatment using a random distribution of penetrable volume elements provides an empirically practical correlation between the molecular radius a (Å) of a given molecular species and its partition coefficient[1]

[15] T. C. Laurent and J. Killander, *J. Chromatogr.* **14**, 317 (1964).

$$a \ (\text{Å}) = a_0 + b_0 \ \text{erfc}^{-1}\sigma \tag{10}$$

the constants a_0 and b_0 are calibration constants of the gel. The quantity ($\text{erfc}^{-1}\sigma$) is the inverse error function complement of σ. More will be said later about the use of column calibration procedures. At this point it is worth noting that the determination of stoichiometries and equilibrium constants by the method to be described here need not depend upon the existence of an accurately known relationship between partition coefficient and molecular weight or size. It is only necessary that the method be sensitive to the changes in these properties which accompany polymerization.

In a gel chromatography experiment the average rate of migration by a particular solute (and hence elution volume) is determined essentially by σ, apart from factors (e.g., void and internal volumes) characteristic of the particular column. In the typical "small zone" experiment the peak elution volume V_e is given (to high approximation) by the relationship

$$V_e = V_0 + \sigma V_i \tag{11}$$

In addition to the average migration rate, a solute zone may be characterized by the extent of spreading (axial dispersion). This dispersion is attributable to (a) nonuniform flow around the beads, (b) linear diffusion within the column, and (c) nonequilibrium exchange between mobile and stationary phases. Under most operating conditions, the last of these processes is dominant with the result that larger solutes generally exhibit the higher rates of dispersion.[2]

2. Associating Solutes. The partition coefficient for a polydisperse system is a weight average of the partition coefficients σ_i of the association complexes M_i

$$\bar{\sigma}_\omega = \sum_i C_i \sigma_i \Big/ \sum_i C_i \tag{12}$$

Here C_i is the weight concentration of species i exterior to the gel, and σ_i is the corresponding partition coefficient.

For noninteracting polydisperse systems the weight average partition coefficient is independent of total solute concentration except for the small concentration dependencies of individual partition coefficients. But if association reactions are taking place in the sample, the various equilibria will be shifted according to the law of mass action so that $\bar{\sigma}_\omega$ will be highly dependent upon total concentration C_T. Measurement of $\bar{\sigma}_\omega$ versus concentration may then be used to resolve the stoichiometry and equilibrium constants for the interactions between species.

The partition isotherm [Eq. (5)] and subsequent relationships have been described in terms of mass concentration units, and it is also useful to consider the mass equilibrium constant K_i' defined by $C_i = K_i'C_1^i$. This constant is related to the molar constant K_i of Eq. (3) by the formula: $K_i' = iK_i/M$ where M is the monomer molecular weight. With these definitions, Eq. (12) may be written

$$\bar{\sigma}_\omega = \frac{\Sigma\sigma_iK_i'C_1^i}{C_T} = \frac{\Sigma iK_i(m_1)^i\sigma_i}{\Sigma iK_i(m_1)^i} \tag{13}$$

Changes in the total concentration C_T bring about shifts in the equilibria so that the average degree of association increases with increasing C_T. This increase is accompanied by a corresponding decrease in the value of $\bar{\sigma}_\omega$ due to the lower values of σ_i for the larger species. The experimental methods described in this chapter are thus aimed at measuring $\bar{\sigma}_\omega$ as a function of C_T and resolving the resulting experimental data into constituent terms σ_i and K_i of Eq. (13).

3. *Types of Experiments.* A sample may be applied to the column to produce either of two distinct types of solute zones, designated the "small zone" and "large zone" methods.

a. SMALL ZONE EXPERIMENTS. Here the sample is introduced at the top of the column in a volume very small (e.g., 0.1–1 ml) compared to the bed volume and is followed by solvent (buffer). The solution may be added (e.g., from a syringe or pipet) by careful layering onto the top of the disc under the solvent which resides over the column bed. This works well when the sample solution is significantly more dense than the solvent. However, in most instances this will not be the case, particularly when the samples are very dilute. In general, we recommend a different procedure in which the upper solvent is first removed (e.g. by pasteur pipet) and the sample is added just as the meniscus of remaining solvent enters the disc. Immediately after sample entry (i.e., when the trailing meniscus enters the disc), a few drops of solvent are added, then several ml more. The volume V_e of solvent which passes through the column between introduction of the sample and the subsequent emergence of its maximum concentration is very closely approximated by Eq. (11). The volumes V_0 and V_i are determined in separate experiments using solutes whose partition coefficients are zero (i.e., a totally excluded molecule) and unity (e.g., glycylglycine), respectively. When this experiment is carried out with a self-associating solute the measured elution volume V_e will shift with variations in the loading concentration, the lowest concentrations corresponding to the largest elution volumes. This experiment is a useful qualitative means of detecting interaction, but there is no quantitative means of using such an experiment to determine association constants. Continuous dilution of the sample (e.g., by axial dispersion) causes the peak to move with decreasing velocity down the column, and the apparent partition coefficient cannot be interpreted in any straightforward way. Calculations have shown that when such experiments are used to estimate apparent equilibrium constants the errors may be as great as several orders of magnitude.[11]

Small zone experiments are nonetheless useful for estimating the minimum subunit partition coefficient and size, by extrapolating the concentration-dependent elution volume to infinite dilution.[12]

In order to obtain a precise determination of $\bar{\sigma}_\omega$ versus C_T a different type of procedure, the large zone experiment must be performed.

b. LARGE ZONE EXPERIMENTS. Here the sample is applied to the column in a sufficient volume that a plateau region of constant concentration is always present as the zone moves through the column. Effects of axial dispersion serve to spread out the leading and trailing boundaries, whereas the plateau between them emerges from the column at the same concentration C_T as that of the initially applied sample. For an interacting system, this plateau constitutes the equilibrium concentration of Eq. (13) corresponding to a value of $\bar{\sigma}_\omega$ which is determined from the elution profile. This weight average partition coefficient is evaluated by first determining the equivalent sharp boundary for the solute zone leading or trailing boundaries, as shown in Fig. 1. These sharp boundaries are first moments (centroids) of the elution profile and satisfy the relationships

$$\bar{V} = \frac{1}{C_T} \int_0^{C_T} V \, dC \qquad \text{leading boundary} \qquad (14)$$

$$\bar{V}^1 = \bar{V} + S \qquad \text{trailing boundary} \qquad (15)$$

where S is the volume of the applied sample. For these determinations the volume coordinate V is assigned a zero value when the leading boundary of the applied sample enters the column bed.

The centroid elution volume is a weight average of elution volumes V_i for the various species:

$$\bar{V} = \frac{\Sigma V_i C_i}{C_T} \qquad (16)$$

Since each species' elution volume satisfies Eq. (11), i.e., $V_i = V_0 + \sigma_i V_i$, it follows that

$$\bar{V} = V_0 + \bar{\sigma}_\omega V_i \qquad (17)$$

Determination of $\bar{V}$ from an experiment performed at a particular plateau concentration C_T permits a determination of $\bar{\sigma}_\omega$ pertaining to the equilibrium distribution at C_T. A series of such experiments are thus required in which C_T is varied. The determination of $\bar{V}$ and hence $\bar{\sigma}_\omega$ as a function of

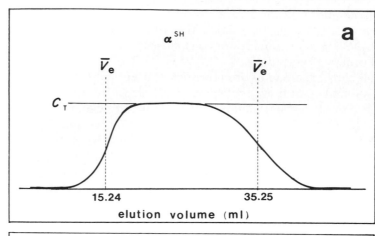

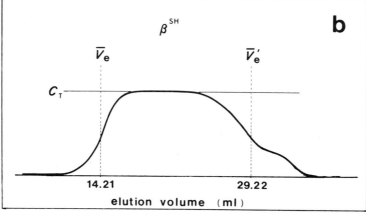

FIG. 1. Elution profiles of large zone experiments for α^{SH} and β^{SH} chains on Sephadex G-100 columns (1×30 cm) at $21.5°$. Buffer was 0.1 M Tris-HCl, 0.1 M NaCl, 1 mM Na$_2$ EDTA at pH 7.4. (a) α^{SH} chains at a plateau concentration of 46.03 μM heme; (b) β^{SH} chains at a plateau concentration of 3.90 μM heme. The profiles exhibit sharp leading boundaries in both cases with diffuse trailing boundaries, characteristic for transport of reversibly self-associating solutes on gel columns. The bimodal character of the β^{SH} chains trailing boundaries is indicative of stoichiometry higher than dimerization. In both cases the centroid volumes for leading and trailing boundaries are indicated by $\overline{V}_e$ and $\overline{V}_e{}'$, respectively. These centroids define equal areas for each boundary below and above the solute profile. (Taken from Valdes and Ackers[10]).

C_T provides a dissociation curve which may be analyzed for σ_i and K_i values of Eq. (13). Methods for doing this will be described in a subsequent section. First, however, we must consider the practical aspects of carrying out these measurements.

II. Experimental Design and Procedures

In this section we will describe experimental methodology for effective application of the theoretical considerations outlined above. Aspects of elution chromatography on associating solutes will be presented with specific details on the equipment and procedures which may be utilized for optimum experimental design.

The most common gel chromatography technique is the elution experiment in which a concentration–volume profile is measured for the sample emerging at the bottom of a column. A wide variety of assays may be performed (enzymatic, radioactive, etc.) if fractions are collected of the eluted solute. However, in this chapter we will be concerned only with the most generally useful approach of continuous spectrophotometric monitoring.

A. Selection of Column

There is no single type of column uniquely suited for studies of protein association. Columns invariably consist of cylindrical tubes into which the preswollen (according to the gel manufacturer's instructions) gel particles or beads are packed. Typical column sizes which are useful in protein association studies are in the range of 25–120 ml total bed volume, i.e., 1 cm diameter by 30–140 cm length. Within this range a great many commercially manufactured columns can be found. A simple glass tube will suffice. For most studies a surrounding thermostated water jacket is also needed. The column is fitted with a suitable arrangement of inlet and outlet devices which retain the gel bed while permitting the flow of solvent through the system. The only important requirements here are that the fittings have minimum "dead space" and provide even flow so that "tailing" and skewing of solute zones is avoided. Some commercial columns are quite bad in these respects. Polyethylene porous discs about 2 mm thickness can be precut from sheets using cork borers to fit snugly into a "homemade" glass column, providing a good definition of the gel bed's top and bottom. The bottom disc should always be placed with its smooth (i.e., least porous) side facing the gel, while the reverse is true for the top disc. Attachments used for inlet and outlet can consist of rubber stoppers through which small (i.e., 1 mm diameter) glass tubes are fitted and polyethylene, Tygon, or Teflon tubing attached on the outside. The bottom stopper is fit snugly against the porous disc, whereas the top of the column should provide several centimeters (5–10) between the disc and top. This permits a few milliliters of sample or solvent to always reside above the top disc during runs. Upward flow is not recommended for analytical purposes, since it is absolutely critical that the packing not change during a series of runs. The temperature of the solvent reservoir above the column may be regulated by having a liquid jacket surrounding the reservoir attached to a suitable circulating thermal regulatory system.

B. Gel Selection and Packing

The criterion used for selection of the gel porosity should be primarily that of maximizing the fraction of penetrable gel volume over the extent of the reaction being measured, i.e., the extremes of species molecular size (see Ackers[1] for a detailed discussion of partition coefficient and molecular size relationships). Packing of the column should be done carefully starting with a gel slurry of approximately twice the final gel volume. Tilting of the column through a small angle, e.g., 10°, will facilitate initial packing and avoid trapped air bubbles from interfering with packing homogeneity. It is desirable to achieve nearly even packing density throughout the length of the column as this minimizes fluctuations in the void volume where extreme variations may interfere with sample profile development. The flow rate during column packing should be maintained nearly constant, and in most cases (smaller columns) a natural flow rate is adequate. In any case the packing flow rate should be greater than the ultimate flow rate used for experiments, as this will avoid any subsequent column compaction. Once the packing is complete, the experimental flow rate should be commenced and maintained throughout.

After several column volumes of equilibration, the top disc may be fitted down snugly against the top of the gel bed. The disc should be tight-fitting so that it will not be subject to displacement during sample loading or other manipulations.

C. Flow Rates

Maintenance of a constant solvent flow rate is most desirable to an elution experiment as this establishes a constant time frame and, therefore, solvent volume reference. Flow rates between 7 and 20 ml/hr are suitable and will depend on the gel type and column size used. For G-100 Sephadex on a 30-ml bed volume, for example, flow rates of 10–14 ml/hr have been successfully used. After the column has been packed it is desirable to reduce the flow rate to a value lower than the natural flow rate of the column. This is accomplished by a pump on the outflow end of the column tubing, which is used as a throttle and constant flow regulator. The importance of using an accurate and constant pump (e.g., a relatively pulseless peristaltic pump) cannot be overemphasized. A very effective method of establishing flow rates and monitoring column elution volumes for recorder calibration is the use of a buret to monitor the eluant volume downstream from the pump. The buret should be kept near the column for ease of visualization during loading of samples.

As the solute is eluted from the bottom of the column, its concentration is monitored as a function of volume of solvent that has passed through the column after introduction of the sample at the top. For chromatography in aqueous solvents, the detector is usually a spectrophotometric device; for organic solvents, it is usually a refractometer. It is, of course, possible to collect fractions and use biological activity or radioactivity to monitor the solute.

D. Spectrophotometers and Flow Cells

The ideal type of spectrophotometer for continuous eluant concentration monitoring is a double-beam instrument. This minimizes instrument drift and therefore yields reproducible and flat base lines throughout an experiment. However, single-beam spectrophotometers have also been successfully used in very precise elution studies. It is simply important to have a flat base line throughout the length of an experiment, or a constant drift which permits accurate corrections to be made. For purified proteins a particularly sensitive and useful wavelength to use is 220 nm where extinction coefficients are generally greater by a factor of 10 than at 280 nm.

The flow cell selected should be one which displays flow characteristics that will accurately reflect the change in concentration with time and will not distort the development and shape of the eluant profile. Flow cells should be tested by passing a "step function" zone through the cell and its tubing while in place and measuring distortion of the zone. The plumbing leading from the column output to the flow cell should be as small as possible in order to minimize "dead space."

E. Recorders

The output signal from the detector is commonly fed into a strip chart recorder and measured as a function of time. If the rate of flow is known, the correlation between concentration and volume is established. The use of an accurate, sensitive, and constant strip chart recorder will greatly simplify both the experimental procedure and the extraction of relevant parameters (see below) from the recording. It is advisable to correlate the strip chart time axis with flow volume as read on the buret and to have an accurate means of indicating on the strip chart the precise "zero" time for entry of the leading edge of a sample into the column bed. A mark introduced at the precise time of sample loading is sufficient.

The sensitivity and pen excursion limits should be set to make use of the maximum span of the concentration. This can be predetermined by an independent absorbance reading prior to sample loading and the recorder span set appropriately. For precise analytical determinations and subsequent automated computational procedures, the spectrophotometer output may also be fed into a digital data acquisition device that interfaces the experiment with a computer. In this article, however, we will concentrate on treating the data manually as this method is readily accessible for most investigations.

III. Analysis of Data

A. Determination of Centroid Boundaries

Various methods are available for calculating the centroid depicted in Fig. 1 obtained in the large zone experiment. As the analytical expressions suggest, calculation of the centroid boundary position is simply a matter of equating for each boundary the respective areas above and below the recorded profile, as shown in Fig. 1. Use of a planimeter is one accurate method. Another useful approach is that of successively minimizing the difference in the two areas by tracing the areas and weighing the paper. Any of these, carefully executed, will yield precise determinations of the desired centroid.

B. Analysis of Weight Average Parameters

Once a determination has been made of $\overline{V}_e$ (or equivalently $\bar{\sigma}_\omega$) as a function of C_T, the resulting dissociation curve may be analyzed according to Eq. (13) or (17). Since the primary data are accumulated as values of $\overline{V}_e$ rather than $\bar{\sigma}_\omega$, it is sometimes desirable to perform as much analysis as possible on these directly measured quantities. Thus a dissociation curve of $\overline{V}_e$ versus C_T may first be analyzed without converting into $\bar{\sigma}_\omega$. It is highly desirable, however, to carry out this conversion at some point in the analysis and also to calibrate the column in order to make correlations between partitioning properties and molecular size and shape. Such correlations can be an obvious aid in making assignments of species property values V_i (or σ_i) to particular aggregates of the solute.

As an example, consider the dissociation curves shown in Fig. 2 for human hemoglobin chains. Here the data cover a wide range of concentration, from 4 μg/ml to 15 mg/ml. Such a wide range is generally neces-

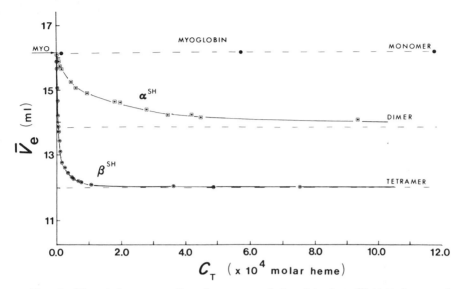

FIG. 2. Dissociation curves from large zone elution data for α^{SH} (dotted squares) and β^{SH} (dotted circles) chains. Myoglobin data points (closed circles) from large zone experiments provide a measure of the nonideality effects as well as a determination of a column calibration end point (monomer). Dashed lines represent elution volumes of monomer, dimer, and tetramer as estimated from calibration of the column. (Taken from Valdes and Ackers[10]).

sary in order to cary out an unequivocal determination of stoichiometry and to evaluate possible effects of nonideality. In systems where the stoichiometry of association is already known, the accuracy and range of data do not have to be as great. From a calibration of the column according to Eq. (10), and a determination of myoglobin elution volumes, the expected elution volumes for monomers, dimers, and tetramers are predicted (horizonal lines, Fig. 2). Such calculations provide a useful guide as to the possible modes of aggregation, and the leveling off at high concentrations of the dissociation curve (or appearance in some cases of positive slope) indicates that association is not of the indefinite type. Additionally, the qualitative shapes of trailing boundaries provide some diagnostic information as to the type of system (cf. Ackers[1,2] for discussion of this point). In general, a sharp unimodal leading boundary and broad trailing boundary are indicative of a homogeneous self-associating solute. For simple two-species systems of the monomer–n-mer type a trailing boundary of the unimodal type (i.e., no inflection points) is expected for a dimerizing system, whereas bimodality (i.e., a bimodal gradient dC_T/dV versus V) is predicted for monomer–n-mer reactions of stoichiometry higher than 2.

Once a dissociation curve such as those of Fig. 2 has been obtained and some suggestive information acquired as regards the possible modes of aggregation, the next step is to carry out least-squares analyses of the data in terms of specific subcases of the general relationships (13) or (17) along with the parametric equations: $C_i = K_i C_1{}^i$. Equation (17) may be conveniently written in terms of weight fractions $f_i = C_i/C_T$

$$\overline{V}_e = \Sigma f_i V_i \tag{18}$$

For the case of simple two-species systems (i.e., $nM_1 \rightleftharpoons M_n$) the formulas assume a particularly simple form:

$$\overline{V}_e = f_1 V_1 + f_n V_n \tag{18a}$$

Solving (18a) for f_1 and noting that $f_1 + f_n = 1$

$$f_1 = \frac{\overline{V}_e - V_n}{V_1 - V_n} \tag{19}$$

and

$$K'_n = \frac{C_n}{C_1{}^n} = \frac{(1 - f_n)}{(f_1)^n C_T{}^{n-1}} \tag{20}$$

Equations (19) and (20) are parametric relationships in the variables V_i, V_n, and K'_n. Thus, in principle, a determination of only three data points ($\overline{V}_e$ versus C_T) would permit estimation of these parameters. In practice a much larger set of data is subjected to least-squares analysis to obtain the best fit. Moreover, the concentration dependence of individual species elution volumes may require at least one additional parameter. If it is assumed that the nonideality of monomer and n-mer are identical then Eq. (19) may be written as

$$f_1 = \frac{\overline{V}_e/(1 - g'C_T) - V_n{}^0}{V_1{}^0 - V_n{}^0} \tag{21}$$

where g' is the coefficient of concentration dependence and superscripted elution volumes for monomer and n-mer refer to values at infinite dilution of these respective solutes.

Numerical analysis of data in terms of these relationships must be carried out by computer, and for this purpose many programs exist. Since they are usually specific to a particular machine or computing center, no detailed programs will be described here. In general it is desirable to use a nonlinear least-squares procedure which affords some estimate of the uncertainty in the best-fit values in addition to the usual estimate of overall variance to the fit. A test for randomness in the distribution of residuals for the best fit provides an overall criterion for absence of systematic error. The reader may wish to consult Box[16] and Fraser and Suzuki[17] for detailed descriptions of applicable methods. Descriptions of the application of these procedures to gel chromatography experiments are contained in Chincone et al.[8] and Valdes and Ackers.[10] By critically testing data against a variety of stoichiometric models a best fitting model, or set of models can be found.

Results of least-squares analyses of the data shown in Fig. 2 are listed in the table.[10] In this analysis the weight average elution volumes were converted into values of $\bar{\sigma}_\omega$ so that estimates of end points and species partition coefficients could be utilized in conjunction with the least-squares fits. Two stoichiometric models were tested: monomer–tetramer and monomer–dimer–tetramer. Under the conditions of these experiments nonideality for myoglobin was apparently absent, and this was verified to be the case for the α and β chains by the fact that no appreciable difference in fitted values was found to arise from neglecting the nonideality terms.

The least-squares analysis was performed using a Gauss–Newton procedure.[16] For each stoichiometric model the fitting was carried out in three consecutive steps: (a) Fitting for the equilibrium constants K_i holding σ_i (end points) constant at the values of $\sigma_{MYO} = \sigma_1$ (the partition coefficient of myoglobin) and the value of σ_i estimated from the column calibration. (b) Next the value of σ_1 alone was held constant and values of K_i and σ_n (high concentration end point) were estimated. (c) Finally, all parameters were floated, σ_1, σ_i, and K_i. Initial guesses were explored over a wide range in order to avoid possible local minima in the residual space. Goodness of fit to the various stoichiometric models was judged on the basis of minimization of the variance of each fit, randomness in the distribution of residuals, and the correspondence of fitted end points to independently determined column calibration parameters.

Table I shows a typical comparison of least-squares fits. The data for α chains exhibited no aggregation beyond dimer so that no critical tests could be made for stoichiometries other than monomer–dimer. This model provided an excellent fit to the data, and it may be noted that the trailing boundary shape (Fig. 1) was also consistent with this stoichiometry. For β chains the comparison is shown between fits for monomer–tetramer, and monomer–dimer–tetramer models. The latter model provided a poorer fit to the experimental data than the monomer–tetramer model. Various attempts to fit this model gave erroneous values for σ_2 (some negative) and K_2 (also negative). The best

[16] G. E. P. Box, Ann. N. Y. Acad. Sci. 86, 792 (1960).
[17] R. D. B. Fraser and E. Suzuki, in "Physical Principles and Techniques of Protein Chemistry" (S. Leach, ed.), Part C, p. 301. Academic Press, New York, 1973.

LEAST-SQUARES FITS TO STOICHIOMETRIC MODELS FOR α^{SH}, β^{SH} CHAIN ASSOCIATION DATA[a,b]

Stoichiometric model	Square root of variance of σ_ω	Estimated partition coefficients[c]			Estimated equilibrium constants[c]		Comments
		σ_1 (limits)	σ_2	σ_4	K_2 (M^{-1} heme × 10^{-3})	K_4 (M^{-3} heme × 10^{-16})	
$2\,\alpha_1 \rightleftharpoons \alpha_2$	0.0043	0.5394 (0.5335, 0.5453)	0.3395 (0.3277, 0.3520)	—	8.422 (6.767, 10.185)	—	Best fit
$4\,\beta_1 \rightleftharpoons \beta_4$	0.0040	0.5224 (0.5159, 0.5286)	—	0.2370 (0.2334, 0.2411)	—	1.406 (1.194, 1.665)	Best fit
$4\,\beta_1 \rightleftharpoons 2\,\beta_2 \rightleftharpoons \beta_4$	0.0081	0.5290 (0.5154, 0.5548)	[0.3720][d]	0.2417 (0.2213, 0.2486)	45.0 (−0.85, 93.9)	2.739 (1.617, 3.346)	Holding σ_2 constant
$4\,\beta_1 \rightleftharpoons 2\,\beta_2 \rightleftharpoons \beta_4$		0.5339	−0.1169	0.2433	Negative value	2.636	All parameters floating

[a] Taken from Valdes and Ackers.[10]

[b] Standard buffer conditions: 0.1 M Tris-HCl, 0.1 M NaCl, 1 mM Na$_2$EDTA, pH 7.4, 21.5°. Sephadex G-100 column calibration parameters: $\sigma_{MYO} = \sigma_{Monomer} = 0.5379$, $\sigma_{Dimer} = 0.3750$, $\sigma_{tetramer} = 0.2430$.

[c] Values in parentheses represent 65% confidence limits for estimated parameters.

[d] Brackets indicate parameter was held constant during fitting routine.

three-species fit could only be obtained by holding σ_2 constant. The square root of the variance for this particular fit was twice the value obtained in the fit for the monomer–tetramer system. The distribution of residuals for this fit are shown in Fig. 3. The distribution of these residuals was essentially random. The variance of this fit was well within the range for meaningful estimation of parameters, as determined by analysis of computer simulated data. The latter procedure consists of (a) assuming a set of parameters for the system, (b) simulating ideal (error-free) data

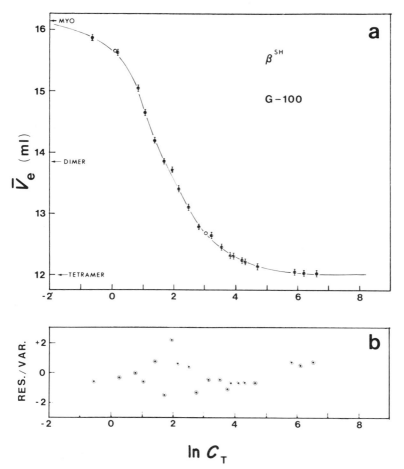

FIG. 3. (a) Large zone G-100 data for β^{SH} of Fig. 2 is here plotted with ln C_T as the independent variable to explicitly show the extent of variation in the measured centroid elution volume over the concentration range studied. Estimated elution volumes of monomer (myoglobin), dimer, and tetramer are depicted by arrows. Error bars are determined by variance of fit. (b) Residuals normalized to the variance for the best fit data of β^{SH} shown in (a) above. The distribution of residuals is essentially random. (Taken from Valdes and Ackers[10]).

over the same range of C_T as the experimental data, (c) assigning random errors to these data corresponding to a range of variances, and (d) analyzing the error-perturbed data in order to assess the reliability with which the originally assumed parameters can be recovered. In addition to the results of least-squares fitting, it may also be noted that the trailing boundary of the large zone elution profile for the β chains (Fig. 1) exhibits a pronounced bimodal character, which is consistent with a monomer–n-mer system having no appreciable population of intermediate species.

In the case of an indefinite (isodesmic) polymerizating system, the dissociation curve will not level off as shown in Fig. 2 over a wide range of concentration, unless the curve is compensated by a high degree of nonideality. The general relationships for isodesmic polymerization are of the form[18].

$$\bar{\sigma}_\omega = \Sigma \, i \, \frac{\sigma_i C_i}{C_T} [1 - (C_1/C_T)^{1/2}]^{i-1} \tag{22}$$

and

$$C_T = \frac{C_1}{(1 - KC_1)^2} \tag{23}$$

Thus if a set of σ_i values is calculated (based upon column calibration and models for the association) the function $\bar{\sigma}_\omega$ versus f_1 (weight fraction monomer) is completely determined. Then for each value of $\bar{\sigma}_\omega$ experimentally determined at a particular C_T, the value of f_1 at that concentration C_T is determined. From such values the complete population of species can be generated since

$$f_i = if_1(1 - f_1^{1/2})^{i-1} \tag{24}$$

An example of the application of these procedures to the analysis of linear indefinite association in L-glutamate dehydrogenase is described in Chun et al.[18]

[18] P. W. Chun, S. J. Kim, C. A. Stanley, and G. K. Ackers, *Biochemistry,* **8,** 1625 (1969).

[10] Multibanded Isoelectric Focusing Patterns Produced by Macromolecular Interactions*

By JOHN R. CANN

Previously (this series, Vol. 25, [11]) we described how a single macromolecule interacting reversibly with a small, uncharged constituent of the

* Supported in part by Research Grant 5R01 HL13909-26 from the National Heart, Lung and Blood Institute, National Institutes of Health, United States Public Health Service. This publication is No. 712 from the Department of Biophysics and Genetics, University of Colorado Medical Center, Denver, Colorado 80262.

buffer medium can give electrophoretic patterns showing two well-resolved zones despite instantaneous establishment of chemical equilibrium. Resolution of the zones is dependent upon generation of stable concentration gradients of unbound small molecule along the electrophoresis column by reequilibration during differential transport of macromolecular reactant and product. In contrast to electrophoresis, steady-state gradients of carrier ampholytes and pH are inherent to the method of isoelectric focusing in natural pH gradients,[1,2] thereby predisposing it to complications arising from macromolecular interactions induced by binding of carrier ampholyte and from pH-dependent conformational transitions.

For ideal situations unencumbered by such interactions the equilibrium isoelectric focusing pattern (plot of macromolecule concentration versus position in the column) of an isoelectrically homogeneous protein will show a single sharply focused peak at that position where the pH corresponds to the isoelectric point (pI). If the protein is isoelectrically heterogeneous, each of the components focuses at its respective pI so that the pattern shows a corresponding number of peaks. Therein resides the power of isoelectric focusing for separating and characterizing proteins and other biological macromolecules. Sometimes, however, the pattern exhibits multiple peaks which do not faithfully reflect the inherent state of homogeneity of the macromolecule. Examples include an acidic protein from wool,[3] some basic microbial proteases,[4] bovine serum albumin,[5,6] myoglobin,[7] and tRNA.[8,9] In each case, fractionation experiments have demonstrated that the multiple peaks arise as a result of essentially reversible, macromolecular interactions. In all but one case[10] interaction with the carrier ampholytes[11] has been implicated, but in no case is the mechanism of interaction fully understood. Thus, for example, there would appear to be at least three possible mechanisms whereby isoelectric focusing of tRNA generates multiple peaks in the pH range 3–5: (a) Strong binding of Ampholines[11] to tRNA may by itself cause changes in

[1] N. Catsimpoolas, ed., "Isoelectric Focusing." Academic Press, New York, 1976.

[2] J. P. Arbuthnott and J. A. Beeley, "Isoelectric Focusing." Butterworth, London, 1975.

[3] R. Frater, *J. Chromatogr.* **50**, 469 (1970).

[4] J. R. Yates, quoted by Frater.[3]

[5] L. J. Kaplan and J. F. Foster, *Biochemistry* **10**, 630 (1971).

[6] K. Wallevik, *Biochim. Biophys. Acta* **322**, 75 (1973).

[7] K. Felgenhauer, D. Graesslin, and B. D. Huismans, *Protides, Biol. Fluids, Proc. Colloq.* **19**, 575 (1971).

[8] J. W. Drysdale and P. Righetti, *Biochemistry* **11**, 4044 (1972).

[9] E. Galante, T. Carvaggio, and P. G. Righetti, *Biochim. Biophys. Acta* **442**, 309 (1976).

[10] The exception is myoglobin for which the appropriate experimentation was not carried out.

[11] The commonly used carrier ampholytes are mixtures of relatively small aliphatic polyaminopolycarboxylic acids with high conductances and closely spaced pI values supplied by LKB, Stockholm, Sweden, under the tradename Ampholine.

the tertiary and secondary structure of the nucleic acid, while at the same time masking phosphate groups and exposing the carboxyl groups of the carrier ampholytes with their higher pK values on the outer surface of the macromolecule.[9] The latter could explain the unexpectedly high pI values observed. (b) Conceivably, tRNA could undergo several sequential pH-dependent conformational transitions as the bases are protonated[8,12] with the carrier ampholytes acting simply as strongly bound counterions,[9] which mask phosphate groups and confer mildly acidic pI values on the different conformations. (c) The interaction may involve elements of both of the foregoing mechanisms, i.e., conformational changes in tRNA may be coupled to both the distribution of carrier ampholytes along the column and the pH gradient.

Recently,[12-15] we have formulated a phenomenological theory of isoelectric focusing of interacting systems which admits certain generalizations, two of which are particularly germaine to routine analytical applications of isoelectric focusing. The first of these is that an amphoteric macromolecule, which complexes reversibly with several species of carrier ampholyte located at different positions along the isoelectric focusing column, can give a multimodal equilibrium pattern. This is based on the assumption that complex formation induces isomerization of the macromolecule with concomitant change in its pI. Each species of ampholyte induces a different isomer. In order to be as general as possible, our model is silent as to whether or not the reversibly bound ampholyte per se might effect the pI of the isomer. It simply assumes that the several isomer–ampholyte complexes have electrophoretic mobilities (at the same pH) and pI's which differ from uncomplexed macromolecule and from each other. The calculations reveal that the number of peaks in the pattern depends upon the number of carrier ampholytes with which the macromolecule can form complexes, the strength of complex formation and the concentrations of the several carrier ampholytes. The illustrative result presented here is for the reaction set

$$M + 4A_1 \overset{K_1}{\rightleftarrows} N(A_1)_4$$

$$M + 4A_2 \overset{K_2}{\rightleftarrows} P(A_2)_4$$

$$P(A_2)_4 + 4A_3 \overset{K_3}{\rightleftarrows} Q(A_3)_4 + 4A_2$$

$$N(A_1)_4 + 4A_4 \overset{K_4}{\rightleftarrows} R(A_4)_4 + 4A_1$$

(1)

[12] D. I. Stimpson and J. R. Cann, *Biophys. Chem.* **7**, 115 (1977).

[13] J. R. Cann and D. I. Stimpson, *Biophys. Chem.* **7**, 103 (1977).

[14] J. R. Cann, D. I. Stimpson, and D. J. Cox, *Anal. Biochem.* **86**, 34 (1978).

[15] D. L. Hare, D. I. Stimpson, and J. R. Cann, *Arch. Biochem. Biophys.* **187**, 274 (1978).

where M is an amphoteric macromolecule; A_1, A_2, A_3, and A_4 are four species of carrier ampholyte; N, P, Q, and R in the four macromolecule–carrier ampholyte complexes, e.g., $N(A_1)_4$, are isomers; and the K_i's are equilibrium constants. The idealized distribution of carrier ampholytes is shown graphically by the dashed curve in Fig. 1; values of the other parameters are given in the legend of the figure. The computed equilibrium isoelectric focusing pattern (solid curve in Fig. 1) exhibits five well-resolved peaks, one corresponding to the uncomplexed macromolecule and four due to the four macromolecule–carrier ampholyte complexes. Since interaction with a single species of ampholyte can give two peaks and with two species, three peaks, etc., inductive reasoning leads to the conclusion that, for interaction with p species, the pattern could show $p + 1$ peaks under appropriate circumstances.

The other generalization is that a macromolecule undergoing a sequence of q reversible pH-dependent conformational transitions can give an equilibrium isoelectric focusing pattern exhibiting $q + 1$ peaks, with the proviso that the sequentially formed conformers have successively lower pI's. Consider the set of two sequential transitions

$$M + nH^+ \overset{K_1}{\rightleftharpoons} NH_n^+$$

$$NH_n^+ + mH^+ \overset{K_2}{\rightleftharpoons} PH_{n+m}^+$$

(2)

in which M, N, and P are conformers with different pI's, and H^+ is hydrogen ion. As shown in Fig. 2, the isoelectric focusing patterns computed for this reaction set exhibit three well-resolved peaks. Thus, the ob-

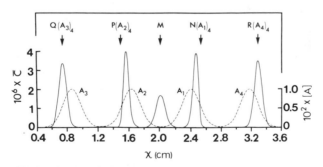

FIG. 1. Equilibrium isoelectric focusing pattern (———) calculated for a four-state, carrier ampholyte-induced isomerization [reaction set (1)], displayed as a plot of molar constituent concentration of macromolecule $\overline{C}$ against position x; distribution of carrier ampholytes (- - - -), plot of molar concentration [A] against x; positions of pI's of indicated macromolecular species are designated by vertical arrows. $K_1 = K_2 = 1.48 \times 10^9 \, M^{-4}$ and $K_3 = K_4 = 0.25$; diffusion coefficient of uncomplexed macromolecule, $3.6 \times 10^{-7} \, \text{cm}^2/\text{sec}$; diffusion coefficients of macromolecule–ampholyte complexes, $2.6 \times 10^{-7} \, \text{cm}^2/\text{sec}$; driven velocities, $V_i(x) = 10^{-4}(a_i - x)$, where a_i is the position of the pI. From Hare et al.[15]

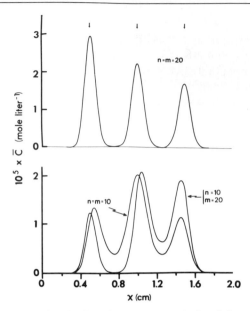

FIG. 2. Equilibrium isoelectric focusing patterns calculated for two sequential pH-dependent conformational transitions [reaction set (2)]. Apparent equilibrium constants, $K_1(H^+)^n$ and $K_2(H^+)^m$, normalized midway between the pΓs of M and NH_n^+ and of NH_n^+ and PH_{n+m}^+, respectively; diffusion coefficients of M, NH_n^+ and PH_{n+m}^+, 3.6×10^{-7} cm^2 sec^{-1}, 3.4×10^{-7} cm^{2}sec^{-1}, and 3.2×10^{-7} cm^{2}sec^{-1}, respectively; driven velocities as given in Fig. 1. When read from right to left, the vertical arrows indicate the positions of the pΓs of M, NH_n^+ and PH_{n+m}^+, respectively. Adapted from Stimpson and Cann.[12]

served behavior of interacting systems, whether they involve carrier ampholyte-induced isomerizations or pH-dependent conformational transitions, is theoretically predicted.

This result has important implications for the many conventional applications of isoelectric focusing to biochemistry and molecular biology. In particular, it underscores the fact that an isoelectric focusing pattern exhibiting two or more peaks need not necessarily be indicative of inherent heterogeneity. Unambiguous interpretation of the pattern is dependent upon fractionation experiments. That is, material must be isolated from each peak; reconcentrated by miniaturized ultrafiltration, for example, and rerun to see if it focuses as a single peak with reproducible pI or shows multiple peaks as in the case of the unfractionated material.[3–8,12–18]

[16] P. Talbot, in "Isoelectric Focusing" (J. P. Arbuthnott and J. A. Beeley, eds.), p. 270, Butterworth, London, 1955.
[17] N. Ressler, Anal. Biochem. 51, 604 (1973).
[18] C. W. Wrigley, in "Isoelectric Focusing" (N. Catsimpoolas, ed.), p. 77. Academic Press, New York, 1976.

Reconcentration of the fractions is necessary because ampholyte-induced association–dissociation reactions can also give isoelectric focusing patterns showing two peaks.[14] When rerunning the fractions, care should be taken that the position of the initial zone of material in the isoelectric focusing column is the same as in the original fractionation. Both theory[12,13] and experiment[16] indicate that the approach to equilibrium can be very slow and the path dependent upon the position of the initial zone relative to the isoelectric region of the interacting system. In fact, rigor requires that both the starting material and its fractions be focused from narrow initial zones located at different positions along the column. The investigation of Talbot[16] on the pH-dependent conformational transition of the 12 S subunit protein of the capsid of foot-and-mouth disease virus is exemplary in this respect.

Last, it is anticipated that the new insights provided by these calculations will also find application in fundamental studies on biochemical reactions such as protein–drug interactions, the interaction of enzymes with cofactors and allosteric affectors, and the interaction of macromolecules with each other. In fact, preliminary findings indicate that isoelectric focusing holds promise for the study of these several classes of interaction.[19] Moreover, the peculiar sensitivity of isoelectric focusing might permit detection of subtle, ligand-induced conformational changes which are beyond the ready reach of other methods. In any case, when an interaction is detected, its actual nature can be determined only by the combined application of isoelectric focusing with one or more other physical methods, such as ultracentrifugation and circular dichroism. Toward this end, Talbot[16] used ultracentrifugation to eliminate association of the 12 S subunit protein of foot-and-mouth disease virus as an explanation for its focusing behavior, and Galante et al.[9] used melting experiments to show that tRNA binds Ampholine. Once the interaction has been classified in general terms (e.g., isomerization, association–dissociation or macromolecular complexing; pH-dependent, ampholyte-induced, or ligand-mediated), the appropriate combination of physical methods can be employed to elucidate its detailed mechanism. The greater the variety of methods brought to bear on the problem, the more precisely can the interaction be described.

[19] J. W. Drysdale, Methods Protein Sep. 1, 93 (1975).

[11] Small-Angle X-Ray Scattering

By INGRID PILZ, OTTO GLATTER, and OTTO KRATKY

I. General Remarks

We will assume that the reader is familiar with the article "Small-Angle X-ray Scattering" by Pessen, Kuminsky, and Timasheff (hereafter P.K.T.), which appeared in Volume XXVII, Part D, p. 151 of this series. It gives a full account of the state of research in the field until 1973. Our main concern will be to give a supplement, covering the progress since 1973. While it is impossible to avoid repetitions with P.K.T. completely, we try to keep them to a minimum and only include duplications with P.K.T. where we consider them essential for the clarity.

Since 1973, important contributions to the evaluation of scattering data (smoothing, desmearing, Fourier-inversion) have appeared. They are presented in Section II,A. In Section II,B a summary of possible interpretation of real space information is given.

Recent contributions concerning instrumentation and experimental methods are covered in Section III. A complete discussion of all existing techniques was not attempted; this chapter is essentially devoted to techniques and instruments which have already been used for the study of proteins. Also covered are instruments whose line of development makes such an application likely or whose suitability for such applications has to be evaluated. The sections on monochromatization and on the determination of absolute intensity are supplements to the corresponding sections in P.K.T.'s article. The camera type developed in Graz will be discussed, as it was recently improved and thereby modified in several of its characteristics. Experimental details concerning the preparation of solutions and measuring strategies to obtain a scattering curve extrapolated to zero concentration will also be covered.

Section IV presents, as far as they have come to our attention, all the small-angle investigations on enzymes since 1973. We have also included several older examples, which were not mentioned by P.K.T.

While P.K.T. use Luzzati's system of notations for the theoretical relationships, we use the one preferred by members of our school. This necessitates a comparison of the two notations and an explanation of their relationship, which will be given in the Appendix. Since the two systems of expressions are the two most frequently used ones, the interested reader will benefit from the correlation between the two systems; their knowledge makes the formalism of practically any publication in the field of small-angle scattering comprehensible.

II. Data Evaluation in Small-Angle Scattering

A. Data Treatment

1. Schematic Description of the Experiment. In this paper we discuss small-angle scattering experiments with particles in solution, i.e., the particles are nonoriented. A large number of particles contribute to the scattering and the resulting spatial average leads to a loss in information. The information contained in the three-dimensional electron density distribution $\rho(\mathbf{r})$ (describing the whole structure of the particle) is thereby reduced to the one-dimensional distance distribution function $p(r)$. This function is proportional to the number of lines with length r which connect any volume element i with any volume element k of the same particle. The spatial orientation of these connection lines is of no account to the function $p(r)$. The connection lines are weighted by the product of the number of electrons situated in the volume elements i and k, respectively. The correlation between the function $p(r)$ and the structure of the particle is discussed in detail in Section II,B. The connection between the distance distribution function $p(r)$ and the measured experimental scattering curve $I_{\exp}(h)$ is shown schematically in the upper half of Fig. 1.

We "observe" the particle indirectly by means of scattering. Each distance between two electrons of the particle, which is part of the function $p(r)$, leads to an angular-dependent scattering intensity as already discussed by P.K.T. This physical process of scattering can be mathematically expressed by a Fourier transformation (transformation T_1), which defines how the information in "real space" (distance distribution function) is transformed into "reciprocal space" (scattering function). In practice, however, the scattering intensity cannot be measured exactly. A monochromatic primary beam with point collimation of very high intensity and an infinitely small detector would be necessary to obtain ideal data. It is impossible to fulfill all these conditions perfectly, but it is possible to reduce particular effects to a negligible degree.

A slightly polychromatic radiation (CuK_{α}- and CuK_{β}-line) is frequently employed in X-ray small-angle experiments (see Section III). In the case of neutron scattering, usually a continous wavelength distribution with a half-width of several percent is used. The effect of any wavelength distribution can be expressed by a linear transformation T_2. The collimation system and the detector have finite apertures in order to provide sufficient scattering intensity. This apertures cause the slit length effect (transformation T_3) and the slit width effect (transformation T_4). The transformations T_2–T_4 can be expressed mathematically by the corresponding integral equations. They cause a "smearing" of the scattering function, i.e., any fine structure of the curve is strongly smoothed.

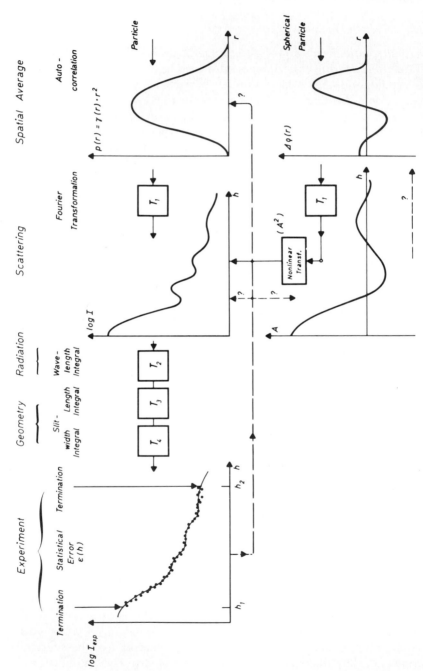

FIG. 1. Schematic representation of the correlation between a particle and its observable experimental scattering data.

The smeared scattering function results from the distance distribution function by performance of the linear transformations T_1 to T_4 in the right order. Two additional experimental factors have to be considered in this context: At small angles the measuring range is limited by the finite dimensions of the primary beam and by the width of the detector window. At large angles the limits are given by the signal-to-noise ratio.

Usually the scattering intensities are registrated with a proportional counter or scintillation counter, and the statistical error is inversely proportional to the square root of counts. This statistical error is superimposed to the smeared scattering function. Subtracting the blank scattering leads to a further increase in the statistical error (see Fig. 35). Together, the limited measuring range and the statistical errors further reduce the significant information. The problem of data evaluation, therefore, lies in the determination of maximum structural information from a given finite number of measured intensity data.

The main problems are the following.

1. Taking into account the statistical accuracy (smoothing, approximation)

2. Desmearing (correction of collimation and wavelength effects, T_2 to T_4)

3. Fourier transformation (termination effect)

These problems can be solved sequentially or in one step. In the sequential procedure smoothing and desmearing operations are performed prior to Fourier transformation. The essential problems involved in the desmearing procedure are the unavoidable smoothing of the statistical fluctuations and the termination effect. The termination effect is also the main problem in Fourier transformation, since the exact inverse transformation T_1^{-1} would require the knowledge of the scattering function over an infinite angular range. Any approximative extrapolation of the scattering function may introduce errors. Nevertheless, this is usually done toward zero angle to enable the estimation of maximum intraparticle distance.

In most of the practical applications that have been reported so far the accumulation of errors made an extensive analysis of the distance distribution function impossible. Thus it was necessary to interpret the data in reciprocal space (scattering curves). The most important examples, as already discussed by P.K.T., are the estimation of the radius of gyration and the scattering intensity at zero angle by means of the Guinier approximation.

The new indirect transformation method solves the three problems mentioned above in one step, using an approximative estimate of the maximum distance of the particle. In the particular case of particles of spherical symmetry there is no loss in information by spatial average. The one-dimensional radial electron density distribution is directly correlated to the scattering amplitude, the square of which is the measurable scattering intensity (see lower half of Fig. 1).

2. *Mathematical Description of the Experimental Effects*. The "characteristic" function of a particle $\gamma(r)$, frequently also termed as correlation function, was introduced in the theory of small-angle scattering by Porod.[1] Multiplying by r^2 we obtain the distance distribution $p(r)$.[2,3]

$$p(r) = \gamma(r) \, r^2 \tag{1}$$

The radius of gyration is defined by

$$R_g^2 = \frac{\displaystyle\int_0^\infty p(r) \, r^2 \, dr}{2 \displaystyle\int_0^\infty p(r) \, dr} \tag{2}$$

The scattering intensity at zero angle is a constant, 4π times the integral of $p(r)$:

$$I(0) = 4\pi \int_0^\infty p(r) \, dr \tag{3}$$

Equations (2) and (3) will be used later to find new approximations for R_g and $I(0)$. We shall now describe the various transformations T_1–T_4 in more detail.

a. FOURIER TRANSFORMATION (T_1). The relation between the scattering intensity and the distance distribution of a particle is given by the Fourier transformation

$$I(h) = 4\pi \int_0^\infty \gamma(r) \, r^2 \, \frac{\sin (hr)}{hr} \, dr = 4\pi \int_0^\infty p(r) \, \frac{\sin (hr)}{hr} \, dr \tag{4}$$

where

$$h = (4\pi/\lambda_0) \sin \theta$$

λ_0 is the wavelength of the monochromatic radiation and 2θ is the scattering angle (see P.K.T.). In P.K.T., s is used as a reduced angular variable.

$$s = (2/\lambda) \sin \theta = h/2\pi \tag{5}$$

[The scattering intensity at zero angle, $I(0)$ in Eq. (3), follows from Eq. (4) by setting h equal to zero.]

[1] G. Porod, *Kolloid-Z.* **124**, 83 (1951).
[2] G. Porod, *Acta Phys. Austriaca* **3**, 255 (1948).
[3] A. Guinier and G. Fournet, "Small Angle Scattering of X-Rays." Wiley, New York 1953.

The corresponding relation between the scattering amplitude $A(h)$ and the radial electron density distribution $\rho(r)$ of a centrosymmetrical particle is given by

$$A(h) = 4\pi \int_0^\infty \rho(r) \, r^2 \, \frac{\sin (hr)}{hr} \, dr \qquad (6)$$

The linear transformation T_1 has an inverse T_1^{-1} given by

$$\gamma(r) = \frac{1}{2\pi^2} \int_0^\infty I(h) \, h^2 \, \frac{\sin (hr)}{hr} \, dh \qquad (7a)$$

or

$$p(r) = \frac{1}{2\pi^2} \int_0^\infty I(h) \, (hr) \sin (hr) \, dh \qquad (7b)$$

and

$$\rho(r) = \frac{1}{2\pi^2} \int_0^\infty A(h) \, h^2 \, \frac{\sin (hr)}{hr} \, dh \qquad (7c)$$

where the limits of integration should be noted.

b. WAVELENGTH EFFECT (T_2). The experimentally observed scattering intensity I_m, measured with polychromatic radiation is related to the theoretical scattering intensity I for monochromatic radiation of wavelength λ_0 as follows

$$I_m(h) = \int_0^\infty W(\lambda')I\left(\frac{h}{\lambda'}\right) d\lambda' \qquad (8)$$

where $\lambda' = \lambda/\lambda_0$, $W(\lambda')$ is the wavelength distribution of the polychromatic radiation. An example for such a wavelength distribution $W(\lambda')$ for neutrons is given in Fig. 2.

c. SLIT LENGTH EFFECT (T_3). The scattering intensity $\tilde{I}$ measured with an infinitely thin slit of finite length can be expressed by the equation

$$\tilde{I}(h) = 2 \int_0^\infty P(t) \, I(h^2 + t^2)^{1/2} \, dt \qquad (9)$$

$P(t)$ is the intensity distribution of the primary beam in its longitudinal direction.

d. SLIT WIDTH EFFECT (T_4). The scattering intensity $\hat{I}$ measured with an infinite short slit of finite width is given by

$$\hat{I}(h) = \int_{-\infty}^\infty Q(x) \, I(h - x) \, dx \qquad (10)$$

where $Q(x)$ is the intensity distribution of the primary beam across its

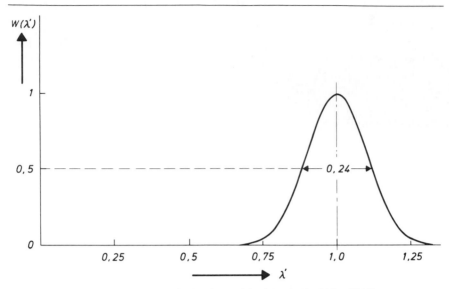

FIG. 2. Wavelength distribution $W(\lambda)$ with a half-width of 24%.

width. Equations (9) and (10) can easily be deduced from Fig. 3 (for further details, see footnote 4).

The two geometrical effects can only be split into the two separate effects T_3 and T_4 if the conditions of Eq. (11) is fulfilled.

$$I_0(x,t) = Q(x)P(t) \tag{11}$$

This condition requires that it be possible to describe the intensity profile of the primary beam $I_0(x, t)$ as a product of the two one-dimensional functions $Q(x)$ and $P(t)$. The sequence of the transformations T_2, T_3, and T_4 must not be changed.[5] The length and width of the detector window can be taken into account by a convolution of $P(t)$ and $Q(x)$ with a step function of adequate length. These convolution operations are already done implicitly, if the profiles $P(t)$ and $Q(x)$ are measured with the same detector window dimensions.

Summarizing all effects (T_1–T_4), the following relation for the experimentally observed scattering intensity, $I_{\exp}(h)$, is obtained

$$I_{\exp}(h) =$$
$$8\pi \int_0^\infty dr \int_0^\infty d\lambda' \int_0^\infty dt \int_{-\infty}^\infty dx \, Q(x) \, P(t) \, W(\lambda') \, p(r) \sin{(\beta)}/\beta + \epsilon(h)$$
$$\beta = r \, [(h - x)^2 + t^2]^{1/2}/\lambda' \tag{12}$$

$\epsilon(h)$ is the statistical error corresponding to the pulse rate of the counter.

[4] O. Kratky, G. Porod, and Z. Skala, *Acta Phys. Austriaca* **13**, 76 (1960).
[5] O. Glatter and P. Zipper, *Acta Phys. Austriaca* **43**, 307 (1975).

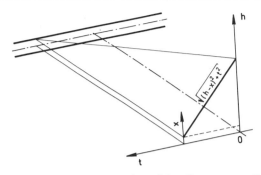

FIG. 3. Schematic representation of the slit geometry effect.

The weight functions $P(t)$, $Q(x)$, and $W(\lambda')$ are normalized usually to one, i.e.,

$$\int_0^\infty W(\lambda') \, d\lambda' = 1 \tag{13a}$$

$$2 \int_0^\infty P(t) \, dt = 1 \tag{13b}$$

$$\int_{-\infty}^\infty Q(x) \, dx = 1 \tag{13c}$$

For the calculation of absolute intensities as used in the Graz group[6] it is necessary to use the following normalization:

$$\tilde{I}(h) = \frac{2}{T_{mh}} \int_0^\infty P(t) \, I(h^2 + t^2)^{1/2} \, dt \text{ where } T_{mh} = \frac{2\pi}{\lambda a} \text{ and} \tag{14}$$

a is the distance between the sample and the plane of recording in centimeters. The height of $P(t)$ is normalized to $P(0) = 1.0$.

3. *Stepwise Procedures.* This section contains the description of the most common methods for solving the desmearing and Fourier transformation problems in two consecutive steps. The elimination of the two collimation effects and the wavelength effect can be performed in one step *or* in individual steps. It should be noted that combined methods do not have the problem of error accumulation. In any case, the right sequence: (1) slit width desmearing, (2) slit length desmearing, and (3) wavelength desmearing has to be followed.

The first important method for slit length desmearing has been given by Guinier and Fournet.[3] It requires a constant primary beam profile $P(t)$ with infinite length. The first derivative of the experimental scattering curve is used for the desmearing procedure. The determination of this first derivative necessitates a smoothing procedure which is the central problem for the application of this method.

[6] O. Kratky, Z. Anal. Chem. **201,** 161 (1964).

A further development to this method was given in our group.[4] It allows the approximative solution of the slit length desmearing problem for $P(t)$ functions of trapezoidal shape. The method was adapted to the use of digital computers by Heine and Roppert.[7] This computer program does not involve a smoothing procedure. An additional method for the preliminary elimination of the slit width effect was given also by Kratky *et al.*[4] The results from this procedure are of sufficient accuracy for small symmetrical $Q(x)$ functions.

A widely used procedure is the method of Schmidt[8] and Taylor and Schmidt[9] which implies the restriction that the $P(t)$ function be a Gaussian. This provides the possibility of partial integration and avoids the computation of the first derivative. The smoothing of the experimental data is performed by a piecewise approximation by polynomials. The experimental data have to be measured at equidistant intervals with the same accuracy. There is some loss of information at the innermost scattering points.

The sampling theorem of the Fourier transformation is used in the smoothing procedures of Damaschun *et al.*[10] As a physical smoothing condition the maximum intraparticle distance is used in their method. The desmearing is done with the method of Heine and Roppert.[7]

An iterative method for desmearing has been developed by Lake.[11] The procedure is very simple and allows arbitrary slit functions $P(t)$ and $Q(x)$. The accuracy of the data is not taken into account and preliminary smoothing of the experimental data is necessary. The quality of the results is not always sufficient.[12,13]

Some new basic aspects are contained in the procedure of Schelten and Hossfeld.[14] A weighted least squares technique, allowing arbitrary primary beam profiles, results in an approximation to the experimental data with minimized second derivative and in the corresponding desmeared scattering curve. Cubic B-Spline functions are used as a basic set. The smoothing effect is excellent, but the minimization of the second derivative involves the possibility of an unverifiable flattening of the maxima, in particular of the maximum at zero angle. This effect increases with increasing statistical error.

[7] S. Heine and J. Roppert, *Acta Phys. Austriaca* **13**, 148 (1962).
[8] P. W. Schmidt, *Acta Crystallogr.* **19**, 938 (1965).
[9] T. R. Taylor and P. W. Schmidt, *Acta Phys. Austriaca* **4**, 293 (1967).
[10] G. Damaschun, J. J. Müller, and H.-V. Pürschel, *Acta Crystallogr., Sect. A* **27**, 11 (1971).
[11] J. A. Lake, *Acta Crystallogr.* **23**, 191 (1967).
[12] P. W. Schmidt, *J. Appl. Crystallogr.* **3**, 137 (1967).
[13] G. Walter, R. Kranold, and G. Becherer, *Stud. Biophys.* **47**, 49 (1974).
[14] J. Schelten and F. Hossfeld, *J. Appl. Crystallogr.* **4**, 210 (1971).

In the procedure of Vonk[15] the integral Eq. (12) is reduced to a linear system of equations. Arbitrary slit length profiles $P(t)$ can be corrected to a weighted approximation of the data. The termination effect and artificial oscillations caused by the statistical error are sometimes not negligible.

The desmearing procedure for the slit length effect developed by Strobl[16] has no restrictions to the primary beam function $P(t)$. A detailed error discussion is given, but it is not a weighted least squares procedure. The method is appropriate for scattering curves containing sharp reflections since the primary result is the integrated scattering function.

The iterative method developed by one of us[17,18] allows the correction of the two geometrical effects and the wavelength effect in one step for arbitrary weighting functions $P(t)$, $Q(x)$, and $W(\lambda')$. The procedure has an implicit smoothing routine. The degree of smoothing is dependent on a freely selectable smoothing parameter and on the accuracy of the data. The iterative process is controlled by a weighted least squares condition. The number of necessary iterations is estimated by a convergence criterion and the termination effect is negligible. The procedure can be used for arbitrary scattering functions, since no special function system is used. The accuracy of the results is comparable to the accuracy of the experimental data. The evaluation of the propagated error is impossible. Slight artificial oscillation may occur in the innermost part of the desmeared scattering function. A special version for scattering curves with discrete reflexes has been written by Jaeneke.[19]

A Fourier transformation can be performed in addition to these desmearing programs in order to compute the distance distribution function. The main problems in this step are caused by the termination effect and by the influence of a remaining background scattering. These effects may cause such strong artificial oscillations in the $p(r)$ function that it becomes useless for the interpretation of the scattering data.

The new indirect transformation method[20,21] makes it possible to carry out the three processes smoothing, desmearing, and Fourier transformation in one step. In order to do this it is necessary to give a rough estimate on the total size of the particle. The method can only be applied to scattering curves from dilute particle systems, i.e., for scattering media whose distance distributions are zero beyond a certain value. Therefore, this method cannot be applied to ordered or densely packed systems.

[15] C. G. Vonk, *J. Appl. Crystallogr.* **4**, 340 (1971).
[16] G. R. Strobl, *Acta Crystallogr., Sec. A* **26**, 367 (1970).
[17] O. Glatter, Monatsh. Chem. **103**, 1691 (1972).
[18] O. Glatter, *J. Appl. Crystallogr.* **7**, 147 (1974).
[19] L. Jaeneke, Diplom Arbeit, Fak. Physik, Univ. Freiburg (1975).
[20] O. Glatter, *Acta Phys. Austriaca* **47**, 83 (1977).
[21] O. Glatter, *J. Appl. Crystallogr.* **10**, 415 (1977).

Here, any remaining background scattering does not cause artificial oscillations in the $p(r)$ function, and the termination effect is minimized.

4. *Indirect Transformation Method.* An important problem in the approximation and smoothing of a function is the choice of the approximation function system to be used. Functions such as polynomals, Hermite functions,[22] Gaussians and Spline functions may be favorable systems for some special problems. An essential improvement can be obtained by defining the function system in real space (distance distribution) taking into account the additional approximative information on the maximum size of the particle. The estimate of this dimension should be an upper limit and is called D_{max}.

$$D_{max} \geqq D \tag{15}$$

where D is the real maximum intraparticle distance. If such an estimate can be given, it follows that

$$p(r) = 0 \quad \text{for} \quad r \geqq D_{max} \tag{16}$$

for scattering curves which are extrapolated to zero concentration (infinite dilution). Equation (16) states that the function $p(r)$ can be different from zero only in the range $0 < r < D_{max}$. For the following it is not necessary that the equality in Eq. (15) holds, i.e., that D_{max} is a perfect estimate of D.

The function $p(r)$ will be approximated by a linear combination of a finite number of linear independent functions $\varphi_\nu(r)$:

$$p_A(r) = \sum_{\nu=1}^{N} c_\nu \varphi_\nu(r) \tag{17}$$

$p_A(r)$ is the approximation to $p(r)$, N is the number of functions, and c_ν are the unknowns. The functions $\varphi_\nu(r)$ are the cubic B Splines.[14,23] These functions are an appropriate system. They are defined only in the range $0 \leqq r \leqq D_{max}$ as multiple-convolution products of a step function and represent curves with a minimum second derivative. They overlap only with a few neighbors as they differ from zero in the range of four knots. This guarantees high numerical stability.

The particular functions φ_ν are transformed according to the transformations T_1 to T_4. The intermediate result after the transformation T_1 represents scattering intensities without collimation effects:

$$\psi_\nu(h) = T_1 \varphi_\nu(r) \tag{18}$$

[22] F. Hossfeld, *Acta Crystallogr.*, Sect. A **24**, 643 (1968).
[23] T.N.E. Greville, "Theory and Applications of Spline Functions." Academic Press, New York, 1969.

Smeared intensities result after execution of all transformations

$$\chi_\nu(h) = T_4 T_3 T_2 T_1 \varphi_\nu(r) = T_4 T_3 T_2 \psi_\nu(h) \tag{19}$$

These functions $\chi_\nu(h)$ are an optimized system for the approximation of scattering data from a particle with a maximum dimension D_{max}, measured under the special conditions represented by T_1–T_4. The coefficients c_ν are determined by a weighted least squares approximation to the experimental data $I_{exp}(h)$. This condition requires that

$$LS = \int_{h_1}^{h_2} [I_{exp}(h) - \sum_{\nu=1}^{N} c_\nu \chi_\nu(h)]^2 / \sigma^2(h) = \text{Min} \tag{20}$$

where h_1 and h_2 are the angles of the first and the last data points and $\sigma^2(h)$ is the variance of the measurements. Instabilities arising from the solution of this minimization problem are eliminated by a new stabilization procedure. This procedure fails if there are systematic errors in $I_{exp}(h)$ and $\sigma(h)$. Another reason for a breakdown of this routine can be the choice of inconsistent parameters (incorrect value for D_{max}) or the use of too many Spline functions, corresponding to an attempt of overinterpretation of the experimental data. For details, see Glatter.[20,21,24]

The coefficients c_ν resulting from Eq. (20) define the solution in real space according to Eq. (17), whereas the approximation curve to the data points is given by

$$\bar{I}_A(h) = \sum_{\nu=1}^{N} c_\nu \chi_\nu(h) \tag{21}$$

The desmeared curve is defined by the expression

$$I_A(h) = \sum_{\nu=1}^{N} c_\nu \psi_\nu(h) \tag{22}$$

(see Fig. 4).

5. *Radius of Gyration and Zero Angle Intensity.* The computation of the radius of gyration according to Eq. (2) from the approximation function $p_A(r)$ gives

$$R_{gA}^2 = \int_0^{D_{max}} p_A(r) \, r^2 \, dr / \left(2 \int_0^{D_{max}} p_A(r) \, dr \right) \tag{23}$$

This approximation has the advantage that the radius of gyration is calculated from the smeared data points, using the whole scattering curve. For this reason, this approximation is often superior to the Guinier approximation.

[24] K. Müller and O. Glatter, in preparation.

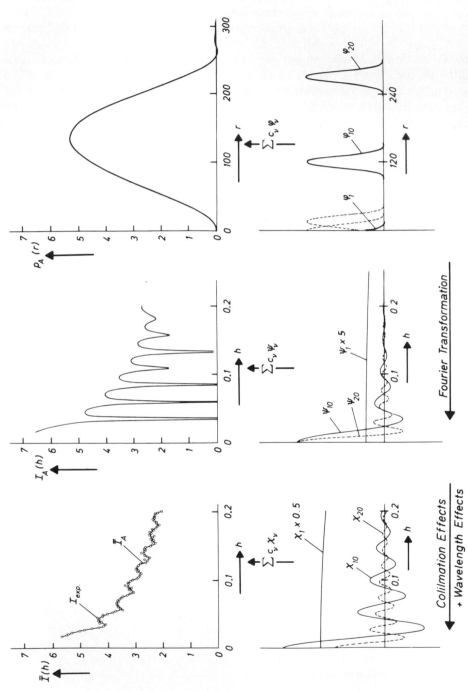

FIG. 4. Function systems $\varphi_\nu(r)$, $\psi_\nu(h)$, and $\chi_\nu(h)$ used for the approximation of the scattering data in the indirect transformation method.

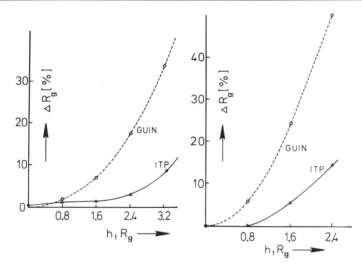

FIG. 5. Dependence of the accuracy of the radius of gyration ΔR_g determined by Guinier's approximation (-- O --) as well as calculated by the indirect transformation method (—△—, ITP) on the angular resolution at small angles for a sphere (left side) and an elongated prism 1 : 1 : 10 (right side).

An approximate value for the zero angle intensity can be computed according to Eq. (3).

$$I_A(0) = 4\pi \int_0^{D_{max}} p_A(r) \, dr \tag{24}$$

Numerical tests have shown that the radius of gyration can be estimated by Eq. (23) with satisfactory accuracy even for such termination values h_1R_g at which the Guinier approximation is no longer valid. A comparison of the approximation methods for two different types of particles is shown in Fig. 5.

An additional advantage of the indirect transformation method lies in the fact that it is possible to calculate the propagated error band in real space. One has to keep in mind, however, that this error band represents only the statistical error. Systematic errors of the measurement cannot be detected by a general treatment. If they are essential they can be signalized by a breakdown of the stabilization procedure. Systematic errors caused by the computational method can be found by means of simulations with theoretical scattering functions.

6. *High Resolution Measurements.* Detailed information about the cross section or the thickness are of particular interest if the particle is elongated in one or two directions (cylindrical particles or lamellae). For that purpose the scattering curve has to be measured to relatively large scattering angles.

The distance distribution of the cross section $p_c(r)$ of a cylinder is correlated to the total scattering intensity $I(h)$ by the equation

$$I(h) = 2\pi^2 L \int_0^\infty p_t(r) \frac{J_0(hr)}{h} \, dr \qquad (25)$$

where $J_0(hr)$ is the zero-order Bessel function and L is the length of the cylinder. A similar relation holds for the correlation function of the thickness $p_t(r)$ for lamellae:

$$I(h) = 4\pi F \int_0^\infty p_t(r) \frac{\cos (hr)}{h^2} \, dr \qquad (26)$$

where F is the area of the plane of the lamella. The indirect transformation method allows the computation of the distance distribution of the cross section $p_c(r)$ and the correlation function of the thickness $p_t(r)$ from the experimental data, if Eqs. (25) or (26) are used instead of Eq. (4) for the transformation T_1.[25]

If the scattering curve is smoothed and desmeared by any other method the following procedure can be taken. The so-called cross section factor $I_c(h)$ and the thickness factor $I_t(h)$ are calculated from

$$I_c(h) = I(h)h/\pi L \qquad (27)$$

and

$$I_t(h) = \frac{I(h)h^2}{2\pi F} \qquad (28)$$

(see Porod et al.[2,26,27]). Now we can calculate the functions $p_c(r)$ and $p_t(r)$ using the equations

$$p_c(r) = \frac{1}{2\pi} \int_0^\infty I_c(h)(hr) J_0(hr) \, dh \qquad (29)$$

(see Fedorov[28]) and

$$p_t(r) = \frac{1}{\pi} \int_0^\infty I_t(h) \cos (hr) \, dh \qquad (30)$$

The quality of such calculations is limited by the finite elongation of the particles and by the termination effect in Eqs. (29) and (30).[29] The correlation function $p_t(r)$ of thin spherical vesicles goes to zero at $r = d$ (d is the

[25] O. Glatter, in preparation.
[26] O. Kratky and G. Porod, *Acta Phys. Austriaca* **2**, 133 (1948).
[27] P. Mittelbach, *Acta Phys. Austriaca* **19**, 53 (1964).
[28] B. A. Fedorov, *Acta Crystallogr.*, Sec. A **27**, 35 (1971).
[29] R. D. Carlson and P. W. Schmidt, *J. Appl. Crystallogr.* **2**, 297 (1969).

thickness of the vesicle) and is different from zero again in the range $D - 2d \leq r \leq D$, where D is the outer diameter of the vesicle.[30]

An exceptional case are particles with spherical symmetry. The radial electron density profile can be computed from the scattering amplitude $A(h) = [I(h)]^{1/2}$ by means of Eq. (7c). Here arises the difficulty of finding the correct signs for the amplitudes. One possibility to solve this problem is given by the variation of the electron density of the solvent.[31,32a]

Experimental curves never show ideal zeros, they only have more or less pronounced minima. This can be due to the following effects: deviations from perfect spherical symmetry, polydispersity, inaccuracy of the desmearing procedure, and experimental errors. One possible procedure used by several groups is the following: The two curves $+[I(h)]^{1/2}$ and $-[I(h)]^{1/2}$ are plotted and the two curves are connected in the region of the minima, i.e., the minima are reduced to zeros and the maxima remain unchanged.[32b]

Another method is the association of the finite intensity at the minima exclusively to the effect of spherical asymmetry.[31] Under this assumption one has to subtract a smooth curve connecting the minima from the whole experimental scattering curve. No detailed comparison of these two approximations has been published till now.

7. Resolution of Small-Angle Scattering Experiments

a. MAXIMUM DIMENSION OF A PARTICLE. It is common practice to define a resolution from the angle of the lowest scattering angle h_1 (see Fig. 1) using the Bragg relation

$$h_1 D_{\text{Bragg}} = 2\pi \qquad \text{or} \qquad D_{\text{Bragg}} = 2\pi/h_1 \qquad (31)$$

In this case, of course, one is not dealing with a periodic arrangement of particles but with scattering functions of isolated particles. It must be emphasized, that D_{Bragg} is a measure for the scattering angle and not a measure for a maximum particle size.

From Fig. 5 we can see that it is possible to determine the radius of gyration with sufficient accuracy if

$$h_1 R_g \leq 1{,}0 \qquad (32)$$

or

$$h_1 D \leq \pi \qquad (33)$$

[30] D. Weick, *Biophys. J.* **14**, 233 (1974).

[31] L. Mateu, A. Tardieu, V. Luzzati, L. Aggerbeck, and A. M. Scanu, *J. Mol. Biol.* **70**, 105 (1972).

[32a] K. Müller, P. Laggner, O. Kratky, G. Kostner, A. Holasek, and O. Glatter, *FEBS Lett.* **40**, 213 (1974).

[32b] J. W. Anderegg, *in* "Small-Angle X-Ray Scattering" (H. Brumberger, ed.), p. 243. Gordon and Breach, New York, 1967.

where D is the maximum dimension of the particle. These relations which were found from simulated experiments are in full agreement with the theoretical limit given by the sampling theorem of the Fourier transformation.[33,34]

Comparing Eqs. (31) and (33) we see that

$$D = D_{\text{Bragg}}/2 \tag{34}$$

Equation (33) could be called a "resolution equation" of small-angle scattering. The value $h_1 D = \pi$ corresponds to the first zero of the sine function in the Fourier integrals in Eqs. (4) and (7).

b. RESOLUTION OF THE FINE STRUCTURE. In order to obtain a good approximation of the scattering curve at large angles up to the last experimental point h_2 (see Fig. 1) it is necessary to adjust the spacing ΔR in real space according to the relation

$$h_2 \Delta R = \pi \tag{35}$$

In the case of the indirect transformation method ΔR corresponds to the distance of the knots of the Spline functions. In practice it is favorable to stay below this limit. Test calculations have shown that

$$h_2 \Delta R \leq 2,5$$

is a practicable limit.

It is difficult to define a lower limit for ΔR. Such a limit would depend on the largest scattering angle h_2, the smearing effect $[P(t), Q(x),$ and $W(\lambda')]$ and the statistical error. No general relation is known for this limit. However, a breakdown of the stabilization routine will indicate that the value of ΔR is too small.

B. Interpretation of Small-Angle Scattering Data

1. *Parameter Estimation.* Some structural parameters can be computed directly from small-angle scattering data, e.g., the radius of gyration R_g, the molecular weight M [using the scattering intensity at zero angle $I(0)$] and the hydrated volume. The maximum intraparticle distance can be obtained from the distance distribution function. These are integral parameters since a part of or the entire scattering curve is used for their evaluation (Damaschun and Pürschel[33] and P.K.T.). These parameters cannot, however, describe the detailed structure of the particle. As mentioned in the beginning of Section II,A. it is impossible to de-

[33] G. Damaschun and H.-V. Pürschel, *Acta Crystallogr., Sect. A* **27,** 193 (1971).
[34] G. Damaschun and H.-V. Pürschel, *Monatsh. Chem.* **102,** 1146 (1971).

termine uniquely a three-dimensional structure from a one-dimensional function.

The methods of interpretation can be classified in three groups:

a. Interpretation in reciprocal space (desmeared scattering functions)

b. Interpretation in experimental space (smeared scattering functions)

c. Interpretation in real space (distance distribution function)

Interpretation in reciprocal space is the most common method. The scattering curves have either to be desmeared or to be measured with point collimation. Several special plots are used to determine the various parameters such as

$\log I(h)$ versus h^2 (Guinier plot)
$I h^4$ versus h^4 (Porod)
$\log (I h)$ versus h^2

and

$\log (I h^2)$ versus h^2 (Guinier plot of the cross section factor and of the thickness factor, respectively)

The comparison of the desmeared experimental scattering curve to theoretical ones of simple triaxial bodies is frequently used to get a rough information on the overall shape. What may essentially be found in this way is a body equivalent in scattering and of uniform electron density. To get a better agreement with the experimental data it is sometimes useful to construct the models by a combination of various simple bodies. For further details see Mittelbach,[27] Kratky and Pilz,[35] and Pilz.[133]

A mathematical procedure for this curve fitting process has been given by Stuhrmann.[36] In this case the shape of the particle is approximated by a finite number of spherical harmonics.

Interpretation in the experimental space does not require any desmearing process. The parameters are calculated from the corresponding plots:

$\log I$ versus h^2

and

$\tilde{I} h^3$ versus h^3 (see P.K.T.)

[35] O. Kratky and I. Pilz, *Q. Rev. Biophys.* **5**, 481 (1972).
[36] H. Stuhrmann, *Acta Crystallogr., Sect. A* **26**, 297 (1970).

If several conditions are fulfilled it is possible to correct these parameters for the collimation effect.[34,37] The process of model fitting is the same as in reciprocal space. The scattering curves of the particular models have to be smeared according to the experimental profiles.[38]

2. *Interpretation of Real Space Information.* Permanent development of experimental equipment and improvement of computer programs make it possible to calculate the distance distribution function (real space) with sufficient accuracy. The interpretation of the functions in real space gives important additional information that can be obtained from small-angle scattering experiments. In this section a summary of the possible structural details that can be extracted from the distance distribution function is given.

a. DEFINITIONS. The distance probability function or "characteristic" function of a particle $\gamma(r)$ [Porod,[1] in the reference termed $H(x)$] can be explained as follows: starting from any point i in the particle, the function $\gamma(r)$ is the probability to find a point k in the distance r which is still inside the particle. This function is usually normalized to unity at the distance $r = 0$, i.e., $\gamma_0(r) = \gamma(r)/\gamma(0)$.

Mathematically the function $\gamma(r)$ is defined as the three-dimensional convolution square of the difference electron density ρ averaged over all directions in space.

The function

$$p(r) = \gamma(r) \, r^2 \tag{36}$$

(multiplied by the factor 4π) is the distance distribution of the particle[26] [in the reference called $\psi(x)$]. As already mentioned in the beginning of this article, this function multiplied by the factor 4π represents the number of lines with lengths r which are found in the combination of any volume element i with any other volume element k. Inhomogeneous particles may have regions with positive and negative difference electron densities. Distances between two regions and with different sign of the electron density give negative contributions to the $p(r)$ function.

A particular case is given by lamellar particles. The number of distances in a plane is equal to $2\pi r \, \gamma(r)$. In analogy to the distance distribution of the whole particle $p(r)$, see Eq. (36), we define the distance distribution of a plane with

$$f(r) = \gamma(r)r = p(r)/r \tag{37}$$

[37] V. Luzzati, *Acta Crystallogr.* **11**, 843 (1958).
[38] B. Sjöberg, *J. Appl. Crystallogr.* **11**, 73 (1978).

Another function in real space, the intersect function or distribution of chords $G(r)$[39-42] can be derived from the characteristic function $\gamma_0(r)$

$$G(r) = \text{const} \times \gamma_0''(r) \tag{38}$$

This function corresponds to the number of connection lines of lengths r between two points at the surface of a homogeneous particle. The condition of homogeneity restricts the range of application.

The characteristic function of a particle and its derivatives at $r = 0$ allow, in principle, the estimation of the surface and the mean radius of curvature of the surface[43] for homogeneous particles with smooth surface. The existence of sharp edges can be recognized if the scattering function has an exact Ih^{-4} tendency at large h values.[1,44]

It is very difficult to fulfill the necessary conditions with experimental data. The distance distributions $p(r)$ and $f(r)$ can be evaluated in most cases with sufficient accuracy. They are important supplementary means for the interpretation of experimental data.[45]

b. INTERPRETATION OF THE DISTANCE DISTRIBUTION FUNCTIONS $p(r)$ AND $f(r)$. The form of these functions enables to distinguish and to recognize directly and rationally the following types of particles:

Compact globular particles
Particles elongated in one direction, with constant cross sections (cylinders and prisms) and with variable cross section (prolate ellipsoids)
centrosymmetrical vesicles

The thickness of flat sheets and vesicles can be determined directly from the $f(r)$ function. This function also yields the thickness and the inner and outer diameter of centrosymmetrical particles.

Inhomogeneous particles with at least two regions of electron densities differing in their signs show typical features in the distance distribution. These characteristics can be found without further assumptions independent from the shape of the particle.

Interparticular scattering contributions (residual concentration effects) cannot be overlooked in real space as it may be the case in reciprocal space (Guinier plot). The formation of dimers can be analyzed without foregoing precise determination of the shape of the monomer.

[39] G. Porod, in "Small-Angle X-Ray Scattering" (H. Brumberger, ed.), pp. 1–15. Gordon and Breach, New York, 1967.
[40] P. W. Schmidt, *J. Math. Phys.* **8**, 475 (1967).
[41] J. Mering and D. Tchoubar, *J. Appl. Crystallogr.* **1**, 153 (1968).
[42] J. Mering and D. Tchoubar, *J. Appl. Crystallogr.* **2**, 128 (1969).
[43] R. Kirste and G. Porod, *Kolloid Z. & Z. Polym.* **184**, (1962).
[44] V. Luzzati, A. Tardieu, and L. Mateu, *J. Mol. Biol.* **101**, 115 (1976).
[45] O. Glatter, *J. Appl. Crystallogr.* (in press).

The distance distribution of a sphere is known analytically.[2]

$$p(r) = \text{const} \times x^2(2 - 3x + x^3) \qquad x = r/D \qquad (39)$$

D is the diameter of the sphere. The maximum is near $r = D/2$ ($x = 0.5$) (Fig. 6). Any deviation from spherical symmetry shifts the maximum to smaller r values if D is kept constant.

Particles elongated in one direction which have a constant cross section of arbitrary shape (long cylinders and prisms) show a linear decrease at large r values. A necessary condition is that the maximum dimension of

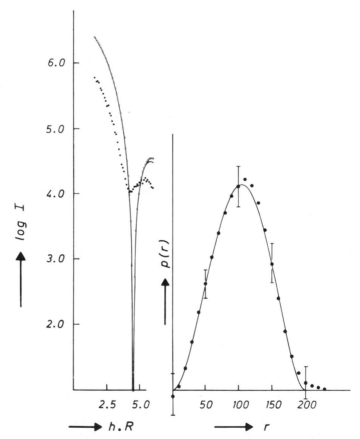

FIG. 6. Indirect transformation. Left: Solid line, theoretical scattering function of a sphere with radius $R = 100$. Closed circles simulated data, smeared according to a slit length effect, 5% statistics, $h_1R = 1.6$, $h_2R = 6.0$, $\Delta hR = 0.1$. Open circles, desmeared scattering function. Right: solid line, theoretical $p(r)$ function. Closed circles, distance distribution with propagated error (error bar) computed from the simulated data using $D_{\max} = 250$. $\Delta R_g = 1.5\%$.

the cross section d be much smaller than the maximum dimension of the particle.

$$D/d \gtrsim 2.5 \tag{40}$$

The slope of this linear region is given by

$$\tan \alpha = - \frac{dp}{dr} = \frac{F^2 \bar{\rho}_c^2}{2\pi} \tag{41}$$

F is the area of the cross section and

$$\rho_c = \frac{1}{F} \int_F \rho(\mathbf{x}) \, df \tag{42}$$

[$\rho(\mathbf{x})$ is the electron density in the cross-section]. Differences in the area of the cross-section result in pronounced differences in the slope than α according to Eq. (41). The $p(r)$ functions of prisms with the same cross section and different lengths are shown in Fig. 7. They show a maximum within the dimensions of the cross section and continue with the linear part at $r > d$.

The $f(r)$ functions of lamellar particles start at $r = 0$ with a linear increase. At $r = d$ (d is the thickness of the particle) run into slightly decreasing, almost linear functions. The form of this part is determined by the shape of the basal plane. The thickness d can be inferred from the

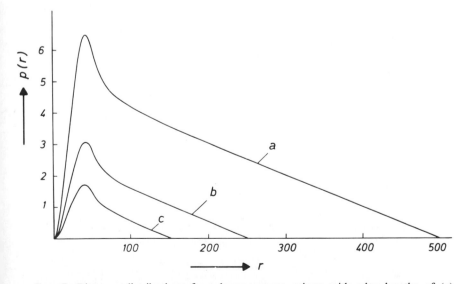

FIG. 7. Distance distributions from homogeneous prisms with edge lengths of (a) 50:50:500; (b) 50:50:250; (c) 50:50:150.

transition point at $r = d$. The limiting value A of the function $f(r)$ resulting from the extrapolation of the quasilinear part following the transition point toward $r = 0$ contains information on the area of the basal plane of the lamella, according to

$$A = f(r) \bigg|_{r \to 0} = \frac{\bar{\rho}_d^{-2} F d^2}{2} = \frac{\bar{\rho}_d^{-2} V d}{2} \tag{43}$$

where

$$\bar{\rho}_d = \frac{1}{d} \int_d \rho(x) \, dx \tag{44}$$

The extrapolation to $r = 0$ is the more accurate the larger the ratio of D/d (see Fig. 8).

In practice the shape of the $f(r)$ function will allow the recognition of lamellar particles and the determination of their thickness. Figure 9 gives a comparison between a sphere, a prolate ellipsoid ($1:1:3$), and an oblate ellipsoid ($1:1:0.2$) having the same number of excess electrons and radius of gyration. The ellipsoids have a larger D than the sphere, and the maximum of the $p(r)$ function is at smaller r values. These properties are

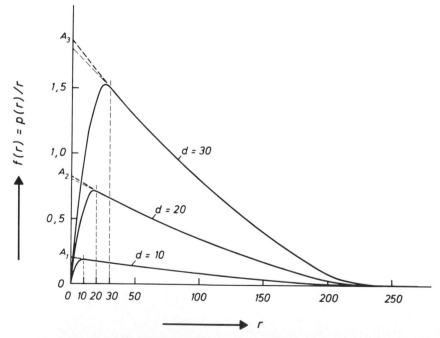

FIG. 8. $f(r)$ functions of lamellar particles with the same plane (100×100) and varying thickness d (10, 20, 30). The transition points are signalized by the vertical dashed lines.

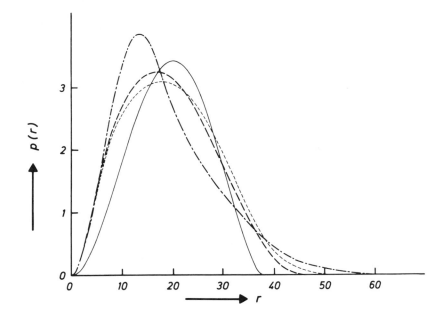

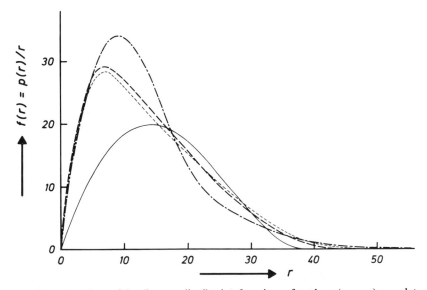

FIG. 9. Comparison of the distance distribution functions of a sphere (———), a prolate ellipsoid of revolution 1:1:3 (—·—·—), an oblate ellipsoid of revolution 1:1:0.2 (- - - -) and a flat prism 1:1:0.23 (- - - -) with the same radius of gyration. Upper: $p(r)$ function; lower: $f(r)$ function.

most significant for the prolate ellipsoid. The varying cross section causes a nonlinear descent as compared to $r = D$. The difference from a particle with constant cross section is obvious. The oblate ellipsoid can be recognized as a flat particle by means of the function $f(r)$.

The limits of interpretation of small-angle scattering results are demonstrated by the distance distribution of a similar flat prism $(1:1:0.23)$. The differences between the prism and the oblate ellipsoid are so insignificant that they may be within experimental errors.

The hollow sphere will be discussed here as the hollow particle of greatest practical importance. The transition from the thin-walled sphere to a full sphere is shown in Fig. 10a and b. The thin spherical shell is a spe-

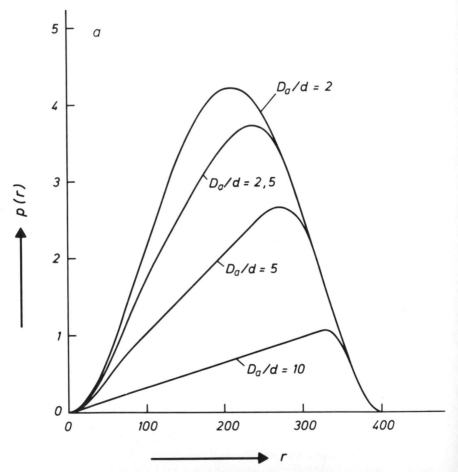

FIG. 10. (a) Hollow spheres, $p(r)$ functions. D_a is the outer diameter, and d thickness of the shell. $D_a/d = 2$ represents a full sphere, $D_a = 400$.

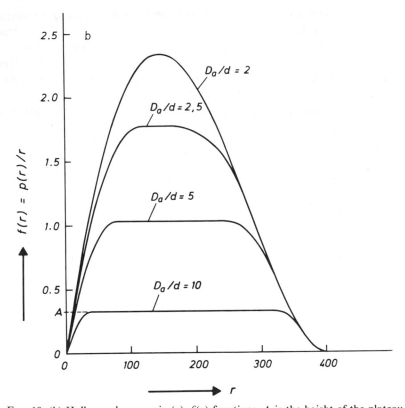

FIG. 10. (b) Hollow spheres as in (a), $f(r)$ functions. A is the height of the plateau.

cial case of a flat particle. The $f(r)$ function shown in Fig. 10b has a transition point at $r = d$ (d is the thickness of the shell). The linear region is horizontal because of the special geometrical shape of this body. This region ends at the inner diameter $r = D_i = D - 2d$. The height of the plateau A is in analogy to Eq. (43) given by

$$A = \bar{\rho}_s^2 2\pi R_m^2 d^2 \simeq \frac{\bar{\rho}_s^2 V d}{2} \tag{45}$$

R_m is the mean radius of the spherical shell and

$$\bar{\rho}_s = \frac{1}{d} \int_{(D/2-d)}^{D/2} \rho(r) \, dr \tag{46}$$

Scattering curves from inhomogeneous particles contain two types of information: the shape and the internal structure of the particle. As an example, the $p(r)$ function of an elongated cylinder with constant electron density along the cylinder axis is compared to one with varying centrosymmetric electron density $\rho(x)$, where x is the distance from the cylinder axis, as shown in Fig. 11. The $p(r)$ function differ within the dimensions of the cross section, there is still a linear descent at large r values, however, its slope varies according to $\bar{\rho}_c$. The comparison to the homogeneous cylinder shows that the maximum of the $p(r)$ function of the hollow cylinder is located at larger distances since the small distances from the centre are missing. The $p(r)$ function of the inhomogeneous cylinder shows a minimum which is negative in this particular example.

Lamellar particles with varying electron density perpendicular to the basal plane, i.e., $\rho = \rho(x)$, where x is the distance from the central plane, show similar properties. The $p(r)$ function deviates from the $p(r)$ function of the homogeneous lamella between $0 \leqq r \leqq d$ and can be negative in this region. The limiting value A and the height of the $f(r)$ function for $r > d$ varies according to ρ_d but the shape does not vary in this range. The $f(r)$ function of inhomogeneous centrosymmetrical thin hollow spheres

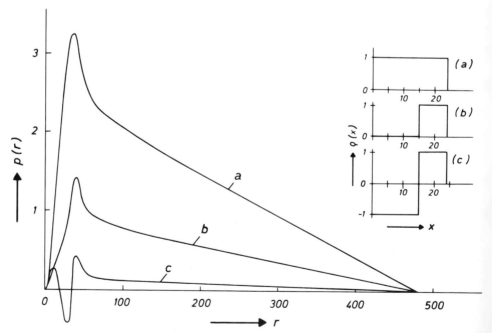

FIG. 11. Circular cylinder with a constant length $L = 480$ and an outer diameter $d = 48$. (a) Homogeneous cylinder, (b) hollow cylinder, (c) inhomogeneous cylinder. The $p(r)$ functions are shown on the left side, the corresponding electron density distributions $p(x)$ on the right side.

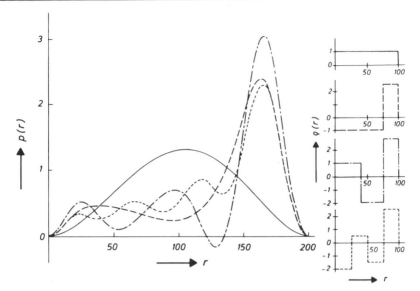

FIG. 12. Spherical multilayer models with constant outer diameter $D = 200$. $P(r)$ functions in the left part, electron density profiles in the right part of the figure.

shows deviations from the homogeneous particles within the two regions $0 \leqq r \leqq d$ and $D_i \leqq r \leqq D$.[46]

The distance distribution functions of spherical multilayer models with alternating signs of the electron density are illustrated in Fig. 12. A direct quantitative analysis of the electron density distribution $\rho(r)$ from the $p(r)$ function is not possible. Under the condition of compact structures the occurrence of minima in the $p(r)$ function signalizes the existence of regions of electron density with alternating sign. The number of such regions is equal to the number of maxima in the $p(r)$ function provided that the particles are spherical (sharp minima in the scattering curve) and that the shells have approximately identical dimensions (thickness) (see Fig. 12). The inner diameter of the outermost shell can be estimated from the point of inflection between the last extrema. The evaluation of the radial electron density can only be performed by Fourier transformation of the scattering amplitude.

The formation of dimers can be analyzed qualitatively by use of the $p(r)$ function. An illustrative example is given in Fig. 13a–d. A prolate ellipsoid with an axial ratio $1:1:2$ is taken as a monomer. The $p(r)$ function of the monomer indicates an elongated particle with decreasing cross section toward the ends. The difference distance distribution of the lines con-

[46] P. Laggner, A. M. Gotto, and J. D. Morrisett, *Biochemistry* **18**, 164–171 (1979).

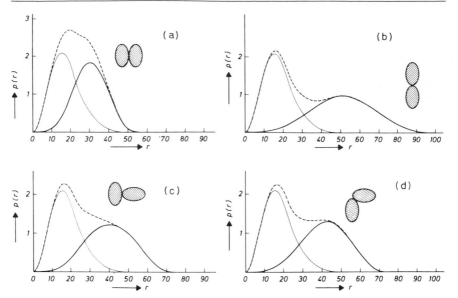

Fig. 13. Distance distribution function $p(r)$ from dimer models built from prolate ellipsoids. Solid line, monomers; dashed line, dimers; heavy solid line, difference between dimers and monomers. (a) Parallel arrangement, (b) linear arrangement, (c) T type, (d) L type.

necting the two subunits in the dimer results from the subtraction of the distance distribution of the two monomers from the distance distribution of the dimer. The difference between the parallel (Fig. 13a) and the linear arrangement (Fig. 13b) is obvious. Nearly the whole difference distribution of the parallel arrangement lies within the distance distribution of the monomer, whereas the linear arrangement contributes up to twice the maximum dimension of the monomer.

The two rectangular configurations of T type (Fig. 13c) and L type (Fig. 13d) lie between the linear and the parallel arrangement. For such a rough analysis it is not necessary to have an exact shape analysis of the monomer.

The interparticle interferences which are not negligible at high concentrations lead to a decrease of the scattering function at small angles. A rough but sufficient approximation is the hard sphere model.[3] The scattering intensity I_{hs} of such a model is given by

$$I_{hs}(h) \simeq \text{const} \times \phi^2(hR) \frac{1}{1 + (8v_0/v_1)\phi(2hR)} \qquad (47)$$

v_0/v_1 is the packing parameter (equal to 0.74 for closely packed hexagonal or cubic systems). R is the radius of the sphere and $\phi(hR)$ is the scattering

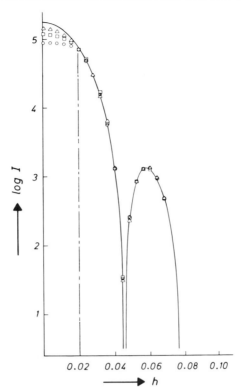

FIG. 14. Interference effect, hard sphere model, radius $R = 100$. Scattering functions for different volume concentrations $c_v = v_0/v_1$. Solid curve, $c_v = 0$; △, $c_v = 1/32$; □, $c_v = 1/16$; ○, $c_v = 1/8$. Dashed vertical line: termination for indirect transformation, first data point at $h R = 2.0$ (see Fig. 17).

amplitude of a sphere with radius R. The approximation for small h values is

$$I_{hs} \simeq \phi^2(hR) \frac{1}{1 + 8c_v} \tag{48}$$

and for low value concentrations $c_v = v_0/v_1$

$$I_{hs} \simeq \phi^2(hR)(1 - 8c_v) \tag{49}$$

A linear extrapolation to zero concentration (see Section IV) is based upon Eq. (49) which is an approximation for Eq. (48).

The scattering curves for several values of $c_v = v_0/v_1$ are shown in Fig. 14. The apparent radius of gyration decreases with increasing concentration (Fig. 15). The length of the linear range increases for particles whose scattering curves deviate from the linear part in upward direction.

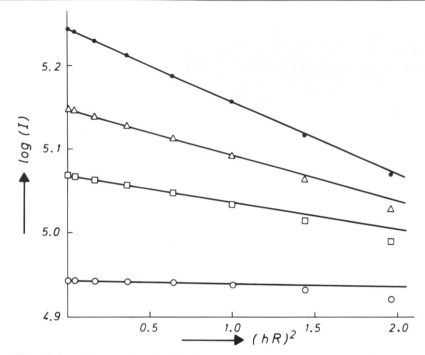

FIG. 15. Interference effect, hard sphere model as Fig. 14, Guinier plot: ●, $c_v = 0$; △, $c_v = 1/32$; □, $c_v = 1/16$; ○, $c_v = 1/8$.

An experimental example is the concentration dependence of the scattering curves (Guinier plot) from hemocyanin of *Astacus leptodactylus*[47] (Fig. 16). The length of the linear range decreases for curves deviating from the linear part downward (see Fig. 15). There exists no method to detect the existence of interparticle interferences directly from the scattering curve or from the Guinier plot. It should be noted that minima at zero angle may also arise from inhomogeneous particles.

The distance distribution is affected considerably by interparticle interferences. It is lowered with increasing distance r, goes to a negative minimum in the region of the maximum distance D of the particle and the oscillations vanish at larger r values. This is shown for the hard sphere model in Fig. 17. The same behavior can be found for the experimental example in Fig. 16. The corresponding $p(r)$ functions are shown in Fig. 18. A similar example for the $\gamma(r)$ function has been given by Damaschun.[48] The essential fact is the negative part in the range of the maximum dimen-

[47] I. Pilz, K. Goral, R. Lontie, and R. Witters, in preparation.
[48] G. Damaschun, H. Damaschun, H.-V. Pürschel, and K. Ruckpaul, *Abh. Akad. Wiss. DDR* p. 289 (1973).

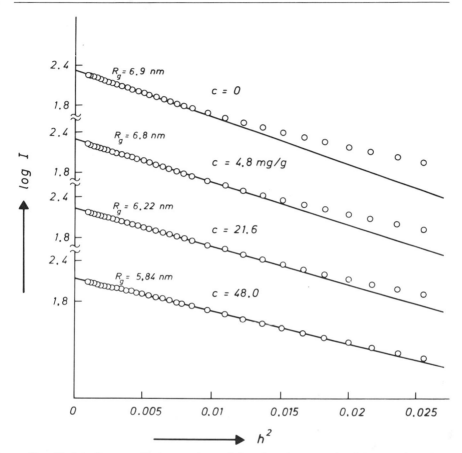

FIG. 16. Interference effect, experimental data from hemocyanin of *Astacus leptodactylus*.

sion *D*. The $p(r)$ function ends with a negative part only for particles having regions with difference electron density of different signs at the distal ends. Such a configuration seems to be a theoretical one in the field of enzymes. Remaining concentration effects can therefore easily be recognized from the distance distribution function.

Sometimes it is impossible to measure a series of concentrations, for example, if the structure of the particle depends on the concentration. In such cases an attempt can be made to reduce the contribution of the interparticular interferences. The main influence from this contribution is at small angles ($h \rightarrow 0$) as can be seen from Fig. 14. If the innermost part of the scattering curve is not used for the evaluation of the experimental data, it may be possible to gain a sufficient reduction of the

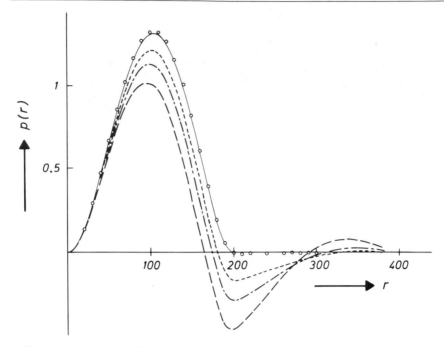

FIG. 17. Interference effect, hard sphere model as Fig. 14, $p(r)$ functions: Theoretical $p(r)$ functions: Solid line, $c_v = 0$; dashed line, $c_v = 1/32$, dot-dashed line, $c_v = 1/16$; long-dashed line, $c_v = 1/8$. Open circles, results from indirect transformation: $c_v = 1/16$, $h_1 R = 2.0$, 2% statistics, $D_{max} = 300$, $\Delta R_g = 0.5\%$, $\Delta I_0 = 1.2\%$.

influence of the concentration effect. On the other hand, the amount of information is lowered and the termination effect may be serious. As shown in Fig. 17 it is possible to obtain useful results with the indirect transformation method. However, if the concentrations are high ($c_v > 0.06$) the interparticle interferences cannot be treated as perturbations that can be eliminated.

3. *Contrast Variation.* The method of contrast variation[49,50] is a valuable tool to increase the information that can be obtained from inhomogeneous particles. This method uses solvents of different electron densities. The difference electron density of the particle $\rho(\mathbf{r})$ is the actual difference between the electron density of the particle and the electron density of the solvent. A basic assumption in this method is that within the volume V of the particle the electron density remains unchanged while the surrounding solvent has a variable homogeneous electron density

[49] H. Stuhrmann, Z. Phys. Chem. (*Frankfurt am Main*) [N. S.] **72,** 177 (1970).
[50] H. Stuhrmann, J. Appl. Crystallogr. **7,** 173 (1974).

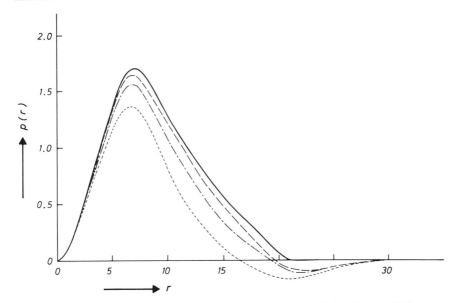

FIG. 18. Interference effect, experimental data as Fig. 16, $p(r)$ functions: solid line, concentration c extrapolated to zero, long-dashed line, $c = 4.8$ mg/g; dot-dashed line, $c = 21.6$ mg/g; dashed line, $c = 48.0$ mg/g.

ρ_L. Under this assumption it is possible to define the difference electron density as the sum of two terms:

$$\rho(\mathbf{r}) = \bar{\rho}\rho_F(\mathbf{r}) + \rho_S(\mathbf{r}) \tag{50}$$

where $\rho_F(r) = 1$ in the particle volume V, $\rho_S(r)$ is the deviation from the mean difference electron density $\bar{\rho}$. It further holds that

$$\int_V \rho_S(\mathbf{r})d\mathbf{r} = 0 \tag{51}$$

$\dot{\rho}$ is the difference between the mean electron density of the particle and the electron density of the solvent ρ_L. The scattering amplitude of the particle is given by

$$A(\mathbf{h}) = \rho A_F(\mathbf{h}) + A_S(\mathbf{h}) \tag{52}$$

and the scattering intensity from diffuse scattering is

$$I(h) = \bar{\rho}^2 I_F(h) + \bar{\rho}I_{FS}(h) + I_S(h) \tag{53}$$

$I_F(h)$ is the scattering function of $\rho_F(r)$ which could be observed if $\bar{\rho}$ approaches infinity. This is the "shape factor" of the volume V and can be determined by extrapolation. $I_S(h)$ originates from $\rho_S(\mathbf{r})$ and is measurable if $\bar{\rho} = 0$. $I_S(0)$ is zero according to Eq. (51). The cross term $I_{FS}(h)$ establishes the correlation between $I_F(h)$ and $I_S(h)$ by the following general inequality

$$I_{FS}(h) \leqq \tfrac{1}{2}\sqrt{[I_F(h)I_S(h)]^{1/2}} \tag{54}$$

Introduction of Eq. (50) into the definition of the radius of gyration R_g yields

$$R_g{}^2 = R_F{}^2 + \alpha/\bar{\rho} - \beta/\rho^2 \tag{55}$$

where

$$R_F{}^2 = (1/V) \int \rho_F(\mathbf{r})r^2 \; d\mathbf{r} \tag{56a}$$

$$\alpha = (1/V) \int \rho_S(\mathbf{r})r^2 \; d\mathbf{r} \tag{56b}$$

$$\beta = (1/V^2) \int\!\!\int \rho_S(\mathbf{r})\rho_S(\mathbf{r}')\mathbf{r}\mathbf{r}' \; d\mathbf{r} \; d\mathbf{r}' \tag{56c}$$

R_F represents the radius of gyration of a homogeneous particle and does not depend on $\bar{\rho}$. α is the second moment of the electron density fluctuations. If the sign of α is positive the positive contributions dominate. β is an asymmetry parameter according to Eq. (56c).

For the determination of α and β it is convenient to plot $R_g{}^2$ versus $1/\bar{\rho}$. The result is a straight line if β is equal to zero. The slope of this line is α and the value of $R_g{}^2$ at $(\bar{\rho})^{-1} = 0$ is $R_F{}^2$. It should be mentioned that the method of contrast variation can be applied with best success for neutron small angle scattering. There, the electron density has to be replaced by the scattering length density for neutrons. This scattering length density differs strongly between H and ^{2}H and therefore the contrast can be varied within a large range by changing the $H_2O:D_2O$ composition in the solvent.

The contrast variation for x-ray small-angle scattering is performed by variations in the concentration of sugar or other low molecular weight solutes. Difficulties may arise from preferential interactions between solvent components and the macromolecule under investigation.

III. Instrumentation and Experimental Methods

A. X-Ray Room

The x-ray room has to meet a number of requirements.

1. Provision for complete darkening, since adjustment of the small angle camera requires maximum adaption of the eye.

2. No strong vibrations, caused for example by traffic, since vibrations may affect the alignment of the camera. If strong vibrations cannot be avoided, one should use a vibration-free base for the generator (commercially available).

3. Air conditioning: constant temperature to $\pm 1°$. Larger fluctuations cause intensity instabilities. The relative humidity should neither be extremely high nor extremely low; stabilization to $\pm 5\%$ is sufficient. Exceedingly low humidity should be avoided because of electrostatic charging, very high humidity may lead to flashovers.

4. Sufficient water supply: direct use of tap water is permissible, as long as its temperature during the whole year (apart from a few days during spring and autumn) fluctuates less than $\pm 1°$. Larger temperature fluctuations affect the intensity too much. A closed cooling water cycle is independent of the public water supply, maintains constant water temperature, and avoids calcification of the tubings and, particularly important, of the anode. The latter effect reduces thermal conduction and decreases the cooling effect. Moreover, a closed water circuit avoids interruptions through dirt which may get into the tap water system. Obviously, one *has* to use a closed cooling water system if the public water supply cannot provide enough water; a 30 kW Rigaku-Denki X-ray generator, for example, needs 25 liters of water per minute. If the public water system is used at the limit of its capacity, pressure fluctuations can create fluctuations in cooling water velocity. This causes fluctuations in the cooling effect and affects the x-ray intensity.

B. X-Ray Tube

1. Choice of Anode Material. For the vast majority of applications, the important question concerning the anode material has been decided in favor of copper, whose K_α wavelength is 0.154 nm. The obvious idea to increase scattering angle (and hence the resolution) by the use of a longer wavelength radiation is not practicable because of the sharp increase of absorption with only moderate increases in wavelength: chromium radiation [λ (CrK$_\alpha$) = 0.229 nm], which brings only a 50% increase in resolution, permits an optimal sample thickness[51] $d_{opt} = 1/\mu$ (μ is the absorption coefficient) which is smaller by a factor of about 3.2 as compared to copper. The recording time (for the same statistical error) increases by the same factor. In addition to that 1, absorption by the windows of x-ray tube and camera and by the tungsten deposits at the surface of the anode (particularly important for older tubes) is much stronger. An even further increase in wavelength (beyond Cr radiation) is thus out of the question,

---

[51] The intensity scattered to small angles is obviously proportional to $de^{\mu d}$. This expression has its maximum at $d = 1/\mu$, where d is the thickness of the sample.

and we know of no serious attempt to use such long-wavelength radiation for the study of biological macromolecules. Indeed, there is little point to put up with all these difficulties, as the resolution attainable with copper radiation is sufficient for practically all applications (see Section,F,1). On the contrary: it is sometimes worth considering the use of shorter wavelength radiation, for example molybdenum radiation [λ (MoK$_\alpha$) = 0.071 nm]. The smaller absorption (by a factor of about 8) allows much shorter exposures. Moreover, a given portion of the scattering curve is compressed into half the angular range when recorded with molybdenum radiation as compared to copper. This can be beneficial when the maximum angular range of a given goniometer is too small to record enough of the scattering curve with copper radiation. In other words, use of a shorter wavelength can be an important tool for the recording of the tail end of the scattering curve up to relatively large angles. However, to record the innermost portion of the scattering curve with Mo radiation would require correspondingly finer entrance and counter slits with considerable loss in intensity; this would compensate or overcompensate the gain in intensity due to the smaller absorption. Moreover, one has to align the camera twice as accurately, and, so far, there is little experience with parasitic scattering of Mo radiation at smallest angles. It is hence generally not advisable to use a wavelength shorter than the CuKα line for the innermost portion of the scattering curve.[52]

2. *Installation of X-Ray Tube and Camera.* We assume that an x-ray tube with line focus and four windows is used. First, one has to decide whether the tube should be mounted vertically or horizontally. Better mechanical stability and access to all four windows calls for the vertical position, but several types of small-angle camera (e.g., the Rigaku Denki goniometer) require a horizontal tube. In the following discussion we assume that the x-ray tube is mounted vertically, as this type of installation is preferred by most workers.

Next, one must decide which of the two types of windows one should use for the camera. Let us assume the size of the focus is 10 × 1 mm and the x-ray path has the usual 6° inclination against the horizontal plane of the focus. Then the projection of the focus into the plane perpendicular to this x-ray path is a line of dimensions 10 × 0.1 mm on the "long" side of the focus. On its "narrow" side, the focus appears as an illuminated square of dimensions 1 × 1 mm. We call these two alternative positions of the camera "line focus" and "square focus," respectively. A great variety of x-ray tubes is commercially available, ranging in focal size from

---

[52] It is a well known fact, however of little interest in the present context, that strongly absorbing inorganic samples or metals require a shorter wavelength than CuKα.

0.4 to 2 mm in width and from 8 to 12 mm in length. This leads to a large manifold of available focal dimensions, as projected into the plane perpendicular to the x-ray path. For our discussion of the most suitable focal size, we assume that a slit camera is used, i.e., a camera whose primary beam has the shape of a narrow band. We call the large dimension of this band the slit length, the narrow one the slit width. It should be noted that, in case the x-ray tube is mounted horizontally, the larger dimension of the focus runs vertically, and the large dimension of the line shaped primary beam is correspondingly often called slit *height* (instead of slit *length* as it will be called in the following). In both cases (i.e., vertical and horizontal x-ray tube), the smaller dimension of the primary beam is referred to as "slit width".

Figure 19a shows a section parallel to the primary beam and perpendicular to the slit length for the collimation system used in our laboratory. *E*, *M*, and *B* have to be imagined as blocks perpendicular to the plane of the paper. The efficiency of this type of collimation in eliminating parasitic scattering is due to the fact that the extension of the upper plane of *M* coincides precisely with the lower one of *B*; the vertical dimensions of the drawing are vastly enlarged. The figure also shows the projection of the focus into the plane perpendicular to the primary beam, drawn to scale.

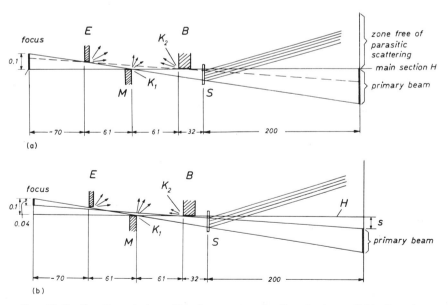

FIG. 19. Section through the collimation system; the direction is parallel to the primary beam and perpendicular to the slit length. Explanation in the text. (a) Completely "illuminated" extrance slit. (b) Primary beam centers "off-center."

We can see that the range of the focal width which can radiate into the collimation system is about twice as wide as the entrance slit E. By far the most applications need an entrance slit between 60 and 100 μm; for lower resolutions one occasionally goes up to 200 μm, only for rare cases requiring ultrahigh resolution one may use entrance slits as narrow as 20 μm.

The projection of a 1 to 2 mm wide focus into the plane perpendicular to the primary beam is, for an inclination of 6°, 100 to 200 μm wide on the "line focus side." Thus, entrance slits of 50 to 100 μm can be fully "illuminated," which illustrates that it is generally favorable to place the camera in front of the line focus, with the added advantage that one can make use of most of the focal length. If one aims an ultrahigh resolution and uses smaller entrance slits, for example, 20 μm, one should use a fine-focus tube with only 0.4 mm focal width corresponding to 40 μm in the plane perpendicular to the primary beam. This type of tube works on a four times higher load per unit area, i.e., produces a brightness which is four times higher than that of the tube with 2mm focal width.

Even for investigations requiring only moderately high resolution, e.g., $E = 50$ μm, it can be advantageous to use a fine focus tube with a 0.4 mm wide focus, although full illumination of the entrance slit would require 1 mm focal width (this means 100 μm in the projections into the plane perpendicular to the primary beam). In that case, the camera should be adjusted as indicated in Fig. 19b, with the primary beam entering "off-center": although no intensity measurements can be carried out in the range s, which is full of parasitic scattering originating from the edges the higher brightness of the fine focus. Thus, given a required resolution, one should always adjust focal width and entrance slit to obtain optimal conditions. Very often, this adjustment cannot be perfect as this may require frequent exchanges of the x-ray tube (with the necessary time-consuming readjustment of the whole camera). However, one should consider very carefully which x-ray tube gives the best overall performance for a given series of measurements. The large difference in brightness between, say, a 0.4 and a 2 mm wide focus makes such considerations very worthwhile.

The "usable" intensity at the square focus is considerably lower than at the line focus: under conditions as discussed above, say, a 1 mm wide projection of the focus can radiate into the collimation system with only 1/10 to 1/5 of its width. The unnecessary large width of the focus is accompanied by its short length (typically about 1 mm), which reduces the available intensity.

It is a matter of fact, on the other hand, that the square focus is a more stable x-ray source than the line focus. This is mainly due to two reasons.

First, the emitted intensity fluctuates within the focus. One easily understands that fluctuations within small regions affect the intensity distribution of the emitted radiation much less for the square focus than for the line focus. Second, intensity fluctuations are frequently caused by thermal and mechanical effects which cause a movement of the focus relative to the entrance slit. Again, it is easy to see that the line focus is more sensitive to such perturbations than the square focus. In cases where one is permitted to use a large entrance slit, i.e., whenever no high resolution it required, one will consider using the square focus and benefit from its superior stability. Thus, a 200 μm wide entrance slit makes use of nearly 50% of the radiation emitted by a 1 mm wide focus.

3. X-Ray Tubes with Rotating Anode. Several companies offer high-power generators with x-ray tubes, which could not be cooled in the conventional way without local overheating at the focus due to the high electron flux. To circumvent this difficulty, the focus is produced at the surface of a water-cooled cylinder, which rotates at high angular velocity to dissipate the heat over the whole of its circumference. More information on the various commercially available models can be obtained from the suppliers.

It is clear that these x-ray sources have a water consumption proportional to their power. This problem was already discussed in Section III,A,4. The power of commercially available instruments ranges from 6 to 60 kW, which is about 3 to 30 times larger than for sealed tubes. The high primary intensity allows correspondingly shorter exposures, which is particularly useful for the study of unstable samples. Moreover, many investigations involving low electron density contrast or high dilutions (for example, if not enough material is at one's disposal) demand high power sources. As with conventional tubes, one can use the square focus or the line focus when working on a rotating anode tube.

Particularly for installations with 30 kW and above, one has to pay for the high primary intensity with several inconveniences: the necessity for periodic services and for periodic replacements of the filament (roughly every 1000 hr of operation); the latter can usually be done without the help of a service representative from the supplier. The stability of the x-ray intensity from rotating anode sources is often comparable to conventional sources.

High power x-ray generators run with the usual voltages (40 to 50 kV) and do not pose excessive radiation protection problems, because their radiation is not more penetrable than the one of conventional sources. Care has to be recommended for adjustments with open shutter, which should only be performed with reduced power. Otherwise, a 2 mm lead protection is always sufficient. The high primary intensity requires great care in design and manufacture of this shield to avoid small leaks which might still be tolerated on conventional sources.

C. Monochromatization

Quantitatively correct interpretation of diffuse small-angle x-ray experiments requires the knowledge of the scattering curve corresponding to *monochromatic* radiation. Polychromatic effects have to be eliminated, either experimentally or computationally. The following methods are in use.

1. Pulse height discriminator combined with a K_β filter.
2. Pulse height discriminator combined with numerical elimination of the K_β contribution
3. Balanced filters
4. Crystal monochromator, either in the incident or in the diffracted beam
5. Totally reflecting mirror, usually consisting of a carefully polished glass plate

Each method is briefly discussed below.

1. Pulse Height Discriminator Combined with a K_β Filter. So far, this seems to be the most widely used method. The pulse height discriminator, which is tuned to the K_α line, is connected to a proportional or scintillation counter. The efficiency in suppressing white radiation is inversely proportional to the channel width. λ (K_β), however, is too close in its wavelength to K_α to be sufficiently attenuated by pulse height discriminator alone (for example, λ (CuK$_\alpha$) = 0.154 nm; λ (CuK$_\beta$) = 0.139 nm). This necessitates the use of a K_β filter. The following discussion will be limited to nickel filters, since so far practically all small-angle investigations on biological systems use copper radiation. The wavelength of the absorption edge of nickel lies between the CuK$_\alpha$ and the CuK$_\beta$ line; thus, it absorbs K_β much stronger than K_α. A 10 μm thick nickel foil is usually quite suitable; it attenuates K_α by a factor 0.6658, and K_β by 0.0865. The intensity ratio γ between K_β and K_α lies (if taken into account that the absorption of the sample enriches the K_β line) roughly between 0.2 and 0.4 for new and old x-ray tubes, respectively. Old tubes suffer from deposits of tungsten at the anode, which absorbs K_α more strongly than K_β. The intensity ratio γ' *behind* the 10μm nickel foil for the two limiting cases (new and old x-ray tube) is, therefore, roughly

$$\gamma' \text{ (new tube)} = \frac{0.2 \times 0.0865}{0.6658} = 0.026$$

and

$$\gamma' \text{ (old tube)} = \frac{0.4 \times 0.0865}{0.6658} = 0.052$$

The influence of the remaining K_β radiation can usually be neglected.

We shall demonstrate below (see Section III,C,4 below and Fig. 22) that 50% channel width in combination with a 10 μm nickel foil indeed has an excellent monochromatization effect, which is sufficient for most applications. It might be noted that a possible source of errors is a shift in channel position which affects intensity and monochromatization. Frequent redetermination of the correct channel position is therefore advisable.

2. *Pulse Height Discriminator Combined with Numerical Elimination of the K_β Contribution.* Zipper[53] has shown how the influence of the β line can be eliminated numerically. This method avoids intensity losses through absorption by the β filter, which amounts to as much as one-third of the CuK_α intensity, as shown in the previous section. The method requires knowledge of the intensity ratio γ between K_β and K_α. Zipper's method involves determination of γ through several intensity measurements. Obviously, the same information can also be obtained with an x-ray spectrometer. It should be noted that γ has to be redetermined for each new sample, because absorption effects have a strong influence on γ. Moreover, γ increases during the lifetime of the x-ray tube, which necessitates periodic redeterminations.

If γ is known, the monochromatization can be performed together with the desmearing procedure. The relevant desmearing programs involving either the iterative method[18] or the indirect transformation method,[20,21] are described in Section II,A.

3. *Balanced Filters.* This monochromatization technique by Ross,[54] involves two exposures with different filters in the primary beam, under otherwise identical conditions. For copper radiation, the two filters consist of nickel and cobalt, respectively. The wavelength of the absorption edge of cobalt ($\lambda = 0.1604$ nm) is slightly above the CuK_α wavelength, the one of nickel ($\lambda = 0.1483$ nm) just below. Let T_{Ni} be the thickness of the nickel filter and T_{Co} the one of the cobalt filter; we call the two filters "balanced," if T_{Ni} and T_{Co} have a certain optimal ratio, namely, $T_{Ni}/T_{Co} = 1/1.0711$. Under this condition, substraction of the two scattering curves leaves only the contribution from radiation whose wavelength lies between the two absorption edges. Contributions from radiation with $\lambda < 0.1483$ nm and with $\lambda > 0.1604$ nm cancel. Since the total intensity of continuous radiation in the narrow range between these two

[53] P. Zipper, *Acta Phys. Austriaca* **30**, 143 (1969).
[54] P. A. Ross, *J. Opt. Soc. Am.* **16**, 433 (1928).

wavelengths is negligible compared to the CuK_α intensity, the difference curve can be regarded as originating from pure K_α radiation.

Monochromatization is guaranteed as long as the two filters have the correct thickness ratio. As one can easily convince oneself, however, the total intensity after subtraction depends on the absolute thickness of the two filters and passes through a maximum when plotted against the thickness. One can show that this maximum is at $T_{Ni} = 6.99$ μm and $T_{Co} = 7.48$ μm. For these values, the K_α intensity in the difference spectrum is 65.2% of the K_α intensity in the unfiltered primary beam. Fortunately, the difference intensity does not depend very sharply on the absolute values of the thickness of the filters, i.e., the maximum is rather flat. This is illustrated in Fig. 20[55]: if both filters have twice their optimal thickness, the difference intensity drops by only 16% from its highest value at the maximum (i.e., to 54.6% of the unattenuated primary intensity). If the thickness *ratio* deviates from its correct value, however, a finite difference intensity remains over the whole spectral range.

However appealing this technique may be conceptually, it is used relatively little. The reason lies mainly in experimental difficulties: two measurements must be performed under strictly identical conditions. In particular, the time integral over the primary intensity (i.e., the total energy) has to be identical for both measurements. Differences in total energy between the two exposures cause a corresponding fraction of white radiation to remain in the difference curve. Moreover, subtraction of the two curves increases the statistical error, which must be compensated by longer exposures. Balanced filters are useful whenever one does not strive for the highest precision. They do have the advantage that no pulse height discriminator is needed. Methods for the preparation of suitable metal foils for copper radiation have been described.[56]

For molybdenum radiation, one needs a pair consisting of one Zr and one Y (or Sr) filter. Practicable methods for the preparation of filters of correct thickness have been described for Zr[56] and Sr.[57]

4. Crystal Monochromator in the Incident or Diffracted Beam. Several types of monochromators were exhaustively described by P.K.T.: the flat quartz monochromator, the bent quartz plate (Johann), and the Johansson monochromator, consisting of a bent quartz plate whose surface is ground to twice the bending curvature. We have to supplement P.K.T.'s discussion by a description of the graphite monochromator.

The biggest disadvantage accompanying the use of quartz monochromators is the considerable loss in intensity upon reflection. Sparks[58] was

[55] Unpublished calculation by O. von Sacken.
[56] C. G. Phelps and B. G. Craig, *Rev. Sci. Instrum.* **34,** 812 (1963).
[57] G. Becherer, G. Herms, and V. Motzfeld, *J. Sci. Instrum.* **42,** 754 (1965).
[58] C. J. Sparks, *Met. Ceram. Div., Annu. Prog. Rep.* **RNL-3970,** 57 (1966).

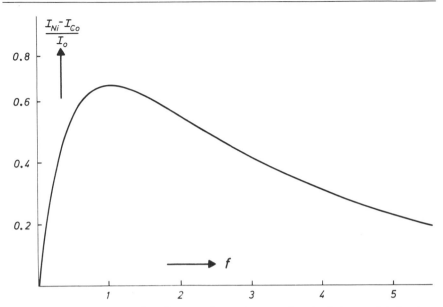

FIG. 20. Intensity of the virtually monochromatic spectral region between nickel and cobalt absorption edge as a function of the thickness of the balanced filters (plotted as multiple f of the optimal thickness).

able to demonstrate that hot-pressed pyrolytic graphite is a highly efficient x-ray monochromator, which overcomes this difficulty. Hendricks[59] has the merit of being the first to introduce this monochromator into the field of small-angle scattering and to test its suitability extensively. This material is so favorable for small-angle scattering because its mosaic spread is roughly equal to the divergence desired for the primary beam in the plane perpendicular to the longitudinal direction of the entrance slit. Even for 100% reflection, an absolutely perfect crystal would give a very poor intensity yield, since it would only reflect rays within a divergence of a few seconds of arc. On the other hand, if the mosaic spread of the monochromator crystal is much larger than the divergence desired for the primary beam, only a small fraction of the mosaic crystals could be used for reflection.[60]

As mentioned above, the situation is most favorable if the mosaic spread is about equal to the maximum desired angular divergence of the

[59] R. W. Hendricks, J. Appl. Crystallogr. 3, 348 (1970).
[60] This can be illustrated by considering the most extreme case, a crystal powder of randomly oriented microcrystals. Only a small fraction of these crystals will be in reflection position for any incoming ray, which will be reflected into a complete Debye–Scherer circle. Depending on the length of the primary beam, only a small fraction of this circle will point into a "useful" direction. In addition, incoming rays will have to penetrate many microcrystals (with a chance to be absorbed) before hitting one which is in reflection position.

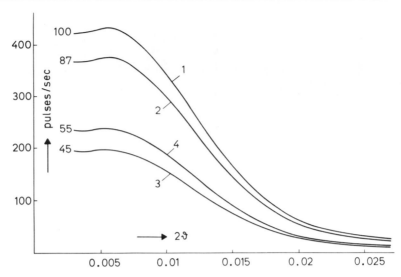

FIG. 21. Scattering curve of Lupolen: curves 1, 2, and 3 were recorded with a 10 μm nickel foil and 100, 50, and 30% channel width, respectively. Curve 4 was obtained with a crystal monochromator and 100% channel width. 2ϑ (=2θ), the scattering angle.[62]

primary beam. A resolution of, say, 20 nm corresponds to the smallest measured scattering angle 2θ given by

$$\lambda = 0.154 \text{ nm} = 20 \times 2\theta$$
$$2\theta = 8 \times 10^{-3} \text{ radians} \simeq 0.4°$$

Commercially available pieces of graphite have a mosaic spread of about 0.45°.[61]

In the following, we give a comparison between the graphite monochromator and a pulse height discriminator in combination with a 10 μm nickel foil. The two monochromatization techniques will be tested on the scattering of a platelet of polyethylene (Lupolen of the Badische Anilin- and Sodafabrik). For this test, the monochromator was arranged in the diffracted beam, following Hendricks'[59] recommendation. Accordingly, the housing for the graphite crystal was mounted on the vacuum tube between the detector slit and the detector in such a way, that it moves together with the detector slit on the perimeter of a circle whose center is the sample axis. The monochromator was adjusted to reflect only the K_α line into the detector.

Figure 21 shows scattering curves of Lupolen measured with entrance slits into the collimation system of 100 μm. Monochromatization for

[61] Commercially available from Union Carbide and Carbon.
[62] Figures 21 and 22 are taken from unpublished measurements by O. Kratky and E. Wrentschur.

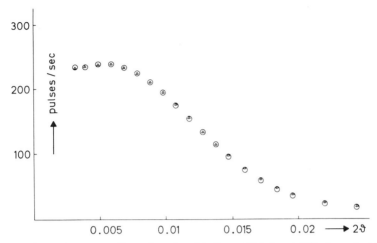

FIG. 22. Scattering curve of Lupolen. △, with 10 μm Ni foil and 50% channel width; ○, with crystal monochromator; $2\vartheta(=2\theta)$, scattering angle.[62]

curves 1, 2, and 3 was performed with pulse height discrimination plus a 10 μm nickel filter. The three curves differ in the channel width, which was 100, 50, and 30%, respectively. The intensity ratio for the three curves is 100:87:45. Curve 4 was measured with the graphite monochromator and 100% channel width. Its intensity (on the same scale) is 55. Thus, the monochromator crystal reduces the intensity attainable with pulse height discriminator plus nickel filter by factors of 55/100 (when compared to 100% channel) and 55/87 = 63/100 (when compared to 50% channel).

The next obvious question, whether it is worth to put up with this loss in intensity, can be answered by considering the polychromatic errors which are introduced when monochromatization is only performed with a pulse height discriminator plus a nickel filter. Figure 22 again shows two scans of the scattering curve of a Lupolen platelet, recorded with pulse height discriminator (50% channel) plus nickel foil (triangles) and with monochromator (circles), respectively. The curves were scaled to coincide in the innermost portion, by application of an appropriate scale factor. It can be seen that they differ by a small but detectable amount further out. These differences are slightly larger for 100% channel width and slightly smaller for 30% channel width; we refrain from giving graphical representations for these cases, since the matter is already quite clear from Fig. 22: if one is interested in the finest details of the scattering curve, one has to accept the longer registration time (almost twice) and use the monochromator. On the other hand, if highest precision in the middle part of the curve is not desired (for example, if the statistical error alone is already larger than the difference between the two curves in Fig. 22), it will be unnecessary to use the monochromator.[62]

This also answers the question concerning the degree of monochromatization attainable with pulse height discrimination plus nickel filter, which was raised in Section III,C,1.

As a matter of fact, most scientists working in the field of small-angle scattering did not use monochromators so far. With increasing refinement of the theory and with the experimental elimination of most classical sources of error (reduction of statistical errors by high power generators and elimination of errors due to intensity fluctuations by monitor cameras) more scientists will choose a complete elimination of polychromatic errors and use monochromators. The more so, as the monochromator (when mounted in the diffracted beam, as described above) automatically eliminates fluorescent radiation originating from the sample. At this point, it should also be noted that with position sensitive detectors (see III,F,4) the loss in intensity caused by the use of monochromatic radiation can be easily tolerated.

5. *Totally Reflecting Mirrors.* Damaschun *et al.*[63,64] have demonstrated, that reflection from a carefully polished glass surface is a practicable monochromatization technique for small-angle investigations. Figure 23 indicates the arrangement of the reflector (combined with the collimation system suggested by us) as described by Damaschun.[63,64] The characteristic radiation retains 90% of its intensity, but reflection does not yield completely monochromatic radiation. For many applications, however, the degree of monochromatization will be sufficient.

D. Absolute Intensity

Experimental determination of molecular weights and related quantities requires knowledge of the absolute intensity, i.e., the ratio of scattered intensity I (or a quantity obtained by multiplication with the scattering angle or its square, i.e., $I\theta$ or $I\theta^2$) to primary intensity (see P.K.T. and Section II). The scattered intensity can be directly determined with commercially available detectors (proportional or scintillation counters) without major difficulties. Direct determination of the primary intensity, on the other hand, is not possible due to the rapid succession of quanta in the primary beam, which cannot be resolved even with the most advanced

[63] G. Damaschun, *Naturwissenschaften* **51**, 378 (1964); *Exp. Tech. Phys.*, **3**, 224 (1965).
[64] G. Damaschun, G. Kley, and J. J. Müller, *Acta Phys. Austriaca* **28**, 223 (1968).

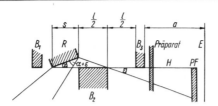

FIG. 23. Totally reflecting monochromator according to Damaschun. R, totally reflecting plane; H, main plane through the camera; PF, primary beam stop; E, plane of registration; B, screens; B_1 avoids penetration of the front edge of the reflector block by x rays.

detectors. P.K.T. have discussed several ways to overcome this difficulty. In the following we will briefly describe the "rotator method."[6,65,66]

This method involves a device which brings about a mechanical attenuation of the primary beam by several orders of magnitude. It works on a most surprising principle: the counter tube is shielded from the direct primary beam by a circular disc of several centimeters diameter. This disc rotates with about 50 revolutions/sec and has a few very small, barely visible holes which cross the line-shaped primary beam upon revolution of the disc. X-ray quanta can only get into the detector while a hole passes through the direct beam. It may come as a surprise, that with a rotator with the above dimensions, only for a small fraction (say one-tenth or one-twentieth) of the passages of a hole through the primary beam, one quantum gets through the hole and enters into the counter. The fraction of passes which let a quantum into the detector allows computation of the primary energy.

The rotator method has the disadvantage that each measurement takes several hours and requires frequent checks to ensure that the primary intensity does not fluctuate in the course of its determination. While the rotator is thus unsuitable for routine applications, it is a very convenient instrument to calibrate *secondary standards.*[67] For this purpose, we use platelets of polyethylene (Lupolen 1811 M of the Badische Anilin- & Sodafabrik) whose absolute intensity (i.e., the quotient of scattered and primary intensity) is determined for an angle of 0.016027 radians (corresponding to a Bragg spacing of 15 nm). Rapid and simple determinations of the primary intensity are then possible in every laboratory with access to a calibrated Lupolen. It should be noted at this point, however, that

[65] O. Kratky, *Makromol. Chem.* **35A,** 12 (1960).
[66] O. Kratky and H. Wawra, *Monatsh. Chem.* **94,** 981 (1963).
[67] O. Kratky, I. Pilz, and P. J. Schmitz, *J. Colloid Interface Sci.* **21,** 24 (1966).

the difference in absorption between secondary standard and sample under investigation has to be allowed for, i.e., the user has to measure the absorption of his sample (determination of the absorption of the Lupolen standard is part of the calibration procedure and each user is supplied with the absorption coefficient of his Lupolen).

The usual calibration also includes the dependence of the scattering power with temperature, which is strictly reversible between 4° and 50°C[68] and the scattering power increases linearly with temperature.

Another study[69] addressed itself to the effect of radiation on the scattering behavior of Lupolen: a Lupolen sample which was subjected to a very high dose of x rays (a high multiple of many years' normal use) did not show any detectable change in scattered intensity.

Our positive results are confirmed and supplemented by an interesting study by Schaffer and Hendricks.[70a] They showed how to determine a calibration constant for Lupolen, independent from collimation system and wavelength. Recently a new technique, the "Moving Slit Method," was described for absolute intensity mesurements[70b]: one vertical slit of 200 μm width is fixed in the plane of registration, while a second one of about the same width, located near the focus, moves perpendicular to its length direction. The primary intensity can be calculated from the width of the two slits, the velocity of the moving slit, and the observed number of counts during one sweep across the whole primary beam.

Hendricks[70c] has recently summarized the theoretical bases of absolute intensity measurements.[70d]

E. Experimental Elimination of the Effect of Intensity Fluctuations, Monitor

No other type of x-ray scattering experiment require as stable x-ray sources as small-angle investigations: small-angle x-ray scattering mea-

[68] I. Pilz, *J. Colloid Interface Sci.* **30**, 140 (1969).

[69] I. Pilz and O. Kratky, *J. Colloid Interface Sci.* **24**, 211 (1967).

[70a] L. B. Schaffer and R. W. Hendricks, *J. Appl. Crystallogr.* **7**, 159 (1974).

[70b] H. Stabinger and O. Kratky, *Makromol. Chem.* **179**, 1655 (1978).

[70c] R. W. Hendricks, *J. Appl. Crystallogr.* **5**, 315 (1972).

[70d] Note added in proof: The "Commission on Crystallographic Apparatus" of the "International Union of Crystallography" has recently issued a report [*J. Appl. Crystallogr.* **11**, 196 (1978)] on an international intercomparison project which was performed to test reproducibility and comparative accuracy of various absolute intensity calibration techniques in current use in small-angle x-ray scattering. Fifteen investigators from eight different laboratories in six countries participated in this study, which was organized by Hendricks and Shaffer.

surements frequently take many hours, in the course of which the scattered intensity is recorded at many different angles; intensity fluctuations during that time lead to a deformation of the observed curve. Subsequently, the scattering curve has to be corrected mathematically for the effect of the collimation system ("desmearing") and errors in the observed curve are propagated through this desmearing procedure in a rather unpredictable way. Moreover, in the case of dissolved macromolecules, the scattering curve of the solvent has to be subtracted from the solution curve. The difference can be very small relative to the observed intensities, especially for low solute concentrations. Small relative errors in one of the two observed curves (solvent and solution) can therefore lead to large ones in the difference curve (see Fig. 35, Section III,G).

P.K.T. have discussed many of the reasons that can lead to fluctuations of the primary intensity, even when well-stabilized x-ray generators are used. Section III,A describes reasons for intensity fluctuations caused by an insufficient environment (x-ray room). Observance of the precautions recommended in that section, however, does not necessarily guarantee perfect stability of the x-ray intensity. We believe the most important cause for troubles to be a relative movement of the camera with respect to the anode; such a movement may either be caused by mechanical or by temperature effects, which cannot be avoided even by careful stabilization of the room temperature: the x-ray tube, which is heated and cooled at the same time, never allows a constant temperature of the entire system. The rigid connection between camera and x-ray tube described in Section III,F is certainly capable of reducing this effect. A *monitor,* however, will be a very useful and important tool if some of the recommended precautions cannot be observed because the necessary equipment is unavailable or if one aims at highest precision.

The basic idea of the monitor is to divide the scattered intensity by the (simultaneously recorded) primary intensity or a quantity proportional to it. This obviously eliminates errors in the scattering curve caused by fluctuations in the primary intensity. Several ways have been suggested and tried for the experimental realization of this idea.

Chipman and Jennings[71] and Jennings *et al.*[72] use the scattering of a mylar foil in the primary beam as a reference. Mylar, however, produces a strong small-angle scattering up to about 0.015 radians, which disqualifies this method if one is also interested in the region below this limit.

[71] D. R. Chipman and L. D. Jennings, *Phys. Rev.* **132,** 728 (1963).
[72] L. D. Jennings, D. R. Chipman, and J. J. De Marco, *Phys. Rev. A* **135,** 1612 (1964).

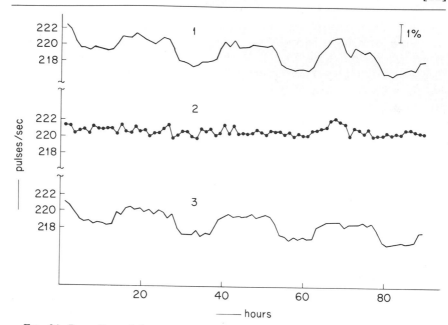

Fig. 24. Recording of the scattered radiation at an angle corresponding to a Bragg spacing of 25 nm. Reflected primary beam (graphite monochromator). Entrance slit, 150 μm; sample, gold sol; recording time, 4000 sec per point. 1, scattering curve; 3, monitor intensity; 2, quotient between 1 and 3.

Recently, Hendricks *et al.*[73] have suggested an arrangement with an ionization chamber within the collimation system. The ionization current caused by the direct beam in this chamber is taken as a measure for the primary intensity. This setup, however, requires full monochromatization of the primary beam prior to its entrance into the collimation system, since the ionization current is proportional to the total energy (and not to the number of CuK_α quanta) and therefore reacts strongly to a change in the spectral composition of the primary beam.

Another suggestion[74] involves the use of a Bragg reflection originating from the primary beam stop made from platinum. Becherer *et al.*[75] use the same physical arrangement as in Kratky *et al.*[74] but record the fluorescence induced by the copper radiation on the iron beam stop. Cobalt is a more suitable material for this purpose than iron,[76] because its absorption edge ($\lambda = 0.1608$ nm) lies closest to the wavelength of the CuK_α radiation

[73] R. W. Hendricks, J. T. De Lorenzo, F. M. Glass, and R. E. Zedler, *J. Appl. Crystallogr.* **6**, 129 (1972).
[74] O. Kratky, H. Leopold, I. Pilz, and H. P. Seidler, *Z. Angew, Phys.* **31**, 49 (1971).
[75] G. Becherer, K. Zickert, and R. Kranold, *Exp. Tech. Phys.* **20**, 259 (1972).
[76] C. Kratky, O. Kratky, and E. Wrentschur, *Acta Phys. Austriaca* **41**, 105 (1975).

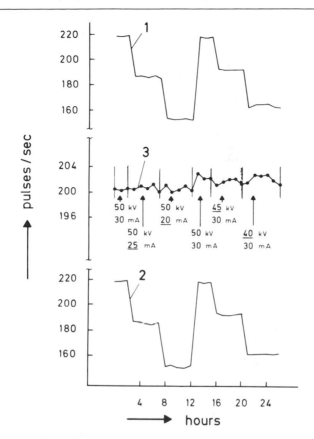

FIG. 25. Recording at an angle corresponding to a Bragg spacing of 30 nm. Effect of artificially induced changes in voltage and current; reflected primary beam (graphite monochromator); entrance slit, 150 μm; sample, gold sol; recording time, 4000 sec per point. 1, scattering curve; 2, monitor intensity; 3, quotient between 1 and 2.

($\lambda = 0.154$ nm) and it therefore yields the most intensive fluorescence. Moreover, the fluorescent radiation originating from cobalt is shorter in wavelength [λ $(CoK_\alpha) = 0.179$ nm] than the one from iron [λ $(FeK_\alpha) = 0.194$ nm] and is therefore less strongly absorbed by the windows which the reference beam has to penetrate before getting into the reference detector. A description of the physical setup of the monitor (according to footnote 76) is given in Section III,F,1.

The above two figures have one thing in common: while scattered intensity and monitor intensity both show large fluctuations, the fluctuations in the quotient are much smaller. This is true for fluctuations by several percent (Fig. 24) as well as for artificially induced fluctuations (change of voltage and current at the generator) of 30 to 40% (Fig. 25).

FIG. 26. Photograph of the new block camera. Explanations in the text.

F. Types of Small Angle Cameras

1. The Block Camera. P.K.T. have described the most frequently used types of small angle cameras, including the one developed in our laboratory by one of the authors.[77,78] In the following discussion, this instrument will be referred to as block camera. The basic idea of its collimation system can be derived from Fig. 19a.

It should be mentioned that housing and installation of the camera were recently considerably modified and improved[79]: to ensure a well-defined positioning of the individual elements, they are now assembled into a stable, evacuated housing from cast brass (Fig. 26). In order to integrate the focus of the radiation source into the system, the housing is directly suspended at the x-ray tube R (in Fig. 26) via a point support. The screw F (in Fig. 26) allows precise adjustment of the main plane into the focus. Two other points of support hold the body on the working table (one of them, S_1, is shown in Fig. 26). The direct connection of camera and x-ray tube avoids relative movements of the focus with respect to the collimation system and considerably improves the constancy of the primary intensity, as could be verified by numerous tests over long periods of time.

Obviously, the continuous vacuum used in this model has the advantage over any other system consisting of separate units with window foils and air paths in between, that the parasitic scattering is greatly reduced. It should be noted at this point that Hendricks[59] has modified an older version of the commercially available block camera to obtain a continuous

[77] O. Kratky, *Z. Elektrochem.* **58**, 49 (1954); **62**, 66 (1958).
[78] O. Kratky and Z. Skala, *Z. Elektrochem.* **62**, 73 (1958).
[79] O. Kratky, H. Stabinger, and J. Schmied, *Makromol. Chem.* (submitted for publication).

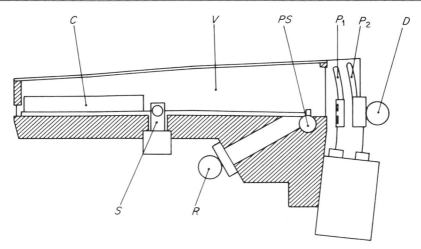

FIG. 27. Schematic representation of the new block camera. Explanation of the camera (including the monitor) in the text.

vacuum: he connected the vacuum tube (containing the counter slit) and the collimation system with a flexible Tombak tube. We prefer an arrangement where the movement of the light-weight counter slit is guided by the slit P_1 (in Fig. 26). The moving device for the counter tube is separated from the one for the counter slit to avoid interference between the supply cables for the detector and the counter slit. The movement of the detector D is guided by the slit P_2 (in Fig. 26) and synchronized with the one of the detector slit, both moving on the circumference of a circle whose center lies in the sample axis.

Figure 27 shows a section through the camera. This drawing also illustrates the physical setup of the monitor, which was discussed above in Section III,E. The primary radiation passes through the collimation system C and hits the sample S. PS, the primary beam stop from cobalt is mounted inside the evacuated housing V. Some of the fluorescent radiation originating from PS leaves the vacuum tube through a window and is recorded by the adjustable reference detector R (monitor).

If one chooses to use film detection, the step scanner, counter, and counter tube have no function. They are replaced by a film cassette which is inserted into a suitable holder. Likewise, a position sensitive detector can be mounted on the same holder.

THE RESOLUTION OF THE CAMERA, EXAMPLE OF A TEST MEASUREMENT. The following example is meant to illustrate the power of the instrument with respect to resolution (i.e., smallest attainable angle) and scattering background.

It follows from the geometry of the camera as shown in Fig. 19a that the distance between the center of gravity of the primary beam and the

(elementary) calculation of the smallest observable scattering angle has to main section H is 2.15E, when measured in the plane of registration. The account for the fact that the broken line connecting the upper edge of the entrance slit with the facing edge K_2 of the bridge hits the plane of registration in the primary intensity maximum. The minimum distance between the center of the counter slit and the center of gravity of the primary beam is therefore 2.15E + 0.5Z (Z equals the width of the counter slit). The theoretically smallest scattering angle observable in the absence of any parasitic scattering is thus given by

$$(2\theta)_{min} = \frac{2.15E + 0.5Z}{200 \times 10^3} \tag{57}$$

(for E and Z in micrometers) which gives the corresponding

$$h_1 \simeq (2\pi/\lambda)(2\theta)_{min} \tag{58}$$

The resolution in terms of the largest particle dimension D is given by h_1 according to Eq. (33) (see Section II), which is rewritten here.

$$h_1 D \leqq \pi \tag{33}$$

Combining Eqs. (57), (58), and (33) leads to

$$D \leqq \frac{\lambda \times 10^5}{2.15E + 0.5Z} \tag{59}$$

The maximum resolution in terms of Bragg spacings (D_{Bragg}) for CuK$_\alpha$ radiation (λ = 0.154 nm) is according to Eqs. (34) and (59) given by

$$D_{Bragg} = 2D \leqq \frac{616 \times 10^3}{4.30E + Z} \tag{60}$$

As follows from Eq. (59), a given resolution can be obtained with different combinations of E and Z. The combination yielding *maximum intensity* is the one obeying the relation $Z = 2.15E$, this means the width of the counter slit equals half the width of the primary beam in the plane of registration. In this case it follows from Eq. (59) that

$$D \leqq \frac{\lambda \times 10^5}{3.22E} \tag{61}$$

Figure 28 shows the scattering curve of polyvinyl chloride, observed with E = 43 μm and Z = 15 μm. The zero point at the abscissa corresponds to the center of gravity of the primary beam, the position of the main section H is indicated by a vertical line. Curve 1 gives the total scattering, and curve 2 the scattering of the sample container alone (Mark capillary), scaled to the absorption of the polyvinylchloride sample. Curve 3 is the blank scattering of the empty camera, again multiplied with

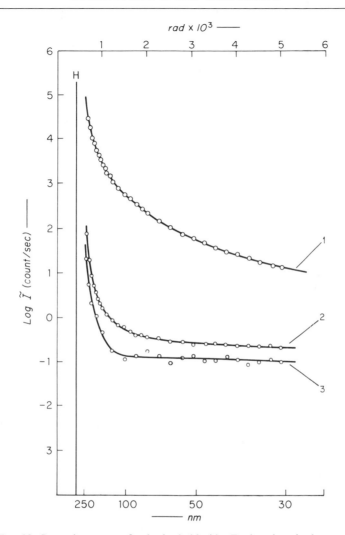

Fig. 28. Scattering curve of polyvinyl chloride. Explanations in the text.

the same scaling factor. The theoretical resolution [from Eq. (60)] for $E = 43$ μm and $Z = 15$ μm is $D_{Bragg} = 308$ nm, the easily attainable one (for the same conditions) is $D_{Bragg}(\exp) = 237$ nm.

The pulse rate at the innermost point of the blank exposure is only 20 counts/sec, a value which can be subtracted without loss in accuracy even for weakly scattering samples. From 110 nm on the blank scattering is as low as 0.1 counts/sec, which is entirely accountable to cosmic radiation. This means that the camera is free of any parasitic scattering beyond angles corresponding to Bragg's value of 110 nm, an important fact for the measurement of the low-intensity tail end of scattering curves.

2. *The Glass Camera of Schnabel, Hoseman, and Röde*[80]. These authors have described a new type of small-angle camera. As suggested by Damaschun *et al.*,[63,64] they use a polished, totally reflecting glass plate as a monochromator; a second glass plate shields the plane of observation from background radiation. The primary beam is limited only by the width of the microfocus and by the slit between the two glass blocks.

The authors report comparative tests between their instrument and the block camera described above, using 20 and 60 μm entrance slits on both instruments. They find that the glass camera yields a 4.2 times more intensive primary beam with a 20 μm entrance slit, while the block camera yields a 2 times brighter beam when a 60 μm entrance slit is used. The conclusion drawn from these test measurements were recently questioned.[81]

3. *The Cone Camera.* Many proteins show very weak subsidiary maxima in the tail end of their scattering curve. Exact knowledge of position and height of these maxima is very important for the reliable determination of particle shape (in the case of nonspherical molecules) or radial electron distribution (for spherical particles). The collimation effect introduced by slit cameras ("slit length smearing") levels such weak maxima, and it is very difficult to obtain them quantitatively correct from the desmearing procedure. This is practically hopeless if the maxima are on top of a strong background, which may either come from parasitic scattering of the camera or from the scattering of solvent and sample container.

For such problems the "cone camera"[82,83] can be very useful. Its cylindrically symmetric collimation system (Fig. 29) masks a primary beam that has the form of a cone-shaped shell. It consists (as shown in Fig. 29) of a hollow, truncated cone HC and a conical needle N, which is concentric with the rotation axis R. The needle is held inside the hollow cone by very small pins and it changes over into the cylindrical body Z, which serves for the elimination of parasitic scattering. The primary beam penetrates the flat sample S adjusted perpendicular to the cone axis and hits the primary beam trap PT consisting of a circular screen. Radiation scattered to the angle 2θ enters into the pinhole diaphragm PD and is recorded by the detector D. 2θ is changed by a shift of the sample along the carrier axis to any other point between the collimation system and the pinhole diaphragm. Two possible positions of the sample and the corresponding scattering angles 2θ are indicated in the figure.

[80] E. Schnabel, R. Hosemann, and B. Röde, *J. Appl. Phys.* **43**, 3237 (1972).
[81] O. Kratky, *J. Makromol. Chem.* (submitted for publication).
[82] O. Kratky, *Monatsh. Chem.* **100**, 376 and 1788 (1969).
[83] O. Kratky, H. Stabinger, E. Wrentschur, and R. Zipper, *Acta Phys. Austriaca* **44**, 173 (1976).

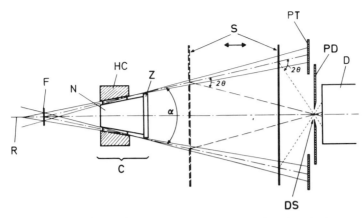

FIG. 29. Schematic drawing of the cone camera. Explanations in the text.

The observed intensity is affected by a moderate collimation effect of the kind known as slit width smearing (in the case of line-shaped primary beams). For its mathematical elimination we refer to the original publication.[83]

At present, this camera is only suitable for measurements in the tail end of the scattering curve, since very small opening angles are mechanically not feasible. The present version allows the movement of the sample within a range of 30 cm, corresponding to a variation of the Bragg spacing between 7.7 and 0.65 nm.

Very encouraging results have already been obtained in the application of this camera to the study of high polymers[84] and biological macromolecules in solution.[85]

4. The Measurement of the Scattered Intensity (The Position Sensitive Proportional Counter). P.K.T. discussed in detail the method of measuring the scattered intensity by means of conventional proportional counters and scintillation detectors. Within the past few years position sensitive proportional counters (PSPC) became available.[85a,85b] PSPC are able to detect simultaneously the scattered intensity at about 1000 locations within the plane of registration covering a range of 5 cm. There exist several commercially available PSPC's, but the systems seem to be still in a stage of rapid evolution. However, a recent systematic test of a PSPC attached to a block camera was very successful.[85c] At present there

[84] K. Lederer, H. Klapp, P. Zipper, E. Wrentschur, and J. Schurz, *J. Polym. Sci., Poly. Chem. Ed.* (in press).

[85] P. Zipper and H. Durchschlag, *Eur. J. Biochem.* **87,** 85 (1978).

[85a] C. J. Borkowski and M. K. Kopp, *IEEE Trans. Nucl. Sci.* **17,** 340 (1970).

[85b] R. H. Hendricks, *Trans. Am. Cryst. Assoc.* **12,** 103 (1976).

[85c] T. P. Russell, R. S. Stein, M. K. Kopp, R. E. Zedler, R. W. Hendricks, and J. S. Lin, "Oak Ridge National Laboratory Report" ORNL/TM-6678 (1979).

are only few reports on applications of PSPC's in the field of biological macromolecules.[85d]

A different problem is the recording of reflections from oriented structures in the small-angle region. Hendricks[85e] reported remarkable innovations in this field by using a very long (10 m) x-ray chamber and a two-dimensional PSPC[85a] according to Borokowski and Kopp.[85b] Since this field of research belongs to the domain of "diffraction" we refrain from a more comprehensive treatment.

G. Remarks to the Working Technique

This chapter concerns itself with the investigation of dissolved biological macromolecules, in particular enzymes, with the x-ray small-angle scattering technique. Different laboratories have developed somewhat different methods, designed to meet different requirements and to use different instruments. We shall confine ourselves, however, to a discussion of the procedures employed in our laboratories in Graz.

1. Sample Preparation. Several questions which are of interest to biochemists who intend to perform a small-angle experiment on an enzyme concern practical details such as total amount of enzyme necessary, special requirements of solvent (buffer), optimal concentration of enzyme, and radiation damage. The first three questions were already discussed by P.K.T. (p. 189), and our treatment will therefore only deal with some supplements.

a. TOTAL AMOUNT OF SAMPLE. As a general rule, a total amount of 50 mg enzyme is the lower limit. Exact measurements, which involve frequent repetitions, require a total amount of 100–200 mg. Twenty milligrams, however, may be enough for a preliminary investigation yielding rough overall parameters for the molecule.

b. CHOICE OF SOLVENTS. Enzymes are usually dissolved in buffers or other low-salt solutions. The only requirement of the SAXS method concerns the electron density of the solvent: as already mentioned by P.K.T., salt concentrations of more than 1 M should be avoided and light ions such as Li^+ and F^- should be preferred to heavy ones such as Na^+, K^+, and Cl^-. Moderately high concentrations of CsCl and high concentrations (8 M) of urea or other organic molecules, mask the scattering of the molecule under investigation.

[85d] A. Tardieu, L. Mateu, C. Sardet, B. Weiss, V. Luzzati, L. Aggerbeck, and A. M. Scanu, *J. Mol. Biol.* **101,** 129 (1976).
[85e] R. W. Hendricks, *J. Appl. Crystallogr.* **11,** 15 (1978).

As already mentioned, a complete sequence of concentrations must be measured. One usually prepares a stock solution and dialyzes against the desired buffer. An important rule pointed out already by P.K.T. should be stressed again: the solvent used as blank solution (for the determination of the background scattering) must be identical in composition and chemical potential to the solvent used for the protein. This requirement is usually best fulfilled by dialysis. Special procedures[86] may be necessary whenever dialysis is unsuitable (such as in the presence of large amounts of sucrose, which is occasionally used to vary the electron density).

Calculation of the molecular weight requires complete knowledge of the buffer composition; the absolute amount of each component added to the buffer when adjusted to the desired pH must, therefore, be carefully recorded.

c. CONCENTRATION OF THE ENZYME SOLUTIONS. All the equations discussed in the Appendix are only valid for infinite dilution. For globular proteins, concentration effects can only be neglected for concentrations below 1 mg/ml. Biological macromolecules, especially proteins in solution, however, produce such a small excess scattering that statistical errors become prohibitively large for concentrations below 5 mg/ml, even when the techniques described in Section III are employed. This makes it necessary to study a concentration series and to extrapolate to zero concentration.

(i) Extrapolation to zero concentration: As a rule, four or five solutions with concentrations in the range between 5 and 30 mg/ml should be measured.

The magnitude of the concentration effect depends on the shape and charge of the macromolecules in solution and on ionic strength and nature of the solvent. No general function exists which would allow a prediction of the magnitude of the concentration effect. A good approximation is given by Guinier.[3]

The procedure currently employed in our laboratory is as follows. The scattering curve of the solvent and of each protein concentration is measured several times. After averaging the intensities separately for each concentration and for the solvent, the blank scattering is subtracted from the scattering of each protein solution and the scattering curves are normalized to unit concentration, i.e., the observed intensities $\tilde{I}$[87] are divided by the corresponding protein concentrations c. All these numerical operations are carried out with the help of a computer program. A plot of the I/c values illustrates the magnitude of the concentration effect, as demonstrated in Fig. 30. The interparticle interference increases with increasing

---

[86] A. Schausberger and I. Pilz, Makromol. *Chem.* **178,** 211 (1977).
[87] $\tilde{I}$ is the slit smeared value of the intensity.

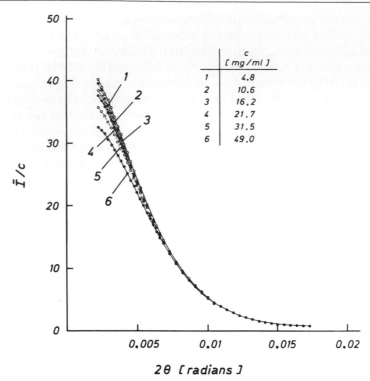

	c [mg/ml]
1	4.8
2	10.6
3	16.2
4	21.7
5	31.5
6	49.0

2θ [radians]

FIG. 30. Slit-smeared scattering curves of a protein (hemocyanin of *Astacus leptodactylus*) for the indicated concentrations c. The curves are normalized to $c = 1$ by plotting $\tilde{I}/c$.

concentration usually with the effect of decreasing the scattering at small angles. The different ways used for the extrapolation to zero concentration are shown in Figs. 31 and 32. The upper thick line (in Fig. 31) is the extrapolated curve; the extrapolation can be carried out either in the normal plot ($\tilde{I}/c$ values versus the scattering angle 2θ) (Fig. 31) or in the Zimm plot (known from light scattering, Fig. 32); at each measured scattering angle, the concentrations are plotted parallel to the abscissa as shown in Figs. 31 and 32. Scattering curves of solutions with relatively low protein concentrations can be extrapolated usually in both plots (normal and Zimm) to zero concentration. If one is forced to use highly concentrated solutions, the more exact extrapolation in the Zimm plot should be preferred.

A third way is to calculate the radius of gyration $\tilde{R}$[88] from the Guinier plot for each concentration (Fig. 33) and to plot the $\tilde{R}$ values versus the

[88] $\tilde{R}$ is the slit smeared value of the radius of gyration, usually called the "apparent radius of gyration."

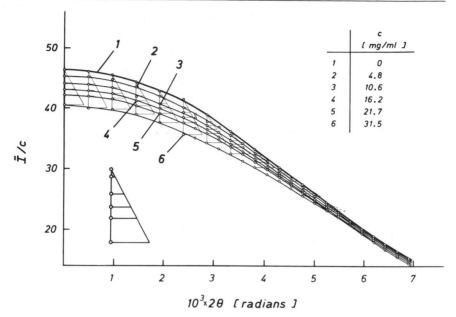

FIG. 31. Innermost portions of the scattering curves shown in Fig. 30. The curves are extrapolated to zero concentration (curve 1) by plotting the corresponding concentration at different scattering angles parallel to the abscissa in arbitrary units, as indicated.

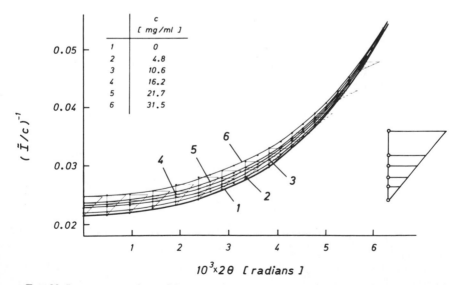

FIG. 32. Innermost portions of the scattering curves shown in Fig. 30 in a Zimm plot. The curves are extrapolated to zero concentration (curve 1, thick line) in the same way as described in Fig. 31.

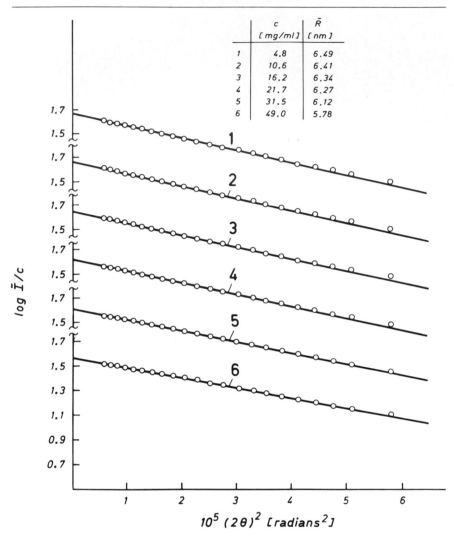

FIG. 33. Guinier plots of the innermost portions of the scattering curves shown in Fig. 30. The values of the slit-smeared radii of gyration $\bar{R}$ obtained for the different concentrations are indicated.

concentration c as shown in Fig. 34. Obviously, this extrapolation is only meaningful when the Guinier plot of the (smeared) scattering curves follows a well-defined straight line, as illustrated in Fig. 33. We routinely perform all three extrapolation procedures on the same system. An agreement of the three extrapolated curves is a good indication that no serious error was introduced by the extrapolation procedure. Another valuable

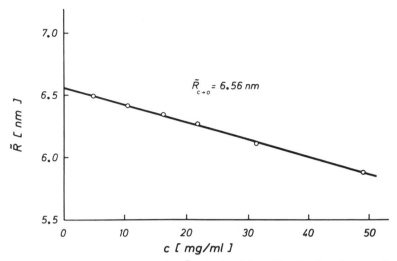

Fig. 34. Slit-smeared radii of gyration $\tilde{R}$ calculated from Fig. 33, plotted versus the protein concentration c.

hint for the correctness of the extrapolation to zero concentration uses the $p(r)$ functions, as already discussed in Section II,B,2,b (Fig. 18).

(ii) Concentration and accuracy: Since the distances between the macromolecules in solution are large as compared to the distances within the macromolecules, the concentration effects due to interparticle interference occur only at small angles, according to the law of reciprocity (Fig. 31). At larger angles the scattering curves at all concentrations should agree in the $\tilde{I}/c$ curves.

While it is necessary to use relatively low protein concentrations (5 to 30 mg/ml) for the extrapolation to zero concentration at small angles, it is a great advantage to use much higher concentrations (50 to 100 mg/ml) for the measurement at larger angles, if enough material is available. The error in the difference between solute and solvent curve becomes very large if the difference is very small. This is illustrated in Fig. 35a, which shows the scattering curves of the solvent and several solutions with different concentrations. The magnitude of the difference decreases linearly with the concentration of the solute, and the errors in the difference curves (Fig. 35b) increase rapidly.

d. DETERMINATION OF THE CONCENTRATION AND ISOPOTENTIAL SPECIFIC VOLUME. The concentrations of the protein solutions under investigation must be determined accurately, as already pointed out by P.K.T.

Apart from the concentration, a volume parameter characterizing the dissolved particle enters into the equation for the molecular weight. This

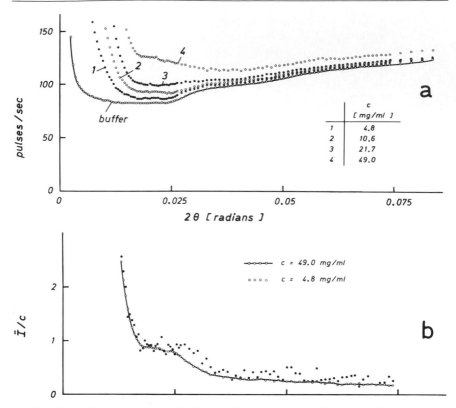

FIG. 35. (a) Scattering curve of buffer and of four protein solutions with different concentrations. (b) Scattering curves for the two indicated concentrations after subtraction of the buffer (blank) scattering and after normalizing to $c = 1$.

parameter is usually called "partial specific volume $\bar{v}$." Kupke[89] pointed out, however, that the quantity entering into the molecular weight equation is not the partial specific volume, but a somewhat different volume parameter called the isopotential specific volume v_1'.

This isopotential specific volume has to be determined with the highest accuracy, since it appears as the square of a difference in the molecular weight equation. Thus, for proteins in the usual dilute buffers, an error of 1% in v_1' leads to errors of 4 to 5% in the molecular weight.

v_1' is usually determined from the densities of solvent and solution. Again, the solvent has to have identically the same composition and chemical potential as the solution (and not *only* the same pH value!). For

[89] D. W. Kupke, *in* "Physical Principles and Techniques of Protein Chemistry" (S. J. Leach, ed.), Part C, p. 1. Academic Press, New York, 1972.

the usual buffers or low salt solvents, dialysis is again the best way to fulfill these requirements. The procedure which has to be applied when the solvent contains large amounts of sugar, etc., is described in a special paper.[86]

Moreover it is essential that the isopotential specific volume of the protein refer to the same solvent which is used for the scattering investigation, since the v_1' value is by no means independent of the solvent.

A special instrument was developed in our laboratory which allows a sufficiently accurate determination of the densities of solution and solvent.[90-92] This instrument requires about 1 ml of protein solution. v_1' is determined from the difference of the density of solvent and solution, and it is clear that the resulting isopotential specific volume will be more accurate for more concentrated solutions (if possible, 10 to 40 mg/ml). Other instruments used to determine v_1' with high precision are mentioned by P.K.T.

e. RADIATION DAMAGE. A question which turns up in the course of every small-angle investigation concerns the possible damage to the enzyme due to the absorbed x rays.

SAXS only records the morphology of the biological macromolecule, independent of its functional activity. That is, a loss in activity can be detected by SAXS only if it is accompanied by a significant change in morphology. A good example for the magnitude of a detectable morphological change is the hemoglobin molecule: the allosteric structural change upon addition and removal of oxygen can be clearly seen by SAXS.[93] The hemoglobin molecule has overall dimensions of about 6 nm, and the maximum movement of a polypeptide chain is about 0.6 nm. Much smaller structural changes, which affect only part of a chain or dislocate only a few atoms are invisible for SAXS.

In other words, damage to the molecule due to radiation or other factors does not influence the small-angle curve as long as there is no structural change of the order of magnitude mentioned above. On the other hand, it cannot be deducted from SAXS measurements that the molecule did not lose its activity in the course of the investigation. This can only be checked with a determination of the activity of each enzyme sample after the x-ray exposure.

As a rule of thumb, enzymes do not suffer much damage if the duration of x-ray exposure does not exceed 8–12 hr (with the usual x-ray generators operated with about 50 kV and 30 mA). In our experience, many

[90] H. Stabinger, H. Leopold, and O. Kratky, *Monatsh. Chem.* **98**, 436 (1967).
[91] O. Kratky, H. Leopold, and H. Stabinger, *Z. Angew. Phys.* **27**, 273 (1969).
[92] O. Kratky, H. Leopold, and H. Stabinger, Vol. 27 [5], 98.
[93] H. Conrad, A. Mayer, H. P. Thomas, and H. Vogel, *J. Mol. Biol.* **41**, 225 (1969).

samples did not display a greater loss of activity after such irradiation than nonirradiated reference samples. Of course, there are enzymes which are much more sensitive to irradiation; an example is discussed in Section IV,D,4,b.

Morphological changes which may occur during irradiation can be monitored by the change of the scattered intensity as a function of time. Repeated recording of the same curve can easily detect and eliminate the effect of such morphological changes.

f. SAMPLE CONTAINER. Cameras with line-shaped primary beam cross sections (which are the only ones discussed in detail in this article) require elongated sample containers to minimize the amount of sample needed. The most important requirements which have to be met by a good sample container were already discussed by P.K.T. (p. 191). Two different types of sample containers are currently used in SAXS: cell containers and glass capillaries. We prefer very thin glass capillaries because of the following advantages: the capillary is easy to clean and its surface does not interact with biological macromolecules, as it is occasionally observed with cells of mica.

Moreover, the capillary absorbs only weakly and produces little disturbing scattering, it has a well-defined thickness and, its most important advantage, it requires only a small amount of sample (about 50 μl). Its only disadvantage is the fact that more time is needed to adjust it precisely in the beam than cell containers.

The cells preferably used by other laboratories have been described in detail by P.K.T. We shall confine ourselves, therefore, to a few remarks concerning the use of capillaries.

The capillary has to fulfill the following requirements: it must be straight and of constant diameter in the region hit by the x-ray beam; the wall thickness in this region should not exceed 1/100 nm. In our laboratory, carefully selected Mark capillaries[94] are mounted on a suitable sample holder. Once adjusted, they can be used for months or years. The capillary is adjusted in the beam by filling it with a strongly absorbing liquid, for instance a saturated cesium chloride solution. With the help of a screen and the x-ray detector, the capillary is first adjusted parallel to the line-shaped primary beam and then exactly in its center, looking for the position with the highest absorption. Once adjusted, the capillary can be used for measurements as long as the camera, entrance slit, and the position of the primary beam remain unchanged.

The diameter of the capillary is chosen to yield maximum scattered intensity as already discussed in Section III,B,1. For dilute aqueous protein solutions, the optical thickness of the capillary (that is, the optimum path

[94] Commercially available for instance from Fa. Hilgenberg, Malsfeld, Germany.

length of the primary beam through the solution) is about 1 mm; capillaries between 0.8 and 1.2 mm diameter can be used without considerable loss of scattered intensity. (The dependence of the scattered intensity on the thickness of the sample shows a relatively broad maximum.)

IV. Selected Examples

Application of small-angle x-ray scattering for the study of enzymes in dilute solution is focussed at the following problems.

Determination of the overall conformation of isolated enzymes

Investigation of structural changes induced in enzymes through interaction with small molecules (effectors)

Investigation of the interaction between enzymes and other biopolymers

Observation of association and dissociation processes

Comparison of structural parameters of native and modified enzymes

These applications are illustrated with selected examples. The first example is presented in some detail to explain the way in which molecular parameters can be obtained from a small-angle experiment. In subsequent examples, observed values for these parameters are quoted and discussed in terms of their structural significance.

Unfortunately, different notations are currently used in the small-angle literature for the same parameters. Whenever the notation in a quoted article deviated from the one adapted in this publication, the symbol of the original article will be given, with the corresponding one from our notation {e.g., correlation function $H(x)$ $[= \gamma(r)]$}.

A. Determination of the Overall Conformation of Isolated Enzymes

1. Pyruvate Dehydrogenase Core Complex from Escherichia coli *K-12.* Durchschlag[95] studied this multienzyme complex in dilute solution. The complex consists of three enzyme components, namely, pyruvate dehydrogenase, dihydrolipoamide transacetylase, and dihydrolipoamide dehydrogenase. It is a good example, as nearly all parameters derivable from SAXS experiments were determined.

The experimental result of any SAXS experiment consists of a scattering curve, i.e., the functional dependence between scattered intensity $I(2\theta)$ and scattering angle 2θ respectively h. The curve for the multienzyme complex as shown in Fig. 36 is already desmeared, monochromatized and extrapolated to zero concentration as described in a previous

---

[95] H. Durchschlag, *Biophys. Struct. Mech.* **1,** 153 (1975).

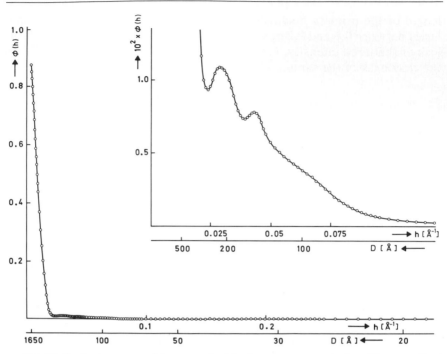

FIG. 36. Scattering curve of the pyruvate dehydrogenase core complex. The normalized scattered intensity $\phi(h)$ is plotted versus the angular argument h; the curve is desmeared, monochromatized and extrapolated to zero concentration. $\phi(h) = I(h)/I_0$, i.e., scattered intensity normalized to 1 for $h = 0$; D (=d_{Bragg}) is the Bragg's spacing. (From Durchschlag[95].)

section. The typical characteristics of this scattering curve are as follows: a steep increase of the scattered intensity toward small angles and a very long outer part with relatively high intensity extending to large angles. The scaled-up drawing (Fig. 36) of the outer section of the curve shows three subsidiary maxima, the first and second well defined, the third one not very distinct.

The correlation function $\gamma(r)$ of the enzyme is shown in Fig. 37.[95a] Intuitively more meaningful is the distance distribution function $p(r)$ (see Section II). Fig. 38 shows this function for the enzyme core complex. The above three functions [scattering curve, $\gamma(r)$, and $p(r)$] can serve as a basis for the computation of the following molecular parameters.

a. RADIUS OF GYRATION. The radius of gyration R_g of a particle can be determined in several different ways.

i. The most common method uses the Guinier approximation[3] described in detail by P.K.T. This method yields $R_g = 16.5$ nm for the

[95a] H. Durchschlag, *Biophys. Struct. Mech.* **1**, 169 (1975).

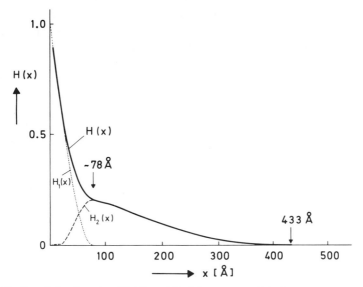

FIG. 37. Correlation function $H(x)$ $[= \gamma(r)]$ of the pyruvate dehydrogenase core complex. The plot suggests a maximum diameter of 433 Å. (From Durchschlag[95a].)

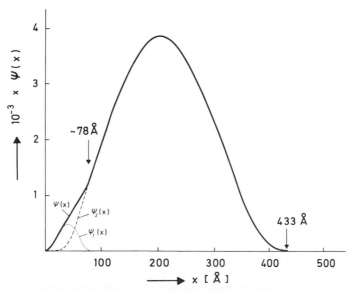

FIG. 38. Distance distribution function $\psi(x)$ $[= p(r)]$ of the pyruvate dehydrogenase core complex. The maximum diameter of 433 Å can be seen more clearly in this plot than in the one of the correlation function shown in Fig. 37. (From Durchschlag.[95a])

multienzyme complex. (A by-product of the linear extrapolation to zero scattering angle is I_0, the scattered intensity at zero angle, a quantity which is necessary for the determination of molecular weight and volume of dissolved particles; see also the Appendix and Figs. 31 and 32).

ii. R_g can also be determined from the distance distribution function p(r) (see Section II,A,2). This method is particularly powerful when used in connection with the indirect Fourier transform described in Section II,A,4. It has the advantage that the whole scattering curve is used for the determination of R_g and not only the innermost portion as is the case with the Guinier method. Here Durschschlag found R_g = 15.65 nm when determined from p(r).

iii. For all homogeneous and isometric particles, whose shape does not deviate strongly from a sphere or cube, R_g can also be determined from the position of the first subsidiary maximum.

$$R_g = 4.5/h \qquad h = (4\pi \sin \theta)/\lambda \qquad (62)$$

This method yields R_g = 15.5 nm. The good agreement with the value obtained from the distance distribution function indicates that the molecules are approximately isometric.

b. MOLECULAR WEIGHT. The molecular weight M of the multienzyme complex was calculated with Eq. (88) (Appendix). The concentration and isopotential specific volume have to be determined as accurately as possible, because the square of a difference enters into the pertaining equation leading to a strong amplification of errors. As usual, I_0 [$= I(0)$] was obtained by extrapolation to zero concentration and zero angle P was determined with a calibrated Lupolen standard sample (see Section III,D). This method yielded a molecular weight of 3.78×10^6.

c. VOLUME. The volume of dissolved macromolecules can be determined either directly from the scattering curve or from its Fourier transform. I_0, the intensity at zero scattering angle (determined from a Guinier plot) and the invariant Q enter into the relevant equation [Eq. (108), Appendix].

Q is given by the integral

$$Q = \int_0^\infty I(2\theta)(2\theta)^2 d(2\theta) = \int_0^{(2\theta)^+} I(2\theta)(2\theta)^2 d(2\theta) + \frac{k_1}{(2\theta)^+} \qquad (63)$$

Obviously, it is not possible to record the scattered intensity up to the angle ∞. The integration is therefore carried out numerically with Simpson's formula up to the relatively large angle $(2\theta)^+$; the remaining tail end of the curve is integrated analytically: according to Porod, the tail end of a scattering curve oscillates about $k_1(2\theta)^{-4}$. The tail end constant k_1 is determined from a suitable plot [$I(2\theta)$ versus $(2\theta)^{-4}$], which should oscil-

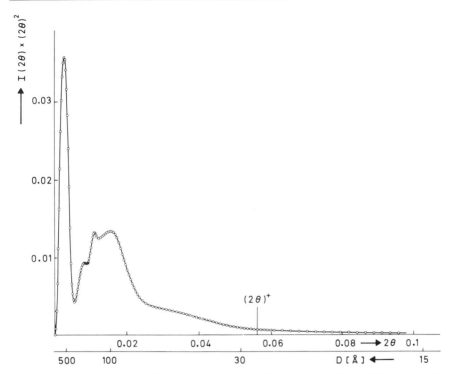

FIG. 39. Plot of $I(2\theta)(2\theta)^2$ versus (2θ), used for the evaluation of the invariant $Q = \int^\infty I(2\theta)(2\theta)^2\, d(2\theta)$ for the pyruvate dehydrogenase core complex $D = d_{Bragg}$. (From Durchschlag.[95a])

late about a straight line with slope k_1. This determination, however, is never very accurate, and the analytically computed portion of the invariant [i.e., the term $k_1/(2\theta)^+$, see Eq. (63)] should be as small as possible. Usually, its contribution to the total of Q is less than 10%. Figure 39 shows the plot $I(2\theta)(2\theta)^2$ versus (2θ). The invariant is equal to the area under this curve. Analytical integration started at the angle $(2\theta)^+$. The value obtained for the volume is 1.075×10^4 nm^3.

The accuracy of such volume determinations is restricted by the fact, that Eq. (108) is exactly only valid for particles with homogeneous electron density. Deviations from complete homogeneity cause a background scattering. Moreover, it is often hard to obtain a reliable estimate for k_1. Due to the limitations mentioned above, the absolute accuracy of the volume determination is not very high. However, if one investigates several very similar systems under strictly identical experimental conditions and if one uses the same techniques for the processing of the observed scattering curves, many of the errors in the volume determination cancel and volume differences can be determined with much higher precision. This has proved very useful for the study of conformational changes of enzymes[96] and immunoglobulins[97,98]

d. MAXIMUM DIAMETER. It follows immediately from its definition that $\gamma(r)$ is zero for all distances larger than the maximum particle diameter D. Figure 37 shows the maximum diameter of the enzyme core complex to be 43.3 nm.

The distance distribution function $p(r) = \gamma(r)\ r^2$, shown in Fig. 38, shows the maximum diameter more clearly than the correlation function $\gamma(r)$. The shape of $p(r)$, especially its step decrease at large distances, suggests again that the enzyme consists of a more or less isotropic body. [To obtain an intuitive understanding of the relationship between particle shape and form of $p(r)$, one should calculate the function for various models of different shape, e.g. spheres, elongated and flat bodies,[21,45] see Section II,B,2b.]

e. OTHER PARAMETERS. The SAXS parameters discussed so far (radius of gyration, molecular weight, volume, and maximum diameter) are the ones most commonly used to characterize dissolved particles. There are a number of other structural parameters applicable to the description of a general colloid distribution of matter.

One of these parameters is the specific inner surface O_s [see appendix Eqs. (115) and (117)], which is identical with S/V, whereby S corresponds to the inner surface of the dissolved particle and V is the volume of the particle. Durchschlag's values for the enzyme core complex are: $O_s = 0.707$ nm^{-1}, $S = 7600$ nm^2.

The *transversal or intersection length* $\bar{l}$ is defined as the average of all intercepts obtained by intersecting the disperse phase in all possible directions. It is proportional to the inverse of the specific inner surface; for a system of identical particles

$$\bar{l} = 4/O_s = 4V/S \qquad (64)$$

This quantity is readily computed as $\bar{l} = 5.66$ nm.

Table I summarizes Durchschlag's values of all parameters discussed so far.

f. OVERALL SHAPE. All parameters discussed so far are directly obtainable from the scattering curve or its Fourier transform. Information on the overall shape of a dissolved particle, on the other hand, can only be obtained by a trial-and-error procedure. This process can be carried out either in reciprocal space (with the scattering curve) or in real space [with

[96] H. Durchschlag, G. Puchwein, O. Kratky, I. Schuster, and K. Kirschner, *Eur. J. Biochem.* **19**, 9 (1971).

[97] I. Pilz, O. Kratky, A. Licht, and M. Sela, *Biochemistry* **12**, 4998 (1973).

[98] I. Pilz, O. Kratky, and F. Karush, *Eur. J. Biochem.* **41**, 91 (1973).

TABLE I

MOLECULAR PARAMETERS FOR THE PYRUVATE DEHYDROGENASE CORE COMPLEX[a]

Parameter	Unit	Calculated from	Experimental curve	Theoretical curve
M	g/mole	I_0, P	3.78×10^6	
R_g	nm	$I(2\theta)$ according to Guinier	16.5	15.4–16.2
R_g	nm	$\gamma(r)$	15.65	15.16
V	nm^3	$I(2\theta)$	1.075×10^4	
S/V	nm^{-1}	$I(2\theta)$	7.07×10^{-1}	7.20×10^{-1}
S	nm^2	$I(2\theta)$	7.6×10^3	7.59×10^3
$\bar{l}$	nm	$I(2\theta)$	5.66	5.56
D	nm	$\gamma(r)$	43.3	43.9

[a] Values obtained from the experimental scattering curve are compared with those from the theoretical curve of the suggested model (Fig. 42).

the distance distribution functions $p(r)$ discussed in Section II]. In the following we will confine ourselves mainly to a description of shape determination in reciprocal space.

The basic idea is to compare the experimental scattering curve with curves calculated for various models. All models not "equivalent in scattering" can be rejected. Models whose scattering curve coincides (within the experimental error) with the observed curve are accepted and presented for discussion.

The general shape, whether the molecule is spherical, flat, elongated, contains hollow spaces, etc., can be determined with certainty from such an analysis. Details of the overall shape on the other hand, can only be suggested if the scattered intensity is known with high precision. It is not possible to obtain positive proof for any detailed model; all one can do safely is to exclude models which are not equivalent in scattering. It is worthwhile to compare the SAXS results on the overall shape with those of the crystal structure analysis, as it was recently done by Kratky and Pilz.[99] In many cases, the agreement is excellent; some of the discrepancies could be plausibly accounted for by solvent effects.

In summary, SAXS provides valuable information concerning the tertiary and quaternary structure of proteins. So far, the investigation of secondary structure was the realm of other methods, but there are first promising attempts to apply small-angle scattering also to problems concerning the secondary structure of proteins: Grigor'ev et al.[100] analyzed subsidiary maxima at larger angles to obtain an estimate of the content of β-structure and α-helix. In that context, it is of interest to note that the course of the scattering curve of a globular, protein is not only a function

[99] O. Kratky and I. Pilz, Q. Rev. Biophys. 11, 39 (1978).
[100] A. I. Grigor'ev, L. A. Volkova, and O. B. Ptitsyn, Mol. Biol. (Moscow) 7, 661 (1973).

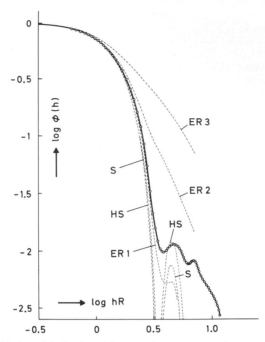

FIG. 40. Double-logarithmic plot of the experimental scattering curve ($\bigcirc$) of the pyruvate dehydrogenase core complex and of theoretical scattering curves of simple full and hollow bodies. S, sphere; HS, hollow sphere of radius ratio $r_i/r_0 = 0.3$ (r_i is inner radius and r_0 is the outer radius); ER1, ER2, ER3: ellipsoids of revolution of axial ratio $a:b:c = \frac{2}{3}:1:1$, $0.1:1:1$ and $10:1:1$, respectively. No satisfactory agreement between experimental and theoretical curves could be obtained in the region of submaxima [$\phi(h) = I(h)/I_0$, $h = (4\pi \sin \theta)/\lambda$, R ($= R_g$) = radius of gyration). (From Durchschlag.[95])

of its overall shape; Fedorov and Denesyuk[101] were recently able to demonstrate that two other factors have an equally strong influence on the curve profile—the course of the backbone and the character of the side chain distribution.

The first idea of the overall shape of a macromolecule is obtained from a comparison of its scattering curve with theoretical curves of simple triaxial full and hollow bodies with various geometries and various axial and hollow-space ratios. It is most convenient to use double-logarithmic plots for this purpose. Since model and protein have to correspond to the same radius of gyration, both curves are usually normalized with respect to R_g by plotting hR_g on the abscissa (instead of the scattering angle 2θ or the angle argument h).

Such a comparison is shown in Fig. 40 for the enzyme core complex, whose experimental curve is compared with those of a sphere, a hollow sphere, and several ellipsoids of revolution with different axial ratios. At

101 B. A. Federov and A. I. Denesyuk, *Acta Crystallogr., Sect. B* **33,** 3198 (1977).

low angles, there is good agreement with the curves of the sphere, the hollow sphere, and the relatively isometric ellipsoid(ER1). This and the steep decrease of the main maximum suggest an isotropic overall shape of the enzyme. The agreement at larger angles, however, is unsatisfactory for any of these simple bodies; the first subsidiary maximum may still be accounted for by regions of low electron density inside the molecule (hollow sphere HS with ratio of inner and outer radius $r_i/r_o = 0.3$), but the remarkably high intensity in the angular range of second and third sub-maximum must be ascribed to more detailed structural elements. As a general rule, the height of subsidiary maxima increases with increasing isometry; regions of low ("hollow particles") or even negative (lipids, etc.) electron density cause a further increase.

Deviations between observed and theoretical curve may be explained through consideration of a more complicated model. For proteins, such deviations from the theoretical curve of a homogeneous, triaxial body can usually be explained by deviations from a simple triaxial shape or in terms of a well-defined substructure, originating, for instance, from the arrangement of subunits in the protein. In other words, this substructure causes deviations from the assumption of homogeneous electron density implicit in the simple triaxial models.

Since there is an infinite variety of possible substructures, one should use any accessible information for the construction of a detailed model; number and size of polypeptide chains obtained from biochemical studies may give valuable hints for the number and size of subunits, and electron microscopy may yield useful information concerning their arrangement. Finally, the SAXS curve itself may contain hints which can be directly used for the construction of a tentative model. This last possibility will briefly be discussed below.

g. SIZE OF SUBUNITS. Debye's[101a] scattering theory of the molecular gas can be applied to a molecule consisting of nearly identical, nearly spherical subunits.[102-104] Since Debye's scattering function always contains the form factor of the spherical subunits, the zero points of the latter also have to appear in the scattering curve of the whole particle. Consequently, the radius of gyration of the subunits and their diameter (assuming spherical shape) can be calculated from the positions of the minima.[105] The above assumption (identical and spherical subunits) are true if

$$h_{01}/h_{02} = 1.73 \qquad (65)$$

[101a] P. Debey, *Ann. Phys. (Leipzig)* [4] **46,** 809 (1915).
[102] This was first carried out by one of us, who approximated the shapes of various anisotropic macromolecules by aggregates consisting of spheres.[103,104]
[103] O. Kratky, *J. Polym. Sci.* **3,** 195 (1948).
[104] O. Kratky, *Monatsh. Chem.* **76,** 325 (1947).
[105] O. Glatter, *Acta Phys. Austriaca* **36,** 307 (1972).

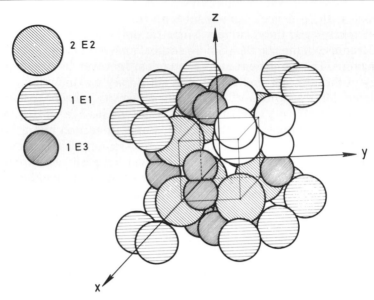

FIG. 41. Model for the pyruvate dehydrogenase core complex consisting of 40 mass centers. Each sphere represents either two poly-peptide chains of the transacetylase (E2) component or one chain of the pyruvate dehydrogenase (E1) or flavoprotein (E3) component. The complex consists of a cubic arrangement of E2 subunits, which are surrounded by E1 and E3 subunits. (From Durchschlag.[95a])

Where h_{01} and h_{02} are the positions of the first two minima. As a consequence of the small size of the subunits, these minima usually appear at very large angles and the scattered intensity is correspondingly weak. To find these minima at all requires highly concentrated solutions (according to our experience at least 40 mg protein per milliliter, usually 60–100 mg/ml; a sufficiently accurate determination is never possible for more than h_{01} and h_{02}. Different size subunits or deviations from the spherical shape cause a flattening of the curve in the region of the long-wave periodicity. Probably because one of these reasons, no minima could be found for the enzyme core complex. For other proteins consisting of many subunits (i.e., hemocyanins), however, this method had proved very successful for the determination of subunit size.[106,107]

To our knowledge, determination of the long-wave periodicity is the only theoretically sound method which yields information on the size of subunits. There are, however, several semiquantitative ways which enabled Durchschlag to estimate the mean diameter of the subunits to about 7.8 nm (see Figs. 37 and 38) and their number to approximately 40.

[106] I. Pilz, O. Glatter, and O. Kratky, *Z. Naturforsch., Teil B* **27**, 518 (1972).
[107] I. Pilz, Y. Engelborgks, R. Witters, and R. Lontie, *Eur. J. Biochem.* **42**, 195 (1974).

h. DETAILED MODEL. Detailed models were calculated for the enzyme core complex using the hints concerning size, number, and arrangement of subunits plus biochemical results of the following kind: molecular weight of the polypeptide chains forming the enzyme components and relative abundance of different polypeptide chains.

Durchschlag calculated the scattering curves of about 150 different models. He varied the number of subunits (between 24 and 48) as well as their relative arrangement. The essential criterion for a satisfying model is "equivalence in scattering." As a first check one generally uses the agreement of radius of gyration and maximum diameter of the enzyme and the model. The best model found in this way (Fig. 41) consists of 40 subunits.

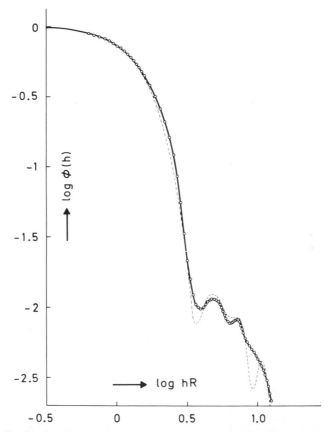

FIG. 42. Double-logarithmic plot of the experimental scattering curve (○) of the pyruvate dehydrogenase core complex and the theoretical curve of the model shown in Fig. 41. This is the model which gives the best fit of all models considered (for explanation of symbols see Fig. 40). (From Durchschlag.[95a])

Its scattering curve fits the experimental curve well (Fig. 42), and its parameters are in satisfying agreement with the experimental ones (Table I).

2. *Further Examples*. Similar studies have been carried out on D-ribulose-1,5-diphosphate carboxylase from *Dasycladies calvaeformis* by Paradies *et al.*[108] on Leucinaminopeptidase from bovine eyes bei Damaschun *et al.*,[109] on phosphofructokinase from baker's yeast by Plietz *et al.*[110,111] and on rabbit muscle phosphofructokinase by Paradies and Vettermann.[112]

B. Investigation of Structural Changes Induced in Enzymes through Interaction with Small Molecules (Effectors)

Small-angle x-ray scattering is well suited for the investigation of conformational changes. While the shape of a dissolved macromolecule can only be deduced indirectly from a comparison with model structures, changes of shape and dimensions can be determined accurately. The allosteric conformational change induced in the hemoglobin molecule by interaction with oxygen (see Section II,G,1,e for a more complete discussion) is a good example; in spite of its small absolute magnitude, it produces a pronounced effect on the scattering behavior of hemoglobin.

This part will enumerate several other examples to illustrate the application of SAXS to problems involving confirmational changes.

1. *Glyceraldehyde-3-Phosphate Dehydrogenase*. This enzyme consists of four chemical subunits which bind one molecule of NAD each. As already mentioned by P.K.T., the structural changes caused by NAD were studied by Durchschlag *et al.*[113] They observed a volume concentration depending on the degree of saturation. Full saturation (i.e., four NAD molecules per enzyme molecule) decreases the volume by 7%. The authors could also answer the question whether the mechanism of binding is sequential or allosteric. For the first case, a strictly linear dependence of the volume contraction on the degree of saturation has to be expected, in the second case deviations from linearity may occur. Since the authors

[108] H. H. Paradies, B. Zimmer, and G. Werz, *Biochem. Biophys. Res. Commun.* **74**, 397 (1977).

[109] G. Damaschun, H. Damaschun, H. Hanson, J. J. Müller, and H.-V. Pürschel, *Stud. Biophys.* **35**, 59 (1973).

[110] P. Plietz, G. Damaschun, H. Damaschun, G. Kopperschläger, R. Gröber, and J. J. Müller, *Stud. Biophys.* **1**, 9 (1977).

[111] P. Plietz, G. Damaschun, G. Kopperschläger, and J. J. Müller, *Acta Biol. Med. Ger.* **37** (in press).

[112] H. H. Paradies and W. Vettermann, *Biochem. Biophys. Res. Commun.* **71**, 520 (1976).

[113] H. Durchschlag, G. Puchwein, O. Kratky, I. Schuster, and K. Kirschner, *Eur. J. Biochem.* **19**, 9 (1971).

found a nonlinear dependence, they could exclude the sequential mechanism.

To prove a deviation from linearity of the dependence between volume and degree of saturation, the volume has to be determined with very high accuracy. As mentioned above (see Section IV,A,1,c), it is impossible to obtain the absolute value of the volume with high accuracy. However, for the present example, only volume changes are of interest, and it was also already mentioned that volume differences between very similar systems can be determined much more accurately. For the present problem, the experimental and data processing conditions were carefully kept identical for all measurements, which indeed permitted a sufficiently accurate determination of the volume changes.

Simon[114] investigated the D-glyceraldehyde-3-phosphate dehydrogenase from mammalian skeletal muscle. He observed a decrease of the radius of gyration by 0.08 nm upon binding of four NAD molecules. To determine such a small change of R_g, Simon had to develop a special method, which allows direct determination of differences in the radius of gyration. In agreement with the results of Durchschlag et al.[113] on yeast glyceraldehyde-3-phosphate dehydrogenase, he found that the decrease of the radius of gyration is not a linear function of the degree of saturation.

Simon was aware of the fact that the change ΔR_g in the radius of gyration can be caused both by structural changes upon binding and by the addition of the mass of the NAD molecules. In spite of the small absolute value of ΔR_g he used a semiquantitative way to estimate both contributions. Subject to the restrictions implied by the twofold symmetry of the enzyme molecule, he calculated the contributions for each of the four binding sites and was able to roughly estimate the most probable positions of the NAD molecules. We should like to point out that according to our experience, evaluations of that kind represents the absolute limit of the SAXS method.

2. Further Examples. Further examples are the studies on yeast pyruvate kinase by Müller et al.,[115] and on pancreatic amylase by Simon et al.[116]

C. Investigation of the Interaction between Enzymes and Other Biopolymers

1. Interaction between Lysine Transfer Ligase and Transfer RNA. Österberg et al.[117] showed by analysis of small angle x-ray scattering data

[114] I. Simon, *Eur. J. Biochem.* **30**, 184 (1972).

[115] K. Müller, O. Kratky, P. Röschlau, and B. Hess, *Hoppe Seyler's Z. Physiol. Chem.* **353**, 803 (1972).

[116] I. Simon, S. Móra, and P. Elodi, *Mol. Cell. Biochem.* **4**, 211 (1974).

[117] R. Österberg, B. Sjöberg, L. Rymo, and U. Lagerkvist, *J. Mol. Biol.* **99**, 383 (1975).

that the predominant complex formed in solution between lysine:tRNA ligase from yeast and tRNA consists of two ligase molecules and one molecule of tRNA.

a. EQUILIBRIUM DATA. The equilibria in the system consisting of lysine:tRNA ligase (A), tRNA (B); and a complex (AB) of both molecules may be described by the general reaction

$$p\text{A} + q\text{B} \rightleftarrows \text{A}_p\text{B}_q \tag{66}$$

which has the association equilibrium constant β_{pq}.

For a sufficiently dilute system, the scattered intensity I at any angle can be written as the sum of intensities scattered from all types of particles in the solution, i.e.,

$$I = I_\text{A} + I_\text{B} + \Sigma\Sigma I_{pq} \tag{67}$$

where I_A, I_B and I_{pq} are the scattered intensities from the molecules A, B and the complexes A_pB_q, respectively. The intensity contributed by each type of particle at a particular small angle depends on its concentration, its molecular weight and the excess electrons Δz [See appendix Eq. (91)]. The scattered intensity increases with an increase in any of these parameters.

Let us compare the scattered intensity I of the following two systems assuming the *same total* concentration of A and B in both cases. The first system consists of a mixture of A and B without any complex between the two. The other system contains free A and B particles plus complexes of the type A_pB_q. As a result of the higher molecular weight of the complexes, the second system will show a higher scattered intensity at sufficiently small angles. The intensity difference ΔI between the two systems depends only on the complex; it is the difference between the intensity I of the system containing components plus complex (i.e., A, B, A_pB_q) and the intensity $I_{\text{A tot}} + I_{\text{B tot}}$ of the system containing only A and B (with the same total concentration of A and B). $I_{\text{A tot}}$ and $I_{\text{B tot}}$ have to be recorded separately for each component (ligase A and tRNA B):

$$\Delta I = I - I_{\text{A tot}} - I_{\text{B tot}} \tag{68}$$

ΔI is divided by the total concentration of the enzyme A and the resulting $\Delta I/\text{A}$ is plotted versus the total tRNA concentration B to obtain the maximum value $(\Delta I/\text{A})_{\text{max}}$. From that, one can compute the normalized difference intensity $(\Delta I/\text{A})/(\Delta I/\text{A})_{\text{max}}$. For the subsequent analysis, the assumption was made that one single complex A_pB_q predominates in the solution. $(\Delta I/\text{A})_{\text{max}}$ was determined in the range where nearly all A particles are bound in order to make the contribution of free A particles negligible.

Comparison of experimental data [plotted as $(\Delta I/\text{A})/(\Delta I/\text{A})_{\text{max}}$ versus B] with theoretical curves generated for different complexes (AB, A_2B,

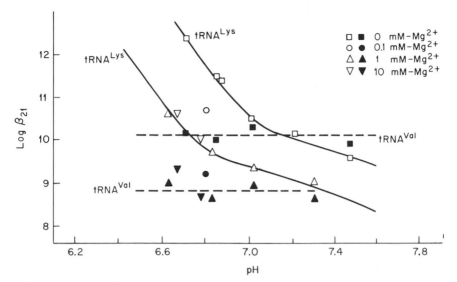

FIG. 43. The stability constants (log β_{21}) of the complexes between lysine:tRNA ligase (yeast) and two different tRNA molecules (tRNA^{Lys} and tRNA^{Val}) as a function of the pH. The experimental points correspond to the following Mg^{2+} concentrations: 10 mM Mg^{2+} (∇, $\blacktriangledown$), 1 mM Mg^{2+} ($\triangle$, $\blacktriangle$) 0.1 mM Mg^{2+} ($\bigcirc$, $\bullet$) and 0 mM Mg^{2+} ($\square$, $\blacksquare$). Open symbols refer to tRNA^{Lys} and the closed symbols for tRNA^{Val}; the solid lines for the tRNA^{Lys} complex, the dashed ones for the tRNA^{Val} comples. (From Österberg et al.[117])

A$_3$B, AB$_2$) and different equilibrium constants yielded the result that A$_2$B (two enzyme + one tRNA molecule) is the predominant complex. Its stability was studied with the cognate tRNA^{Lys} and the noncognate tRNA^{Val}. The authors found that the stability of both complexes decreases with increasing Mg^{2+} concentration. The two complexes, however, differed in the pH dependence of their respective equilibrium constant: While the stability of the noncognate complex was found to be independent of the pH, the stability of the cognate complex decreases with increasing pH as shown in Fig. 43.

b. MOLECULAR PARAMETERS OF THE COMPLEX LYSINE: tRNA LIGASE AND tRNA. The radius of gyration, volume and molecular weight of the complex were calculated as usual, after a correction to subtract the contribution of free ligase and free tRNA. To make this correction small, the experimental condition were chosen in a range where the complex has a high stability. Subsequently, the (corrected) experimental curve was compared with theoretical curves of different models. The two ligase molecules were assumed to have the same shape as unliganded native ligase[118] (i.e., an ellipsoid with semiaxes: a = 6.27 nm, b = 5.01 nm, and c = 2.35

[118] R. Österberg, B. Sjöberg, L. Rymo, and U. Lagerkvist, J. Mol. Biol. 77, 153 (1973).

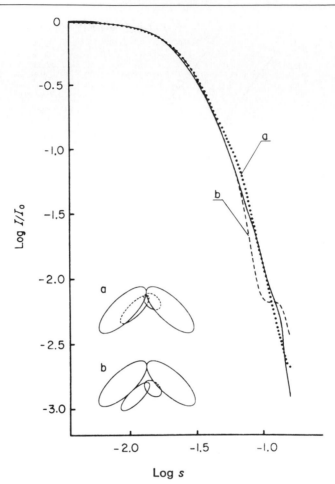

FIG. 44. Experimentally observed scattering curve from the complex of lysine: tRNA ligase with tRNA[Lys] from yeast (————), compared with theoretical curves calculated for two models. The models consist of two large ellipsoids at right angles (representing the ligase) and two smaller ones (representing the tRNA) in the hinge region. (a) About 50% of the tRNA ellispoid volume is within the enzyme ellispoid volue. (b) The tRNA allipsoids barely touch the enzyme ellipsoids. $s \ (= h) = (4\pi \sin \theta)/\lambda$. (From Österberg et al.[117])

nm). The tRNA molecule was approximated by two ellipsoids with the semiaxes: $a = 3.85$ nm, $b = c = 1.3$ nm and $a = 2.0$ nm, $b = c = 1.3$ nm, similar to the model suggested in 1970 by Pilz et al.[119] and in rough approximation to the x-ray crystal structure.[120]

[119] I. Pilz, O. Kratky, F. Cramer, F. von der Haar, and E. Schlimme, Eur. J. Biochem. 15, 401 (1970).
[120] S. H. Kim, J. L. Sussman, F. L. Suddath, G. J. Guigley, G. J. McPherson, A. H. J. Wang, N. C. Seeman, and A. Rich, Proc. Natl. Acad. Sci. U.S.A. 71, 4970 (1974).

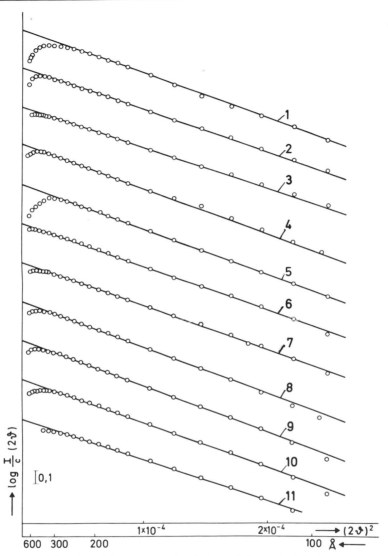

FIG. 45. Guinier plot of the cross section factors of glutamate dehydrogenase recorded for eleven different concentrations. The resulting values for the radius of gyration of the cross section R_q ($= R_c$) and the mass per unit length ($M/1$ Å) are plotted versus the concentration in Fig. 60 ($2\vartheta = 2\theta$). (From Sund et al.[123])

The best agreement was obtained for model a from Fig. 44, which has the two large ligase ellipsoids arranged at right angles and the tRNA molecule (represented by the small ellipsoids) in between.

2. *Further Examples.* Similar studies have been carried out on phenyl-alanine synthetase from yeast and the synthetase: tRNA[Phe] complex, by

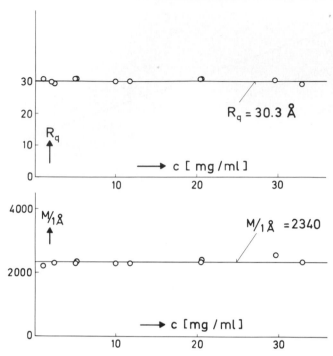

FIG. 46. Radius of gyration of the cross section R_q ($= R_c$) and mass per unit length ($M/1$ Å) ($= M_c$) of glutamate dehydrogenase as a function of enzyme concentration c. (From Sund et al. [123])

Pilz et al., [121] and on human antithrombin III and its complex with heparin by Furugren et al. [122]

D. Observation of Association and Dissociation Processes

Usually, a homodisperse solution of enzyme particles is a necessary prerequisite for SAXS investigations. There are, however, special circumstances where valuable information can be obtained even from polydisperse systems. Some of these cases are discussed in the following examples.

1. *Glutamate Dehydrogenase.* In the range of concentrations suitable for small-angle scattering (1 to 30 mg/ml), beef liver glutamate dehydrogenase exists as a mixture of aggregates, whose size depends on the concentration. Thus, the observed values for the radius of gyration and

[121] I. Pilz, K. Goral, and F. von der Haar, *Z. Naturforsch.* **34c**, in press.
[122] B. Furugren, L. O. Andersson, and R. Einarsson, *Arch. Biochem. Biophys.* **178**, 419 (1977).

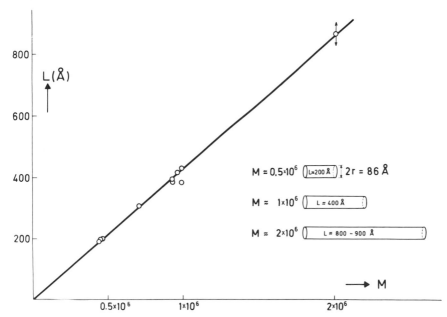

FIG. 47. Average length L of glutamate dehydrogenase particles in phosphate buffer, pH 7.6, as a function of the average molecular weight M. (From Sund et al.[123])

the molecular weight are only averages[123]; the observed molecular weight, for instance, varies (in the above concentration range) between 0.5×10^6 and 2×10^6. The cross-section curves, however, showed almost identical slopes in the Guinier plot, independent of concentration and average molecular weight (Fig. 45). Figure 46 shows the radius of gyration of the cross section ($R_c = 3.03$ nm) and the mass per unit length ($M/1$ Å ($=M_c$) = 2340 [see Appendix Eq. (99)] to be completely independent of the size of the associated molecule. This implies that this enzyme aggregates in the direction of the longest axis, while the cross section remains unchanged. Comparison of the experimental cross-section curves with theoretical curves for elliptical cross sections of different axial ratios indicated a more or less isotropic cross section, either circular or slightly elliptical, with a diameter of 8.6 nm.

While the parameters for the cross section are accurate, those obtainable for size and shape of the entire molecule are much less accurate due to the polydispersity of the solutions investigated.

Nevertheless, it could also be shown from overall parameters that the enzyme molecules associate in a linear fashion. Thus, the quotient of observed radii of gyration R_g and R_c yields the average length L of the ag-

[123] H. Sund, I. Pilz, and M. Herbst, Eur. J. Biochem. 7, 517 (1969).

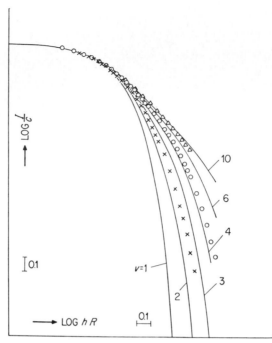

FIG. 48. Theoretical scattering curves of circular cylinders (full lines) in log-log plot; the ratios ($v = L/2r$) of length L to diameter $2r$ of the cylinder are 1, 2, 3, 4, 6, and 10. These curves are compared with experimental scattering curves for glutamate dehydrogenase of the following average molecular weights: 0.5×10^6 (×), 1×10^6 (○), and 2×10^6 (△); ($hR = hR_g$). (From Sund et al.[123])

gregates. A plot of these lengths against the average molecular weight is linear; that is, the length increases in proportion to the molecular weight of the aggregates (Fig. 47). This fact is also illustrated by a comparison of the experimental scattering curves of different concentrations with theoretical curves for spherical cylinders of different axial ratios (Fig. 48). Increasing aggregation causes an increase in anisotropy which leads to a flattening of the curves. To a first approximation, the axial ratios correspond to the models shown in Fig. 47.

Precise information on the shape of the glutamate dehydrogenase molecule can only be obtained from monodisperse solutions. Such solutions are prepared by addition of NADH and GTP, which cause a splitting of the enzyme molecule into its smallest enzymatically active units. These units[124] have a radius of gyration of 4.7 nm, a molecular weight of 300,000 and a ratio of length to diameter of 1.5:1. Their scattering curve is in fairly good agreement with the theoretical curve for the model suggested by

[124] I. Pilz and H. Sund, Eur. J. Biochem. 20, 561 (1971).

Eisenberg[125] from electron microscopic studies. This model consists of six subunits which are arranged in two layers with three subunits each.

2. *Malate Synthase.* Zipper and Durchschlag[126–128] studied malate synthase from baker's yeast in the absence and presence of its substrates (acetyl-CoA, glyoxylate) and the substrate analog pyruvate. The results showed two interesting effects. First, it was found that the enzyme molecule undergoes slight changes in the overall structure upon substrate binding. Second, an aggregation of the enzyme caused by x-ray exposure was observed.

a. MOLECULAR PARAMETERS OF ENZYME AND ENZYME–SUBSTRATE COMPLEXES. The enzyme and the enzyme–substrate complexes were studied in a Tris buffer (5 mM Tris-HCl, pH 8.1, containing 10 mg/ml MgCl$_2$, 1 mg/ml MgK$_2$EDTA, 0.2 or 2 mM dithiothreitol) at 4° using enzyme concentration between 2.5 and 60 mg/ml. The usual molecular parameters (such as radius of gyration R_g, molecular weight M, maximum particle diameter D, radius of gyration of the thickness R_t, and particle volume V) were determined for the enzyme and the complexes. The values for the free enzyme are $R_g = 3.95$ nm, $M = 18700$, $D = 11.2$ nm, $R_t = 1.04$ nm, and $V = 338$ nm^3. The parameters of the different enzyme–substrate complexes are very similar to those of the free enzyme. Only the radii of gyration and the maximum diameters of the various enzyme–substrate complexes are slightly smaller and the radii of gyration of the thickness are somewhat larger than the corresponding values observed for the apoenzyme. This indicates small structural changes upon substrate binding.

A comparison with theoretical scattering curves of simple triaxial bodies showed the enzyme to be an anisotropic flat particle with an axial ratio of about 1 : 0.3.

Good agreement was obtained with the curves for two kinds of models—oblate circular cylinders and oblate ellipsoids of revolution. Discrimination between both models was aided by the use of a camera with the new cone-collimation system. This collimation system (for a full discussion see Section III,F,3) was used to remeasure the outer part of the scattering curve with high accuracy. Thus it was possible to show that an ellipsoid of revolution (axes $a = b = 6.06$ nm, $c = 2.21$ nm) fits the experimental curve better than a circular cylinder.

The very similar scattering curves and molecular parameters for the

[125] H. Eisenberg, in "Pyridine Nucleotide Dependent Dehydrogenase" (H. Sund, ed.), p. 293. Springer-Verlag, Berlin and New York, 1970.
[126] P. Zipper and H. Durchschlag, *Biochem. Biophys. Res. Commun.* **75**, 384 (1977).
[127] P. Zipper and H. Durchschlag, *Z. Naturforsch.* **33c**, 504 (1978).
[128] P. Zipper and H. Durchschlag, in preparation.

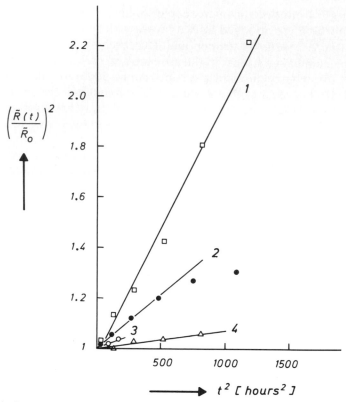

FIG. 49. Dependence of the apparent radius of gyration $\bar{R}$ ($=\bar{R}_g$) of various enzyme-substrate complexes of malate synthase on the time of irradiation t. 1, Substrate-free enzyme; 2, enzyme–glyoxylate; 3, enzyme–acetyl-CoA; 4, enzyme–pyruvate. (From Zipper and Durchschlag.[137]

enzyme–substrate complexes imply that they are also very similar in shape.

b. X-RAY-INDUCED AGGREGATION. Usually, enzymes are not affected by the exposure to x rays in the course of a small-angle investigation. Frequently, the biological activity of an enzyme solution is nearly identical before and after irradiation. Apart from that, a decrease of the enzymatic activity does not necessarily imply a change in the morphological parameters determined in a small-angle x-ray scattering experiment.

Malate synthase showed a clearly different behavior. Zipper and Durchschlag observed considerable changes of the scattering behavior after prolonged irradiation. The radius of gyration and the molecular weight increased as a function of the time of irradiation indicating an aggregation to larger particles. This effect was studied systematically by varying the time of exposure to x rays and the intensity of the incident (primary) radiation. For this purpose, some of the samples were irradiated considerably longer than in a typical small angle experiment.

The rate of aggregation was found to be a function of the intensity of the incident radiation. Interruption of irradiation was accompanied by an interruption of the aggregation. Figure 49 shows a plot of the square of the slit-smeared radius of gyration R_g versus the square of the irradiation time t, for the enzyme and some enzyme–substrate complexes. The data can be approximated by straight lines whose slopes represent a measure of the rate of aggregation. The rate of aggregation was found to be considerably reduced in the presence of a substrate or substrate analog. A similar effect was observed after adding excess dithiothreitol. All these substances appear to stabilize the enzyme against x-ray damage.

This small-angle investigation did not only yield information on the rate of aggregation but also on the aggregation process itself; it turned out that the thickness factor remains unchanged during aggregation (Fig. 50a). This rules out a stacking of the disc-like particles in the direction of the ro-

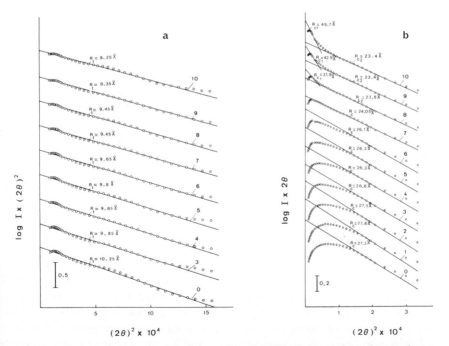

FIG. 50. Thickness plots (a) and cross-section plots (b) of aggregating substrate-free malate synthase. Time interval between subsequent curves: 5.7 hr; R_t, radius of gyration of the thickness; R_c, radius of gyration of the cross section. (From Zipper and Durchschlag.[128])

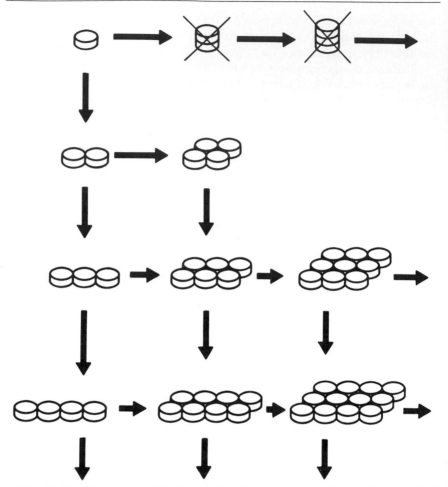

FIG. 51. Tentative scheme of the X ray-induced aggregation of malate synthease, derived from small-angle x-ray scattering measurements. (From Zipper and Durchschlag.[128])

tation axis (such as stacked coins). Small aggregates show one, and higher aggregates show two cross-section factors (Fig. 50b), which led the authors to suggest the association pattern indicated in Fig. 51. The first several molecules aggregate in one direction forming a linear row of side by side particles. This leaves the thickness of the flat particle unchanged and one cross-section factor is found, corresponding to the cross section of the linear row of particles. For larger oligomers, aggregation in a second direction has to be assumed, indicated by the existence of a second cross-section factor. This aggregation pattern is also supported by the distance distribution function $p(r)$ (Fig. 52b). The maximum distance in-

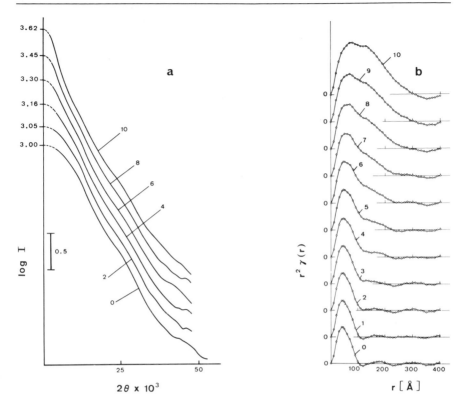

FiG. 52. Scattering curves (a) and distance distribution functions $p(r)$ (b) of aggregating substrate-free malate synthase. Time interval between subsequent curves: 5.7 hr.

creases stepwise, each step being roughly equal to the diameter of a single enzyme molecule.

All samples investigated became more and more polydisperse in the course of the aggregation. The authors performed a computer simulation of the assumed aggregation model, calculating distance distribution functions and cross-section curves for the various states of aggregation (Fig. 53). The agreement with the corresponding experimental curves (Figs. 52b and 50b) is quite striking. It illustrates again that even quite complex polydisperse systems can be successfully investigated with the small-angle technique.

3. Rabbit Muscle Glycogen Phosphorylase Dimer b and Tetramer b. Many enzymes exist in different states of aggregation, such as the present example which coexists as dimer and tetramer. Frequently, only one of these aggregation states can be obtained in a pure form in solution. The other component can only be studied in a paucidisperse system containing both aggregation states. If sufficiently accurate structural parameters are

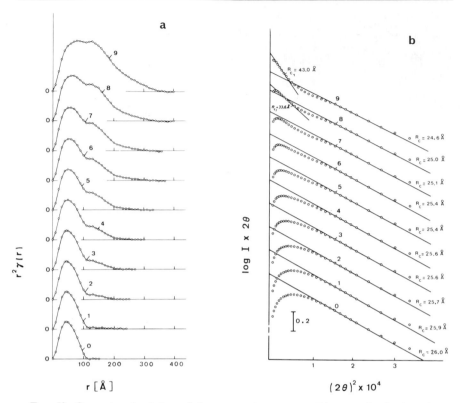

Fig. 53. Computer simulation of the aggregation process. Distance distributions functions $r^2\gamma(r)$ $[=p(r)]$ (a) and cross-section plots (b) calculated for various stages of aggregation. R_c, radius of gyration of the cross-section. (From Zipper and Durchschlag.[128])

obtainable for one of the components from studies on the homodisperse system, it may be possible to calculate the structural parameters of the other component from the scattering curve of the paucidisperse solution. This was done by Puchwein et al.[129] for the glycogen phosphorylase aggregates.

4. *DNA-Dependent RNA Polymerase.* Depending on the ionic strength of the solution, DNA-dependent RNA polymerase[130] from *Escherichia coli* K12 exists of a monomeric and a dimeric form. In this case, both forms can be obtained in a homodisperse solution, and the usual SAXS analysis was carried out.

At high ionic strength, the enzyme exists as a monomer, and at low

[129] G. Puchwein, O. Kratky, C. F. Gölker, and E. Helmreich, *Biochemistry* **9**, 4691 (1970).
[130] I. Pilz, O. Kratky, and D. Rabussay, *Eur. J. Biochem.* **28**, 205 (1972).

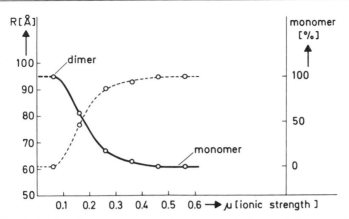

FIG. 54. Equilibrium between monomeric and dimeric forms of the DNA-dependent RNA polymerase (holoenzyme) as a function of the ionic strength μ of the buffer; solid line, experimentally observed radii of gyration R $(=R_g)$; dashed line, calculated amount of monomer. The total enzyme concentration was always around 9 mg/ml. (From Pilz *et al.*[130])

ionic strength as a dimer. The mixture of monomer and dimer existing at intermediate ionic strength was also studied to obtain information about the equilibrium between the two forms. Figure 54 shows the observed radius of gyration as a function of ionic strength.

E. *Comparison of Structural Data of Native and Modified Enzymes*

1. Native and Trypsin-Modified Methionyl-tRNA Synthetase from E. coli. The comparison between these two molecules is of special interest since so far only modified methionyl-tRNA synthetase could be crystallized in a form suitable for x-ray diffraction analysis.

The structural parameters found for the native and trypsin modified enzyme[131] are summarized in Table II.

Apart from these parameters Gulik *et al.* also determined the so-called distribution of cords $G(r)$. This function is equal to the second derivative of $\gamma(r)$. For homogeneous particles, it has the intentive meaning of a distance distribution function of the particle surface. $G(r)$ can be calculated from the scattering curve according to the following relationship.

$$G(r) = \int_0^\infty \left(1 - \frac{s^4 I(s)}{\lim_{s \to \infty} s^4 I(s)}\right) 8 \left(-\frac{d^2(\sin 2\pi rs/2\pi rs)}{d(2\pi rs)^2}\right) ds \qquad (69)$$

$$s = \frac{2 \sin \theta}{\lambda}$$

[131] A. Gulik, C. Monteilhet, P. Dessen, and G. Fayat, *Eur. J. Biochem.* **64,** 295 (1976).

TABLE II
MOLECULAR PARAMETERS FOR NATIVE AND TRYPSIN-MODIFIED
METHYONYL-tRNA SYNTHETASE

Parameter	Unit	Value for	
		Modified enzyme	Native enzyme
R_g	nm	2.48	4.3
M	g/mole	66,400	147,000
V	nm³	90	244
Solvation	grams H_2O per gram of protein	0.09	0.11
S	nm²	135	298

$G(r)$ is zero at the origin for particles with a smooth surface; a nonzero value indicates discontinuities of the particle surface ("sharp-edged particle"). The function becomes zero at the maximum particle diameter, which can thus be determined from $G(r)$. The only problem with this function is the fact that it is often hard to determine it with sufficient accuracy, as it is necessary to know the scattered intensity to very large angles. As mentioned above (compare Section IV,A,1), the $(2\theta)^{-4}$ extrapolation of the tail end is usually difficult to find and very unreliable.

The $G(r)$ functions for the native and modified enzyme are shown in Fig. 55. They yield a maximum diameter of 9 nm for the modified enzyme and the rather high value of 15 nm for the native synthetase.

The authors approximated the *modified enzyme* by a prolate ellipsoid with dimensions 9.4 × 4.1 × 4.1 nm, in excellent agreement with the

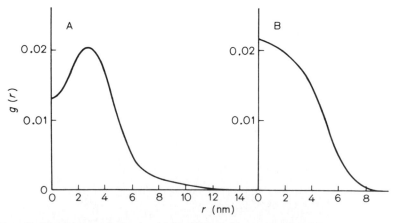

FIG. 55. Distribution of chords $g(r)$ [$= G(r)$], $\int g(r)\,ds = 1$. (A) Native methionyl-tRNA synthetase. (B) Trypsin-modified methionyl-tRNA synthetase. (From Gulik et al.[131])

crystal structure. The nonzero value of $G(r)$ at the origin (Fig. 55) was explained by the presence of edges or crevices, presumably related to the cleft shown in the crystal structure.

For the *native enzyme*, $G(r)$ shows a much slower decrease at large distances. The authors conclude that the native enzyme cannot consist of a compact structure formed by one globule; it consists rather of two (or more) globular domains linked by a flexible hinge. Such a model would also agree with other data found for the native enzyme.

APPENDIX

Comparison of the Notations of the Graz School and of Luzzati's School

With regard to summarizing papers relating predominantly to biological macromolecules we refer to the following articles.

Graz school	Luzzati school
Kratky and Pilz[35,132,133]	Luzzati[44,134]

Scattering angle 2θ (see P.K.T.)

ANGULAR VARIABLE

$$\boxed{h = 2\pi s} \qquad (71)$$

$$s = (2/\lambda)\sin\theta \qquad (72)$$

$$h = (4\pi/\lambda)\sin\theta \qquad (70)$$

λ is the wavelength of the radiation in nanometers.

$$m = a\sin 2\theta \qquad (73)$$

$$h \simeq (2\pi/\lambda a)\,m \qquad (74)$$

a is the distance between the sample and the plane of registration in centimeters.

$$\boxed{m/\lambda a \simeq s} \qquad (75)$$

FOURIER TRANSFORMATION

$$hI(h) = 4\pi \int_0^\infty \gamma(r)r\sin(hr)\,dr \qquad (76)$$

or

$$si(s) = 2\int_0^\infty rp(r)\sin(2\pi sr)\,dr \qquad (78)$$

$$I(h) = 4\pi \int_0^\infty \gamma(r)^2\,\frac{\sin(hr)}{hr}\,dr \qquad (77)$$

or

$$i(s) = \frac{2}{s}\int_0^\infty rp(r)\sin(2\pi sr)\,dr \qquad (79)$$

Graz school		Luzzati school
$I(h)$	Scattering intensity of the particle	$i(s)$
$\gamma(r)$	Characteristic function or correlation function of the particle	$p(r)$

$$A(h) = 4\pi \int \rho(r) r^2 \frac{\sin(hr)}{hr}\, dr \tag{80}$$

$$\sigma(s) = \frac{2}{s} \int r\rho(r) \sin(2\pi sr)\, dr \tag{81}$$

$A(h)$ | Scattering amplitude of a particle of spherical symmetry | $\sigma(s)$

$\rho(r)$ | Electron density distribution | $\rho(r)$

RADIUS OF GYRATION

$$R_g^2 = \frac{\int p(r) r^2\, dr}{2\int p(r)\, dr} = \frac{\int \gamma(r) r^4\, dr}{2\int \gamma(r) r^2\, dr} \tag{82}$$

$$R^2 = \frac{\int p(r) r^4\, dr}{2\int p(r) r^2\, dr} \tag{83}$$

R_g | Radius of gyration | R

$p(r)$ = distance distribution function

$$p(r) = \gamma(r) r^2 \tag{84}$$

$$I(h) = I(0) e^{-h^2 R_g^2/3} \tag{85}$$

$$i(s) = i(0) e^{(-4\pi^2/3) s^2 R^2} \tag{86}$$

MOLECULAR WEIGHT

m = number of electrons of the particle

$$\boxed{Mz = m} \tag{87}$$

$$M = \frac{I(0)}{P} \frac{21.0 a^2}{\Delta z^2 dc} \tag{88}$$

$$m = \frac{I(0)}{E_0 \eta v} \frac{1}{c_e(1 - \rho_0 \Psi)^2} \tag{89}$$

M = molecular weight

z = mole electrons per gram solute

$I(0)$ = scattering intensity at zero angle, i.e., scattered energy per second and unit area in the distance a of the sample

$I(0)$ = scattering intensity at zero angle, i.e., scattered energy per second and unit area in s scale (energy nm²/sec).

P = energy per second of the incident beam

E_0 = energy per second of the incident beam.

Δz = excess electrons (mole-electrons per gram)

d = thickness of the sample (cm)

c = concentration (g/cm³)

(Continued)

APPENDIX (continued)

Graz school	Luzzati school

$$d = \eta / \bar{\rho} \quad (90)$$

21.0 = $1/TN$

T = Thomson's constant 7.90×10^{-26}

N = Avogadro's number 6.03×10^{23}

In practice $I(0)$ (often written as I_0) and P are related to a segment of unit length of the incident beam. The scattering intensity per unit length of the primary beam is calculated with Eq. (14) in Section II,A,2. The corresponding energy of the incident beam can be detected by the rotator method or by a secondary standard (Section III,D).

$$\Delta z = (z_1 - v_1' \rho_2) \quad (91)$$

v_1' = isopotential specific volume of the solute in cm^3

ρ_2 = electron density of the solvent in mole-electrons per cm^3

$Z_1 = z_1 N$ (electrons per gram solute).

$\bar{\rho}$ = averaged electron density of the sample in electron/cm^3

η = "thickness of the sample" in electrons/cm^2

$\nu = T\lambda^2 = \lambda^2 \, 7.90 \times 10^{-26}$

λ = wavelength of the radiation in nm

c_e = concentration, measured by the ratio of the numbers of electrons (solute/solution)

Ψ = partial electronic volume of the solute in the solvent (nm^3/electron)

ρ_0 = electron density of the solvent (electrons/nm^3)

$$\rho_2 \frac{v_1'}{Z_1} = \Psi \rho_0 \quad (92)$$

$$Z_1 c = c_e \bar{\rho} \quad (93)$$

PARTICLES ELONGATED IN ONE DIRECTION, CHAIN-LIKE PARTICLES

SCATTERING INTENSITY

$$I_c(h) = \frac{I(h)h}{\pi L} \quad (94)$$

$$q(s) = \frac{2si(s)}{L} \quad (95)$$

$I_c(h)$ Scattering intensity of the cross section $q(s)$

L Length of the particle in nm L

R_c Radius of gyration of the cross section R_c

$$I_c(h) = I_c(0)e^{-h^2 R_c'^{2}/2} \quad (95)$$

$$q(s) = q(0)e^{-2\pi^2 h^2 R_c'^{2}} \quad (97)$$

MASS PER UNIT LENGTH

$\boxed{zM_c = \mu \quad (98)}$

M_c = mass (of the cross section) per unit length

μ = number of electrons (of the cross section per unit length)

$$M_c = \frac{[I(h)\cdot h]_{h\to 0}}{P}\,\frac{6.68a^2}{\Delta z^2 dc} \quad (99)$$

$$\mu = \frac{1}{c_e(1-\rho_0\Psi)^2}\,\frac{2[si(s)]_{s\to 0}}{E_0\eta v} \quad (100)$$

$$6.68 = \frac{1}{\pi}\,\frac{1}{TN}$$

PARTICLES ELONGATED IN TWO DIRECTIONS, LAMELLAR PARTICLES SCATTERING INTENSITY

$$I_t(h) = \frac{I(h)h^2}{2\pi F} \quad (101)$$

$$t(s) = \frac{2\pi s^2 i(s)}{F} \quad (102)$$

$I_t(h)$ Scattering intensity of the thickness $t(s)$

F Area of the plane of the lamella F

R_t Radius of gyration of the thickness R_t

$$I_t(h) = I_t(0)e^{-h^2 R_t^2} \quad (103)$$

$$t(s) = t(0)e^{-4\pi^2 h^2 R_t^2} \quad (104)$$

(Continued)

APPENDIX (*Continued*)

Graz school	Luzzati school

MASS PER UNIT AREA

M_t = mass (of the thickness) per unit area μ_t = number of electrons (of the thickness) per unit area

$$z M_t = \mu_t \quad (105)$$

$$M_t = \frac{[I(h)h^2]_{h \to 0}}{P} \frac{3.34 a^2}{\Delta^2 dc} \qquad (106)$$

$$\mu_t = \frac{1}{c_e(1 - \rho_0 \Psi)^2} \frac{2\pi[s^2 i(s)]_{s \to 0}}{E_0 \eta \nu} \qquad (107)$$

$$3.34 = \frac{1}{2\pi} \frac{1}{TN}$$

Eqs. (103), (105) and (107) are not given directly by Luzzati,[44] but they can be derived from the previous formulas.

VOLUME OF THE PARTICLE

V

$$V = 2\pi^2 \frac{I(0)}{Q} \qquad (108)$$

v_1

$$v_1 = \frac{i(0)}{4\pi \int_0^\infty s^2 i(s)\,ds} = \frac{i(0)}{2\pi \int sj(s)\,ds} \qquad (109)$$

where

$$Q = \int_0^\infty I(h)h^2\,dh = \frac{1}{2}\int_0^\infty \tilde{I}(h)h\,dh \qquad (110)$$

$\tilde{I}(h)$ Smeared scattering intensity (infinite length) $j(s)$

Specific Inner Surface

$$O_s = S/V \tag{111}$$

$$O_s = s_1/v_1 \tag{112}$$

S   Inner surface of a particle s_1

V   Volume of a particle v_1

$$v_1 = v/N \tag{113}$$

$$s_1 = S/N \tag{114}$$

v = total volume of the dissolved molecules in the sample

N = number of particles in v

S = interface surface of v

$$O_s = w_2\pi \frac{\lim_{h\to\infty} h^4 I(h)}{\displaystyle\int_0^\infty h^2 I(h)\, dh} \tag{115}$$

$$\frac{8\pi^3 \lim_{s\to\infty} s^4 i(s)}{\displaystyle\int_0^\infty 4\pi s^2 i(s)\, ds} = \frac{S}{V} = \frac{\rho_1 - \rho_0}{\rho_1 - \rho_0[1 - c_e(1 - \rho_0\Psi)]} \tag{116}$$

or

$$O_s = w_2 4 \frac{\lim_{h\to\infty} h^3 I(h)}{\displaystyle\int_0^\infty h\tilde{I}(h)\, dh} \tag{117}$$

Extrapolation to $c_e = 0$ and introduction of Eqs. (113), (114) gives

$$O_s = \frac{s_1}{v_1} = \frac{8\pi^3 \lim_{s\to\infty} s^4 i(s)}{\displaystyle\int_0^\infty 4\pi s^2 i(s)\, ds} \tag{118}$$

$$O_s = \frac{16\pi^2 \lim_{s\to\infty} s^3 j(s)}{\displaystyle\int_0^\infty 2\pi s j(s)\, ds} \tag{119}$$

w_1 = volume fraction of the solute
w_2 = volume fraction of the solvent
Extrapolation $w_1 \to 0$ gives $w_2 \to 1$

[132] O. Kratky, Prog. in Biophys. 13, 105 (1963).
[133] I. Pilz, in "Physical Principles and Techniques of Protein Chemistry" (S. J. Leach, ed.), Part C, p. 141. Academic Press, New York, 1973.
[134] V. Luzzati, Acta Crystallogr. 13, 939 (1960).

[12] Practical Aspects of
Immune Electron Microscopy

By JAMES A. LAKE

Knowledge of ribosome structure and function has increased at a remarkable rate as a result of the very fruitful combination of the techniques of biochemistry, immunology, and electron microscopy. Immune electron microscopy has provided structural information about the three-dimensional arrangements of ribosomal proteins and the factors participating in protein synthesis in detail that was almost unthinkable five years ago. It has led to tentative locations being assigned for the codon–anticodon interaction, the tRNA binding sites, initiation factor binding sites, and the peptidyl transferase.[1] The locations of these sites have, in turn, provided important clues to some of the molecular events that occur during the elongation cycle of protein synthesis, and this has recently led to the development of the concept of aminoacyl-tRNA binding at the recognition (R) site.[2] This chapter explains in detail the techniques of electron microscopy of antibody labeled ribosomes and ribosomal subunits in anticipation that they can be usefully extended to the analysis of other cell organelles and supramolecular structures, such as microtubules, flagella, nuclear pores, thick and thin muscle filaments, gap junctions, and enzyme complexes.

The most widely used method of negatively staining macromolecules and macromolecular assemblies is the droplet method described by Brenner and Horne.[3] The simplicity of this technique has tended to obscure the important advantages of other negative staining methods. Although the droplet technique is well suited for resolving structural details of many objects (such as viruses), smaller molecules, (such as antibodies) cannot be reliably contrasted by this method. It was, therefore, considered to be a significant advance when Valentine and Green,[4] used another negative staining technique to reproducibly resolve the shape of IgG molecules. The technique of immune electron microscopy was initially used, in combination with the droplet method, by Yanagida and Ahmad-Zadeh[5] to map the locations of specific proteins in bacteriophage

---

[1] J. A. Lake, *FEBS*, "A Chapter in Gene Expression" (B. F. C. Clark, H. Klenow, and J. Zeuthen, eds.), pp. 121–130. Pergamon Press, Oxford, 1978.

[2] J. A. Lake, *Proc. Natl. Acad. Sci. U.S.A.* **74,** 1903 (1977).

[3] S. Brenner and R. W. Horne, *Biochim. Biophys. Acta* **34,** 103 (1959).

[4] R. C. Valentine and M. Green, *J. Mol. Biol.* **27,** 615 (1967).

[5] M. Yanagida and C. Ahmad-Zadeh, *J. Mol. Biol.* **51,** 411 (1970).

METHODS IN ENZYMOLOGY, VOL. 61

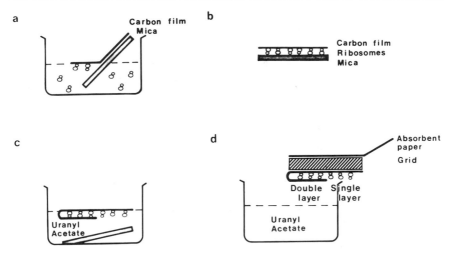

FIG. 1. A diagram of the technique of preparing single carbon layer and double carbon layer negatively stained specimen grids. (a) Adsorbing the ribosomes to the carbon film. (b) The carbon film after withdrawing the mica. (c) Floating the carbon film off the mica and onto the surface of the negative stain. A single carbon layer region is illustrated at the right and a double carbon layer region at the left. (d) Lifting the grid from the negative stain.

T4. It was first applied to large ribosomal subunits, small ribosomal subunits, and monomeric 70 S ribosomes by Wabl,[6] Lake et al.,[7] and Lake,[8] respectively. Subsequently, extensive mapping of both small and large subunits has been done in Germany by Tischendorf, Zeichardt, and Stöffler and in the United States by Kahan, Strycharz, Nomura, and Lake. Antibodies were first visualized by electron microscopy (Valentine et al.,[9] 1968) in negatively stained regions that are currently referred to as *single-carbon-layer* regions. This is in order to distinguish them from the *double-carbon-layer* regions that are primarily used in antibody labeling studies of ribosomal subunits (Lake et al.,[10] 1974).

The method used to produce both types of negatively stained regions is illustrated in Figure 1. In this procedure, a small carbon film that has been previously deposited onto the surface of freshly cleaved mica is partially floated onto a solution of ribosomes (or other macromolecules) being stained, as shown in Fig. 1a. The carbon film is then withdrawn by lifting the mica from the solution, and the attached layer of carbon traps

[6] M. R. Wabl, *J. Mol. Biol.* **84**, 241 (1974).

[7] J. A. Lake, M. Pendergast, L. Kahan, and M. Nomura, *J. Supramol. Struct.* **2**, 189 (1974).

[8] J. A. Lake, *J. Mol. Biol.* **105**, 131 (1976).

[9] R. C. Valentine, B. M. Shapiro, and E. R. Stadtman, *Biochemistry* **7**, 2143 (1968).

[10] J. A. Lake, M. Pendergast, L. Kahan, and M. Nomura, *Proc. Natl. Acad. Sci. U.S.A.* **71**, 4688 (1974).

some of the ribosome solution between it and the mica, as shown in Fig. 1b. In the third step of the process (Fig. 1c), the mica is inserted into a solution of uranyl acetate so that the carbon film is completely floated off the mica substrate. If the mica is inserted gently, then the carbon floats off as a single, unfolded sheet. These regions of the film, illustrated at the right in Figure 1c, produce *single carbon layer* images. On the other hand, if the mica is plunged into the uranyl acetate, the carbon film folds back onto itself as shown at the left in Fig. 1c. In these circumstances, in many regions the ribosomes are sandwiched in a shell of negative stain between two carbon films. These regions produce *double-carbon-layer* images. Next, a grid[11] is placed on top of the carbon and then the grid and attached carbon film are lifted from the uranyl acetate solution by placing a piece of absorbent paper[12] on top of the grid, waiting for the staining solution to begin wetting the absorbent paper and lifting the grid from the solution, as shown in Fig. 1d. Alternatively, good results can also be obtained if the completed grid is directly lifted from the surface by forceps (K. Leonard, personal communication). Neither technique produces negative staining exclusively of the single or of the double carbon layer type. Just as the carbon film often folds back on itself forming small regions of double layer images when the single layer technique is used, likewise, sizable regions of single carbon layer images will still be found on the grid in addition to the desired double layer regions when the double layer technique is used.

The first small subunit protein to be mapped using immunoelectron microscopy was S14,[6] and it will be used to illustrate the procedure of mapping protein locations. Protein S14 is a functional protein and is incorporated into the small subunit relatively late in the assembly process. An electron micrograph of a field of small subunits reacted with antibodies against S14 is shown in Fig. 2A. Pairs of subunits connected by either one or by two IgG's are indicated by arrows. In the central pair of subunits the Fab region attached to each subunit is visible as well as the Fc region of the IgG.

---

[11] A 400 mesh grid coated with adhesive is preferred. Adhesive covered grids can be made by dissolving the adhesive from a ⅛ in. strip of Scotch Brand transparent tape (not magic mending tape) in 5 ml of chloroform, pouring the solution over grids arranged on the bottom of a petri dish, and allowing the chloroform to evaporate. After drying thoroughly, the grids should be easily removed from the bottom of the dish using finely tipped forceps. If too much adhesive was used and the grids are firmly attached to the bottom of the dish, the dish can be rinsed with chloroform to remove the excess adhesive.

[12] A good adsorbent paper is newsprint covered with very dark printing so that the paper fibers are flattened.

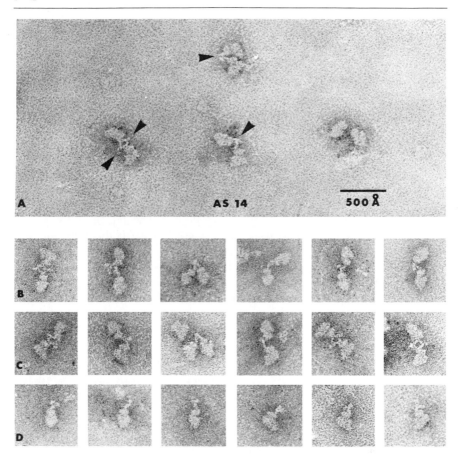

FIG. 2. Electron micrographs of small subunits reacted with AS14 antibodies. (A) A field of subunits. Pairs of subunits connected by either one or two IgG's indicated by arrows. (B) A gallery of subunits in the quasisymmetric projection. (C) A gallery of subunits in the asymmetric projection. (D) A gallery of single subunits with attached antibodies.

The small subunit profiles observed in electron micrographs correspond to the small subunits being positioned on the carbon in different orientations. Two of the most distinctive views are shown in Fig. 2b and 2c. The view illustrated at the left in Fig. 3, the quasisymmetric projection, is characterized by a line of approximate mirror symmetry that is coincident with the long axis of the subunit. Pairs of subunits with the lower subunit of the pair in this orientation are shown in Fig. 2b. The lower subunit of each pair in Fig. 2c is oriented in the "asymmetric" view (the view shown

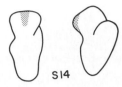

FIG. 3. A diagrammatic representation of the region of attachment (stippled area) of antibodies to protein S14.

on the right in Fig. 3). Antibodies against S14 thus were mapped at a single, unique three-dimensional site shown in Fig. 3.

As is the case with the droplet technique, both negatively and positively contrasted images are seen in single carbon layer contrasted regions. Figure 4 illustrates the transition between a region of single carbon layer (at the top) and a double carbon layer (at the bottom) in a negatively contrasted field of small ribosomal subunits. Similar fields are frequently observed in which the subunits in the single layer region are positively contrasted. Both types of negative staining are useful for structural studies on ribosomes. The double layer carbon image offers the advantage of providing a very uniform background of thin stain in the regions between subunits and hence makes it easier to observe antibodies. This is also possible, but more difficult, in single carbon layer regions. Tilting experiments suggest that subunits contrasted in double layer regions are flattened between the carbon layers. The flattening, however, does not appear to cause significant distortion of the structure in the plane of the carbon films, since similar structural features of ribosomes are revealed by the droplet, the single carbon layer, and the double carbon layer techniques.[8] The dimensions of ribosomes contrasted by these techniques do vary considerably, however. In the table the lengths of the longest dimension of the small subunit of the *E. coli* ribosome is listed for each of the three staining techniques, together with the length measured by X-ray scattering from solutions of small subunits.[13]

The largest dimension of the 30 S subunit as measured by X-ray scattering involves no assumptions about the shape of the subunit and is a relatively reliable physical measurement of the maximum distance within the subunit in solution. Hence, for ribosomal subunits at least, the double layer carbon method gives images with dimensions that are quite close to those measured in X-ray studies. The single carbon layer method gives dimensions that are about 10% less than the X-ray results, and the droplet method gives dimensions that are about 30% less than the X-ray results.

A major advantage of the double carbon layer technique is that the areas of negative contrast are frequently much larger than those typically

[13] W. E. Hill, J. D. Thompson, and J. W. Anderegg, *J. Mol. Biol.* **44**, 89 (1969).

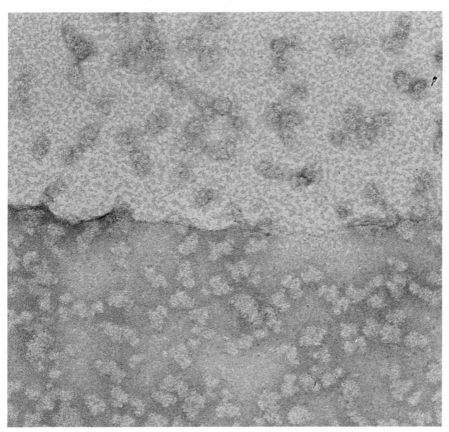

FIG. 4. A negatively contrasted field of *E. coli* small ribosomal subunits. The lower one-half of the field is a double layer carbon image and the upper one-half is a single carbon layer image.

TABLE I
COMPARISON OF 30 S RIBOSOMAL SUBUNIT DIMENSIONS DETERMINED
BY DIFFERENT TECHNIQUES

Technique	Longest dimension (Å)
Electron microscopy (droplet)	180
Electron microscopy (single layer)	220
Electron microscopy (double layer)	250
X-Ray scattering from solutions	250

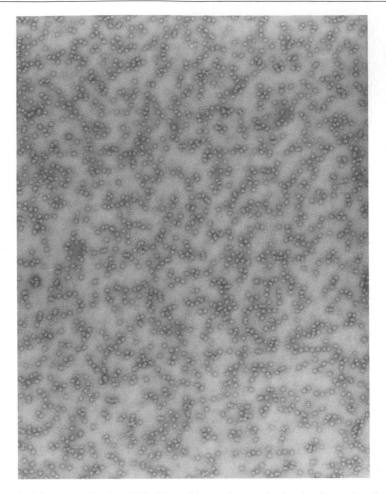

Fig. 5. A low magnification field of *E. coli* large ribosomal subunits, illustrating the large regions of double carbon layer negatively stained images that can be obtained.

obtained with the droplet method. The double layer images can, in favorable circumstances, extend over several squares of a 400 mesh grid with only slight differences in the negative staining occurring throughout the entire area. A typical large area of double layer contrasted 50 S subunits is shown in Fig. 5.

Although the double layer technique is obviously the method of choice for antibody labeling experiments on large and small ribosomal subunits, it is not a panacea. For macromolecular assemblies that are more than 250–300 Å thick, the double layer technique tends to flatten the specimens to such an extent that structural details are obscured. For this reason the single layer technique is, in general, superior to the double layer techique for contrasting larger structures, such as monomeric ribosomes. In other cases, the droplet method is superior to both the single and double carbon layer methods. For example, in investigating the structure of the capsid of bacteriophage ϕCbK,[14,15] it was found that droplet negative staining produced electron micrographs in which structural details were preserved to 20 Å resolution as measured by optical diffraction, whereas capsids negatively stained by the other two methods did not diffract. The tail of the same phage, however, produced optical diffraction patterns extending to 16 and 20 Å resolution when contrasted by the single carbon layer and the droplet techiques, respectively.[16] Hence, the staining properties of a particular specimen cannot be predicted in advance, and in any new situation it is sensible to try all three techniques.

[14] K. R. Leonard, A. K. Kleinschmidt, N. Agabian-Kashishian, L. Shapiro, and J. W. Maizel, Jr., *J. Mol. Biol.* **71,** 201 (1972).
[15] J. A. Lake and K. R. Leonard, *J. Mol. Biol.* **86,** 499 (1974).
[16] K. R. Leonard, A. K. Kleinschmidt, and J. A. Lake, *J. Mol. Biol.* **81,** 349 (1973).

Section III

Conformation and Transitions

[13] Microcalorimeters for Biological Chemistry: Applications, Instrumentation and Experimental Design[1]

By NEAL LANGERMAN and RODNEY L. BILTONEN

Introduction

During the past decade, answers have been sought to increasingly difficult questions concerning energy changes in biological systems. In many instances, methods such as equilibrium dialysis or spectrometry are sufficient to provide the necessary data. However, many problems, especially those involving the entropy and enthalpy changes occurring during a reaction, are best approached by direct calorimetric methods.

Until very recently, calorimetric techniques have not been widely used in the study of biological systems due to the lack of commercial instruments applicable to the small sample size generally required. The problem of sensitivity has now been essentially solved. Many theoretical and practical aspects of biological reaction calorimetry have been recently reviewed.[1a-7] A variety of calorimeters particularly suited to the study of biological samples are now available. In this chapter we will attempt to provide a guide to (1) establishing the need for a calorimeter, (2) determining which instrument is best suited to the problem, and (3) setting up a calorimetry laboratory which can operate in the manner necessary to answer thermodynamic and other questions. In Biltonen and Langerman [14], this volume, we will examine various methods of data analysis, interpretations of the data, and some of the problems currently being studied calorimetrically.

[1] This research was supported in part by United States Public Health Service Research Grants GM22049, GM 20637, and GM 22676 from the National Institutes of General Medical Sciences, and National Science Foundation Grant BMS75-03030 and BM 75-23245.

[1a] J. M. Sturtevant, Vol. 26, Part C, p. 227.

[2] J. M. Sturtevant, *Annu. Rev. Biophys. Bioeng.* **3,** 35 (1974).

[3] C. Spink and I. Wädso, *Methods Biochem. Anal.* **23,** 1 (1976).

[4] I. Wädso, *Pure Appl. Chem.* **38,** 529 (1974).

[5] M. A. Marini and C. J. Martin, Vol. 27, Part D, p. 590.

[6] H. D. Brown, ed., "Biochemical Microcalorimetry." Academic Press, New York, 1969.

[7] D. J. Eatough, J. J. Christensen, and R. M. Izatt, "Experiments in Thermometric Titrimetry and Titration Calorimetry." Brigham Young Univ. Printing Serv., Provo, Utah, 1973.

I. Establishing the Need for a Calorimeter

Techniques, Concepts and Terms. A calorimeter can measure the quantity of heat associated with a particular process in three distinct ways. The first of these is to measure the temperature rise associated with the heat production.[7a] In this case, assuming a purely adiabatic system, the rise in temperature will be equal to the heat input divided by the heat capacity of the system. Therefore, the heat associated with the process can be calculated if the temperature rise is measured and the apparent heat capacity of the system is known. The calorimeter used in this type of process is usually referred to as (pseudo)adiabatic or isoperibolic. The Tronac 450 Titration calorimeter and the LKB 8700 calorimeter system are examples of this type.

The second method involves the determination of the rate of heat flow resulting from a heat effect. There are two basic procedures to measure this heat flow. The first involves the *active* use of a Peltier heating-cooling device which is controlled by a feedback system. (A thermistor, or other type of temperature sensor, monitors changes in the temperature of the solution. As the temperature is changed, a current is applied to the Peltier device to maintain the solution at constant temperature.) The total power input to the device during the course of the experiment is equal to the reaction heat effect. This type of calorimeter, an example of which is the Tronac 550 Titration calorimeter, is strictly isothermal. The second procedure is to use a Peltier device, or other type of differential temperature sensor, in a passive mode. Here the device produces a voltage proportional to the temperature difference between the heat sink and the calorimeter cell, which according to Newton's law of cooling is directly proportional to the rate of flow between the cell and the sink. In a twin or differential calorimeter, the sensors in the two cells are connected in electrical opposition so that the voltage measured is proportional to the difference in temperature between the sample and reference cell. Integration of the voltage–time curve provides an estimate of the total heat effect. Although in such a device, the temperature of the reactant solution is identical at the beginning and end of the experiment, the temperature within the reaction cell changes during the experiment. This type of calorimeter is thus not strictly an isothermal device and is generally referred to as a heat-leak calorimeter. The LKB batch and flow microcalorimeters are of this type.

It is possible to initiate reactions or to disturb their thermal equilibrium position by changing the temperature of the system. For example, if the temperature is raised by the introduction of heat, the system will ad-

---

[7a] We have treated all discussion of heat effects as if they were exothermic. For an endothermic process, the appropriate changes may be made throughout the discussion.

just to a new equilibrium position. The temperature dependence of this shift is a quantity that can be directly related to the heat capacity of the system. Direct measurements of the heat capacity can provide information about changes in chemical equilibria and allow evaluation of the enthalpy change for the process. The calorimeter used in this type of process is usually referred to as a differential scanning calorimeter.

The heat quantity associated with all processes occurring can be obtained either by direct measurement of the heat effect or by measurement of the change in the heat capacity of the system. One obtains from calorimetric experiments either the net heat energy change associated with all processes which are occurring in the system or the heat capacity of the system. Analysis of the data and their subsequent interpretation is dependent upon the detailed design of the experiment and knowledge of the system under consideration.

Two basic types of experiments can be performed. The first type of measurement involves the determination of the heat change associated with the mixing of two, or more, reactive components in the calorimeter. The second type of measurement employs the use of differential scanning calorimetry. In this, and the following chapter, we will consider only the first type of experiment; differential scanning calorimetry is discussed in detail in Krishnan and Brandts.[7b]

Before initiating a discussion regarding the detailed use of reaction microcalorimeters in the study of biological systems it is useful to summarize, in a general way, the types of information which can be obtained from such studies. We will use a number of examples from the recent literature to do this.

Calorimetry of Enzyme-Catalyzed Reactions. Hexokinase reactions have been carefully studied by Goldberg.[8,9] The enthalpy change of hexokinase-catalyzed phosphorylation of glucose, mannose, and fructose to their respective hexose 6-phosphates have been measured calorimetrically in Tris/Tris-HCl buffer at various temperatures and ionic strengths. The data were corrected to the thermodynamic state of unit activity of the reactants and products. Rigorous analysis of the data demonstrated that the reactions are strongly dependent on pH below pH 8, but independent of ionic strength.

Rothman et al.[10] have examined the conversion of deoxyuridine monophosphate to deoxythymidine monophosphate by the enzyme thymidylate synthetase. They were able to estimate the magnitude of the entropy

[7b] K. S. Krishnan and J. Brandts, in "Methods in Enzymology" (C. H. W. Hirs and S. N. Timasheff, eds.), Vol. XLIX, p. 3. Academic Press, New York, 1978.
[8] R. N. Goldberg, *Biophys. Chem.* **4**, 215 (1976).
[9] R. N. Goldberg, *Biophys. Chem.* **3**, 192 (1975).
[10] S. W. Rothman, R. L. Kisliuk, and N. Langerman, *J. Biol. Chem.* **248**, 7845 (1973).

change for the reaction as well as point out that, on thermodynamic grounds, it is impossible to reverse the direction of the reaction without a significant amount of Gibb's energy being utilized. A heavy emphasis was placed on model studies in order to arrive at this conclusion.

Wädso[4] and Beezer et al.[11,12] have demonstrated that flow calorimetric techniques can also be used under certain conditions for enzyme assay and for the determination of the apparent rate of enzyme-catalyzed reactions. More recently, Johnson and Biltonen[13] has shown that the enthalpy change, the Michaelis constant, and the maximum velocity of the chymotrypsin and ribonuclease systems can be accurately determined calorimetrically. This latter work examined in detail the methods of data analysis necessary to derive the information. The use of calorimetric techniques to obtain kinetic information will be discussed in detail in Biltonen and Langerman [14], this volume.

Many more examples of calorimetric studies of enzyme catalyzed reactions are reviewed in Brown[6] which should be consulted for further information.

Calorimetry of Ligand Binding. Biltonen and co-workers[14-18] conducted an extensive study of the binding of 3′-CMP to ribonuclease over a broad range of ionic strength and pH and obtained an extensive set of values for $\Delta G^{0'}$, $\Delta H^{0'}$, and $\Delta S^{0'}$. The results were interpreted as demonstrating that van der Waals' interactions between the riboside moiety and the protein and that electrostatic interaction between the negatively charged phosphate group of the inhibitor and the positively charged histidine residues on the protein were the primary driving forces for binding. Of special interest was an estimate of the actual magnitude of the electrostatic contribution to the interaction and its importance in the catalytic process.

Hemoglobin is currently one of the more thoroughly studied proteins. Critical calorimetric determinations of the heat of oxygenation were first performed by Chipperfield et al.[19] and later extended in the recent study of Atha and Ackers.[20] The latter workers demonstrated that a model

[11] A. E. Beezer, T. I. Steenson, and H. J. V. Tyrrell, *Thermochim. Acta* **9**, 447 (1974).

[12] A. E. Beezer, T. I. Steenson, and H. J. V. Tyrrell, *Talanta* **21**, 467 (1974).

[13] R. E. Johnson and R. L. Biltonen, *J. Am. Chem. Soc.* **97**, 2349 (1975).

[14] D. W. Bolen, M. Flogel, and R. L. Biltonen, *Biochemistry* **10**, 4136 (1971).

[15] M. Flogel, D. W. Bolen, and R. L. Biltonen, *Protides Biol. Fluids, Proc. Colloq.* **20**, 521 (1973).

[16] M. Flogel and R. L. Biltonen, *Biochemistry* **14**, 2603 (1975).

[17] M. Flogel and R. L. Biltonen, *Biochemistry* **14**, 2610 (1975).

[18] M. Flogel, A. Albert, and R. L. Biltonen, *Biochemistry* **14**, 2616 (1975).

[19] J. R. Chipperfield, L. Rossi-Bernardi, and F. J. W. Roughton, *J. Biol. Chem.* **242**, 777 (1967).

[20] D. H. Atha and G. K. Ackers, *Biochemistry* **13**, 2376 (1974).

which distributes the total heat evenly over each oxygenation step could not be distinguished from other distributions of the total enthalpy change at each step of the reaction. A similar study of the oxygenation of hemoglobin was performed by Gill and his co-workers[21-23] who confirmed this result.

Rehfeld *et al.*[24] have used titration methods to evaluate both the stoichiometry and energetics of binding of human albumin and concanavalin A to normal and sickled erythrocytes. They were able to deduce the number of albumin molecules bound to the cell, the apparent binding constant, and the binding enthalpy change. These data revealed a significant difference between the enthalpy change of binding of the two types of cells and suggested a potential method for identifying abnormal cell lines.

Barisas *et al.*[25] and Biltonen and co-workers[26,27] have determined the enthalpy change associated with the binding of dinitrophenyl-ϵ-lysine to anti-DNP antibodies. In these studies, the association constants were not determined by calorimetry because of the very high binding affinity. Both groups provided evidence that the variation in affinity within an antibody population was not due to a difference in enthalpy change but rather to a change in the entropy change associated with binding. The latter group further showed that the difference in affinity appeared to be a result of the degree of interaction between the protein and the nonpolar side chain of the hapten which could be correlated with the ''depth'' of the hapten binding site.

A few studies have focused on the thermodynamics of proton binding to proteins. Shaio and Sturtevant[28] determined the apparent pK values and associated enthalpy change for protonation of chymotrypsin, lysozyme, and ribonuclease. Flogel and co-workers[15-18] studied in detail the titration behavior of the four histidine residues of ribonuclease. An interesting result of the latter study was the demonstration that at low ionic strength histidine 48 titrated with a normal pK but an extremely large enthalpy change which was interpreted to indicate a conformational change in the protein upon protonation.

[21] S. A. Rudolph and S. J. Gill, *Biochemistry* **13**, 2451 (1974).
[22] H. T. Gaud, B. G. Barisas, and S. J. Gill, *Biochem. Biophys. Res. Commun.* **59**, 1389 (1974).
[23] H. T. Gaud, S. J. Gill, B. G. Barisas, and K. Gersonde, *Biochemistry* **14**, 4588 (1975).
[24] S. J. Rehfeld, D. J. Eatough, and L. D. Hansen, *Biochem. Biophys. Res. Commun.* **66**, 586–591 (1975).
[25] B. G. Barisas, J. M. Sturtevant, and S. J. Singer, *Biochemistry* **10**, 2816 (1971).
[26] J. F. Halsey and R. L. Biltonen, *Biochemistry* **14**, 800 (1975).
[27] J. F. Halsey, J. Cebra, and R. L. Biltonen, *Biochemistry* **14**, 5221 (1975).
[28] D. D. F. Shiao and J. M. Sturtevant, *Fed. Proc., Fed. Am. Soc. Exp. Biol.* **29**, 335 (1970).

Vander Jagt *et al.*[29] determined the microionization constants and enthalpy changes associated with protonation of glutathione and confirmed the original estimates for these constants reported by Benesch and Benesch.[30]

Calorimetry of Conformational Changes. Reaction calorimetry generally has not been used for direct study of conformational changes of biological macromolecules. This is more generally done with scanning calorimetric techniques and are discussed in Krishnan and Brandts.[7b] Sturtevant and co-workers[31] did study the helix forming reaction poly(A) + poly(U) → poly(A + U) in the presence of Na^+ ion and found $\Delta H = -7$ kcal/mole of base pairs.

More recently, Atha and Ackers[32] determined the enthalpy change associated with the acid denaturation of lysozyme at a number of concentrations of guanidine hydrochloride. Their calorimetrically obtained value was in very close agreement with that obtained by van't Hoff analysis of the temperature dependence of the apparent equilibrium constant. This agreement demonstrated that to a good approximation the unfolding reaction of lysozyme was of the two-state type.

Two final examples which may broadly be considered in this category are the study of the self-association of proteins. An early effort in this area was conducted by Langerman and Sturtevant[33] on hemerythrin. By direct calorimetric measurement, they confirmed the enthalpy change for association was zero, which agreed with previous van't Hoff estimates. Thus, octamer formation of hemerythrin is completely entropically driven.

Valdes and Ackers[34] have recently examined the polymerization of bacterial flagellin. They report an enthalpy association (-12.7 kcal/mole) at 25°, which contrasted greatly with that observed at 40° ($+38$ kcal/mole). If the same mechanism of association is operating at both temperatures, this implies a heat capacity change of 3.3 kcal mole^{-1} deg^{-1} associated with polymerization!

These examples should provide an overview of the types of information regarding biological systems which can be obtained using reaction calorimetric techniques. It must be noted, however, that the acquisition of data relating to any particular system depends upon the nature of the question being asked, the estimated heat effect, and the availability of

[29] D. L. Vander Jagt, L. D. Hansen, E. A. Lewis, and L. B. Han, *Arch. Biochem. Biophys.* **153**, (1972).

[30] R. E. Benesch and R. Benesch, *J. Am. Chem. Soc.* **77**, 5877 (1955).

[31] M. A. Rawitscher, P. D. Ross Ross, and J. M. Sturtevant, *J. Am. Chem. Soc.* **85**, 1915 (1963).

[32] Atha, D. H. and G. K. Ackers, *J. Biol. Chem.* **246**, 5845 (1971).

[33] N. Langerman and J. M. Sturtevant, *Biochemistry* **10**, 2809 (1971).

[34] R. Valdes and G. K. Ackers, *Biochem. Biophys. Res. Commun.* **60**, 1403 (1974).

material, and this dictates the type of reaction calorimeter required for the study. Section III will describe in some detail different types of calorimeters and how they should be judged as regards their applicability.

II. Instrumentation

Currently, LKB Productor, Inc. (Rockville, Maryland) and Tronac, Inc. (Orem, Utah) account for the majority of all commercial microcalorimeters sold in the United States. While other instruments are also available,[34a] our remarks will be directed primarily toward the LKB and Tronac instruments, since these are the instruments with which we are most familiar. Direct evaluation of the other instruments should be possible based on this section.

An option is to construct rather than purchase a calorimeter. Several review articles describe the details of the design and construction of a calorimeter.[35–38] Our comments here will be primarily directed toward the persons interested in purchasing and using a commercially available instrument. We will, however, briefly consider various modifications of such instruments which have been reported in the literature. In general, our discussion will be limited to experimental work requiring an accuracy of at least 10% and carried out at 1 atm pressure in the temperature range of 0°–60°.

Type of Calorimeter. There exist three fundamental types of microcalorimeters, and the operating characteristics of each are different. These three types have been designated as batch, flow, or titration calorimeters. Figure 1 illustrates the signal (thermogram) obtained from each type of unit, and the solution mixing system employed.

The batch calorimeter produces a signal such that the total area under the thermogram is proportional to the heat of the reaction. The heat q is dependent upon the total number of moles n of limiting reagent reacting in the calorimeter. A single loading of a batch calorimeter results in a single value for the heat of reaction at the concentration of the experiment.

[34a] In addition to the Tronac and LKB instruments, the Picker Flow Calorimeter is also currently capable of meeting the operating specifications to which we have limited this report. We have not commented on the Picker instrument, since neither of us have direct experience with it. The manufacturer's claimed operating specifications are listed in Table I. Other instruments which might also be considered are Microsal, Ltd. Mark 2V Flow Calorimeter (London, England); Setaram CALVET (batch) microcalorimeter (Lyon, France); and the American Instrument Co. Enthalpimeter (Silver Springs, Maryland).

[35] J. M. Sturtevant, *in* "Physical Methods of Chemistry" (A. Weissberger and B. W. Rositer, eds.), p. 523. Wiley (Interscience), New York, 1971.

[36] E. Calvet and H. Prat, "Recent Progress in Microcalorimetry." Pergamon, Oxford, 1963.

[37] I. Wadso, *Acta Chem. Scand.* **22,** 927 (1968).

[38] P. R. Stoesser and S. J. Gill, *Rev. Sci. Instrum.* **38,** 422 (1967).

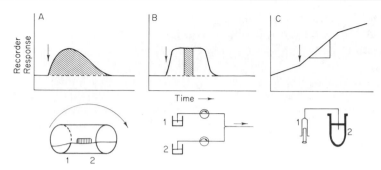

FIG. 1. Thermograms produced by various type calorimeters (upper) and schematic outline of each reactant handling system (lower). (A) Batch calorimeter thermogram and mixing cell. The small numbers 1 and 2 represent the two reactants. The area under the thermogram is proportional to the total heat of reaction. Mixing is effected by rotating and inverting the entire reaction vessel, as indicated. (B) Flow calorimeter thermogram and flow system. The area swept per second under the steady state thermogram is proportional to the heat of reaction. Mixing is effected by constant reactant flow into the "Y" mixing junction. (C) Titration calorimeter thermogram and titration cell (Dewar vessel). The indicated slope is proportional to the heat of reaction. Mixing is effected by flowing the contents of the syringe (1) into the well-stirred reaction vessel which contains the other reactant (2).

A flow calorimeter produces a signal which is proportional to the rate of heat production by the reaction $\dot{q}$ (mcal/sec). This heat flux is dependent upon the rate of consumption of limiting reagent, i.e., the mole flux n. A particular mole flux will produce a single value for the heat of the reaction; however, the continuous flow characteristics of the system make changing the concentration of limiting reagent, and thus changing of the mole flux, very simple.

TABLE I
TYPE OF CALORIMETERS USED IN STUDIES CITED

Study (Reference)	Type of calorimeter used
Hexokinase[8,9]	Batch
Thymidylate synthetase[10]	Batch
Kinetic studies[3,11,13]	Flow
Ligand binding to RNase[14–18]	Batch, Flow
Ligand binding to hemoglobin[19–23]	Flow, modified Batch
Protein binding to cells[24]	Titration
Antigen–antibody interaction[25–27]	Flow, Batch
pH titration of proteins[15,18,28]	Flow
pH titration of glutathione[29]	Titration
Nucleic acid interactions[31]	Batch
Self-association of hemerythrin[32]	Batch
Self-association of flagellin[33]	Batch

A titration calorimeter is loaded in a manner similar to a batch device with a single concentration of each reagent, but, because addition of the titrant is slow and continuous, the resulting signal is a heat flux. Knowledge of the concentration of each reactant, flow rate of titrant, and total volume of the system allows the calculation of the heat effect at a series of different total concentrations of titrant.

The goal of the research, the availability and stability of the material, and the magnitude of the heat change will dictate the type calorimeter used. In many cases more than one type of calorimeter will satisfy the experimental needs. We have tabulated (Table I) the type instruments used in the various studies previously described.[34a]

Specifications. Several operating specifications must be understood in order to select the appropriate calorimeter for a given type problem. Values for the parameters defined below are given in Table II. Useful sensitivity is the minimum instantaneous heat effect which will cause a response in the heat sensor from which the sign of the enthalpy change [exothermic $(-)$ or endothermic $(+)$] for the process can be determined. This parameter is based on an instantaneous heat effect and does not include data which must be deconvoluted from a complex signal. Type reaction refers to the kinetics of the reaction being studied. An instantaneous reaction such as an acid–base reaction is considered "fast" and readily handled by all calorimeters. The heat effect associated with a slow reaction, such as many enzyme-catalyzed reactions, is most precisely measured by either a batch or an isothermal device. Proper experimental design, as will be discussed [14], this volume will also permit the use of a flow calorimeter for this purpose. The isoperibol calorimeter will exhibit significant drift if reactions lasting much more than 20 min are attempted. Reaction vessel materials are the materials from which the calorimetric cell is constructed. Operating volume is the volume of solution necessary to determine an enthalpy of reaction. For a flow calorimeter, the operating volume is greater than the volume necessary to simply fill the reaction vessel. Operating volumes are given for both reactants A and B in Table II. Operating temperature is the temperature range available for routine operation. Sample delivery system is the device used to contain the reactants and initiate mixing. The dimensions refer to the size of the laboratory space needed to set up the system.

Times Required to Perform an Experiment (Table III). Time can be a very important limitation when working with biological materials. Biological samples are often unstable at a given temperature, so it is necessary to know how much time an experiment requires. Table III indicates typical times required to (a) equilibrate the samples with the calorimeter thermostat, (b) perform an experiment which will allow the calculation of an

TABLE II
SPECIFICATIONS[a]

	LKB flow	Tronac isothermal[a]	LKB batch	Tronac isoperibol
Useful sensitivity (μcal/sec)	0.05	0.05	0.05[b]	8[c]
Type reaction	Fast	Fast/slow	Fast/slow	Fast
Reaction vessel materials	Gold	316 stainless steel or Tantalum	Glass or gold	Glass
Operating volume	5 ml + 5 ml	2.5 ml + 1 ml	2 ml + 4 ml	2.5 ml + 1 ml
Operating temperature (°C)	0–60	0–100	0–60	0–100
Sample delivery system	Peristaltic pumps	Buret syringe	Batch mixing	Buret syringe
Dimensions (W × D × H) (meters)	Tronac 0.8 × 0.6 × 0.5 counter top location		LKB 1.5 × 1.4 × 0.25	Buret syringe / free standing

Specifications for Picker calorimeter as published by manufacturer

Heat of mixing
Sensitivity: ±1 × 10^{-6}°C
Precision: ±0.5% up to the limit of sensitivity
Reproducibility: ±0.2% under the best conditions
Response time: 6 sec or less

Specific heat
  Sensitivity: 5×10^{-5} JK^{-1} cm^{-3}
  Precision: $\pm 0.3\%$ up to the limit of sensitivity
  Reproducibility: $\pm 0.2\%$ under the best conditions
  Response time: 2 sec or less
  Overall volume of solution required per measurement: 5 ml
  Overall time per measurement: 5 min
Heat exchangers
  Length: 20 cm
  Average residence time of reactants in exchanger: 10 sec
  Coil, material: external, PTFE (Teflon)
                  internal, stainless steel
  Cross section: 0.5 mm^2
Temperature control
  Small volume of circulating liquid (less than 0.1 liters)
  Flow of circulating liquid: 2 liters/m
  Stability: better than 0.001°C
  Linear programming capacity (0°–80°C) with rates between 0.005 and 1°C/min

^{a} The figures reported in this table are based upon our experience with the calorimeters, and do not necessarily reflect the manufacturers' claims.
^{b} This corresponds to an impulse of 50 μcal.
^{c} This corresponds to a total heat input of about 0.1 mcal.
^{d} The maximum sensitivity possible is attained by placing an RC filter in front of the recorded input. A time constant of about 1 min is appropriate, but this markedly slows the instrument response time.

TABLE III
TIME REQUIRED TO PERFORM AN EXPERIMENT

	Times (min)		
	Equilibration[a]	Running[b]	Turn-around
Tronac Isothermal	45–60	20	30
LKB Batch	60–180[c]	20	45
Tronac Isoperibol	45–60	10	30
LKB Flow	5–10	15	10

[a] It is assumed that the sample is approximately at the thermostat temperature when the reaction vessel is filled.
[b] The reaction is assumed to be instantaneous.
[c] To obtain maximum useful sensitivity, up to 24 hr equilibration may be required.

enthalpy change, (running time), and (c) empty the calorimeter, clean the reaction vessel, and set up for another experiment (turn-around time). The time required to analyze the single experimental datum is about 15 min in all cases. Equilibration time is very important to avoid experimental artifacts and will vary depending upon the calorimeter used and the magnitude of the heat effect. This is discussed in detail in [14], this volume.

Experimental Milieu. The milieu in which the experiment is to be performed is also critical in choosing a calorimeter. The Tronac 180 thermal calorimeter is particularly suitable for reactions to be run in an atmosphere other than air. It is our experience that this is the simplest system to maintain anaerobic. We have had good success working in a variety of atmospheres including argon, oxygen, and nitrogen. Flow systems usually require complicated jackets or stainless steel tubing to maintain an anaerobic situation and are often difficult to use. Any solution which contains 2-mercaptoethanol will present problems if the calorimeter cell or flow tubes are made of platinum[10] which is the metal often used in flow systems. The Tronac isothermal reaction vessel may be purchased in 316 stainless steel or tantalum. The tantalum cell seems to function well with solutions containing 0.5 M in mercapatoethanol which cannot be used with platinum flow tubing. The Tronac isoperibol reaction vessel is made of glass and presents no difficulty. The LKB instruments, both flow and batch, have gold cells which are also acceptable.

These calorimeters function well with both aqueous and nonaqueous solvents. If volatile solvents are to be mixed, the flow calorimeter is the one of choice since no vapor space exists in the calorimeter cell. In all cases, care must be taken to use a pumping system which is inert to the solvents. For example, the silicon rubber tubing used in the head of peristaltic pumps is not inert to most organic solvents. Sturtevant and co-workers[39,40] have reported on the use of flow calorimeters to measure the heat of mixing of volatile organic solvents.

In summary, the questions which must be carefully considered prior to acquiring a calorimeter are (1) the mode of operation of the device, i.e., batch, flow, or titration; (2) the expected minimum heat effect; (3) the kinetics of the reaction to be studied; (4) the concentration of reactants available; (5) the stability of the material to be studied; and (6) the milieu of the experiments.

We will now turn our attention to the problem of developing a calorimeter laboratory and then describe some useful experiments for testing the system and experimental procedures.

III. Laboratory Requirements

The establishment of a calorimetric laboratory facility requires consideration of a number of things in addition to the purchase of a calorimeter. Several pieces of supporting equipment, such as amplifiers, recording devices, and auxillary thermostats are required. The location, space, and power requirements must also be carefully considered.

Location. The ideal location for a calorimeter is a constant temperature room which is maintained at the temperature of the instrument. A small constant temperature room, 3.5×4 m, is adequate. Recognizing that most laboratories do not have the space or money to install a constant temperature room, both Tronac and LKB have devised excellent constant temperature baths for their calorimeters. These are part of the purchase price of the instrument. If the calorimeter is contained within a water bath, the control of the external temperature is not critical. However, if it is contained within an air bath, the external temperature should be maintained to $\pm 0.2°$ for optimal operation. An auxillary thermostat is not included in the basic purchase but is required to regulate the main calorimetric thermostat. We have had good success in placing our instruments either in a central instrument laboratory or in a small, individual research laboratory. The latter is preferred, as it minimizes both physical and electrical interference. Setting up a calorimeter in the main research areas of a laboratory is not suggested, as this will lead to several problems, such as using the calorimeter top as a storage shelf.

Space. The space requirements are not extreme. The Tronac instrument requires 2.5 m of bench top including rack mounting several compo-

[39] J. M. Sturtevant and P. A. Lyons, *J. Chem. Thermodyn.* **1**, 201 (1969).
[40] K. J. Breslauer, B. Terrin, and J. M. Sturtevant, *J. Phys. Chem.* **78**, 2363 (1974).

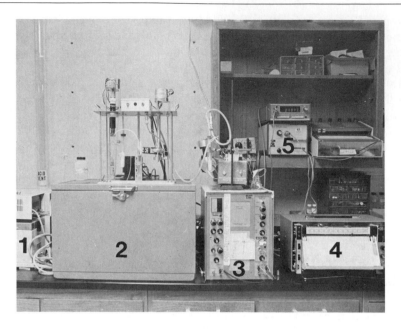

FIG. 2. Photograph of actual arrangement of components for a Tronac 550 titration calorimeter. The temperature bath (1) is a Lauda K2/R Super Constant Temperature Circulator used to provide a constant cooling source to the calorimeter bath. The titration calorimeter consists of a controlled temperature water bath (2) inside of which is the reaction vessel and an electronics console (3). The recorder (4) is a Leeds and Northrup XL 600 two-pen recorder and Disc Integrator (5). The digital multimeter (6) is a Keithley 190. Other equipment visible in the photograph is part of a photomultiplier and fiber optic system plus support equipment for the calorimeter.

nents. A photograph of the actual arrangement is shown in Fig. 2. We have found this amount of space very adequate. The LKB unit requires about the same space, but is a floor unit. A photograph of a typical arrangement of two LKB calorimeters is shown in Fig. 3.

All calorimeters require several electrical outlets. Sources of air, tap and distilled water, and other services are also useful. The most serious problem is the availability of 120 V outlets. They should all be on the same circuit to minimize ground loops, and that circuit should be "clean" of electrical noise. Tronac has greatly helped by providing 12 service outlets on the rear panel of their calorimeter.

Recorders, Amplifiers, and Data Acquisition Systems. Analog recorders are usually used to record the experimental data, although there is an increasing trend to digitize the data. The best possible recorder available for the money at hand should be obtained. Experience tends to suggest recorders such as the Hewlett-Packard 7100, the Leeds and Northrup XL 600, and the Sargent SRG recorders.

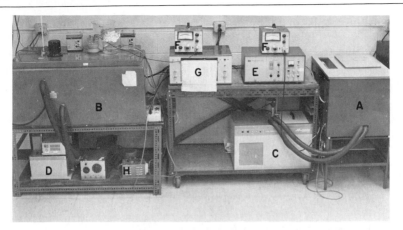

FIG. 3. A photograph of a laboratory arrangement of the components of the LKB batch and flow microcalorimeters. The batch calorimeter is contained in the air bath (A). The flow calorimeter is contained in a submersible submarine in water bath (B). The constant temperature cooling sources for A and B are a Forma (C) and Lauda (D) refrigerated constant temperature circulators. The control unit (E) for both calorimeters contains a constant current supply for the calibration heaters and a timer. The signals from each calorimeter are amplified by means of Kiethly 150B microvoltmeter (F) and the output recorded on a Sargent TRG potentriometer recorder (G). Other equipment visible in the photograph include a 120/210 V transformer (H) and two LKB peristaltic pumps (I) used to introduce the two solutions into the flow calorimeter cell. The height of this set-up is approximately 4.5 ft and occupies a floor area of about 8 × 2.5 ft.

The signals from the LKB calorimeter are generally in the 1–100 μV range, which requires the output to be amplified prior to being recorded. The Keithley 150B (Keithley Instruments, Cleveland, Ohio) has been found to be satisfactory for this purpose. We suggest that it be maintained at a constant temperature ($\pm 1°$) and never turned off except for servicing.

Several methods of converting the thermogram into numerical form are possible. Direct digitization using an analog to digital converter or similar device is very convenient. Integration of the signal using a DISC integrator provides an inexpensive and convenient method of data conversion. For experiments in which a rate of heat production is measured (i.e., flow calorimetry), a DISC printer is also very useful. We have incorporated a simple timer into the printer to allow automatic periodic printing of the area under the thermogram as well as resetting of the integrator accumulator.

Auxillary Thermostats. Precise temperature control is best obtained by providing a source of constant cooling to the calorimetric thermostat and using a controller to carry out the fine regulation. We currently use a Lauda K2/R Super refrigerated circulating bath (4 liter capacity) to provide the necessary cooling on a Tronac 550 Titration and the LKB batch and flow calorimeters. This performs adequately. The only requirements are that the circulator be capable of controlling to better than ± 0.1° over the temperature range − 30° to 60°. (We found that this same circulator, placed on a Beckman flow calorimeter, was not adequate. However, when replaced with a Forma bath having a 75-liter capacity, a remarkable improvement in performance was observed. We strongly suggest using a large thermal reservoir type circulating cooler.)

Other Incidental Items. Most of the other equipment needed to set up a calorimetry laboratory is commonly available. The only other very useful item is a digital voltmeter (DVM) especially useful in adjusting the thermistors in the Tronac unit and for monitoring a run. It is also convenient to use a DVM with BCD output to directly digitize continuous data obtained from a flow calorimeter.

Estimated Cost for Purchasing a Commercial Calorimetry Facility. Table IV outlines the itemized, estimated cost of purchasing a complete TRONAC or LKB calorimeter, including items not supplied by the calorimeter manufacturers. The total cost is in the range of $15,000–$20,000. Individual options and special equipment are not included in this estimate.

The Calorimetric Experiment. A detailed description of the design of a calorimetric experiment and data analysis will be given for various types of calorimeters in [14], this volume. At this time, the relevant parameters which influence the design and execution of a calorimetric experiment will be considered.

It is assumed that mixing of the two solutions results in an instanta-

TABLE IV
APPROXIMATE COSTS OF A COMMERCIAL CALORIMETER FACILITY[a]

Tronac system[b]		LKB system[b]	
550 Microcalorimeter	$11,000	10700 Microcalorimeter[c]	$19,100
Leads and Northrup recorder with disc integrator and printer	3,600	Sargent RS potentiometric recorder (without integrator)	1,125
Keithly 190 digital multimeter	700	Keithly 150 B microvoltmeter	1,495
Lauda K²/R refrigerated circulating bath	745	Lauda water bath	745
	$16,045		$21,765

[a] These prices reflect the manufacturers suggested 1977 retail price.

[b] The accessory equipment listed in this table are those items used in the author's laboratories and are not the only ones which will suffice. Several other brands of recorders, multimeters, and water baths will also perform adequately.

[c] For the batch colorimeter. The basic flow unit, including two peristaltic pumps, costs $18,475. The calorimeter blocks only can be obtained at a cost of about $10,500.

neous reaction in which heat is liberated. In a batch experiment the amount of heat produced Q is

$$Q = \Delta H\,[P] \cdot V + Q_{dil} + Q_{mix}$$

where ΔH is the enthalpy change per mole of product formed, $[P]$ is the concentration of product, and V is volume of the mixed solutions. Q_{dil} is the heat of dilution of all of the components and Q_{mix} is the heat associated with frictional effects of mixing. In a flow experiment the heat flux is

$$\frac{dq}{dt} = \dot{q} = f\,\Delta H\,[P] + \dot{Q}_{dil} + \dot{Q}_{mix}$$

where f is the flow rate of solution through the calorimetric cell. In general Q_{dil} and Q_{mix}; or $\dot{Q}_{dil}$ and $\dot{Q}_{mix}$, will be small compared to the heat of reaction. This is not always true, however, and, in any case, these quantities must be determined in separate experiments.

In general, Q or $\dot{Q}$ should be at least greater than ten times the useful sensitivity of a given instrument if reliable data are to be obtained. For the LKB and Tronac instruments this means that $Q \simeq 1$ mcal or $\dot{q} \simeq 5$ μcal/sec or greater. In Table V we have tabulated values for the various parameters for a hypothetical reaction whose $\Delta H = 10$ kcal/mole.[40a] In general, we find that the batch-type instrument is more sensitive (in terms of required material) than the flow instrument. However, the latter device offers certain advantages which will be discussed later.

As experience with a specific calorimeter is gained, it will become possible to perform experiments using significantly smaller heat effects and/or amounts of material. For example, Halsey and Biltonen[26] have reported data obtained in an LKB batch calorimeter using heat effects as low as 0.15 mcal. Beaudette and Langerman[41] have performed complete thermal titrations with the Tronac isoperibol calorimeter on about 0.3 μmole of material.

Calibration. Regardless of the type of calorimetric experiment, it is necessary to convert the observed signal to a heat quantity by a predetermined calibration constant. This factor is experimentally determined and should be frequently and carefully assessed. Two methods of calibration exist: electrical and chemical. Both should be used whenever possible. Electrical calibration is performed with an electrical resistance heater placed in the calorimeter cell. The total power dissipated in the resistor is given by the relation

$$P = i^2 R$$

[40a] Throughout this chapter 1 cal = 4.184 J.
[41] N. V. Beaudette and N. Langerman, *Anal. Biochem.* **90,** 693 (1978).

TABLE V

Typical Requirements for a Microcalorimetric Experiment for a Hypothetical Instantaneous Reaction A + B → C for which $\Delta H° = 10^4$ cal/mole[a]

Batch instrument

$Q = 1$ mcal $= V \, \Delta H$

$V = 4$ ml

$[P] = 2.5 \times 10^{-5} \, M$

Total amount of material $= V[P] = 0.1 \, \mu$mol

Flow instrument

$\dot{Q} = 5 \times 10^{-6}$ cal/sec $= \dot{V} \, \Delta H = f[P] \, \Delta H$

$f = 5 \, \mu$l/sec

$[P] = 0.1 \, mM$

$t = 10$ min, which is the total time during which protein must be flowing through the calorimeter

Total amount of material required $= ft[P] = 0.3 \, \mu$mole

[a] These numbers represent typical amounts of material required for a single determination of Q or $\dot{Q}$. The flow calorimeter operated in the exponential dilution mode will allow a complete thermal titration using about 1 μmole of material. The Tronac Isoperibol Calorimeter used in the titration mode also requires about 1 μmole for a complete thermal titration.

where P is in watts (joule/sec) and i is the current through the resistor of resistance R. The power in watts, or more commonly in cal/sec; is determined using the relations[40a]

$$P = V^2/R \text{ (watts)}$$

and

$$P = 4.184 \, V^2/R \text{ (cal/sec)}$$

The Tronac calorimeter provides a variable voltage source and the LKB calorimeter provides a variable current source for performing those calibrations.

Chemical calibration is done using one or more of three "standard" reactions ($T = 25.00°$)[42]

Tris + HCl → Tris H$^+$Cl$^-$	$\Delta H = -11,340$ cal/mole	(Ref. 1)
HClO$_4$ + KOH → H$_2$O + KClO$_4$	$\Delta H = -13,430$ cal/mole	(Ref. 1)
Sucrose (m) → sucrose (m')	$\Delta H^{43} = +128.9$ cal/mole Δm	(Ref. 44)

[42] See Sturtevant[1a] for a listing of the ΔH of neutralization of Tris and KOH as a function of temperature.

[43] The expected heat effect for diluting sucrose from a molal concentration of m to a new concentration m', is calculated using the relations q(cal) $= n_{sucrose} \Delta H = n_{sucrose} \, 128.9$ (cal/mole) ($m - m'$), where $n_{sucrose}$ is the number of moles of sucrose being diluted.

[44] F. T. Gucker, H. B. Pickard, and R. W. Planck, *J. Am. Chem. Soc.* **61**, 454 (1938).

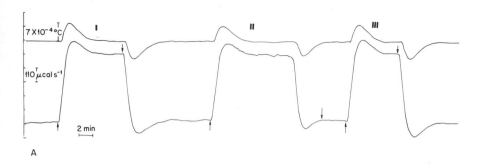

A

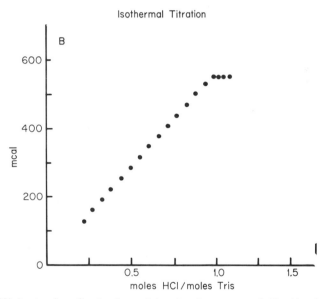

Isothermal Titration

B

FIG. 4. (A) A retracing of an Isothermal titration thermogram obtained by titrating 2.0 ml of 0.0241 M Tris with 0.135 M HCl. The delivery syringe for the acid was operated at 0.325 μl/sec. The lower tracing (10 V full scale) across the figure represents the energy being removed from the reaction vessel each second by the Peltier cooler while the upper tracing (1 mV full scale) represents the actual temperature in the reaction vessel. The figure is divided into three parts, with parts I and III being calibration experiments using a heater (resistor) within the reaction vessel. The heater was operated at a power setting of 540 μcal/sec. Part II is the signal due to the slow addition of HCl to Tris. The arrows pointed up indicate the points in time at which the calibration heater was turned on (parts I and III) or the syringe containing the HCl (part II) was started. The arrows pointed down indicate the points in time at which these functions were turned off. During periods of large changes in the rate of heat production in the reaction vessel, i.e., immediately after turning on or off a function, the temperature changes as indicated and the isothermal system is out of thermal control. The net heat, however, under this "uncontrolled" portion of the thermogram is correct. These data may be used to construct the titration curve shown in (B). The average cali-

(continued)

An example of an actual chemical calibration using the Tris-HCl reaction is shown in Fig. 4A using the Tronac Isothermal Calorimeter. The steady state signal can be used to calculate the number of recorder divisions (or volts) which correspond to 1 mcal/sec of heat. Conversion of these data into a titration curve, as shown in Fig. 4B, is useful for evaluating the precision of the data and the data treatment. The precision is best tested by evaluating the stoichiometry from the experimental data. In Fig. 4B, the apparent stoichiometry is 1:1, as expected for this reaction.

Except for a few details, specific to the actual instrument used, the calculation outlined in Table V is correct for designing a calibration experiment. The actual concentrations to use, and the expected heat effects, for a batch, flow, or titration calorimeter are indicated in Table VI. Note that some neutral salt, Tris H^+Cl^-, is added to the standard Tris solution to suppress the hydrolysis of Tris. All solutions should be made up in freshly boiled distilled water which was cooled under N_2 and stored under N_2.[45]

V. Operation

Rather than attempt to explain how to run each calorimeter switch by switch, we will outline a useful thermal titration which we use to test experimental design. The instruction manuals for the various instruments are suggested as sources for detailed operating instructions. A laboratory manual, such as "Experiments in Thermometric Titrimetry and Titration Calorimetry" by Eatough et al.[7] is also a valuable source of detailed experiments designed for the Tronac calorimeter.

In planning an experiment, we usually calibrate the instruments as explained previously with Tris, and then use the reaction

---

[45] We have recently used 0.02 mM HCl plus 0.1 mM Tris to calibrate a Tronac Isoperibol Calorimeter. We find that the water used to prepare our solutions caused a 15% error in the derived stoichiometry for the neutralization.

---

bration factor (area/mcal) is determined from the calibration experiments by determining the area under the thermogram for a period of time, e.g., 60 sec, and dividing this value by the total millicalories produced by the heater during this same period. The evaluation of the heat effect for the titration proceeds in a similar manner. The thermogram is divided into equal time increments and the area under the thermogram during each increment is determined. If the base line shifts during the reaction, an appropriate correction must be made. The corrected area for each time interval is converted to a heat effect in mcal by use of the calibration factor. The data contained in the "uncontrolled" portion of the thermogram are handled as a single point. The mole ratio of acid to base is determined at each point using the known concentration of the base and acid plus the known flow rate of the acid. A point on the tiration curve in (B) represents the total number of millicalories evolved versus the total number of millimoles of HCl per millimoles of Tris. Thus, the heat effects are additive and errors can propogate rapidly if due caution is not observed.

TABLE VI
CALIBRATION SOLUTIONS[a]

	Flow calorimeter	Expected heat
Solution 1	$[\text{Tris}] = 2.5 \times 10^{-3}\ M$ $[\text{Tris-HCl}] = 1 \times 10^{-3}\ M$	0.24 mcal/sec
Solution 2	$[\text{HCl}] = 4 \times 10^{-3}\ M$	
Solution 3	$[\text{HClO}_4] = 2.5 \times 10^{-3}\ M$	0.28 mcal/sec
Solution 4	$[\text{KOH}] = 5 \times 10^{-3}\ M$	
	Batch calorimeter	
Solution 1	$[\text{Tris}] = 1.1 \times 10^{-3}\ M$ $[\text{Tris HCl}] = 0.6 \times 10^{-3}\ M$	50 mcal
Solution 2	$[\text{HCl}] = 1.1 \times 10^{-2}\ M$	
Solution 3	$[\text{HClO}_4] = 4.65 \times 10^{-4}\ M$	25 mcal
Solution 4	$[\text{KOH}] = 4.65 \times 10^{-3}\ M$	
	Isoperibol titration calorimeter	
Solution 1	$[\text{Tris}] = 4 \times 10^{-3}\ M$ $[\text{Tris HCl}] = 2 \times 10^{-3}\ M$	136 mcal
Solution 2	$[\text{HCl}] = 0.20\ M$	
Solution 3	$[\text{HClO}_4] = 0.115\ M$	92 mcal
Solution 4	$[\text{KOH}] = 2.29 \times 10^{-3}\ M$	

[a] All solutions should be prepared on $0.1\ M$ KCl, using CO_2 free deionized water. In each example, solutions 1 and 2 or 3 and 4 are mixed to generate the indicated heat effect. The valves given are useful for learning to operate a calorimeter, as well as for calibration. As one becomes more proficient with the operation of the instrument, a significant reduction in required heat effect is possible for precise experimental work.[45]

$$\text{RNAse} + 3'\text{-CMP} \rightarrow (\text{RNase-CMP})$$

to test the actual planned experiment. For example, we have 1 μmole of a protein of molecular weight 13,800 g/mole and we plan to carry out a thermal titration to evaluate ΔH and K_{eq} for a ligand binding process. Assuming that the enthalpy for the reaction is 10^4 cal/mole and the equilibrium constant is greater than 10^3, we can easily calculate all of the conditions necessary to obtain the data. This calculation is illustrated in Table VII. The calculations outlined in this table use "reasonable" assumed parameters. The experiment should require about 10 minutes to complete after equilibration plus calibration time. The expected heat flux, about 70 μcal s^{-1} is easily detected.

Using this planned experiment as a guide, a similar experiment with ribonuclease is performed. An actual thermogram using the Tronac Isoperibol titration calorimeter is shown in Fig. 5A. For this experiment, 2 ml of $4.58 \times 10^{-4}\ M$ RNAse was titrated with $2.78 \times 10^{-2}\ M$ 3'-CMP in pH 5.6 acetate buffer at 25°. The delivery rate of titrant was 0.325 μl/sec. Also shown as an insert to Fig. 5B is the derived titration curve (binding isotherm) based on this thermogram.

TABLE VII

DETERMINATION OF CONDITIONS FOR MEASURING ΔH USING
A TITRATION CALORIMETER[a]

Reaction: Protein (P) + Ligand (L) = P—L
ΔH (assumed) $= -10,000$ cal/mole
K (assumed) $= 10^5 \, M^{-1}$ (see Chapter [14], this volume, for details of using K)

1. $[P] = \dfrac{1 \, \mu\text{mole}}{2 \, \text{ml}} = 5 \times 10^{-4} \, M$

2. Using a 2.5 ml syringe operated at a flow rate of 1.67×10^{-3} ml/sec, calculate concentration necessary to deliver 1 μmole (1 equivalent) of L in 2.5 min
   $[L] \times$ time(t) $\times$ flow rate(f) $= 1 \times 10^{-3}$ mmole
   $[L] = 4.0 \times 10^{-3} \, M$

3. Calculate $\dot{n}$ for addition of L
   $\dot{n} = 1.6 \times 10^{-2}$ mmole/ml $\times 1.67 \times 10^{-3}$ ml/sec
   $\quad = 6.7 \times 10^{-6}$ mmole/sec

4. Calculate the expected heat flux $\dot{q}$
   $\dot{q} = \Delta H \dot{n}$
   $\quad = 10^4$ mcal/mmole $\times 2.7 \times 10^{-5}$ mmole/sec
   $\quad = 67 \, \mu$cal/sec

[a] Using the instrument shown in Fig. 2, a 1 mV full scale recorder setting would be appropriate for this experiment. A complete experiment performed as outlined in this table would provide sufficient information to calculate $\Delta H'$, K, $\Delta G'$, $\Delta S'$, and the stoichiometry for the reaction.

Special Applications. In this final section, we have brought together several applications of calorimetry which require either modification of the equipment and its operation or special treatment of the data. The purpose of this discussion is to indicate the versatility and power of calorimetry.

Exponential Dilution. The flow calorimeter possesses several advantages including the possibility of continuously changing the concentration of one or more reactants. The time dependent signal, which must be corrected for the time response of the system, is now a continuous heat effect which can be correlated with the changes in concentration of the reactant.

Mountcastle *et al.*[46] have developed a constant volume device in which the concentration of solute [S] is an exponential function of time. These workers have used this device to obtain continuous heat affect versus log of concentration curves using an LKB flow calorimeter. For example, the heat of reaction between excess NaOH and HCl, whose concentration is continuously varied, will be an exponential function of time.

If such an experiment is performed with a system in which a ligand binds to a macromolecule, and the ligand concentration is continuously varied, an enthalpy-binding isotherm is obtained. One system which has

[46] D. B. Mountcastle, E. Freire, and R. L. Biltonen, *Biopolymers* 15, 355 (1976).

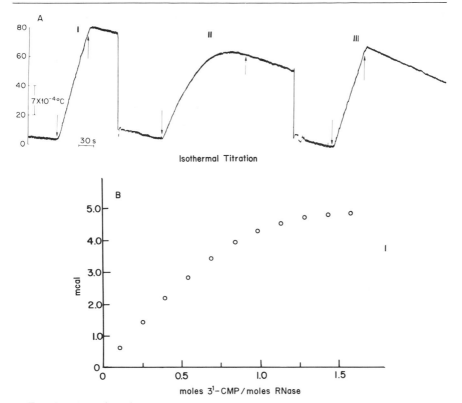

FIG. 5. A retracing of an Isoperibol titration thermogram obtained by titrating 2.0 ml of 0.458 mM RNase with 27.8 mM 3′-CMP. The delivery system for the CMP was operated at 0.325 μl/sec. The tracing represents the net temperature change occurring in the reaction vessel as a function of time. The recorder was operated at 40 μV full scale and the Wheatstone bridge voltage on the Tronac 450 was reduced to 2.000 V for this experiment. Parts I and III of the figure represent calibration experiments with the calibration heater operated at 87μ cal/sec. Part II is the signal due to the slow addition of CMP to RNase. The arrows pointed down indicate the points in time at which the calibration heater (parts I and III) or the CMP syringe was turned on, while the arrows pointed up indicate when these functions were turned off. These data, after correcting for heat loss from the reaction vessel, may be used to construct the titration curve shown in (B). (A detailed description of the calculation is contained in Beaudette *et al.*[42]). The calibration experiments are used to evaluate a calibration factor α, chart divisions mcal^{-1}, by dividing the net deflection of the recorder pen by the net heat input. The titration thermogram is divided into equal time increments, e.g., 15 sec, and the observed recorder deflection at each time is recorded. This value is corrected for heat loss by using the initial and final slopes and the heat effect at each time point to evaluate the rate of heat loss and then integrating these rate data from $t = 0$ to t. The net heat loss is then added to each observed data point to obtain the corrected heat effect and this is converted to energy units using α. The mole ratio of CMP to RNase is determined at each point using the known concentrations of these reagents and the flow rate of the CMP. The titration curve is constructed by plotting the corrected energy term obtained above versus the corresponding mole ratio term. Note, the heat effects are not additive as was the case with the Isothermal titration.

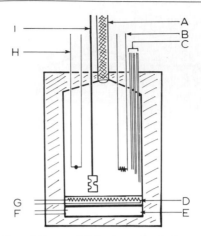

FIG. 6. Schematic outline of Tronac 550 isothermal calorimeter cell modified with fiber optic light guide. A, Light guide; B, calibration heater; C, sample inlet tubes; D, controlled heater; E, peltier cooler; F and G, electrical connections; H, control thermistor; I, stirrer.

been successfully studied in this manner is the binding of FMN to bacterial luciferase.[47] This type of experiment can be performed in about 1 hr and is the equivalent of an infinite number of discrete experiments. Such an experiment requires computer analysis of the data and will be discussed in Chapter [14], this volume.

Heats of Solution. The heat of solution of pure solid materials into various solvents can be of use as thermodynamic descriptions of model reactions. For example, Halsey and Biltonen[48] determined the heat of solution of dinitrophenyl-ϵ-lysine into ethanol and water at several temperatures. The data obtained were used to calculate the thermodynamics of transfer of the hapten from water to ethanol. Using this transfer reaction as a model for the thermodynamics of hapten binding to its antibody, they concluded that the Δc_p for the binding reaction could be at least partially rationalized in terms of transfer of the hapten from the aqueous solvent to a hydrophobic binding region on the protein.

Of the calorimeters discussed in this chapter the Tronac Model No. 550 with an ampoule attachment is best for this type of experiment. The LKB 8700 Precision calorimetric system can also be used for these types of experiments. In such a device, the solid reactant is sealed within a thin-walled glass ampoule which immersed in the solvent into which it is to dissolve. After the system is thermally equilibrated, the ampoule is broken and the heat effect evaluated.

[47] N. Langerman, *in* "Flavins and Flavoproteins" (T. P. Singer, ed.), p. 82. Elsevier, Amsterdam, 1976.
[48] J. F. Halsey and R. L. Biltonen, *J. Solution Chem.* **4**, 275 (1975).

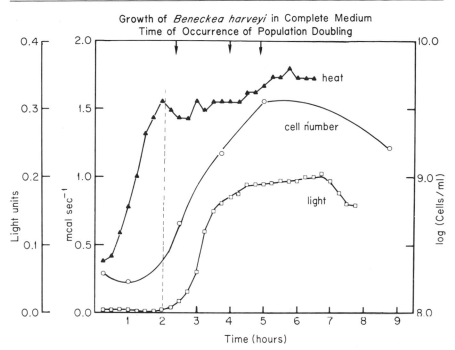

FIG. 7. Combined thermogram, light production (expressed in light units, where 1 light unit = 1.44 × 10⁻² protons/sec), and viable cells/ml (obtained by periodic sampling of the contents of the reaction vessel, plating, and counting for the marine luminescent bacteria, *Beneckea harveyi*). The scale at the top represents the points in time at which the total number of cells in the reaction vessel doubled. The bacteria were grown in a seawater medium containing 5 g/liter Bacto-Tryptone, 0.5 g/liter Bacto-yeast extract, and 2 ml/liter glycerol. The temperature was held constant at 25.00°.

The Simultaneous Detection of Heat and Light. A Tronac calorimeter was modified by the manufacturer to accept a fiber optical bundle positioned, as shown in Fig. 6 above the reaction vessel. This device has been used by McIlvaine and Langerman[49] to simultaneously monitor the metabolic heat production and the light production from luminescent bacteria. A set of data obtained using the isothermal calorimeter are shown in Fig. 7. This system has also been used to study the growth of photosynthetic bacteria under anaerobic illuminated conditions. Cooper and Converse[50] have used a similar arrangement in an LKB batch calorimeter to study the energetics of rhodopsin in the rod outer segment membranes.

Determination of Reaction Kinetics by Calorimetry. Historically, the determination of rate parameters has been well separated from thermo-

[49] P. McIlvaine and N. Langerman, *Biophys. J.* **17**, 17 (1977).
[50] A. Cooper and C. A. Converse, *Biochemistry* **15**, 2970 (1976).

TABLE VIII

CALCULATED SECOND-ORDER RATE CONSTANTS FOR THE ALKALINE
HYDROLYSIS OF ETHYL ACETATE AT 25°

[EtAc] or [KOH]	k (min^{-1})	Literature value[13]
0.01–0.03	6.74 ± 0.07	6.86
		6.69
		6.5
0.03–0.06	6.75 ± 0.05	6.76
		6.66
0.06–0.13	6.95 ± 0.06	7.22
0.13–0.20	6.94 ± 0.05	6.96

[EtAc] or [KOH]	Flow rate (μl/sec)	k (min^{-1})
0.012–0.02	21.5	6.78 ± 0.19
0.017–0.05	9.07	6.92 ± 0.22
0.017–0.05	5.48	6.63 ± 0.21
0.038–0.06	2.73	6.67 ± 0.17

dynamic approaches. However, several attempts have been made to couple reaction calorimetry to kinetics. Recently, a very successful approach has been demonstrated by Johnson and Biltonen[13] using an LKB flow calorimeter. Their analysis is explained in detail in Chapter [14], this volume, but their results for the hydrolysis of ethyl acetate are reproduced in Table VIII. These results demonstrate the accuracy which can be achieved using a calorimeter to obtain kinetic information. A similar approach has been made using the Tronac instrument. Eatough *et al.*[7] have published a detailed experimental procedure, also using the hydrolysis of ethyl acetate by base.

This chapter is intended to provide an overview of concepts and techniques and to establish the proper perspective for evaluating the feasibility of calorimetric experiments. Chapter [14] will examine specific aspects of experimental design and interpretation.

[14] Microcalorimetry for Biological Chemistry: Experimental Design, Data Analysis, and Interpretation[1]

By R. L. BILTONEN and NEAL LANGERMAN

Introduction

The application of calorimetric techniques to the study of biological systems has been quite limited. The major reason for this limited application has been the unavailability of commercial instrumentation. This particular problem has been alleviated in the past few years, and now several different instruments can be obtained. In the preceding chapter[1a] the basic operating characteristics of some of these instruments were described and the general questions relating to application of calorimetric techniques to the study of biological systems addressed. That chapter should provide for the biological scientist the proper perspective regarding application of the technique to his particular needs.

Although in recent years a number of publications[2-6] concerned with calorimetry and biology have appeared, it does not seem that the technique has been fully exploited. The general applicability of calorimetric techniques resides in the ubiquitous nature of heat changes associated with all chemical reactions, but a great deal more information than enthalpy changes can be obtained regarding the nature of specified chemical reactions. For example, Gibbs energy changes, stoichiometry, differential proton uptake, and other thermodynamic as well as kinetic details of reactions can be obtained from appropriate sets of calorimetric data.

In this chapter we shall describe and discuss the design of specific types of calorimetric experiments. It is suggested that the reader consult the preceding chapter[1a] before proceeding with this one. It is hoped that the interested reader will be able to seriously consider application of cal-

[1] This work was supported in part by United States Public Health Service Research Grants GM22049, GM20637, and GM22676 and National Science Foundation Grants BMS75-03030 and PCM75-23245-A01

[1a] N. Langerman and R. Biltonen, this volume [13].

[2] J. M. Sturtevant, *Annu. Rev. Biophys. Bioeng.* **3,** 35 (1974).

[3] C. Spink and I. Wadso, *Methods Biochem. Anal.* **23,** 1 (1976).

[4] J. M. Sturtevant, Vol. 26, p. 227.

[5] I. Wadso, *Pure Appl. Chem.* **38,** 529–538 (1974).

[6] G. Rialdi and R. Biltonen, *in* "International Review of Science, Physical Chemistry," Series Two (H. A. Skinner, ed.), Vol. 10, p. 147. Butterworths, London, 1975.

orimetric techniques to his particular problem after the study of these two chapters.

The Basic Experiment

The basic calorimetric experiment consists of mixing two solutions which contain reactive components. The heat effect associated with the overall process is

$$Q = Q_x + Q_{dil} + Q_{mix} \tag{1}$$

where Q_x, Q_{dil}, and Q_{mix} are the heat of reaction, the heat of dilution of the components of the solutions, and the heat effect associated with the physical process of mixing, respectively. Q_{mix} may include the heat of ampoule breaking, the viscous heat associated with flowing solutions, or the heat associated with the wetting of the calorimeter cell as well as the heat of mixing.

In a differential, or twin, batch calorimeter the mixing effects are, to a large extent, canceled by mixing solvent in the reference cell. However, because the calibration constants of the sample and reference cells are not identical and because the nature of the solutions in the sample and reference cells (e.g., viscosity) may not be identical, a separate experiment is required to determine Q_{mix}. Similarly, Q_{mix} for the mixing process in ampoule, titration, and flow calorimeters must also be determined in separate experiments. In most cases Q_{mix} is small compared to Q_x. The exception occurs when Q_x is small or the solutions being mixed are very viscous.

In the batch calorimeter a particular problem is the heat associated with the wetting of the calorimeter cell upon mixing. The calorimeter cell is usually dried prior to loading the sample in order to remove all solvent. After loading the cell, the upper portion of the cell remains dry. Upon rotation of the calorimetric unit to effect mixing, the solution wets the dry portion of the cell. A significant heat effect can be associated with this wetting or adsorption process. The magnitude of the heat effect is a function of the material from which the cell is made, the solvent, and the composition of the solution; water by itself does not appear to have any significant effect on gold or glass cells, but salt solutions can produce large heats of wetting. In certain cases dilute solutes can also adsorb to the cell surfaces. Examples of this include surfactants and enzymes. Since no general rules regarding this phenomenon can be established, it is best for the experimentalist to be prudent.

In the flow calorimeter no specific heat effects associated with wetting of the cell are usually observed, since the cell is always saturated. However, in some cases, solute may be specifically adsorbed to the cell surface. For example, Pain and Larner[7] discovered that dilute solutions of penicillinase (10^{-9} M) were almost totally adsorbed onto the gold flow cell. In addition components of the solution may be adsorbed to the tubing of a flow calorimeter. These phenomena do not produce any measureable heat effects, but can seriously effect the concentration of the components within the calorimeter cell. In the case of penicillinase, for example, the effective concentration of the enzyme within the cell under steady-state conditions was about tenfold greater than the actual enzyme concentration in solution.

The heat associated with ampoule breaking in an ampoule calorimeter is a very serious problem and is probably the limiting factor in the useful sensitivity of such an instrument. The heat associated with breaking glass ampoules is of the order of several millicalories and generally not very reproducible. For this reason, the heat of ampoule breaking should be determined several times and an average value used for all calculations. The reason for this poor reproducibility is that the thickness of the glass diaphram varies from ampoule to ampoule. In general, this method of mixing the components should be avoided unless the anticipated heat effect for the reaction is on the order of 100 mcal or more.

The heat of dilution Q_{dil} for all components is

$$Q_{dil} = \Sigma n_i Q_{d,i} \tag{2}$$

where n_i is the number of moles of component i in the reaction mixture and where $Q_{d,i}$ is heat associated with the process of diluting component i from its initial concentration to its final concentration. (For experiments in the flow calorimeter n_i is the number of moles per unit volume of the final reaction mixture.) The heat of dilution for each of the components whose concentration changes upon mixing must be measured in separate experiments. Ideally this should be done at the exact initial and final concentrations of these components. If, however, Q_{dil} is small compared to Q_x, it is not necessary to perform the experiment at exactly identical concentrations.

In all but the batch calorimeter, each dilution experiment must be done separately. In the batch calorimeter, it is possible to perform one dilution in the reference cell with its associated heat effect canceling out the similar heat effect occuring upon mixing in the sample cell. However, this simultaneous experiment should only be performed if Q_{dil} is small, since the calibration constants of the two cells are not identical.

The necessary experiments to determine Q_{mix} and Q_{dil} should be performed with the same care that is given the actual mixing experiment. The

[7] R. Pain and A. Larner, unpublished observations.

reliability of Q_x is equally dependent upon the accuracy of both the actual and the control experiments.

Thermal Equilibration

The success of a calorimetric experiment depends upon a number of factors, but perhaps the most commonly ignored is the achievement of thermal equilibrium of the solutions prior to mixing. It is important to realize that during the calorimetric experiment temperature changes on the order of 10^{-4} deg or less are being measured. For example, in the batch-type calorimeter a heat effect of 1 mcal[7a] produces a maximal temperature change of less than 0.1 millideg. Therefore, if good thermal equilibration is not achieved severe artifacts will occur.

In the batch-type calorimeter loading of the solutions into the reaction vessels requires that the calorimetric heat sink and air bath be exposed to the environment, thus disturbing the existing equilibrium between the cells, heat sink, and bath. The time required to achieve satisfactory thermal equilibrium after the calorimeter is loaded depends upon the magnitude of the heat effect being measured. At least 1 hr is sufficient if Q is on the order of 3 mcal or more. If Q is about 1 mcal it is necessary to wait 2 or 3 hr, and if Q is only a few hundred microcalories, up to 24 hr may be required. Recently, Prosen and co-workers[8] have designed a batch microcalorimeter with a "disposable" cell. In this instrument the loaded sample cell is inserted into the calorimeter through a small slot, thus not severely exposing the heat sink to the environment. It has been reported that thermal equilibration can be achieved within 15 min.

The symptoms of nonequilibrium in the batch calorimeter are signals substantially different than zero prior to mixing and substantial drift in the signal. Upon mixing, the signal will not return to the zero (or base line) value. This latter effect always occurs to some extent and will be discussed later, but if the deviation between the initial and final base line is greater than 10% of the peak value of the signal, it can be assumed that the experiment is unsatisfactory.

In the flow calorimeter, thermal equilibrium between the solutions and the calorimeter heat sink is achieved by flowing the solution through one or more heat exchangers which are in equilibrium with the heat sink. The degree to which thermal equilibrium is achieved depends upon the initial temperature of the solutions and the rate at which the solutions are being

---

[7a] Throughout this chapter, 1 cal = 4.184 joules.

[8] E. Prosen, R. N. Goldberg, B. P. Staples, R. N. Boyd, and G. T. Armstrong, *in* "Thermal Analysis: Comparative Studies on Materials" (H. Kambe and P. D. Garn, eds.), p. 253. Wiley, New York, 1974.

pumped. The closer the initial temperature is to the calorimeter temperature and the slower the pumping rate, the better the equilibrium. Usually the solutions are incubated at a temperature close to that of the calorimeter. However, it is possible to incubate the solutions at temperatures far from the calorimeter temperature (e.g., in an ice bath) and still achieve good preequilibration if the pumping rate is sufficiently slow. The degree of equilibration required depends upon the magnitude of the signal.

The symptoms of nonequilibrium in the flow calorimeter are substantial base line drift and periodic fluctuations in the base line. In general, the base line drift over a 15-min interval should be less than 10% of the measured heat effect for a successful experiment.

In the isoperibol and isothermal calorimeters the solutions are equilibrated within the calorimeter prior to mixing. The time requirements for equilibration are similar to that required for the batch calorimeter. The symptoms associated with nonachievement of thermal equilibrium are substantial base line drift in the isoperibol calorimeter. With the isothermal calorimeter, nonequilibrium is indicated by a large signal as well as substantial base line drift.

Since good thermal equilibrium is an absolute requirement for a successful calorimetric experiment, great care is required. Although the general comments made above offer some advice regarding this problem, experience with any particular calorimeter is the best teacher. As one gains this necessary experience, the probability of success of an experiment increases and, as well, it becomes possible to do successful experiments under more difficult conditions, e.g., measurement of smaller heat effects.

Calculation of Heat Effects

The calculation of the heat effect from the experimental data is straightforward, but the exact procedure depends upon the type of calorimeter being used. In all cases to be discussed, it will be assumed that the calorimeter has been appropriately calibrated as described in the previous chapter[1a] and that the proportionality constant between observed signal and the rate of heat production has been determined.

The flow and batch-type calorimeters are referred to as heat-burst or heat-leak calorimeters. This means that the thermal contact between the calorimeter cell and heat sink is maintained. In such a system, any process initiated in the calorimeter which either absorbs or releases heat produces a temperature differential between the heat sink and the cell. The temperature difference causes a voltage to be generated by the thermal elements, which form the thermal barrier between the cell and heat sink. This voltage is directly proportional to the temperature difference. According to Newton's law of cooling, the rate of heat flow is also directly proportional to this temperature difference. Therefore

$$\dot{Q} = \frac{dQ}{dt} = \epsilon V \tag{3}$$

where ϵ is the calibration constant.

In a batch-type experiment the initiation of the reaction generates a temperature gradient. Initially the temperature difference achieves a maximum and then returns to thermal equilibrium as depicted in Fig. 1. The area under the curve is

$$Q = \int dQ = \epsilon \int_0^\infty V \, dt \tag{4}$$

This area can be easily measured in a various number of ways: cutting out and weighing the tracing, using a planimeter, or using an automatic integrator attached to a recorder.

In some cases, the final and initial base lines are not identical. This base line displacement is the result of not achieving good thermal equilibrium prior to mixing or the result of physical effects on the thermal elements. If it is the result of the former situation, the base line displacement can be very large and the experiment can only provide an approximate value of the heat effect. If the base line displacement is small, it is prob-

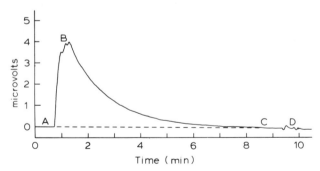

FIG. 1. A typical experiment with an LKB batch microcalorimeter. A designates the initial base line, and C the final base line. The fine structure in this thermogram (labeled B) is the result of reproducible mechanical effects on the thermal elements and do not represent distinct chemical (or physical) processes occurring within the calorimetric cell. The total heat associated with mixing the reactants were calculated from the area between the voltage–time curve and the extrapolated base line (dotted line) as described in the text. The heat effect associated with the physical mixing process ($\sim 50 \, \mu$cal) is shown at D. This heat effect must be subtracted from the measured heat. In this case a solution of transfer RNA was mixed with spermine and produced a total heat of 1.67 mcal.

ably the result of the latter phenomenon. In this case, it is possible to obtain a relatively precise estimate of the heat effect by assuming the base line displacement occurred at the instant of mixing and extrapolating the final base line to zero time and calculating the area between this base line and the observed signal.

The measured heat effect from a batch experiment includes a contribution from mixing which must be subtracted from the observed effect. This contribution can be determined by simply rotating the calorimeter after the thermal equilibrium between the reaction vessel and heat sink has again been achieved. This remixing is shown in Fig. 1 (area of curve labeled D). The area of this curve should be subtracted from the total area of the initial experiment. It is generally found that this mixing heat is very reproducible from experiment to experiment and on the order of 50 μcal. If the apparent heat of mixing in such a remixing experiment is significantly larger than this value, then complete mixing did not occur in the initial effort. This lack of complete mixing is unusual. It can be the result of having very viscous solutions or the adsorption of the solute onto the calorimeter cell surface. In any case, this possible effect should be carefully assessed in all experiments.

The signal from a typical flow calorimeter experiment is shown in Fig. 2. The zero signal from the calorimeter (i.e., the voltage measured without any solution flowing through the calorimeter) is indicated (A in Fig. 2). This zero value is always close, but not identical, to zero. At the point marked B in Fig. 2, the pumping of solvent through the calorimeter is initiated. At the point marked C in Fig. 2, a steady-state value for the signal has been obtained. The difference between this signal and the zero value is the result of viscous flow through the calorimeter. This latter signal is the base line for the experiment. At the point marked D in Fig. 2, the two solutions to be mixed in the calorimeter are being pumped through the reaction vessel and a new steady state value of the signal is obtained. The difference between this signal (D in Fig. 2) and the baseline signal (C in Fig. 2) is proportional to the heat effect $\dot{Q}$. It includes the heat of dilution of all the components and the heat of reaction and is equal to

$$\dot{Q} = f(C \ \Delta H + Q_{\text{dil}}) \tag{5}$$

where f is the flow rate, C the concentration of product, ΔH is the heat of the reaction, and Q_{dil} the apparent heat of dilution per unit volume of solution.

The above description of the flow calorimetric experiment is based upon the assumption that the reaction is instantaneous. This is not always the case. Normally, using a pumping rate of about 10 ml/hr, the sample is retained within the calorimeter cell for about 100 sec. If the reaction is not

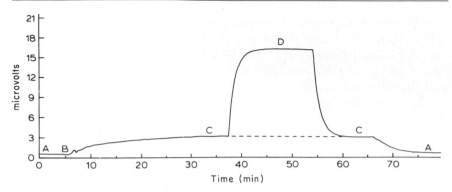

Fig. 2. A typical experiment with an LKB flow microcalorimeter. Zero microvolts is the zero point of the recorder. A represents the signal (~ 0.5 μV) from the calorimeter thermal elements with no solution flowing. At B, pumping of a solution at the rate of ~ 20 ml/hr is initiated. After about 20 min a steady-state base line (labeled C) is obtained and represents the heat effect due to viscous heating within the calorimeter. At about 37 min two reactant solutions are first mixed within the calorimeter cell and a new steady heat effect (labeled D) is obtained in about 7 min. These reactant solutions are then replaced with water and base line C is obtained. At about 66 min the pumps are turned off and base line A is again obtained. The steady-state heat effect due to the reaction is proportional to the difference between D and C, which in this case is about 14 μV or 56 μcal/sec.

completed within this residence time, the measured heat refers only to the extent to which the reaction occurred during that time. It is, therefore, necessary to determine the rate of the reactions in an approximate manner. This can be done with usual kinetic experiments or determined by varying the flow rate in the calorimetric experiment.[9] If the reaction is instantaneous, the signal should be a linear function of the flow rate as indicated in Eq. (5). If the signal does not behave properly as a function of flow rate, it means that the reaction is slow compared to the residence time in the calorimeter. Analysis of such data will be discussed in detail in a later section of this chapter.

With isoperibol calorimeters, which operate in a quasi-adiabatic mode, equilibrium with the environment is not maintained during the experiment. Furthermore, stirring of the solution and dissipation of heat by the thermistor detector during the experiment generates heat at a presumed constant rate. An experimental signal profile for a slow reaction is shown in Fig. 3. The initial base line reflects a continuous change in the temperature of the solution. The slope of this base line is essentially constant and proportional to the heat leakage constant κ. At about the 10-min point on the abcissa, an ampoule containing the solute was broken and a change in

[9] R. E. Johnson and R. Biltonen, J. Am. Chem. Soc. 97, 2349 (1975).

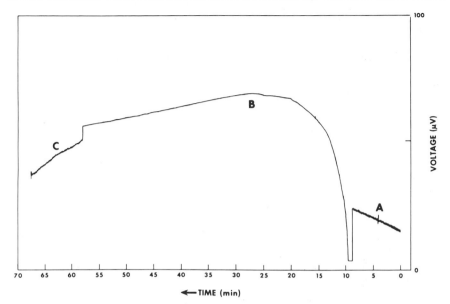

FIG. 3. A typical signal obtained for a heat of solution experiment with an LKB 8700 iso-peribol calorimeter. The time dependent temperature change is monitored by the voltage output from a Wheatstone bridge in which a thermistor forming one arm of the bridge is used for monitoring. The before period (A) and after period (C) are at higher amplifier gain to obtain more precise estimates of their slopes (λ_1 and λ_2). At about 10 min the ampoule, containing about 30 mg of thymine, was broken and the curve labeled B obtained. The area under this curve was used to calculate the heat of solution as described in the text. The total heat for this experiment was about 0.5 cal.

the temperature of the solution occurs until at about 35–40 min the change in temperature again becomes a linear function of time.

During the experiment it is assumed that heat exchange with the environment obeys Newton's cooling law.[10] The rate of change of temperature of the solution due to this heat exchange is

$$\frac{dT}{dt} = \kappa(T_\infty - T) \qquad (6)$$

where κ is the heat leakage constant, T_∞ the temperature of the vessel at infinite time (i.e. the temperature of the calorimeter thermostat) and T the temperature of the calorimetric vessel at any time. Letting

$$T_\infty - T = (T_\infty - T_f) + (T_f - T)$$

[10] I. Wadso, *Sci. Tools* **13**, 33 (1966).

and after integration it follows that the total change in temperature during the experiment due to heat loss to the environment is

$$\Delta T = \kappa (T_\infty - T_f)t + \kappa \int_0^t (T_f - T) \, dt \qquad (7)$$

where T_f is the temperature of the reaction vessel upon termination of the experiment. The first term on the right-hand side of Eq. (7) is given by the slope of the final base line. The integral is given by κ times the area under the curve as indicated in the figure. The value of κ can be obtained from the difference in the slopes of the initial and final base lines divided by the temperature difference

$$\kappa = (\lambda_2 - \lambda_1)/(T_2 - T_1)$$

where λ_1 and λ_2 are the slopes of the initial and final base lines (labeled A and C, respectively, in Fig. 3). The true difference in temperature of the solution due to the reaction if the calorimeter were operating in a truly adiabatic mode is given by

$$\Delta T_{true} = \Delta T_{ob} - (T_2 - T_1) \qquad (8)$$

The true value of the heat effect for the reaction is given by $Q = \epsilon \, \Delta T_{true}$, where ϵ is the calibration constant for the reaction vessel and its contents and is an "effective heat capacity" of the system. Since the heat capacity of the system will vary from experiment to experiment, it is necessary to determine ϵ after each experiment. This is usually done electrically using a resistance heater placed within the reaction vessel.

In most isoperibol calorimeters the temperature of the solution is not directly measured, but rather the resistance of a thermistor is measured. In this case, a proportionality constant relating temperature to resistance is required. This aspect does not affect the procedure for calculating the heat effect, however, and is described in detail elsewhere.[10]

These considerations of the calculation of the heat effect associated with reactions occurring in isoperibol calorimeters are exact and necessary for reactions which are not instantaneous. In the case of essentially instantaneous reactions, a useful approximation can be made which reduces the complexity of the calculation of the true heat effect and does not introduce serious error. The integral in Eq. (7) can be approximated using the mean value theorem so that

$$\frac{dT}{dt} = \kappa (\overline{T_\infty - T}) \qquad (9)$$

where $(\overline{T_\infty - T})$ is the mean value of $(T_\infty - T)$ during the experiment. Now assuming that the temperature changes during the before and after periods of the experiment are small compared to the total temperature change, Eq. (9) can be rewritten in terms of the slopes of the initial and final baseline signals such that

$$\frac{dT}{dt} = \tfrac{1}{2}(\lambda_2 - \lambda_1) \quad \text{and} \quad \Delta T = (\lambda_2 - \lambda_1)\Delta t / 2 \tag{10}$$

Thus, the true temperature change for the reaction is simply the difference in the extrapolated values of T of the two base lines to the midpoint of the signal change. It is necessary to correct these values of the heat effect for the heats of mixing and the heats of dilutions to obtain the true heat for the reaction.

The isothermal calorimeter is maintained at constant temperature by means of a Pelitier device which can add or remove heat as required. The amount of energy input is thus equal to the heat effect. Prior to mixing, the rate of heat input (or extraction) is equal to the rate of heat exchange with the surrounding and the rate of joule heating associated with stirring of the solution. This steady rate is indicated in Fig. 4. The heat associated with mixing causes a temperature change in the cell, and, via a feedback mechanism, current is applied to the Pelitier element to maintain constant temperature. Because of some sluggishness in the systems, the temperature is not maintained exactly constant, and temperature fluctuations in the system may occur; this is indicated in Fig. 4. After the reaction is

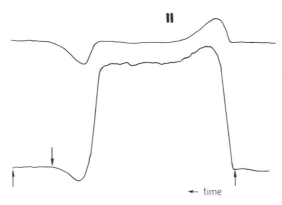

FIG. 4. Thermogram obtained from an isothermal titration calorimeter. The lower trace represents the power requirements to a Peltier device to maintain isothermal conditions. The upper signal is proportional to the absolute temperature within the reaction vessel. Details of these signals were discussed in Chapter [13], this volume. The area under the steady-state signal per unit time is proportional to the heat effect. Dividing this area, by an appropriate calibration factor, expressed in area per millicalorie, for example, will provide the number of millicalories produced per unit time. The enthalpy of the reaction may be determined by dividing this heat effect by the number of millimoles being added to the reaction vessel per unit of time.

complete the rate of power input returns to its original value. The total power input, proportional to the area under the curve indicated in Fig. 4, is the measured heat effect. The calculation of the heat effect is thus similar to the procedure used with the batch calorimeter.

Determination of the Heat of Reaction

In the following discussion it is assumed that the total heat effect associated with the mixing of two reactive components in the calorimeter has been appropriately corrected for the heat of mixing and the heats of dilution of the components. This corrected heat now corresponds to the heat associated with the reaction that has occurred. For purposes of discussion we shall assume that the reaction can be written in terms of clearly defined products as follows.

$$A + B \rightarrow P$$

The apparent heat of reaction can thus be used to determine the apparent enthalpy change for the reaction if [P] is independently known or [P] can be estimated if ΔH is known. In the following discussion, examples of the uses of Q_x or $\dot{Q}_x$ to study particular reactions will be given.

The rate of heat production within the calorimetric vessel is

$$\dot{Q}_x = \Delta H^{0'} \frac{dn_p}{dt} \tag{11}$$

where $\Delta H^{0'}$ is the apparent enthalpy change associated with the formation of one mole of product. In the batch, isothermal, and isoperibol calorimeters the magnitude of heat effect is

$$Q_x = \int dQ = n_p \, \Delta H^{0'} = V[P] \, \Delta H^{0'} \tag{12}$$

$n_p = V[P]$ is the total amount of product formed, where V is the final volume of the mixed solutions and [P] the final product concentration. If the reaction is instantaneous, the analysis can be readily extended to the flow calorimeter. In the flow or titration calorimeter under steady-state conditions the heat effect is

$$Q_x = \frac{dQ}{dt} = \Delta H^{0'}[P] \frac{dV}{dt} = \Delta H^{0'}[P]f \tag{13}$$

where f is the total flow rate of the mixed solution through the flow cell.

Determination of Apparent Gibbs Energy and Enthalpy Changes

Although the calorimeter is primarily intended to measure enthalpy changes associated with chemical reactions, proper experimental design can provide additional information relating to the thermodynamics of the reaction. Let us assume that the reaction occurring is the reversible association of two reactants as follows.

$$M + L \rightleftarrows ML$$

and that Q_x is measured at a constant total concentration of M as a function of the concentration of L. At any given final concentration of $[L]$, the free ligand concentration, the fractional degree of saturation of the macromolecule is given by

$$F = \frac{[ML]}{[M] + [ML]} = \frac{K[L]}{1 + K[L]} \tag{14}$$

where $K = [ML]/[M][L]$, the association constant for the reaction. Thus,

$$Q_{app} = \frac{Q_x}{M_t} = F \Delta H' \tag{15}$$

where M_t is total number of moles of M in the reaction mixture. If Q_{app} is determined as a function of ligand concentration all the necessary information required to determine both $\Delta H'$ and K is available. An example of this type of experiment is shown in Fig. 5 for the system of 3'-cytosine monophosphate and ribonuclease where a one to one complex is known to form. A plot of Q_{app} versus [3'-CMP] is the familiar rectangular hyperbola. These data, in double reciprocal form, are shown in Fig. 6. In this form Eq. (15) transforms to

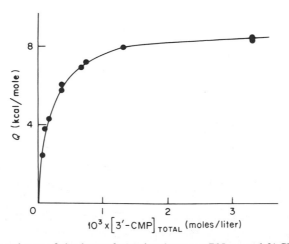

Fig. 5. Dependence of the heat of reaction between RNase and 3'-CMP on the total 3'-CMP concentration. [RNase] = 6.63×10^{-5} M; pH 5.52; $T = 25°$. Reproduced from Bolen et al.,[11] with permission of the publishers.

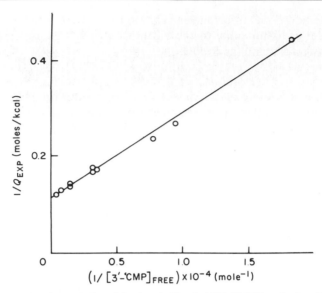

FIG. 6. Double reciprocal plot of the data in Fig. 5. $1/[3'-CMP]$ is calculated from the free 3'-CMP concentration by an iterative least-square technique. The original paper should be consulted for details. Reproduced from *Bolen et al.*,[11] with permission of the publishers.

$$\frac{1}{Q_{app}} = \frac{1}{\Delta H'} + \frac{1}{\Delta H' K}[L]^{-1} \tag{16}$$

Thus the intercept of Q_{app}^{-1} as the magnitude of $[L]^{-1}$ approaches zero is the reciprocal of the molar enthalpy change for complex formation and the slope of the line is $(\Delta H' K)^{-1}$. Since $\Delta G^{0'} = -RT \ln K$, both the apparent enthalpy change and apparent Gibbs energy change for the reaction can be obtained from such a set of calorimetric experiments.

It is to be noted that precise estimation of these thermodynamic parameters depends upon the method of analysis of the data. First, the mathematical analysis must be performed as a function of the *free* ligand concentration. Further, if double reciprocal plots are used with a least squares procedure, it is necessary that the data be properly weighted. However, the purpose of this chapter is not to consider details of numerical analysis and other references should be consulted.[11-13]

The apparent thermodynamic quantities associated with a number of reactions involving ligand binding to biological macromolecules are sum-

[11] D. W. Bolen, M. Flogel, and R. Biltonen, *Biochemistry* **10**, 4136 (1971).
[12] C. Bjurulf, J. Laynez, and I. Wadso, *Eur. J. Biochem.* **14**, 47 (1970).
[13] P. R. Bevington, "Data Reduction and Error Analysis for the Physical Sciences." McGraw-Hill, New York 1969.

marized in the table.[12,14-32] In many biological systems the stoichiometry of ligand binding to the macromolecules is not 1 : 1. If all the sites on the macromolecule are identical and independent, analysis of calorimetric data as described above will provide the correct K' for the reaction but $\Delta H'$ will now refer to the enthalpy change associated with saturation of the n binding sites. Thus $\Delta H' = n\langle\Delta H\rangle$, where $\langle\Delta H\rangle$ is the enthalpy change for binding to 1 mole of sites. If the experiment is performed at constant ligand concentration and the total concentration of macromolecule varies, the calculated Q_{app} is per mole of ligand bound and the extrapolated value of $\Delta H'$ is equal to $\langle\Delta H\rangle$. A comparison of the two values of $\Delta H'$ thus obtained provides an excellent estimate for the stoichiometry of the reaction. It is to be emphasized that estimation of the stoichiometry in this manner requires that binding of ligand to each site occur with identical values of the enthalpy change.

Determination of Heat Capacity Changes

The constant pressure heat capacity, C_p, of a system is defined as

$$C_p = \left(\frac{\partial H}{\partial T}\right)_p \tag{17}$$

Similarly the heat capacity change, $\Delta C_p^{0'}$, associated for a chemical process is written as

[14] B. G. Barisas, S. J. Singer, and J. M. Sturtevant, *Biochemistry* **11**, 2741 (1972).
[15] D. H. Atha and G. K. Ackers, *Biochemistry* **13**, 2376 (1974).
[16] M. F. M. Johnston, B. G. Barisas, and J. M. Sturtevant, *Biochemistry* **13**, 390 (1974).
[17] S. F. Velick, J. P. Baggot, and J. M. Sturtevant, *Biochemistry* **10**, 779 (1971).
[18] H. Hinz, D. F. Shaio, and J. M. Sturtevant, *Biochemistry* **12**, 2780 (1973).
[19] B. Hedlund, C. Damelson, and R. Lorren, *Biochemistry* **11**, 4660 (1972).
[20] M. Flogel, A. Albert, and R. Biltonen, *Biochemistry* **14**, 2616 (1975).
[21] Y. Kuriki, J. Halsey, R. Biltonen, and E. Racker, *Biochemistry* **15**, 4956 (1976).
[22] R. P. Hearn, F. M. Richards, J. M. Sturtevant, and G. D. Watts, *Biochemistry* **10**, 806 (1971).
[23] D. D. F. Shaio and J. M. Sturtevant, *Biochemistry* **21**, 4910 (1969). D. D. F. Shaio, *Biochemistry* **22**, 1083 (1970).
[24] R. A. O'Reilly, S. I. Ohms, and C. H. Motley, *J. Biol. Chem.* **244**, 1303 (1969).
[25] R. Valdes, Jr. and G. Ackers, *J. Biol. Chem.* **252**, 88 (1977).
[26] R. J. Baugh and C. G. Trowbridge, *J. Biol. Chem.* **247**, 7498 (1972).
[27] G. Rialdi, J. Levy, and R. Biltonen, *Biochemistry* **11**, 2472 (1972).
[28] T. Sturgill, Ph.D. Dissertation, University of Virginia, Charlottesville (1976).
[29] F. Schmid, H. J. Hinz, and R. Jaenicke, *Biochemistry* **15**, 3052 (1976).
[30] H. J. Hinz and R. Jaenicke, *Biochemistry* **14**, 24 (1975).
[31] J. Donner, C. H. Spink, B. Borgström, and I. Sjöholm, *Biochemistry* **15**, 5413 (1976).
[32] H. J. Hinz, K. Weber, J. Flossdorf, and M. R. Kula, *Eur. J. Biochem.* **71**, 437 (1976).

A SUMMARY OF THE THERMODYNAMIC QUANTITIES ASSOCIATED WITH INTERACTIONS INVOLVING BIOLOGICAL MACROMOLECULES[a]

Macromolecule	Ligand	$-\Delta G^{0\prime}$ (kcal/mole)	$-\Delta H^{0\prime}$ (kcal/mole)	$-\Delta S^{0\prime}$ (kcal/mole deg)	$-\Delta C_p^{0\prime}$ (cal/mol deg)	Reference
Anti-DNP antibody (rabbit)	DNP-Lys	10.8	15.3	15	—	14
	TNP-Lys	9.1	11.1	7	—	14
Anti-TNP antibody (rabbit)	DNP-Lys	9.9	17.8	27	155–185	14
	TNP-Lys	12.2	21.3	31	205	14
Anti-DNP antibody (guinea pig)	DNP-Lys	10.0	13.9	13.1	—	43
	DNP-Lys	12.1	8.7	11.4	—	43
MOPC 315	DNP-Lys	8.1	16.6	32	180	16
	DNP-glysine	5.4	19.9	48	291	16
	DNP-aminocaproate	7.9	14.4	21	111	16
	TNP-Lys	8.7	18.1	31	276	16
	Dinitronaphthol	6.5	20.2	46	301	16
	Menadione	6.7	12.1	18	201	16
MOPC 460	DNP-Lys	6.4	15.8	31	270	16
	Menadione	6.7	9.3	9	249	16
Glyceraldehydephosphate dehydrogenase (yeast)	Nicotinamide adenine dinucleotide	7.3	12.4	17	520	17
(rabbit)	Nicotinamide adenine dinucleotide	12.2	16.0	13	300	17
Rabbit muscle aldolase	Hexitol 1,6-diphosphate	7.2	1.3	−20	1100	18
Lysozyme	N-Acetyl-D-glucosamine	2.2	5.8	12	53	12
	Tri(N-acetyl-D-glucosamine)	6.9	13.6	22	−41	12

Protein	Ligand					Ref.
Hemoglobin	ATP	6.5	6.5	0	85	19
	DGP	4.9	9.1	14	−135	19
	O$_2$	7.6	14.1	22	—	15
Ribonuclease A^{b,c}	3′-CMP	10.8	7.1	12	—	20
	2′-CMP	12.0	6.6	18	—	20
	Cytidine	5.1	6.1	−3	—	20
	Orthophosphate	10.5	0.3	36	—	20
Na$^+$-K$^+$ ATPase	Mg^{2+}	4.2	43	128	—	21
	Orthophosphate	3.5	50	154	—	21
S-Protein	S-Peptide	12	24	42	700	22
Chymotrypsin	Indole	4.3	15.2	37	—	23
	N-Acetyl-D-tryptophan	3.3	19.0	52	—	23
	Benzoate	2.6	18.2	52	—	23
	Hydrocinnamate	3.7	15.7	40	—	23
	β-Napthoate	5.3	17.5	40	—	23
	N-Acetyl-L-tryptophan	3.0	21.2	61	—	23
	Phenol	3.1	13.5	35	—	23
Human Plasma albumin	Warfurin	7.4	3.1	−14.2	—	24
Hemoglobin	2α → α$_2$	5.4	4.3	−4	—	25
	4β → β$_4$	22.3	24	6	—	25
	2(αβ) → (α$_2$β$_2$)	8.3	4	−14	—	25
Trypsin	Soybean trypsin inhibitor	12.3	−8.6	−70	440	26
	Ovamucoid inhibitor	10.2	−5.6	−53	270	26
	Lima bean inhibitor	12.7	−2.1	−49	430	26
tRNA^{phe}	Mg^{2+}	8.2	0	28	—	27
		5.5	0	18	—	27
	Ethidium bromide^b	10.3	12	−6	—	28
		8.9	9	0	—	28

(Continued)

TABLE (*Continued*)

Macromolecule	Ligand	$-\Delta G^{0\prime}$ (kcal/mole)	$-\Delta H^{0\prime}$ (kcal/mole)	$-\Delta S^{0\prime}$ (kcal/mole deg)	$-\Delta C_p^{0\prime}$ (cal/mol deg)	Reference
Lactate dehydrogenase (heart)	NADH	7.4	10.6	11	170	29
	NAD	4.8	6.1	5	85	29
Lactate dehydrogenase, NAD complex (heart)	Oxalate	7.7	7.8	0	340	29
Lactate dehydrogenase (Muscle)	NAHD	6.9	7.6	2	325	30
Lipase	Colipase	8.6	6.9	6	313	31
Isoleucine: tRNA ligase	L-Isoleucine	7.2	3.7	-12	430	32
E. coli MRE 600	L-Leucine	3.5	3.7	0	—	32
	L-Valine	4.4	3.7	-2	—	32
	L-Norvaline	4.1	3.7	-1	—	32
	L-2-Amino 3S,4-dimethyl-pentanoic acid	5.8	3.7	-7	—	32
	L-Isoleucinol	2.8	-0.9	-12	—	32

[a] Except where noted all quantities refer to a standard state of 1 *M*. Most quantities also refer to a temperature of 25°. The original references should be consulted for a more complete description of the conditions of the experiment.

[b] These quantities refer to a standard state of unit mole fraction.

[c] These quantities refer to the hypothetical reaction of the dianionic ligand with the fully protonated enzyme. The original paper should be consulted for a description of the calculations.

$$\Delta C_p^{0\prime} = \left(\frac{\partial \Delta H^{0\prime}}{\partial T}\right)_p \tag{18}$$

Thus the calorimetric measurement of $\Delta H^{0\prime}$ at two or more temperatures can yield precise estimates of average values of $\Delta C_p^{0\prime}$. For example,

$$\Delta C_p^{0\prime} \simeq \frac{\Delta H_2^{0\prime} - \Delta H_1^{0\prime}}{T_2 - T_1} \tag{19}$$

where $\Delta H_1^{0\prime}$ and $\Delta H_2^{0\prime}$ are the enthalpy changes measured at temperature T_1 and T_2, respectively. Estimates of $\Delta C_p^{0\prime}$ determined from the temperature variation of calorimetrically determined $\Delta H^{0\prime}$ values for several reactions involving biological macromolecules are also summarized in the table.

Calorimetry of Coupled Reactions

The heat effect observed for complex reactions includes the heat changes associated with all steps of a reaction. Consider the sequential reaction

$$M + L + B^- \rightleftarrows ML + H^+ + B^- \rightarrow ML + BH$$

where M and L have been previously defined, B^- and BH represent the unprotonated and protonated forms of a buffer, and H^+ is a proton released upon binding of ligand. The overall heat effect is

$$Q = F(\Delta H_L + \Delta n \, \Delta H_B) \tag{20}$$

where ΔH_L is the intrinsic heat of ligand binding (per mole of complex) including release of Δn protons, ΔH_B is the molar enthalpy change associated with protonation of the buffer, and F is the fractional degree of binding which is a function of [L]. ΔH_L will be pH dependent and the magnitude of Q, at any pH, will depend upon the enthalpy of protonation of the buffer. This situation complicates the interpretation of Q, and further experimentation is required to resolve the situation. If the solvent system is well defined (in terms of ΔH_B of the buffer) the magnitude of Q will be directly proportional to F, and the apparent association constant K_L' and $Q_{max} = \Delta H_L + \Delta n \, \Delta H_B$ can be obtained from analysis of the data as previously discussed. If Δn and ΔH_B are known from separate experiments then ΔH_L can be determined. For example, if Q_{max} is determined in two different buffer systems (identified by subscripts 1 and 2) with different ΔH_B

$$Q_{max,1} = \Delta H_L + \Delta n \, \Delta H_{B,1} \tag{21a}$$

$$Q_{max,2} = \Delta H_L + \Delta n \, \Delta H_{B,2} \tag{21b}$$

and

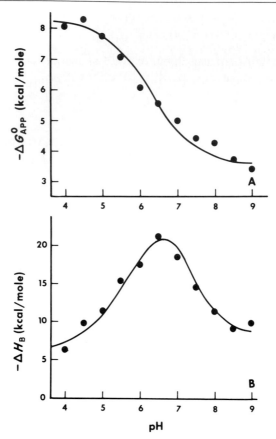

FIG. 7. (A) The apparent Gibbs energy change associated with the binding of 3'-CMP to ribonuclease A as a function of pH, $T = 25°$, ionic strength $= 0.05$. The solid curve was calculated according to a model described in the original paper.[34] Reproduced from Flogel and Biltonen,[34] with permission of the publishers. (B) The apparent heat of binding of 3'-CMP to ribonuclease A as a function of pH, $T = 25°$, ionic strength $= 0.05$. The solid curve was calculated according to a model described in the original paper.[34]

$$Q_{max,2} - Q_{max,1} = \Delta n (\Delta H_{B,2} - \Delta H_{B,1}) \tag{21c}$$

These two experiments thus provide an estimate of Δn and the equations can then be solved for ΔH_L. This analysis requires that neither buffer influence the interaction between M and L.

If the buffering system is not well defined then ΔH_B may be a complex function of the reacting system. For example, the buffering capacity, and hence ΔH_B, may vary with composition. In such cases, the analysis of the data to yield K'_c and ΔH_L is complicated and requires knowledge of Δn.

Flogel and Biltonen[33] have outlined a procedure by which analysis of such a system can be performed provided that Δn is known and the variation of the characteristics of the buffering system can be estimated.

If protons are either released or absorbed upon ligand binding, then both K'_L and ΔH_L will be pH dependent. In order to attempt to interpret thermodynamic quantities ($\Delta G^{0'}$, $\Delta H^{0'}$, $\Delta S^{0'}$), it is absolutely necessary that they refer to well-defined reactions. Therefore, a complete characterization of the pH dependence of these quantities is required. This usually requires a great number of experiments, but can be done. Such a detailed study has been performed for the binding of 3'-CMP to Ribonuclease A.[33,34] The pH dependence of $\Delta G^{0'}$ and $\Delta H^{0'}$ for this system is shown in Fig. 7. These quantities which are a function of seven ionization processes have been analyzed to yield the thermodynamic changes associated with the binding of the dianionic ligand to the fully protonated protein. These same workers[34] extended the studies to include a number of other inhibitors of the enzyme. From these results they were able to deduce the magnitude of specific interactions between ligands and the protein surface.[20]

The Use of Coupled Reactions in Studying Ligand Binding for Which $\Delta H_L = 0$

Many reactions involving ion binding to biological macromolecules exhibit a $\Delta H^{0'}$ which is approximately zero. In such cases, the binding cannot be directly monitored using calorimetric techniques. However, the total concentration of specific ions in solution can be easily measured using calorimetry. For example, the heat change associated with the interaction of Mg^{2+} with EDTA (ethylenediaminetetraacetic acid) is about 8 kcal/mole at 25°.[35] Such uses of calorimetry as an analytic tool are well documented.[36]

This heat effect has been used by Rialdi et al.[27] to study the binding of Mg^{2+} to tRNA[phe] for which the heat of binding was found to be zero. The experiments were carried out in the following way: A buffered solution of tRNA[phe] was extensively dialyzed at 4° against a large excess volume of buffer solution containing a known concentration of $MgCl_2$. Thus, except

[33] M. Flogel and R. Biltonen, Biochemistry 14, 2603 (1975).
[34] M. Flogel and R. Biltonen, Biochemistry 14, 2610 (1975).
[35] J. Jordan and T. G. Alleman, Anal. Chem 29, 9 (1957); P. T. Priestly, Analyst 18, 194 (1963).
[36] H. J. V. Tyrrell and A. E. Beezer, "Thermometric Titrimetry." Chapman & Hall, London. 1968.

for a small Donnan membrane correction,[37] the concentration of free Mg^{2+} is identical in both solutions. The total concentration of Mg^{2+} in the $tRNA^{phe}$ solution is

$$[Mg^{2+}]_t = (1 + \delta)M + nC_t \tag{22}$$

where M is the total concentration of Mg^{2+} in the dialysate, C_t is the concentration of $tRNA^{phe}$, n is the average number of moles of Mg^{2+} bound per mole of $tRNA^{phe}$, and δ is the Donnan correction factor which is small and can be approximately calculated knowing the salt concentration. Equal volumes of the $tRNA^{phe}$ solution and the dialysate were then separately mixed with buffer solution containing an excess of EDTA in the calorimeter. The two respective heats of mixing are

$$Q(tRNA) = Q_{dil} + (1 + \delta)Q_E + nVC_t(\Delta H_E - \Delta H_L) \tag{23}$$

and

$$Q(buffer) = Q'_{dil} + Q_E$$

Q_{dil} and Q'_{dil} are the sum of the heats of dilution of the components in each solution and were measured separately. V is the volume of the $tRNA^{phe}$ solution and dialysate used. ΔH_E is the molar heat of reaction between free Mg^{2+} and EDTA, and ΔH_L is the molar heat of binding of Mg^{2+} to $tRNA^{phe}$

$$Q_E = MV \, \Delta H_E$$

It thus follows that

$$\begin{aligned}\Delta Q &= Q(tRNA) - Q_{dil} - [Q(buffer) - Q'_{dil}] \\ &= \delta Q_E + nVC_t(\Delta H_E - \Delta H_A) \end{aligned} \tag{24}$$

Since ΔH_A, the intrinsic heat of Mg^{2+} binding to tRNA, is zero

$$n = \frac{\Delta Q - \delta Q_E}{VC_t\Delta H_E} \tag{25}$$

Thus a knowledge of Q, V, C_t, ΔH_E, and δ allows a determination of n as a function of Mg^{2+} concentration. V, C_t, and ΔH_E were all determined in separate experiments.

Values of n as a function of free $[Mg^{2+}]$ obtained in this fashion are shown in Fig. 8. Scatchard analysis of these results indicated that at least two distinct sets of binding loci exist: A strong set of four sites characterized by a binding constant of 10^6 M^{-1} and a weaker set of about twenty sites characterized by a binding constant of 1.1×10^4 M^{-1}.

[37] C. Tanford, in "Physical Chemistry of Macromolecules." Wiley, New York, 1961.

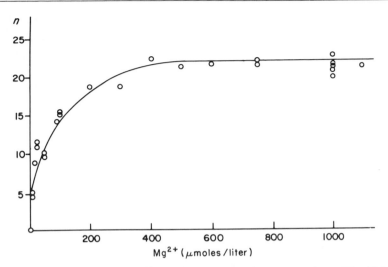

FIG. 8. Binding of Mg^{2+} to RNA^{Phe} as a function of Mg^{2+} concentration at 25°, pH 7.2, 10 mM Na phosphate buffer, 5 mM NaCl. The solid line has no theoretical significance but the data were analyzed to yield the binding parameters. See text for details. Reproduced from Rialdi *et al.*,[27] with permission of the publishers.

The Use of a Flow Microcalorimeter as a Kinetic Instrument

Johnson and Biltonen[9,38] have demonstrated that a flow microcalorimeter can be used to determine rates of reactions in solution that have half-lives between a few seconds and several hours. Under steady-state conditions the rate of heat production within the calorimetric cell ($\dot{Q}$) is equal to the heat flux out of the cell. The total heat flux is composed of three components: that heat conducted to the heat sink via the thermal elements ($\dot{Q}_t$), that heat transferred via the air gap and tubing at the edges of the gold calorimetric cell ($\dot{Q}_a$), and that heat transported out of the cell by the flowing solution ($\dot{Q}_s$). From Newton's law of heat transfer it follows that

$$\dot{Q}_t = \alpha \Delta T_{hs} \quad \text{and} \quad \dot{Q}_a = \beta \Delta T_{hs} \quad (26)$$

where α and β are proportionality constants and ΔT_{hs} is the temperature difference between the calorimetric cell and the heat sink. Assuming that the existing solution is in thermal equilibrium with the body of the calorimetric cell it follows that

$$\dot{Q}_s = C_p \Delta T_{hs} f \quad (27)$$

where C_p is the heat capacity of the solution per unit volume and f is the total flow rate through the calorimetric cell. Thus

[38] R. E. Johnson, Ph.D. Dissertation, Johns Hopkins University, Baltimore, Maryland, 1974.

$$\dot{Q} = (\alpha + \beta + C_{\mathrm{p}}f)\,\Delta T_{\mathrm{hs}} = \epsilon V \tag{28}$$

These conditions have been shown to be valid for fast reactions.[39]

With slow reactions, a significant amount of reactants may remain unreacted when the solution exits the mixing cell. The rate of heat generation within the calorimetric cell in such cases (W_r) is proportional to the amount of product formed in the time the solution residues within the calorimeter cell:

$$W_{\mathrm{r}} = f\,\Delta H[\mathrm{P}]_r \tag{29}$$

If thermal equilibration between the solution and the body of the calorimetric cell is achieved, then the apparent rate of heat generation $Q = W_r$. It is this proportionality which makes the flow microcalorimeter potentially capable of measuring reaction rates.

The amount of product formed within the calorimetric cell $[\mathrm{P}]_r$ is related to the residence time τ of the solution within the cell. The residence time is proportional to the "effective volume" v such that

$$v = \tau f \tag{30}$$

where τ is the residence time of a volume element in the calorimeter cell from the point of mixing to the point of exit.

Knowing the flow rate, the effective volume and ΔH for the overall reaction, the average velocity of the reaction r over the time interval τ can be calculated;

$$r_\tau = \frac{W_{\mathrm{r}}}{f\,\Delta H} \tag{31}$$

r can then be used to calculate the rate constant by solving the integral equation

$$r\tau = k \int_0^\tau (\Pi_i\,[R_i]^n)\,dt \tag{32}$$

where R_i is the time-dependent concentration of reactant i and n_i is the order of the reaction which respect to that component. Thus, it is, in principle, possible to determine the rate constant for any biomolecular or higher molecularity reaction using this technique.

For a first-order or pseudo-first-order reaction the integrated rate expression is

$$W_{\mathrm{r}} = \Delta H f(1 - e^{-k_1\tau})[R]_0 \tag{33}$$

where $[R]_0$ is the initial concentration of the reactant R and k_1 is the first-order rate constant. Thus the average rate and hence the signal for

[39] P. Monk and I. Wadso, *Acta Chem. Scand.* **22**, 1842 (1968).

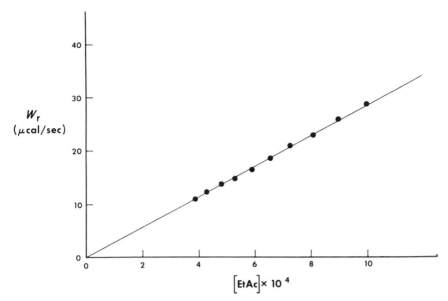

FIG. 9. Heat effect W_r produced by the alkaline hydrolysis of ethyl acetate under pseudo-first-order conditions: $[KOH] = 0.1\ M, f = 2.73$ liters/sec; the line is the calculated best least-squares fit to the data. See text for details. Reproduced from Johnson and Biltonen,[9] with permission of the publishers.

first-order reactions under constant flow rate conditions is a linear function of the variable reactant.

For higher-order reaction schemes the relationship between the average rate and the initial concentration of reactants is more complex. However, if conditions can be established such that the rate of the reaction is essentially constant over the time τ (i.e., initial velocity conditions) then

$$W_r \simeq \Delta H\, fk(\Pi_i\, [R_i]^n) \tag{34}$$

and the average rate is equal to the initial rate of the reaction.

Using the hydroxide-catalyzed hydrolysis of ethyl acetate as a model reaction, Johnson and Biltonen[9] have shown that precise rate constants could be obtained if either of the above two conditions could be met. The type of data obtained for the reaction under pseudo-first-order and constant velocity conditions are shown in Figs. 9 and 10. The analysis of such data to yield rate parameters requires evaluation of the enthalpy change for the reaction and the effective volume of the calorimeter. The determination of the effective volume is straightforward and described in detail by Johnson and Biltonen.[9,38]

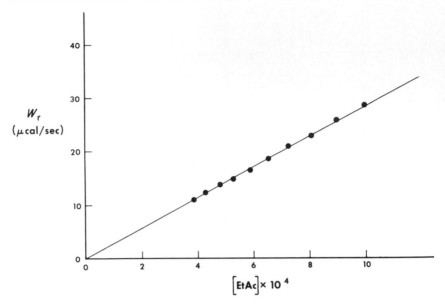

FIG. 10. Heat effect w_r produced by the alkaline hydrolysis of ethyl acetate under conditions where the reaction rate remains constant during the calorimeter transit time τ. $[KOH] = [EtAc]$, $f = 21.5$ $\mu l/sec$, extent of reaction $= 2$–4%, the line is the calculated best least-squares fit to the data. See text for details. Reproduced from Johnson and Biltonen,[9] with permission of the publishers.

Determination of the Enthalpy Change for Slow Processes

Although the flow microcalorimeter was initially designed to measure enthalpy changes of very fast reactions, it is also possible to determine enthalpy changes for slow reactions by appropriate experimental design. A particularly convenient scheme is available for second-order reactions for which the integrated rate expression is

$$\frac{1}{W_r} = \frac{1}{\Delta H\, f[R]_0} = \frac{1}{\Delta H\, Vk_2[R]_0^2} \tag{35}$$

If the reaction is run at several different initial concentrations of reactant at constant flow rate, the data can be analyzed according to Eq. (35) to yield estimates of ΔH and k_2. Data obtained for the alkaline hydrolysis of ethyl acetate under constant flow rate conditions where $[EtAc]_0 = [OH^-]_0 = [R]_0$ are shown in Fig. 11.

ΔH values can also be determined by measuring W_r under second-order conditions as a function of the flow rate. Data obtained under constant initial concentration of reactants as a function of flow rate are shown

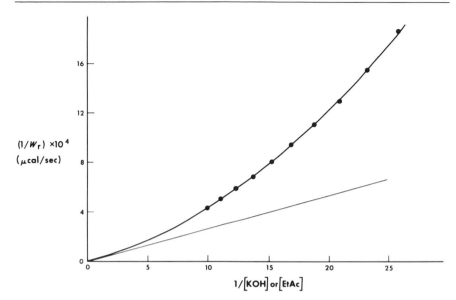

FIG. 11. Heat effect W_r produced by the alkaline hydrolysis of ethyl acetate as a function of initial concentration when $[KOH]_0 = [EtAc]_0$ plotted in the form of Eq. (35); $f = 2.73$ $\mu l/sec$; the curved line is the calculated best least-squares fit to the data, and the straight line is the calculated linear term of Eq. (35). Reproduced from Johnson and Biltonen,[9] with permission of the publishers.

in Fig. 12. The two determinations of ΔH agreed well with the values obtained by Papoff and Zambonin[40] and Becker and Spalink.[41]

Thus, the enthalpy change and rate constant for a second-order reaction can both be estimated from the variation in heat effect as a function of the initial concentration of reactants at constant flow rate or from the variation in the heat effect as a function of the flow rate at constant initial concentration of reactants.

The apparent second-order rate constant can also be calculated using an appropriate integral form of the rate equation for a second-order reaction. In the case where the two reactant concentrations are equal

$$k_2\tau = \frac{1}{[R]} - \frac{1}{[R]_0} \tag{36}$$

where $[R] = [R]_0 - \dot{Q}_r/\Delta H f$, the concentration of each reactant at time τ. In cases where the reactant concentrations are not equal, k_2 can be estimated by iterative stepwise numerical integration of Eq. (34) over the

[40] P. Papoff and P. G. Zambonin, *Talanta* **14**, 581 (1967).
[41] F. Becker and F. Spalink, *J. Phys. Chem.* **26**, 1 (1960).

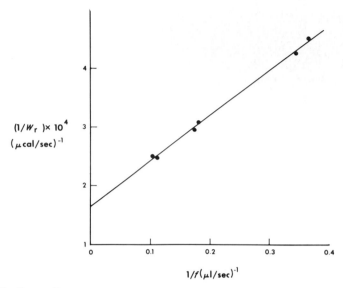

FIG. 12. Heat effect W_r produced by the alkaline hydrolysis of ethyl acetate when $[KOH]_0 = [EtAc]_0 = 0.1$ M as a function of flow rate; the line is the calculated best least-squares fit to the data. Reproduced from Johnson and Biltonen,[12] with permission of the publishers.

time interval $t = 0$ to $t = \tau$ assuming that the reaction was first order with respect to each reactant. This procedure is repeated until a value of k_2 was obtained which provided a calculated W_r in agreement with the experimental W_r.

These procedures were used by Johnson and Biltonen[9] to estimate the apparent second-order rate constant for the alkaline hydrolysis of ethyl acetate over a limited concentration range at different flow rates. Their results, listed in Table IV in Chapter [13], this volume, demonstrate that the calculated value of the rate constant is independent of the flow rate.

Thus, a flow microcalorimeter can be used to measure reaction velocities which in turn can provide reliable values for reaction order, rate constant, and enthalpy change for the reaction. This application of the instrument requires that thermal equilibrium between the reacting solution and the body of the calorimeter cell be at least approximately attained, and that the residence time of the solution in the calorimeter cell be well defined.

In order to obtain a high degree of accuracy in rate determinations using a flow microcalorimeter, rapid thermal equilibration between the flowing solution and the body of the calorimetric cell and efficient mixing of the two solutions is required. The former problem may become serious at high flow rates. The latter problem will become serious when the mixing time is a significant fraction of the residence time.

The use of flow microcalorimetry to measure reaction rates in the liquid phase has wide potential in all areas of organic, inorganic, and biological chemistry because it measures a ubiquitous property of chemical reactions. Its particular advantages include the high degree of attainable precision and accuracy, its high sensitivity, the wide range of reaction times which can be resolved, and the fact that it can provide both thermodynamic and kinetic information.

An area of special interest is that involving enzyme-catalyzed reactions. Since this technique relies only on heat generation by the reaction, it is thus not necessary to develop complicated assay procedures to monitor the reaction. In addition, full recovery of the enzyme after reaction is feasible. Assuming a useful sensitivity of 0.1 μcal/sec and a reaction enthalpy of 10 kcal/mole, it is possible to detect enzymatic activity at concentrations as low as 10^{-15} M for those enzymes which exhibit the highest turnover numbers (e.g., carbonic anhydrase). Johnson[38] has used this technique to determine V_{max} and K_m for reactions catalyzed by α-chymotrypsin and ribonuclease A.

The Use of Continuous Gradients in a Flow Calorimeter

The types of calorimetric data analysis discussed thus far can require a great amount of experimental data. The usual procedures for data acquisition will consume a great amount of time and material. It is, therefore, useful if techniques or procedures can be developed which allow rapid data generation with a minimum consumption of material, while still maintaining normal precision and accuracy. Flow systems offer such a potential and can be applied to the calorimetric technique.

If two instantaneously reacting solutions are continuously mixed in a flow heat-conduction calorimeter, a steady-state signal S, corrected for the heat of dilution of the components, is given by Eq. (37). If the amount of product formation is a function of time, the calorimeter is not operating under steady-state conditions and the true heat effect signal S_t is

$$S_t = S + t_r \frac{dS}{dt} \tag{37}$$

where t_r, the ratio of the heat capacity to the thermal conductance of the calorimeter mixing cell, is the response time of the instrument.

If two reactants, A and B, are mixed in the calorimeter and the concentration of one of the reactants continuously changes with time t, the rate of heat production within the cell is also a function of time. Therefore, the observed signal S can be "corrected" to S_t and hence related to the concentration of the variable component if the latter quantity is known as a function of time.

Mountcastle and co-workers[42] have shown that such a procedure is valid in a flow calorimeter using an exponential dilution device in the following way. If solvent is continuously added to a constant volume initially containing solute at a concentration C_0, the concentration of solute as a function of time will be

$$[C] = C_0 \, e^{-at}$$

where $a = f/v$, where f is the flow rate of the solution and v the volume of the vessel.

In order to judge the applicability of continuous concentration gradients to flow calorimeters, a solution of HCl continuously diluted in a constant volume device, was mixed with a constant excess concentration of NaOH. Thus $[HCl]_t = C_0 \, e^{-at}$ and S_t is the signal associated with the heat of neutralization corrected for a *constant* signal associated with the heat of dilution of the components

$$S_t = \epsilon f_c \, \Delta H \, C_0 \, e^{-a(t-t_0)}$$

ΔH is the molar heat of neutralization and t_0, the "zero time," is the time at which the front of the gradient reaches the mixing cell. The HCl solution, at its initial concentration, was pumped through the calorimeter for a period of time to establish true steady-state conditions for the calorimetric signal. Therefore, at $t < t_0$, $S = S_t = \alpha$ (where $\alpha = \epsilon f_c \, \Delta H \, C_0$) and the solution of the differential Eq. (27), assuming $a \neq t_r^{-1}$, is

$$S = \frac{\alpha}{(1 - a\tau)} \left(e^{-a(t-t_0)} - a \tau e^{-[(t-t_0)/t_r]} \right)$$

If $a < 1/t_r$

$$S \simeq \frac{\alpha}{1 - at_r} \, e^{-a(t-t_0)}$$

at sufficiently long time. t_r was experimentally determined to be 75 sec for the LKB calorimeter used, and $a < 0.004$ sec^{-1} was used for these experiments.

S_t was obtained as a function of time, and data were collected in digital form on paper tape and then differentiated and corrected according to Eq. (37). Logarithmic representations of S_t and S versus time are presented in Fig. 13. Overall time $S_t = \alpha e^{-a(t-t_0)}$ and at long time

[42] D. Mountcastle, E. Freire, and R. Biltonen, *Biopolymers* **15,** 355 (1976).
[43] J. Halsey, J. Cebra, and R. Biltonen, *Biochemistry* **14,** 5221 (1975).

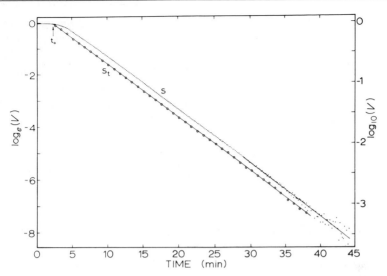

Fig. 13. Logarithmic representation of the recorded voltage minus base line (S) every 10 sec for an exponential dilution experiment in which a solution of 0.1 N HCl is continuously diluted with H_2O and mixed with excess NaOH in a flow calorimeter at 25°. The gradient mixer, an open bottom portion of a test tube, used magnetic stirring and had a nominal volume of 3.4 ml. The diluting flow rate was 0.7755 ml/min. S_t is every fifth value of the voltage corrected by equation $S_t = S + \tau(ds/dt)$ using $\tau = 75$ sec. t_0 marks the beginning of the HCl gradient at the calorimeter mixing cell. Reproduced from Mountcastle et al.,[42] with permission of the publishers.

$$S \simeq \alpha e^{-a[t-t_0]} + (1/a) \ln (1 - at_r)$$

Therefore the logarithmic representations of the S and S_t curves is displaced by an amount $\Delta t = -(1/a) \ln (1 - at_r)$ on the time axis. The intersection of the S_t curve and the initial plateau region of the S curve define the gradient zero time t_0 on the real time axis.

Linearity of the S_t curve over more than three decades of concentration was observed. At extreme dilution ($\Delta \log C > 3$), random deviation of the logarithm of the corrected signal from linearity was observed and is due to instrument noise. Except at early times, the S curve was observed to be linear and parallel to the S_t curve.

It is clear that precise continuous concentration gradients can be useful in monitoring any chemical reaction as a continuous function of the concentration of one or more reactants or perturbants. The technique can be used successfully with any flow calorimeter with sufficiently low base line drift to be compatible with a flow system. In such a configuration the continuous gradient is transformed into an automatic discrete sampling device. The exponential gradient is particularly suitable for thermodynamic studies on equilibrium reactions, since the logarithm of the gradient reactant concentration is a linear function of time. The technique is applicable to the study of such equilibria as ligand binding and self-associating systems.

Continuous concentration gradients are also useful in kinetics studies. Johnson[38] has demonstrated its usefulness in the study of steady-state enzyme kinetics. In such an experiment with an enzyme, which exhibits Michaelis–Menten behavior, the directly measured initial velocity as a continuous function of the logarithm of the substrate concentration will exhibit the same characteristics as an equilibrium binding isotherm. Such data can be directly analyzed to yield the parameters K_m and V_{max}.

Concluding Remarks

We have attempted to survey the basics of experimental design and data analysis required for microcalorimetric experimentation. We have also attempted to demonstrate the versatility and potential of this technique in studying biological systems. It is hoped that the reader will have gained some insight into the methodology and that application of calorimetric techniques to experimental biology will become more prevalent.

[15] Conformational Changes in Proteins by Low Temperature–Rapid Flow Analysis[1]

By D. S. AULD

Introduction

The detection of conformational changes which might accompany the binding of a substrate to an enzyme and the subsequent catalytic and product release steps has not been readily possible due to the high rate of enzymatic reactions. The relationship between protein conformation and function in biochemical reactions has therefore provided an impetus for the development of rapid and sensitive methods for the observation of reaction intermediates. The achievement of the objective probably will require integration of several different areas of current research. These include (1) the design of rapid mixing devices, (2) their adaptation to spectral methods which allow identification of the chemical nature of intermediates, (3) the design of rapid scanning detectors (1--10 msec time range)

[1] This work was supported by Grant-in-Aid GM-15003 from the National Institutes of Health of the Department of Health, Education and Welfare.

which have excellent spectral resolving powers, (4) the placement of chromophoric groups within the enzyme and/or substrate which do not interfere with the enzyme reaction but rather act as monitors of its progress and (5) the use of subzero temperatures to decrease the rate of reaction.

Rapid Flow Methodology

There are several excellent reviews dealing with continuous and stopped-flow methodology and their adaptation to absorption and fluorescence spectrometry.[1a-3]

The resolution time of these techniques, the so-called dead time, now in the range 0.3 to 5 msec, is not likely to be decreased further. Even within this range a trade-off must be made between time resolution, optical path length, and volumes of sample needed. Thus, the original apparatus of Gibson had an optical path length of 2 mm, required 0.3 ml of fluid per determination, and had a dead time of 1.5 msec.[4] Increasing the optical path length to 20 mm required 0.6 ml of fluid, and yielded a dead time of 3.5 msec. An apparatus has been described that produces both high acceleration and flow velocity of liquids and extremely rapid stopping of the liquid after mixing has been achieved.[5] It has a dead time of approximately 0.3 msec for a 3 mm optical path length and requires 4 ml of solution per determination.

Stopped-flow absorbance and fluorescence instrumentation has been used frequently in the study of both models for enzyme reactions and enzyme-catalyzed reactions themselves. Only a few examples will be cited. Such instrumentation has been essential to mechanistic studies of ester hydrolysis,[6] of imine formation,[7] of ligand interactions with peroxidase[8] and hemeprotein[9] as well as investigation of intermediates in flavoproteins[9] and alcohol dehydrogenase reactions.[10,11] Pre-steady-state studies using stopped-flow absorbance and fluorescence techniques have

[1a] Q. H. Gibson, Vol. 16, p. 187.
[2] B. Chance, *Tech. Chem. (N.Y.)* **6**, Part II, 5 (1974).
[3] F. J. W. Roughton and B. Chance, *Tech. Or. Chem.* **8**, Part II, 704 (1963).
[4] Q. H. Gibson and L. Milnes, *Biochem. J.* **91**, 161 (1964).
[5] R. L. Berger, B. Balko, W. Borcherdt, and W. Friauf, *Rev. Sci. Instrum.* **39**, 486 (1968).
[6] B. Holmquist and T. C. Bruice, *J. Am. Chem. Soc.* **91**, 2982 (1969).
[7] T. C. French, D. S. Auld, and T. C. Bruice, *Biochemistry* **4**, 77 (1965).
[8] B. Chance, *J. Biol. Chem.* **151**, 553 (1943).
[9] Q. H. Gibson, *Anal. Rev. Biochem.* **35**, 435 (1966).
[10] H. Theorell and B. Chance, *Acta Chem. Scand.* **5**, 1127 (1951).
[11] M. Dunn, *Struct. Bonding (Berlin)* **23**, 61 (1975).

also been critical to the demonstration of the presence of an intermediate in addition to the Michaelis complex of many hydrolytic enzymes.[12,13]

While the use of absorption and fluorescence optics with stopped-flow systems is therefore well established, these spectral properties of molecules are not as exquisitely sensitive to conformation changes as is the optical activity of molecules. The conformational sensitivity of optical rotatory dispersion (ORD) and circular dichroism (CD) is well founded both in theory[14] and in experiment,[15,16] and CD in particular has been used widely to study the equilibrium properties of proteins.[17] Measurement of optical activity has been especially useful for probing active site interactions of inhibitors and pseudosubstrates with enzymes which exhibit chromophoric properties.

In the past two years, there have been several reports of the successful adaptation of CD to stopped-flow instrumentation. Such instruments have been used to monitor binding of a small molecule to a macromolecule, such as a chromophoric sulfonamide to bovine carbonic anhydrase,[18] crystal violet to calf thymus DNA,[19] and carbon monoxide to hemoglobin.[20]

The application of stopped-flow CD to the visualization of enzyme–substrate (ES) complexes should aid greatly in the establishment of details involved in the mechanism of action of enzymes. Thus, changes in optical activity of a spectral probe attached to either the enzyme or the substrate should allow not only monitoring of substrate binding but also changes in spatial orientation of the substrate within different ES complexes. It may be anticipated that stopped-flow CD will yield a great deal of sorely needed information regarding local conformational changes occurring at the active site of enzymes in the course of catalysis.

Rapid Scan Spectrometry

Kinetic analysis of intermediates in enzyme reactions based on observation at only a single spectral wavelength makes it difficult to detect the

[12] K. J. Laidler and P. S. Bunting, in "The Chemical Kinetics of Enzyme Action," 2nd ed., p. 312, and references therein. Oxford Univ. Press (Clarendon), London and New York, 1973.

[13] D. S. Auld, in "A Survey of Contemporary Bioorganic Chemistry" (E. E. van Tamlen, ed.), Vol. 1, p. 1, and references therein. Academic Press, New York, 1977.

[14] H. Eyring, H.-C. Liu, and D. Caldwell, Chem. Rev. 68, 525 (1968).

[15] D. D. Ulmer and B. L. Vallee, Adv. Enzymol. 27, 37 (1965).

[16] A. J. Adler, N. J. Greenfield, and G. D. Fasman, Vol. 27, Part D, p. 675.

[17] S. N. Timasheff, in "The Enzymes" (P. D. Boyer, ed.), 3rd ed., Vol. 2, p. 371. Academic Press, New York, 1970.

[18] P. Bayley and M. Anson, Biochem. Biophys. Res. Commun. 62, 717 (1975).

[19] P. Bayley and M. Anson, Biopolymers 13, 401 (1973).

[20] F. A. Ferrone, J. J. Hopfield, and S. E. Schnatterly, Rev. Sci. Instrum. 45, 1382 (1974).

presence of multiple intermediates characterized by closely spaced or overlapping spectral bands. The advantages to be gained from the capacity to scan a large spectral region (200–400 nm) in a few milliseconds have recently led to the development of fast scanning spectrometry. Two different technical approaches have been chosen which may be characterized as "array detector" and "scanned spectrum" techniques.[21]

The "array detector method" requires illumination of the sample with a high-intensity, broad-band wavelength source followed by the collection, dispersion, and detection of the existing radiation by a photodiode array detector. At present, scan rates of 100 spectra/sec for a 400 nm range can be achieved.[22] This approach has the advantage of being readily adaptable to a variety of existing stopped-flow instruments, such as the Durrum Gibson instrument. However, the array detector cannot as yet match all the performance parameters of photomultipliers.[21] This approach also precludes the analysis of photo-sensitive and photo-activated systems.

In the "scanned spectrum approach," radiation from the source is first dispersed, and the dispersed light is swept across the exit slit of the monochromator, allowing monochromatic light in a sequence of wavelengths to pass through the sample onto a single photomultiplier detector. It has the advantage of a wider choice of wavelength scanning ranges, and it is applicable to photo-sensitive reactions. Its disadvantage lies in the movable parts which are subject to mechanical wear.

Dye has designed a double beam stopped-flow instrument based on a Perkin Elmer Model 108 scanning monochromator.[23] A speed of 75 spectra/sec over a 250 nm range was found suitable for most applications, but scan speeds of from 3 to 150 spectra/sec are possible. A commercial spectrophotometer based on the scanned spectrum approach covers the electromagnetic spectrum from 200 nm to 3 μm, depending on suitable gratings and detectors, with rates of 200 spectra/sec for a 300 nm range.[24] This instrument follows initial developments by Kuwana and associates.[25] An electronically tunable acoustooptic filter serving as a monochromator and modulator for incorporation into a rapid scanning spectrophotometer for the wavelength range 450–750 nm constitutes a very recent innovation.[26] Scan rates for a 50 nm range as high as 12,000 spectra/sec have been reported for it.

[21] R. E. Santini, M. J. Milano, and H. L. Pardue, *Anal. Chem.* **45**, 915A (1973).
[22] P. Burke, *Res. Dev.* **24**, 24 (1973).
[23] J. L. Dye and L. H. Feldman, *Rev. Sci. Instrum.* **37**, 154 (1966).
[24] "Harrick Rapid Scan Spectrophotometer," Data Sheet 5. Harrick Sci. Corp., Ossining, New York, 1974.
[25] J. W. Strojek, G. A. Gruver, and T. Kuwana, *Anal. Chem.* **41**, 481 (1969).
[26] W. S. Shipp, J. Biggens, and C. W. Wade, *Rev. Sci. Instrum.* **47**, 565 (1976).

It is anticipated that the further development of these different rapid scanning approaches will provide the spectral resolution needed for studying intermediates in enzyme reactions.

Spectrokinetic Probes

The determination of the rates of interconversion of different intermediate states in an enzyme reaction can be aided greatly by the placement of appropriate chromophoric groups either into the enzyme, the substrate, or both. When these can be monitored without interfering with the reaction, they act as spectrokinetic probes.

Organic chemical modification of protein side chain groups can yield sensitive probes of the enzyme active site. Monoarsanilazotyrosine 248-modified carboxypeptidase A serves as an example of the usefulness of this approach. The intensely chromophoric intramolecular coordination complex between arsanilazotyrosine 248 and the zinc atom of azocarboxypeptidase A has proved particularly informative in probing vicinal perturbations and mutual orientation of and distance between these two constituents of the active site of this enzyme. Stopped-flow studies have shown that rapidly turned over ester and peptide substrates disrupt the azo Tyr-248 · Zn complex before hydrolysis occurs.[27] Stopped-flow pH-jump studies have also revealed the existence of rapidly interconverting substructures of carboxypeptidase A.[28]

A spectral probe in a substrate can also allow monitoring of the progress of an enzymatic reaction. Resonance energy transfer between fluorescent enzyme tryptophanyl residues as intrinsic donors and an extrinsically placed acceptor in the substrate, e.g., a dansyl group, has allowed direct visalization of ES complexes.[29] The spectral overlap between the absorption of the dansyl group and the emission of tryptophan is excellent, and the spectrum is red-shifted far enough not to overlap with its own absorption spectrum, properties which make these an exceptionally good donor–acceptor pair. Quantitatively, the degree of energy transfer is sensitive to the distance between and mutual orientation of the donor–acceptor pair and to the environment of the acceptor. Differences in tryptophan to dansyl transfer efficiencies, and/or dansyl quantum yields can characterize the ES species formed. Thus, if a set of reversible ES complexes, ES and/or covalent intermediates, EA are

[27] L. W. Harrison, D. S. Auld, and B. L. Vallee, *Proc. Natl. Acad. Sci. U.S.A.* **72**, 3930 (1975).
[28] L. W. Harrison, D. S. Auld, and B. L. Vallee, *Proc. Natl. Acad. Sci. U.S.A.* **72**, 4356 (1975).
[29] S. A. Latt, D. S. Auld, and B. L. Vallee, *Proc. Natl. Acad. Sci. U.S.A.* **67**, 1383 (1970).

formed in the course of an enzyme catalyzed reaction it should be possible to determine the minimal number of significantly populated states, the rates of interconversion of molecules among these states, and the equilibrium constants determining the relative properties of the populations [Eq. (1)].

$$E + S \rightleftharpoons (ES)_1 \rightleftharpoons (ES)_2 \rightleftharpoons (ES)_n \rightarrow EA + P_1 \rightarrow E + P_2 \qquad (1)$$

This approach has been applied successfully to carboxypeptidase A and has allowed delineation of mechanisms of inhibition,[30] enzymatic consequences of chemical modification,[31] syncatalytic measurement of distances between the active site metal atom and the dansyl blocking group of peptide substrates,[32] and differentiation of the mechanism of ester and peptide hydrolysis.[31] Most recently, it has been applied to a number of other hydrolytic enzymes, such as yeast carboxypeptidase, chymotrypsin, and alkaline phosphatase.[13] A similar approach has been used for pepsin.[33] The analysis of the rate of formation and breakdown of the intermediates have allowed quantitative assessment of kinetic schemes for these enzymes and determination of rates of conformational changes occurring during catalysis.

Low Temperature Biochemistry

As has been stated earlier, the time resolution of rapid mixing instruments, now reliably 1–2 msec, is not likely to be decreased substantially. However, if a time resolution of 2–5 msec can be obtained at subzero temperatures, reaction intermediates should become much more accessible to observation.

The relationship between the rate constant of any chemical reaction, k, and the absolute temperature, T, is given by the Arhenius relationship

$$k = A e^{-E_a/RT} \qquad (2)$$

which indicates that a rate constant in general will decrease with temperature, and in any particular case, the extent of such decrease is governed by the activation energy E_a. Thus, if the activation energies for three different reactions are 5, 10, and 15 kcal/mole, a reduction in temperature from $+25°$ to $-55°$ will decrease the rate constant by 22-, 511-, and 10,260-fold. Therefore if one step of an enzyme reaction has a rate constant of 10^4 sec^{-1} at 25° it will occur within the mixing time of the instru-

[30] D. S. Auld, S. A. Latt, and B. L. Vallee, *Biochemistry* **11**, 4994 (1972).
[31] D. S. Auld and B. Holmquist, *Biochemistry* **13**, 4355 (1974).
[32] S. A. Latt, D. S. Auld, and B. L. Vallee, *Biochemistry* **11**, 3015 (1972).
[33] G. P. Sachdev and J. S. Fruton, *Proc. Natl. Acad. Sci. U.S.A.* **72**, 3425 (1975).

ment. However, if the energy of activation is 10 kcal/mole, lowering the temperature of $-55°$ would decrease this rate constant to ~20 sec^{-1}. If the dead time is less than 10 msec at $-55°$ this step of the reaction can then be monitored easily. Since temperature-jump studies show the rate constant for dissociation of substrates and of pseudosubstrates for a number of enzymes to be in the range from 10^2 to 10^4 sec^{-1} [34] it may be possible to monitor even the substrate binding step by lowering the temperature.

If different elementary steps have different energies of activation, it may be anticipated also that the rate determining step in the reaction may differ at $-55°$ and at $+25°$. Intermediates present only at low steady state concentration at $+25°$ may then increase in concentration as the temperature is lowered, thus allowing much easier identification of their chemical nature.

Douzou and his associates during the past decade have delineated the fundamental physical chemistry of solutions at subzero temperature. In 1966 they initiated an investigation of the temperature dependence of physical chemical parameters of solutions such as density, viscosity, dielectric constant, and dissociation constants of neutral salts, acids, bases, and buffers to provide the background necessary for experimentation with biochemical compounds at subzero temperatures.[35] They have shown that these parameters vary widely on addition of organic solvents and lowering of temperature. However, more importantly, this vast body of information now allows the empirical prediction of the values of these parameters for a wide range of subzero temperatures, given any set of operational conditions. Thus, it has been demonstrated that the dielectric constant can be held constant by synchronizing the progressive addition of an organic solvent with cooling of the solution. Moreover, buffer solutions can be prepared for which the pH is practically invariant over a broad range of subzero temperatures. Table I exemplifies the type of information available from these studies. The physical parameters of a methanol–H$_2$O 70:30 mixture which has a freezing point of $-115°$, are compared at $+20°$ and $-60°$. Much more extensive tables are given in the review by Douzou.[35] Such studies have shown that proton activity and ionic strength are critical factors for biological reactions in mixed solvents at subzero temperatures.[36] The methodology for obtaining subzero temperatures has been described.[35] The equipment used has three main components: (a) a cryogenic temperature production unit, (b) a temperature regulation device and (c) a cell holder and sample compartment adapted for low temperature studies. The adaptation of the Beckman Acta III,

[34] G. G. Hammes and P. R. Schimmel, in "The Enzymes" (P. D. Boyer, ed.), 3rd ed., Vol. 2, p. 67. Academic Press, New York, 1970.

[35] P. Douzou, Methods Biochem. Anal. 22, 402 (1974).

[36] P. Douzou, Biochimie 53, 1135 (1971).

TABLE I

COMPARISON OF SOLUTION PARAMETERS AT +20° AND −60°
IN A METHANOL–H$_2$O 70 : 30 MIXTURE[a]

Parameter	+20°	−60°
Volume mass (g/cm³)	0.883	0.941
Viscosity (cP)	1	37
Dielectric constant	46.3	74.9
Tris buffer pH	8.0	11.5
Phosphate buffer pH	8.6	9.0

[a] The data are taken from figures and tables compiled by Douzou and co-workers.[35]

Aminco Chance DW2, and Cary 15 instruments for low temperature studies has been reported.[35]

Douzou has shown that polyalcohols, e.g., ethylene glycol or glycerol, when mixed with water (50:50, v/v) are far less effective denaturing agents than simple aliphatic alcohols, e.g., methanol. In such mixtures the native conformations of many proteins are stable, making it possible to analyze conformational changes of proteins at subzero temperatures.[37] Intermediates have been stabilized at low temperatures for time periods that enable measurements of their spectra in conventional, i.e., slow scanning, spectrophotometers. The reaction of horseradish peroxidase with H$_2$O$_2$ to yield complex I occurs over a time interval of 3 sec at −33°C compared to 10 msec at 25°.[38] A 100% yield of complex I is achieved, and it is stable in the presence of electron donors at temperatures < −30°. The transition from complex I to complex II can be achieved by rapidly raising the temperature from −35 to −25°, giving a 100% yield of complex II. Optical rotatory dispersion spectra of complexes I and II at −60° suggest there is a change in conformation during this process.[38]

The reaction of α-chymotrypsin with N-acetyl-L-phenylalanine methyl ester has been studied at subzero temperatures using 65% dimethyl sulfoxide as a solvent.[39] The reaction was monitored by changes in enzyme tryptophan fluorescence at 330 nm. After initiation of the reaction at −90°, three sequential reactions, which suggests two intermediates, could be detected as the temperature was raised progressively. The first reaction was attributed to binding of substrate, the second to the acylation of the enzyme and the third to deacylation. The reaction of chymotrypsin and N-acetyl-L-phenylalanine-p-nitroanilide can be observed under similar conditions either by liberation of the p-nitroanilide or by changes occurring in the enzyme tryptophan fluorescence. In the latter case after ini-

[37] F. Travers, P. Debey, P. Douzou, and A. N. Michelson, *Biochim. Biophys. Acta* **19,** 265 (1969).
[38] P. Douzou, R. Sireux, and F. Travers, *Proc. Natl. Acad. Sci. U.S.A.* **66,** 787 (1970).
[39] A. L. Fink and E. Wildi, *J. Biol. Chem.* **249,** 6087 (1974).

tiation of the reaction at $-90°$, a series of four reactions occurred prior to release of the product p-nitroaniline.[40] Again reaction 1 was attributed to formation of the initial Michaelis complex. Reactions 2 and 3 are pH independent, and reaction 4 is pH dependent. The chemical nature and mechanistic significance of these intermediates have yet to be established. However, both studies indicate that intermediates can be stabilized long enough at subzero temperatures so that they could be observed by even slow scanning spectrophotometers.

The rebinding of oxygen or carbon monoxide to myoglobin (Mb) has been examined over an extremely wide temperature range, from $40°$ to $350°K$, using photodissociation of the ligand to initiate the reaction.[41] The rebinding process can be monitored at 436 nm since the Soret absorption bands of Mb, MbCO, and MbO_2 are 434, 423, and 418 nm, respectively. If these bands do not change appreciably with temperature, changes in the transmission at 436 nm should largely reflect the rebinding process. Four different processes have been found. Between $40°$ and $160°K$ a single process is observed which is not exponential with respect to time and independent of CO concentration. Processes having similar kinetic characteristics occur at approximately $170°$ and $200°K$, but at $210°K$ a CO concentration-dependent process occurs. When myoglobin is imbedded in a glass, only the first three nonexponential processes are observed.

In a glycerol–water fluid, rebinding is exponential. These processes have been interpreted to represent four successive energy barriers to ligand rebinding. The nonexponential changes of absorbance at low temperature in solid samples has been attributed to the heme existing in many different conformational states which are thought to be characterized by a distribution of activation enthalpies and entropies.[41] The exponential behavior of the absorbance above $230°K$ in the liquid state might then be attributed to rapid interconversions of molecules among these states. The solid state in this system would then represent stabilization of the "fluctuant configurations" of the protein known to exist in the liquid state.[42] Studies on carboxypeptidase A in the solution and crystalline states have similarly suggested that in crystals the rearrangement of molecular structure may be severely impaired or restricted, and crystallization might single out either active or inactive conformations.[28] The results of both of these studies suggest that low temperature biochemistry is likely to result in increasingly valuable insight regarding the nature of conformational adaptability of proteins.

[40] A. L. Fink, *Biochemistry* **15,** 1580 (1976).
[41] R. H. Austin, K. W. Beeson, L. Eisenstein, H. Frauenfelder, and I. C. Gunsalus, *Biochemistry* **14,** 5355 (1975).
[42] K. U. Linderstrom-Lang and J. A. Schellman, *in* "The Enzymes" (P. D. Boyer, H. Lardy, and K. Myrbäck, eds.), 2nd ed., Vol. 1, p. 743. Academic Press, New York, 1959.

Low Temperature-Rapid Flow Instrumentation

An important aspect of this field is the combined use of rapid flow and rapid freezing methods for the study of chemical reactions by electron paramagnetic resonance (EPR) spectroscopy. Bray developed the rapid freezing method which stops reactions in 3–5 msec.[43] This rapid quenching technique permits the EPR method to be used at a variety of low temperatures to observe components of the reaction mixture at intermediate stages of the reaction. Quenching of the reaction allows signal averaging by slow repetitive scanning techniques, thus increasing the sensitivity of the EPR method. Several excellent sources of information are available on this topic.[2,43,44]

Allen et al.[45] designed the first low temperature stopped-flow device which consisted of a stainless-steel block machined to provide all the reservoirs, gas-driving and stopping systems, mixing chamber, observation tube, as well as to accept the optical windows. The entire block was immersed in a thermostated bath. The system was capable of operating at temperatures at low as −120°. However, stainless steel is not compatible with a number of reactive systems and might cause problems for studies of those enzymes which require metal-free conditions. The volume required for each determination (3 ml) was also excessive for enzymological work.

Dewald and Brooks[46] described the construction of an all-Pyrex stopped-flow apparatus which was capable of functioning at −40° and serviceable with reactive systems, e.g. metal–ammonia solutions. The thermostating bath fluid was an ethylene glycol–water mixture which was cooled by a refrigerated bath. The low temperature limit of the instrument was determined by the greases used to lubricate the syringe and stopcocks. The system was reported to be suitable for reactions with half-lives of 2 msec or greater but the volume needed per run was not reported.

Another all-glass stopped-flow apparatus has been described which can be used under anaerobic conditions over the temperature range from −110 to +100°C.[47] The total volume of solution used in each run was usually in the range of from 0.2 to 0.5 ml, and the dead time of the instrument was 3 to 4 msec at room temperature. Light is transmitted to and from the observation cells by flexible Pyrex light guides. The system was

[43] R. C. Bray, in "Rapid Mixing and Sampling Techniques in Biochemistry" (B. Chance et al., eds.), p. 195. Academic Press, New York, 1964.
[44] D. P. Ballou, Doctoral Dissertation, University of Michigan, Ann Arbor (1971) (University Microfilms No. 72-14796).
[45] C. R. Allen, A. J. W. Brook, and E. F. Caldin, Trans. Faraday Soc. 58, 788 (1960).
[46] R. R. Dewald and J. M. Brooks, Rev. Sci. Instrum. 41, 1612 (1970).
[47] D. Michael, D. O'Donnell, and N. H. Rees, Rev. Sci. Instrum. 45, 256 (1974).

TABLE II
VALUES OF THE RATE CONSTANTS FOR THE REDUCTION OF
2,6-DICHLOROPHENOL–INDOPHENOL (DCPIP) BY ASCORBATE[a]

T (°C)	k_{obs} (sec^{-1})
+20	5.56
+20[b]	0.25
+11	3.33
0[b]	0.091
−8[b]	0.047
−22	0.26
−33	0.17

[a] The concentrations of DCPIP and ascorbate are $10^{-5} M$ and $5 \times 10^{-2} M$, respectively.[48]
[b] Ascorbate concentration $2 \times 10^{-3} M$.

reported to hold a vacuum of better than 10^{-3} mm Hg and thus was particularly useful for oxygen-sensitive solutions, such as that of the radical ion of triphenylmethyl sodium in tetrahydrofuran.

Hui Bon Hoa and Douzou have recently reported the construction of a stopped-flow apparatus which is similar in principle to that designed by Gibson[4] but which can function over the temperature range from + 40° to − 35°.[48] The mass of the mixer, the observation chamber, and driving syringes were reduced to provide rapid cooling and rapid changing of temperature by a thermostated bath of methanol. The mixing and observation chambers are made of stainless steel. The apparatus has an optical path length of 10 mm and requires 0.2 ml of solution for each run. The mixing is complete at the moment flow stops, 20 msec. The performance of the instrument is reported to be practically unchanged over the temperature range from +20° to −33°.[48] Rate constants for the reduction of 2,6-dichlorophenolindophenol by ascorbate have been obtained over this temperature range (Table II). These studies have shown that it is essential to mix solutions of identical viscosity at low temperatures in order to avoid inhomogeneities. In addition rigorous temperature control and homogeneity of the solutions in both chambers are necessary to avoid optical artifacts due to thermal effects in the observation tube.

Design of a New Low Temperature Stopped-Flow Instrument

We have considered a number of criteria in the design of a new low temperature stopped-flow instrument and the resultant design considerations are listed in Table III. This new instrument incorporates some of

[48] G. Hui Bon Hoa and P. Douzou, *Anal. Biochem.* **51,** 127 (1973).

TABLE III
DESIGNS CONSIDERATIONS FOR THE SUB-ZERO STOPPED-FLOW

Criterion	Design consideration
1. Temperature control to ±0.2°.	Thermostat drive syringes and observation chamber uniformly
2. Minimal leakage over wide temperature range (i.e., +50 to −50°)	Avoid interfaces of different materials
3. Resistance to organic and corrosive fluids	Construct components of glass and/or inert plastics
4. Minimal contamination by metals	Construct components of plastic and/or glass
5. Minimal artifacts from air bubbles at low temperature	Use vertically aligned drive syringes and inclined observation chamber
6. High fluid economy	Minimize volume of delivery tubes, mixing jet and observation cell
7. Low dead time (<10 msec)	Same as 6, plus pneumatic drive controls
8. Flexibility in accepting variety of light sources and detectors	Use fiber optics for transmitting light

the features of previous stopped-flow devices and *some novel* features which makes it particularly suitable for fluorescence and absorption spectrometry at sub zero temperatures while retaining adaptability for future modifications.[49] The instrument consists of four major components: the environmental chamber containing the stopped-flow module, the temperature control unit, the drive control unit, and the light supply and detector system (Fig. 1).

The environmental chamber is a heavy, double walled, air-insulated plexiglass assembly which serves as a stable thermal environment, allows easy viewing of nearly all the components and is sufficiently large to allow future modifications. The coolant medium is nitrogen gas which has been passed through a heat exchanger coil immersed in liquid nitrogen and then past a heating tape controlled by a temperature regulator. A nitrogen gas coolant was chosen primarily because it eliminated many of the problems related to fluid thermostating baths. Since these fluid baths usually contain organic solvents, they dissolve lubricating greases, cements, and often penetrate the reaction vessels. A large liquid nitrogen tank provides both the liquid nitrogen and the nitrogen gas the flow of which is controlled by a constant flow regulator to minimize heating fluctuations caused by flow rate changes. The temperature is monitored by a digital thermometer which has thermocouple probes mounted at the exit of the N_2 supply line, within the solution exiting the reaction cell, inside the bodies of the drive syringes and observation cell and in the environmental chamber. Tests

[49] D. A. Hanahan and D. S. Auld, in preparation.

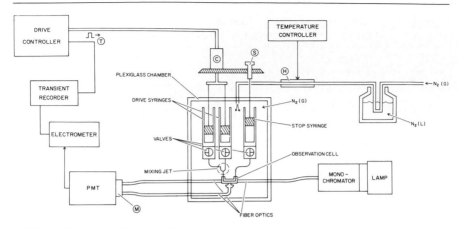

FIG. 1. Schematic diagram of the low-temperature stopped-flow apparatus showing temperature regulation of cold N_2 gas by a heating tape (H) and circulation of thermostated gaseous nitrogen into a plexiglass environmental chamber which contains two brass jacketed teflon drive syringes and one stop syringe, Teflon selector valves, a Kel-F mixing jet, a quartz observation cell and Teflon tubing for connections. Quartz fiber optics are used for transmitting light to the cell from the lamp source and from the cell to the cooled photomultiplier tube (PMT). A mode selector (M) allows one to view either the transmittance or the fluorescence of the sample. Upon initiation of the reaction the drive controller simultaneously triggers (T) the transient recorder to start recording and sends an air pulse to the pneumatic cylinder (C). The amount of solution mixed is controlled and monitored by the adjustable stop (S).

have indicated that the chamber can be maintained at a temperature of $-55° \pm 0.2°$ for several hours.

The supply syringes are driven by a pneumatic cylinder. A controller has been constructed that triggers a transient recorder or scope and activates the pneumatic cylinder. A time delay for the triggers of either the scope or the cylinder driver control is incorporated. It is, therefore, possible to observe the absorbance or fluorescence of the final reaction mixture prior to flow, as well as the flow process, by triggering the scope before initiation of flow. The time during which the pneumatic cylinder is activated, i.e. the drive time, is also adjustable to minimize the period during which the flow system is subjected to high pressure.

The flow system must be inert, leak-free, durable, and bubble-free to eliminate artifacts. Although metal parts will aid in rapid thermostating of vessels, they are corroded by the salt solutions required in many enzyme reactions. Since many of our studies pertain to metalloenzymes which are sensitive to the presence of adventitious metal ions, a nonmetallic flow system was imperative. The use of organic solvents in low temperature studies exclude most plastics, leaving glass, Teflon, and Kel-F as potentially satisfactory materials. In a flow system experiencing large temperature changes, material interfaces between different substances are particularly troublesome since the substantial differences in thermal expansion can cause leakage. The elimination of all material interfaces in the flow system would constitute the most effective solution. Glass is a logical choice since the observation cell must be quartz, in order to efficiently pass uv and visible radiation. However, previous low-temperature stopped-flow instruments using all-glass flow systems were subject to breakage due to mounting strain and nonuniform expansion of bends where glass thickness varied.[46]

Availability of Teflon tubing, valves and fittings, suggested Teflon as applicable for all but the observation cell, minimizing but not completely eliminating material interfaces. In the present instrument, the high thermal expansion property of Teflon (Table IV) has been exploited in the Teflon–quartz cell interface by press-fitting the input/output tubes of the cell into Teflon blocks, which actually clamp the quartz more firmly the lower the temperature. The Teflon syringes, constructed of a brass-jacketed Teflon barrel and a Teflon-tipped steel plunger, are mounted directly onto Teflon selector valves. The resulting flow system is all-Teflon, with the exception of the quartz cell and a Kel-F mixing jet. Hence, the flow system is inert, metal-free, and has been found to be leak-free at temperatures as low as −55°C.

In order to minimize artifacts arising from the presence and movement of bubbles after flow stops, the flow system has been aligned on a vertical axis (Fig. 1). Any bubbles introduced will collect at the top of the syringes, away from the exit port, and will not be pumped through the system unless the syringes are emptied completely.

Quartz fiber optics were chosen to interface the observation cell in the low temperature chamber with the external light supply and detection systems. Fiber optics are flexible and do not require a precisely defined optical axis for the interface, which greatly simplifies the use of a nonhorizontal cell axis. In our studies a variety of detection capabilities are desired, such as rapid scanning, absorbance, fluorescence, or circular

TABLE IV
LINEAR THERMAL EXPANSION COEFFICIENTS (K_t) OF MATERIALS[a]

Materials	K_t (cm/cm °C)
Quartz	5×10^{-7}
Teflon (polytetrafluoroethylene)	1×10^{-4}
Kel-F (polytrifluorochloroethylene)	7×10^{-5}

[a] Values of K_t taken from Chemical Rubber Handbook.

dichroic systems. In general, this requires a stopped-flow module which can be interfaced easily with various light supplies and detectors while the reaction is being carried out at subzero temperatures. In the present design, flexible optical fibers allow alternate monitoring of either the fluorescence or the absorbance of the sample solution (Fig. 1). A single detector is employed with a shutter assembly selected for the desired measurement. Optical alignment is performed easily. At the present time a low noise–high sensitivity detector system is comprised of a trialkali photomultiplier in a cooled housing ($-30°$) and a very high quality high voltage power supply. A low noise electrometer with reference voltage offset facilitates the observation of small changes in light levels above a large background. A high intensity 200 W xenon light supply, used in conjunction with this high sensitivity detector, counterbalances losses at optical interfaces and should also extend significantly the lower limit of detection of ES complexes.

Scarcity and expense of certain enzymes and/or substrates makes it desirable to minimize the instrument's dead volume, which is defined as the fluid volume that must be expended in order to completely fill the observation cell with freshly mixed solution. The dead volume is comprised of the volume of the supply channels from the selector valves to the mixing jet and the observation cell. The dead volume of the present flow system is 0.250 ml, with the volume of the supply channels between selector valves and the mixing jet comprising one-half of it. At room temperature mixed reactants diffuse back into this section of the supply channels over a period of minutes, necessitating using about 0.3 ml of solution per

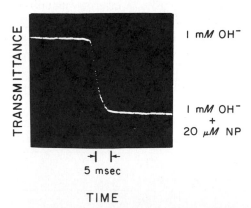

TIME

FIG. 2. Displacement of sodium hydroxide ($10^{-3} M$) by o-nitrophenolate ($10^{-5} M$) both in a 60:40 methanol–H_2O solution at $-56.3°$. A pretrigger record mode is used in the experiment. The upper portion of the oscilloscope trace indicates the transmittance of the sodium hydroxide solution in the cell prior to the initiation of the flow, and the lower portion of the trace indicates the transmittance of the o-nitrophenolate solution after flow stops.

experiment. However, at low temperatures the back diffusion effects from the mixing jet to the selector valves are less significant, reducing the minimum volume needed for a reaction to about 0.15 ml.

Performance Tests for a Low-Temperature Stopped-Flow Instrument

The performance of the instrument at low temperature has been evaluated with a methanol–H_2O solution by measuring the time to displace the dead volume of the system and the rate of the base-catalyzed hydrolysis of the ester, o-nitrophenyl chloroacetate (NPCA) at 412 nm. At $-56.3°$, sodium hydroxide is displaced from the cell by the o-nitrophenolate ion in ca. 5 msec (Fig. 2). It should be readily possible therefore to measure reactions having half times of 5 msec (first order rate constants of about 100 sec^{-1}). The time resolution of the instrument even may allow monitoring of the events occuring during the binding of substrates to an enzyme.

At 30°C, the ester (NPCA) is hydrolyzed by hydroxide ion with a second-order rate constant of $5.5 \times 10^3 M$ sec^{-1}.[6] Pseudo-first order rate constants of 5–550 sec^{-1} can be achieved readily by varying the hydroxide concentration between 10^{-3} and 10^{-1} M at 30°. First order kinetics were obtained for the reaction of 10^{-5} M ester and 5×10^{-3} M hydroxide ion in a 70:30 methanol–H_2O mixture over the temperature range of $+19$ to $-52°$. The oscilloscope traces for a reaction at $-53°$ is shown in Figure 3.

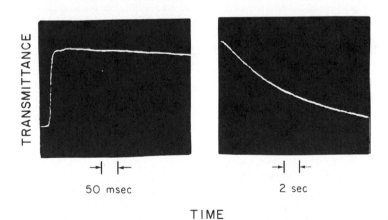

TIME

FIG. 3. Oscilloscope traces of the hydrolysis of 10^{-5} M o-nitrophenyl chlorocetate by $5 \times 10^{-3} M$ hydroxide ion in an 80:20 methanol–H_2O mixture at $-53°$. The reaction is monitored by the change in transmittance at 412 nm. Upon mixing the two reagents, the signal rises rapidly yielding a transmittance equivalent to the unreacted ester (left panel). The subsequent decrease in transmittance reflects the hydrolysis of the ester (right panel).

Upon mixing the ester and hydroxide ion, the observed transmittance at 412 nm rapidly increases to that of the unreacted ester. Hydrolysis then taked place, signaled by the decreasing transmittance, with a first-order rate constant of 0.17 sec^{-1}. Over the temperature range $+19°$ to $-52°$ the first-order rate constants decreased ~150-fold, from 29 to 0.19 sec^{-1} (Table V).

Aqueous salt solutions can also be used with the low-temperature stopped flow instrument. The hydroxide ion catalyzed hydrolysis of NPCA in 4.5 M NaCl and 3.5 M $CaCl_2$ has been studied over the temperature range $+20°C$ to $-30°C$ (Table V). First order kinetics were obtained at all conditions tested. Since the freezing points of 4.5 M NaCl and 3.5 M $CaCl_2$ are $-23°$ and $-55°$, respectively, such solutions may be particularly useful for studying the dynamics of enzyme reactions where aqueous organic solutions cause enzyme denaturation or weakening of substrate binding.

The flexible design of the instrument should allow its use with ultraviolet, visible, infrared and circularly polarized light sources, as well as a variety of detectors and rapid scanning devices. The low-temperature conditions do not exclude the use of temperature or pH-jump perturbations to study reactions at equilibrium. As has been pointed out earlier in

TABLE V

VALUES OF THE RATE CONSTANTS FOR THE HYDROXIDE CATALYZED HYDROLYSIS OF O-NITROPHENYLCHLOROACETATE[a]

T (°C)	k_{obs} (sec^{-1})		
	70/30 MEOH/H$_2$O	4.5 M NaCl	3.5 M CaCl$_2$
+18.7	28.8	5.3	0.63
+9.8	19.2		
+0.6	11.7		
−2.0			0.17
−3.0		1.19	
−18.0		0.43	
−20.6	2.72		
−28.6			0.025
−32.2	1.19		
−43.6	0.29		
−52.0	0.19		

[a] NaOH, 5 millimoles per liter, was added to the aqueous organic and salt solutions. The concentration of the ester, NPCA, was 10^{-5} M.

this article the integration of these different approaches under low-temperature conditions should give access to greatly expanded information about the chemical nature of reaction intermediates as well as the role local conformational changes play in enzyme catalysis. In turn these results should assist in discernment of the details involved in the mechanism of action of enzymes.

Section IV

Conformation: Spectroscopic Techniques

[16] The Interpretation of Near-Ultraviolet Circular Dichroism[1]

By PETER C. KAHN

Optical activity is exquisitely dependent upon conformation. This awareness has led since the earliest studies in the field[1a] to the tantalizing hope for direct conformational interpretation of CD data. Given a spectrum and the covalent structure of a molecule, one asks, "What can we say of its three-dimensional conformation?" The answer at present, unfortunately, is, "Not enough." In many cases, we can estimate what fraction of the peptide bonds are in α-helical, β-pleated sheet, and aperiodic conformations,[2] and we can sometimes make statements about the conformational freedom available to aromatic side chains, disulfides, and prosthetic groups. That is all that an inspection of spectra presently reveals. There are formidable computational and experimental problems which must be solved before our interpretive abilities will improve. Some of them are discussed below.

The theory of optical activity is sufficiently well developed that given the three-dimensional structure of a molecule, it is possible to calculate its CD spectrum. Calculated and experimental curves can then be compared to validate the theoretical work. Moreover, by omitting terms due to particular classes of amino acid—all tyrosines, for example—or by omitting single residues, one can pinpoint the contribution to the spectrum that arose from the presence of the omitted groups. Such a procedure is particularly important in the near uv, where individual side chains can act as sensitive probes of their local environments. In favorable cases, this will enable us to monitor simultaneously several separate and distinct parts of the protein molecule. Having thus correlated circular dichroism (CD) contribution with molecular structure for known conformations, one is then in a position to examine the spectral effects of the crucial conformational changes that accompany substrate and inhibitor binding, subunit assembly, and other biological phenomena. As our understanding of the detailed relationship between conformation and spectrum grows, so will our ability to make direct structural inferences from spectroscopic data. In this chapter, therefore, we describe a scheme for calculating the CD spec-

---

[1] Paper of The Journal Series, New Jersey State Agricultural Experiment Station.

[1a] V. L. Rosenfeld, *Z. Phys.* **52**, 161 (1928).

[2] J. T. Yang, G. C. Chen, and B. Jirgensons, *Hand. Biochem. Mol. Biol. 3rd Ed.* **3**, p.3 (1976).

METHODS IN ENZYMOLOGY, VOL. 61

trum of proteins of known three-dimensional structure. Rotating about covalent bonds to modify the conformation with recalculation of the spectrum for the new three-dimensional structure gives the dependence of the optical activity upon conformation.

The theory of optical activity has been studied intensively for over forty years. An extensive and fascinating literature has arisen in that time. Because the purpose here is to describe how to do and to interpret optical activity calculations, we use the theory in its current state and can only allude to this literature. A number of good reviews exist.[3-9] For the near-uv the reader is referred especially to Sears and Beychok[6] and to Strickland.[7] In the following discussion equations describing spectroscopic properties are usually presented without derivations, for these may be found both in the review literature and in the original work. We concentrate instead on explaining the physical significance of the terms in the equations and on detailing the manipulations that carry one from a set of atomic coordinates to a CD spectrum. Some familiarity with molecular orbital theory and with elementary matrix algebra is assumed.

The Representation of CD Data

Circular dichroism is an absorptive phenomenon. It is defined as the *difference* between the absorbances of left and right circularly polarized light:

$$\Delta\epsilon(\lambda) = \epsilon_L(\lambda) - \epsilon_R(\lambda) \tag{1}$$

$\epsilon_L(\lambda)$ and $\epsilon_R(\lambda)$ are the molar absorbances of the sample for left and right circularly polarized light at wavelength λ. The difference in Eq. (1), $\Delta\epsilon(\lambda)$ is formulated separately at each wavelength. (For a description of how circularly polarized light is produced and of its electromagnetic nature, see Velluz *et al.*[8].) The absorbances $\epsilon_{L,R}$ are given, like ordinary absorption, by the Beer-Lambert law:

$$\epsilon_{L,R}(\lambda) = A_{L,R}(\lambda)/(lc) \tag{2}$$

[3] C. R. Cantor and S. N. Timasheff, *in* "The Proteins" (H. Neurath and R. L. Hill, eds.), 3rd ed. Academic Press, New York (in press).

[4] C. W. Deutsche, D. A. Lightner, R. W. Woody, and A. Moscowitz, *Annu. Rev. Phys. Chem.* **20**, 407 (1969).

[5] M. Goodman and C. Toniolo, *Biopolymers* **6**, 1673 (1968).

[6] D. W. Sears and S. Beychok, *Phys. Princ. Tec. Protein Chem.*, Part C p. 446 (1973).

[7] E. H. Strickland, *Crit. Rev. Biochem.* **2**, 113 (1974).

[8] L. Velluz, M. Legrand, and M. Grosjean, "Optical Circular Dichroism." Academic Press, New York, 1965.

[9] I. Tinoco and C. R. Cantor, *Methods Biochem. Anal.* **18**, 81 (1970).

where l is the path length of the cell in centimeters, and c the concentration in moles/liter if molar units are desired. L,R means that Eq. (2) of identical form are written separately for left and right circular polarizations. Substitutions of these into Eq. (1) produces a Beer–Lambert law for $\Delta\epsilon$

$$\Delta\epsilon_{L,R}(\lambda) = \Delta A_{L,R}(\lambda)/(lc) \tag{3}$$

where $\Delta A_{L,R}(\lambda)$ represents $A_L(\lambda) - A_R(\lambda)$.

For historical and instrumental reasons CD data are often presented not as $\Delta\epsilon(\lambda)$ but as the ellipticity, $[\theta(\lambda)]$, which is related to $\Delta\epsilon(\lambda)$ by

$$[\theta(\lambda)] = 3298[\epsilon_L(\lambda) - \epsilon_R(\lambda)]$$
$$= 3298\,\Delta\epsilon(\lambda) \tag{4}$$

The units of $[\theta(\lambda)]$ are deg cm²/*decimole*. The specific ellipticity $[\psi(\lambda)]$ is analogous to the specific absorbance and is obtained from the raw ellipticity $\psi(\lambda)$ as measured by the spectrometer from

$$[\psi(\lambda)] = \psi(\lambda)/dc' \tag{5}$$

d is the cell path in *decimeters* and c' the concentration in grams per milliliter. The units of $[\psi(\lambda)]$ are deg cm²/*dekagram*. The *molar* ellipticity is related to the raw data by

$$[\theta(\lambda)] = 100\psi(\lambda)/lc \tag{6}$$

with l and c as in Eq. (3). The factor 100 in Eq. (6) is "Biot's contribution to the confusion in units"[9] and is responsible for the use of decimoles rather than moles.

A cautionary word on molar units is in order. For studies in the far-uv (185 − 240nm) the molar unit is generally based on the mean residue weight (MRW), which is the molecular weight of the protein divided by the number of amino acid residues per molecule. This choice is made because the principal chromophore of interest in the far-uv is usually the peptide bond.[10] In the near-uv, where chromophoric side chains and prosthetic groups are of interest, one's choice of molar units must be guided accordingly. MRW is probably a poor choice here, for the number of nonpeptide chromophores is usually much smaller than the number of peptide bonds. The use of MRW leads to inconveniently small values of the ellipticity.

---

[10] Note that while the MRW is based on the number of amino acids per molecule, it is the molarity of peptide bonds that is actually of interest. The number of peptide bonds per molecule is one less than the number of amino acids. For proteins the discrepency is not serious—one part in a hundred for a small protein of 100 amino acids and less for larger species. For smaller compounds such as peptide hormones having 10–20 residues, one must be careful in setting up the comparison of theory with experiment.

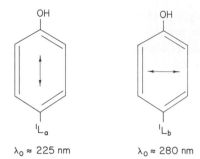

$\lambda_0 \approx 225$ nm $\lambda_0 \approx 280$ nm

FIG. 1. Polarization directions of the longest wavelength tyrosyl transitions.

The Physical Basis of Optical Activity: Rotational Strength, R, and
    Dipole Strength, D

If one were to "average" the motions of an electron in a molecule, one
would obtain a point representing the "center of gravity" of its charge.
Charge distributions in excited states differ, often greatly, from those of
ground states. The promotion of an electron from a ground state 0 to an
excited state a thus involves a net, if transient, displacement of charge
from the "center of gravity" of 0 to that of a. A linear charge displacement is a dipole. It is a vector quantity whose magnitude and direction depend upon the particular electronic states involved, i.e., it is a property of
the chromophore. The transient dipole moment which is induced by
virtue of the excitation is called the electric transition moment, and for the
$0 \rightarrow a$ transition, we represent it by $\boldsymbol{\mu}_{0a}$. By way of example, the orientation of two transitions of the tyrosine phenolic group are illustrated in Fig.
1. They are perpendicular to one another, and both are in the plane of the
ring.

The transition moment can be calculated from molecular orbital
theory. It is, however, an experimentally observable quantity. Its direction can be obtained from polarized absorption measurements on single
crystals of the chromophore if three-dimensional structure of the crystal
is known. Its magnitude is related to the area under the transition's absorption band by[11,12]

$$D_{0a} = \boldsymbol{\mu}_{0a} \cdot \boldsymbol{\mu}_{0a} = |\boldsymbol{\mu}_{0a}|^2 = 9.180 \times 10^{-39} \int_0^\infty \frac{\epsilon(\lambda)\, d\lambda}{\lambda} \qquad (7)$$

D_{0a} is called the dipole strength. It is equal to the vector scalar product
(dot product) of the electric moment with itself, i.e., to the square of its

[11] R. S. Mulliken, *J. Chem. Phys.* **7**, 14 (1939).
[12] A. Moscowitz, in "Optical Rotatory Dispersion" (C. Djerassi, ed.), p. 269. McGraw-Hill,
    New York, 1960.

magnitude. The constant 9.180×10^{-39} is an amalgamation of unit conversion factors and fundamental constants (see Sears and Beychok[6]). $\epsilon(\lambda)$ is, as before, the absorption curve in molar absorbance units as a function of wavelength λ which is given in nanometers. The cgs units of D_{0a} in Eq. (7) are esu cm. The small magnitude of the proportionality constant in Eq. (7) is sometimes inconvenient, especially in computer calculations. Dipole moments μ_{0a} are often, therefore, represented in Debye units: 1 Debye = 10^{-18} cgs. The magnitude of μ_{0a} may therefore be given as

$$
|\mu_{0a}| = (D_{0a})^{1/2} = \left(\frac{9.180 \times 10^{-39}}{(10^{-18})^2} \int_0^\infty \frac{\epsilon(\lambda)\, d\lambda}{\lambda} \right)^{1/2}
$$

$$
= 9.581 \times 10^{-2} \left(\int_0^\infty \frac{\epsilon(\lambda)\, d\lambda}{\lambda} \right)^{1/2} \tag{8}
$$

which is merely a conversion of Eq. (7). We shall use Eq. (7) and Eq. (8) interchangeably, the conversion being implied as needed.[13]

With the exception of the disulfide, the chromophores present in proteins are symmetric and, as a consequence, optically inactive. They become optically active by virtue of their presence in the asymmetric environment provided by the rest of the protein molecule. In classical terms the effect of asymmetry upon chromophoric electrons is to impart a circular component to their motion of transfer from ground state 0 to excited state a. It will be recalled that the circular motion of charge produces a magnetic moment. Associated with an optically active transition $0 \rightarrow a$, there is thus a magnetic transition moment $\mathbf{m}_{0a}$. Like the electric transition moment μ_{0a} to which it is analogous, it is a vector quantity. Its direction is perpendicular to the average plane of the circular motion, and it reverses sign if the direction of motion about the circle reverses. (See Kauzmann[14] for a lucid and full development of this point and for some excellent interpretive diagrams.) Opposite handedness of molecular structure produces opposite circularity of motion.

If μ_{0a} is zero, there is no net linear displacement of charge, and, as a result, no absorption. The transition is said to be "electrically forbidden," and it is optically inactive. If μ_{0a} is not zero but $\mathbf{m}_{0a}$ is zero, the transition is "electrically allowed," and absorption occurs, but since there is no circular component to the charge displacement, there is no optical activity. The transition is "magnetically forbidden." For optical

[13] A quantity that is often reported in place of the dipole strength, D_{0a}, is the oscillator strength, f_{0a}. The two are proportional:

$$
f_{0a} = 1.207 \times 10^{-4}\, D_{0a}/\lambda_{0a}.
$$

where λ_{0a} is the wavelength of the transition in nm.

[14] W. Kauzmann, "Quantum Chemistry," pp. 613ff. Academic Press, New York, 1957.

activity to arise both $\boldsymbol{\mu}_{0a}$ and $\mathbf{m}_{0a}$ must be nonzero. The quantitative expression of this is due to Condon,[15] and is similar in form to Eq. (7) for the dipole strength

$$R_{0a} = Im(\boldsymbol{\mu}_{0a} \cdot \mathbf{m}_{a0}) \tag{9}$$
$$= Im(|\boldsymbol{\mu}_{0a}||\mathbf{m}_{a0}| \cos \phi). \tag{9a}$$

where R_{0a} is the rotational strength and I_m denotes the "imaginary part of." R_{0a} is analogous to D_{0a} in that it measures the dichroic intensity of the $0 \rightarrow a$ transition. In a manner strictly analogous to Eq. (7) for D_{0a}, R_{0a} can be obtained from the area under a CD curve:

$$R_{0a} = 0.696 \times 10^{-42} \int_0^\infty \frac{[\theta(\lambda)] \, d\lambda}{\lambda} \tag{10a}$$

The cgs units of R in Eq. (10a) are usually converted to the more convenient form of Debye-Bohr magnetons (D-BM). The Debye unit, 10^{-18} cgs, converts $\boldsymbol{\mu}_{0a}$ in Eq. (9). The corresponding unit of magnetic moment is the Bohr magneton, 0.9273×10^{-20} cgs. R_{0a} in D-BM is then given by Eq. (10b):

$$R_{0a} = 0.751 \times 10^{-4} \int_0^\infty \frac{[\theta(\lambda)] \, d\lambda}{\lambda} \tag{10b}$$

The vector scalar product in Eq. (9a) is written out in Eq. (9b) to emphasize that there is one further condition in which the optical activity can be zero. Even when both $\boldsymbol{\mu}_{0a}$ and $\mathbf{m}_{0a}$ are nonzero, if they are perpendicular to one another, R_{0a} vanishes, for cos ϕ, the angle between them, is then zero. The physical meaning is that for a transition to be optically active, the vectors describing the linearity and circularity of its charge displacement must be at least partly parallel to one another. The intensity of optical activity depends thus upon the magnitude of linear charge displacement $|\boldsymbol{\mu}_{0a}|$ and the magnitude of the circular displacement whose vector direction is that of the linear displacement $|\mathbf{m}_{0a}|\cos \phi$.

The notation Im in Eq. (9) means that we take the imaginary part of $\boldsymbol{\mu} \cdot \mathbf{m}$. In principle, $\boldsymbol{\mu} \cdot \mathbf{m}$ is a complex number of the form $x + iy$, $i = (-1)^{1/2}$. This introduces a computational matter only and not a matter of interpretation. It arises from the nature of the magnetic moment operator involved in calculating $\mathbf{m}_{0a}$. For all cases of interest in proteins, the real part of $\boldsymbol{\mu} \cdot \mathbf{m}$ is zero in any case, and we take only y, the remaining imaginary part of the number.

[15] E. U. Condon, *Rev. Mod. Phys.* **9**, 432 (1937).

Curve Resolution and Data Handling

Implicit in the discussion thus far is the availability of experimental spectra for single electronic transitions, $0 \rightarrow a$. Data seldom come that way, though. One generally obtains instead an envelope consisting of the summed contributions of several transitions whose bands overlap one another. For n such bands the ellipticity and absorbance at each wavelength, λ, are given by

$$[\theta(\lambda)] = \sum_{a=1}^{n} [\theta_{0a}(\lambda)] \tag{11}$$

$$\epsilon(\lambda) = \sum_{a=1}^{n} \epsilon_{0a}(\lambda). \tag{12}$$

(see Fig. 2). From the curves in Eq. (11) and Eq. (12) one must resolve the individual components. For each transition one seeks the following.

1. The wavelength position of the band extremum λ_a
2. The band shape as a function of wavelength (see below)

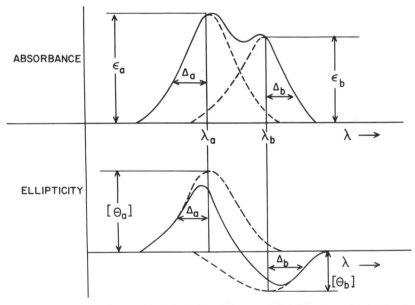

FIG. 2. Gaussian resolution of absorption and corresponding CD spectra for two bands, a and b. Parameters as defined in text. Note difference in absorption between resolved and apparent peak positions due to overlap of bands. Note also difference in apparent shape and position of longest wavelength absorption and CD bands.

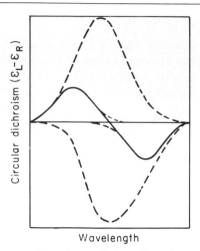

FIG. 3. The uniqueness problem in resolution of overlapping CD bands. Either the dashed lines which follow the spectrum (solid line) or the large dashed bands which nearly cancel will yield the observed spectrum. (From Tinoco.[15a])

3. In the case of CD, the sign of the band
4. The extremum value of the ellipticity or absorbance

Because of the complexity of most protein spectra, resolution is often difficult and may not produce a unique set of bands. The problem is illustrated in Fig. 3[15a] in which either the intense closely spaced pair of bands or the weaker pair at the greater separation sum to give the experimental curve. Fortunately, the situation is not hopeless. Because CD and absorption arise from the same phenomenon, namely, the promotion of electrons from their ground states to excited states, features that are visible in one may help in the resolution of the other. For this reason, among others, the CD and absorption spectra of a compound are generally resolved together as a single problem as illustrated for the disulfide of cystine in Fig. 4.[15b]

The most common procedure is to fit the spectra with a sum of Gaussian functions. One thus assumes that actual CD and absorption bands are Gaussian in shape within the error of the data.

$$[\theta_{0a}(\lambda)] = [\theta_{0a}] \exp \left(-(\lambda - \lambda'_a)/\Delta'_a\right)^2 \tag{13}$$

$$\epsilon_{0a}(\lambda) = \epsilon_{0a} \exp \left(-(\lambda - \lambda_a)/\Delta_a\right)^2 \tag{14}$$

ϵ_{0a} and $[\theta_{0a}]$ are the molar absorbance and ellipticity, respectively, at the peak extrema, and λ_a and λ'_a are the wavelengths at which the absorption and CD extrema occur. The half-widths Δ_a and Δ'_a govern the breadth of

[15a] I. Tinoco, Jr., in "Molecular Biophysics" (B. Pullman and M. Weissbluth, eds.), p. 274. Academic Press, New York, 1965.
[15b] P. C. Kahn, Ph.D. Thesis, Columbia University, New York (1972).

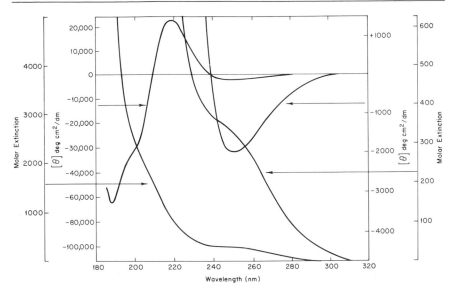

FIG. 4. Absorption and CD spectra of L-cystine in 1 M HClO$_4$ set up for curve resolution. (From Kahn.[15b])

the bands and equal one-half the band widths at the point where the curves reach exp (-1) times their extremum values. Although the positions of band centers and the half-widths of CD and absorption bands are not, in principle, the same, Moscowitz[12] has shown that for electrically allowed transitions ($\boldsymbol{\mu}_{0a} \neq 0$, $\epsilon_{0a} > {\sim}1000$) they are (i.e., $\lambda_a = \lambda'_a$ and $\Delta_a = \Delta'_a$). Nearly all of the near-uv transitions of proteins fall into this category. We therefore drop the prime notation in Eq. (13).

Substituting Eq. (13) and Eq. (14) into Eq. (11) and Eq. (12), respectively, and integrating yields

$$R_{0a} = 0.696 \times 10^{-42} \pi^{1/2} [\theta_{0a}] \Delta_a / \lambda_a \tag{15}$$

$$D_{0a} = 9.180 \times 10^{-39} \pi^{1/2} \epsilon_{0a} \Delta_a / \lambda_a \tag{16}$$

provided that $\lambda_a \gg \Delta_a$, which for uv bands is generally the case. If now, one divides Eq. (15) by Eq. (16) and solves for $[\theta_{0a}]/\epsilon_{0a}$ one obtains

$$\frac{[\theta_{0a}]}{\epsilon_{0a}} = (3298)(4) \frac{R_{0a}}{D_{0a}} \tag{17}$$

and dividing Eq. (13) by Eq. (14) allows the replacement of the left-hand side of Eq. (17), yielding

$$\frac{[\theta_{0a}(\lambda)]}{\epsilon_{0a}(\lambda)} = (3298)(4) \frac{R_{0a}}{D_{0a}} \tag{18}$$

$4R_{0a}/D_{0a}$ is called the anisotropy [15] and is often tabulated in optical activity investigations. Its meaning is deduced by dividing Eq. (9b) by Eq. (7). Except for the factor 4 one obtains the ratio of the component of the magnetic moment that is parallel to μ_{0a} to the magnitude of μ_{0a}

$$4\frac{R_{0a}}{D_{0a}} = \frac{4|\mathbf{m}_{0a}| \cos \phi}{|\mu_{0a}|} \tag{19}$$

i.e., the ratio of the "effective circularity" of charge displacement to the magnitude of linear displacement for the electronic transition.

From the standpoint of curve resolution, however, Eq. (18) expresses the fact that at any wavelength λ the ellipticity is proportional to the absorbance when $\lambda_a = \lambda'_a$ and $\Delta_a = \Delta'_a$ for Gaussian bands. This is of utmost importance, for it enables us to use a single set of bands to fit both absorption and CD spectra. The bands must have the same wavelengths at their extrema and the same half-widths in the former as in the latter. They may differ only in sign and in peak magnitude. In imposing these constraints one is fitting two observables rather than one, and the problem of uniqueness is much reduced. Horwitz et al. [16] have made elegant use of this procedure in their study of ribonuclease (see below).

We will not discuss mathematical methods of curve resolution here. They have been reviewed carefully by Tinoco and Cantor. [9] In addition, two lucid and practical discussions of nonlinear least-squares procedures with detailed examples are available. One, by Frazer and Suzuki, [17] contains a thoroughly documented Fortran program which can be keypunched and used directly. The other, by Daniel and Wood, [18] offers Fortran programs through both the I.B.M. SHARE and C. D. C. VIM libraries. These programs, too, are fully described. All three sources should be consulted.

We deal here instead with problems that may arise. The most serious is obtaining a unique set of bands that resolves the CD spectrum. The problem of closely spaced bands was raised in Fig. 3. If the separation between two bands is much less than the smaller of their half-widths, their separate rotational strengths cannot be determined. If the corresponding absorption bands are of equal size, then neither the rotational strengths nor the wavelengths separating their band centers (the band splitting) can be obtained. The observable in this case is the rotatory couple, which is the product of the band splitting and the rotational strength. [19] When closely spaced bands differ greatly in size, the observable is the difference in their rotational strengths. Account must be taken of such anomalies in setting up the comparison of calculated with experimental spectra.

[16] J. Horwitz, E. H. Strickland, and C. Billups, J. Am. Chem. Soc. **92**, 2119 (1970).

[17] R. D. B. Fraser and E. Suzuki, Phys. Princ. Tech. Protein Chem., Part C p. 301 (1973).

[18] C. Daniel and F. S. Wood, "Fitting Equations to Data." Wiley, New York, 1971.

[19] P. M. Bayley, E. B. Nielsen, and J. A. Schellman, J. Phys. Chem. **73**, 228 (1969).

There are two other constraints that must be imposed. First, one's choice of bands (λ_a, Δ_a) must be consistent with all available data, particularly model compound absorption data, for the chromophores present. Second, one must use the minimum number of bands that will fit the experimental data. The fit could be made as good as one wishes by adding more and more bands, but the result would make little spectroscopic sense. The principle of Occam's razor should be applied where there is doubt.

The result of curve resolution is a set of bands each of which corresponds to a single electronic transition from the ground state 0 to an excited state a. Each is characterized by λ_a, the wavelength position of its extremum, by $[\theta_{0a}]$ or ϵ_{0a}, the ellipticity and absorbance at the extremum, and by a shape. If Gaussian bands are used, the shape is governed by the half-width Δ_a. From these D_{0a} and R_{0a}, the dipole and rotational strengths, are obtained, ultimately to be compared with the interpretive calculations described below. If Gaussian functions are used, R_{0a} and D_{0a} are obtained directly from Eqs. (15) and (16). Similar equations may be derived for other analytical functions (See Frazer and Suzuki,[17] for Cauchy-type bands). If nonanalytical empirical functions are used, the areas are obtained by numerical integration or by cutting out the bands and weighing the paper. In this case R_{0a} and D_{0a} are obtained from Eqs. (10) and (7).

Basic Theory: Introduction *

Most calculations of optical activity have utilized the perturbation theory approach developed by Tinoco[20] and his co-workers (see also Sears and Beychok[6]). Tinoco's formalism is complete in that it contains terms for all mechanisms by which measurable optical activity is known to arise,[6] and it has been used with success in a variety of computations.[21] For proteins, however, perturbation theory has two significant disadvantages. It is quantitative only when the perturbations of individual chromophores by their surroundings in the protein are small, i.e., their electronic properties remain very similar to those found in simple model compounds. Because of the close juxtaposition of residues within a folded protein, however, the electron clouds of unsaturated groups can and do overlap. The perturbations thereby introduced are no longer small. To cite a single example, the indole rings of tryptophans 63 and 64 of chicken egg lysozyme are so close in the crystal structure of the native enzyme that neither can move very much without a corresponding motion of the other. One of them, Trp 63, is replaced in human lysozyme by a tyrosine

[20] I. Tinoco, Jr., *Adv. Chem. Phys.* **4**, 113 (1962); *Radiat. Res.* **20**, 133 (1962).
[21] M.-C. Hsu and R. W. Woody, *J. Am. Chem. Soc.* **93**, 3515 (1971).

with profound spectroscopic consequences,[22] and in both proteins these aromatics are involved in substrate binding,[23] also with profound effects upon the spectra. A further disadvantage of perturbation expressions is that some of their terms contain $(\lambda_b - \lambda_a)$ in denominators, where λ_b and λ_a are the wavelengths of electronic transitions b and a, respectively. When these wavelengths are the same, as would be the case for two identical side chains, e.g., Trp 63 and Trp 64 of lysozyme, the denominator containing $(\lambda_b - \lambda_a)$ goes to zero, and the expression is discontinuous.

For these reasons, among others, the recent trend is to use the matrix formalism pioneered by Bayley *et al.*[19] and by Pysh[24] and exploited successfully by Hooker and his co-workers (see below).

Methods of Calculation

In considering electronic transitions, we are dealing with changes in the spatial distribution of electrons. That distribution, it will be recalled, is defined by the wave function ψ. The physical significance of ψ is that its square, $\psi^2(x,y,z)$ gives the probability of finding the electron at the point (x,y,z). There is thus a wave function for the ground (unexcited) state, ψ_0, and one for each excited state, ψ_i, $i = 1, n$ if there are n possible excitations. The energy difference between ground and excited states ΔE_{0i} is given by the wavelength λ_i at which the transition occurs through $\Delta E_{0i} = E_i - E_0 = h\nu_i$, where h is Planck's constant and ν_i is the frequency of the light (cycles/second). $\lambda_i = c/\nu_i$, c being the velocity of light.

Corresponding to each physically observable quantity such as the energy E_i, there is a quantum mechanical operator by means of which it is possible to calculate the value of the observable provided that the wave function is known. In the case of the energy, for example, the operator is the familiar Hamiltonian, symbolized by $\mathcal{H}$, and the energy is given by

$$E_\psi = \int \psi^* \mathcal{H} \psi \, d\tau \tag{20}$$

Wave functions are continuous, differentiable functions of space coordinates $(x,y,z$ or r, θ, ϕ, etc.), and in Eq. (20) and henceforward, we assume that they are normalized. An operator, such as $\mathcal{H}$, is an explicit set of mathematical instructions to be carried out on whatever is written to its right. One "operates" on ψ in Eq. (20) with $\mathcal{H}$. Having done so, one then

[22] J. P. Halper, N. Latovitzki, H. Bernstein, and S. Beychok, *Proc. Natl. Acad. Sci. U.S.A.* **68**, 517 (1971).

[23] J. A. Rupley and V. Gates, *Proc. Natl. Acad. Sci. U.S.A.* **57**, 496 (1967).

[24] E. S. Pysh, *J. Chem. Phys.* **52**, 4723 (1970).

multiplies the result by the complex conjugate of ψ, which is ψ^*. If the wave function is real, as is often the case, $\psi^* = \psi$. In either case, the result is still a continuous analytic function of space coordinates, and the integration $\int d\tau$ is a shorthand notation for integration over all space, since the electron must be somewhere. In Cartesian coordinates,

$$\int d\tau = \int_{-\infty}^{+\infty} \int_{-\infty}^{+\infty} \int_{-\infty}^{+\infty} dx\ dy\ dz \tag{21}$$

In the discussion above on the nature of the electric transition moment μ, it was described as a dipole representing the displacement of charge from the center of gravity of the initial (usually ground) to the center of gravity of the final (usually excited) state. Like any other dipole moment it is defined in classical mechanics by the product of the charge displaced and the vector $\mathbf{r}$ giving the direction and distance of the displacement, i.e., $e\mathbf{r}$, where e is the electronic charge. The classic definition of an observable (here the dipole moment) becomes the quantum mechanical operator, and the electric transition moment is thus given by

$$\boldsymbol{\mu}_{0a} = \int \psi_0(e\mathbf{r})\psi_a\ d\tau \tag{22}$$

The "operation" in this case is simply to multiply the excited state wave functions ψ_a by the electric charge e and by $\mathbf{r}$, which in the operator is the vector from the origin to the electron. If the wave function is expressed in Cartesian coordinates, then $\mathbf{r} = xi + yj + jk$, where i, j, k are unit vectors in the x, y, and z directions.

The operator for the magnetic moment is derived from the circularity of the motion, i.e., from the angular momentum associated with the charge displacement.

$$\mathbf{m} = -\frac{\hbar e}{2im_e c}\ \mathbf{r} \times \mathbf{p} \tag{23}$$

where $\hbar = h/2\pi$, e and c are as above, m_e is the mass of the electron, and $i = (-1)^{1/2}$, making the magnetic moment a pure imaginary. $\mathbf{r}$ is as above, $\mathbf{p}$ is the operator for the linear momentum of the electron, and their vector cross product is taken. $\mathbf{m}$ is discussed in greater detail later (see also Kauzmann[14] for a full development.) The magnetic transition moment is then

$$\mathbf{m}_{0a} = \frac{\hbar e}{2im_e c} \int \psi_a^*(\mathbf{r} \times \mathbf{p})\psi_0\ d\tau \tag{24}$$

which is similar in form to Eqs. (20) and (22). The rotational strength R_{0a} for the $0 \rightarrow a$ transition is obtained from Eq. (9).

The similarity of form of Eqs. (20), (22), and (24) is not a coincidence. It is an expression of the quantum mechanical fact that any physically definable quantity can be calculated if the wave functions are known, for the operators of the observables can be written down directly from classical mechanics. The expression for the energy in Eq. (20) is perhaps more familiar, but it is the same in principle as the transition moments of Eqs. (22) and (24). This point is crucial to the interpretation of any spectroscopic data. It means that if we can write a suitable Hamiltonian for the protein's chromophoric electrons, the wave functions found by solving the resulting secular equation may be used directly to obtain the electric and magnetic transition moments and from these, the rotational strengths. It also means that when transition moments and energies for the individual chromophoric groups of a protein are used to compute the protein's CD (see below), experimental data obtained with model compounds may be used instead of calculated values. This point is crucial to the accuracy with which protein optical activity can be interpreted.

The protein Hamiltonian that we wish to construct has as its basis set the wave functions of the separate, individual chromophores that are present in the macromolecule. In the wave functions that we seek these isolated chromophoric states are "mixed" with one another, expressing the fact that the electrons of each chromophore may perturb those of all others. To formulate the Hamiltonian that yields these interactions, we first establish a naming and ordering convention for the states of the macromolecule. Following Bayley et al.[19] the ordering is: ground state, all singly excited states of group 1, all singly excited states of group 2, . . . , all singly excited states of group n, all doubly excited states, etc. Each state is designated by a single integer, I^0, J^0, . . . , the superscript 0 indicating the state of nonperturbed isolated groups. Capitals I, J, . . . , without superscripts indicate the final macromolecular state. The Hamiltonian is then given by

$$\mathscr{H} = \sum_i \mathscr{H}_i^0 + V \tag{25}$$

in which $\mathscr{H}_i^0$ is the local Hamiltonian of the ith group and V is the intergroup interaction potential of the rest of the protein with group i. Diagonal elements of the Hamiltonian matrix take the form

$$\mathscr{H}_{I^0,I^0} + V_{I^0,I^0} = E_{I^0} + V_{I^0,I^0} \tag{26}$$

Here, E_{I^0} is the energy of the Ith state as it occurs in an isolated chromophore, for example, in model compounds, and V_{I^0,I^0} is the shift in this energy that occurs by virtue of its incorporation into the protein. Numerical values for the diagonal elements can usually be obtained from resolved spectra of the chromophore in a series of suitable solvents. Alternatively, the resolved absorption spectrum of the protein itself may be used. Off-diagonal elements have the form V_{I^0, J^0}, representing the interaction of states I^0 and J^0. Expressions for the V_{I^0, J^0} are discussed briefly below and in detail by Tinoco,[20] Woody and Tinoco,[25] Hohn and Weigang,[26] Sears and Beychok,[6] and Bayley et al.[19]

A unitary matrix C that diagonalizes $\mathcal{H}$ can be computed:

$$\mathcal{H}^{\text{diag}} = C^{-1} \mathcal{H} C \tag{27}$$

The diagonal elements of $\mathcal{H}^{\text{diag}}$ give the final energies of the states available to the protein's chromophoric electrons, and the columns of C are the eigenvectors that express the merging of the individual chromophoric groups into protein wave functions.

We now construct matrices $\boldsymbol{\mu}$ and $\mathbf{m}$ for the electric and magnetic transition moments as for $\mathcal{H}$ in Eqs. (25)–(27). It is at this point that the similarity of Eqs. (20), (22), and (24) is utilized, for a unitary matrix that yields one physically observable property of a molecule from the properties of its constituent groups yields any other observable as well. (Mathematically, their operators must commute. Properties for which this is true are constants of the motion.)

$$\boldsymbol{\mu} = C^{-1} \boldsymbol{\mu}^0 C \qquad \mathbf{M} = C^{-1} \mathbf{M}^0 C \tag{28}$$

where $\boldsymbol{\mu}^0$ and $\mathbf{M}^0$ are matrices of the electric and magnetic transition moments of the isolated, unperturbed groups as they occur in suitable model compounds. $\boldsymbol{\mu}$ and $\mathbf{M}$ are matrices of the transition moments of the protein. For the transition to protein state K from the ground state 0:

$$\boldsymbol{\mu}_{0K} = \sum_{I^0, J^0} C_{0I}^{-1} \boldsymbol{\mu}_{I^0}^0 C_{J^0 K} \tag{29}$$

$$\mathbf{M}_{K0} = \sum_{L^0, M^0} C_{KL^0}^{-1} \mathbf{M}_{L^0, M^0} C_{M^0 0} \tag{30}$$

The corresponding rotational strength is then

$$R_{0K} = \text{Im} \ (\boldsymbol{\mu}_{0K} \cdot \mathbf{M}_{K0}) \tag{31}$$

$$= \sum_{\substack{I^0, J^0 \\ K^0, M^0}} C_{I^0 0}^* C_{M^0 0} C_{L^0 K}^* C_{J^0 K} (\boldsymbol{\mu}_{I^0, J^0}^0 \cdot \mathbf{M}_{L^0, M^0}^0) \tag{32}$$

where $C_{I^0 0}^* = -C_{0 I^0}^{-1}$. The rotatory strength of each transition R_{0K} thus involves "coupling" of all possible excited states of all the individual

[25] R. W. Woody and I. Tinoco, J. Chem. Phys. 46, 4927 (1967).
[26] E. G. Hohn and O. E. Weigang, Jr., J. Chem. Phys. 48, 1127 (1968).

groups. The coupling depends upon the properties of the individual transitions of the isolated groups ($\boldsymbol{\mu}_{I^0,J^0}{}^0$ and $\mathbf{M}_{L^0,M^0}{}^0$), upon the extent to which they affect one another (the elements of C), and upon the three-dimensional geometry of the groups in the protein (both the elements of C and the scalar products). The elements of C, in turn, depend upon the nature of the intergroup interaction potentials specified in V of Eq. (25).

The dipole strength, D_{0K}, is obtained from Eqs. (7) and (29) and yields intensity transfer values (hyper- and hypochromicity and exciton selection rules).

The formalism thusfar developed is complete but computationally and experimentally intractible. The elements of $\boldsymbol{\mu}^0$ and $\mathbf{M}^0$ for all transitions could, in principle, be obtained from theory if group wave functions of sufficient accuracy were available, but unfortunately they are not. Neither, and for the same reasons, are all possible transition energies available for use in $\mathcal{H}$. Experimentally determined values are to be preferred in any case, but available data are generally limited to a few transitions from the ground state to singly excited states. Bayley et al.[19] therefore make two simplifying approximations.

1. The basis set of individual group excited states is restricted to transitions that are characterizable by ordinary ground state spectroscopy.

2. The ground state and all states involving the simultaneous excitation of two or more groups are dropped from the secular equation. This leaves only singly excited states.

These approximations are standard in many fields of spectroscopy. The Hamiltonian then reduces to

$$
\begin{pmatrix}
E_0 & \cdots\cdots\cdots\cdots\cdots\cdots\cdots\cdots \\
\cdot & (1-1)(1-2) & \cdots\cdots(1-N) \\
\cdot & (2-1)(2-2) & \cdots\cdots(2-N) \\
\cdot & \cdots & \cdots \\
\cdot & (N-1)(N-2)\cdots\cdots(N-N)
\end{pmatrix}
$$

which is Eq. (7) of Bayley et al.[19]. E_0, the ground state energy, is taken as zero by convention. The partitioning of $\mathcal{H}$ is by groups, $(1-1)$ being a submatrix representing interactions among excited states of group 1, and $(1-2)$ giving interactions between states of group 1 and those of group 2.

The matrix C which converts the representation in terms of isolated groups to that of the assembled protein now has all elements of its first row and first column equal to zero except $C_{00} = 1$. Equations (29) and (30) then reduce to

$$\mu_{0K} = \sum_{I^0} \mu_{0I^0}{}^0 C_{I^0K} \tag{34}$$

$$M_{K0} = \sum_{J^0} M_{j^00}{}^0 C_{j^0K}^* \tag{35}$$

and the rotational and dipole strengths of the Kth transition become [cf. Eq. (32)]

$$R_{0K} = \mathrm{Im} \sum_{I^0,J^0} C_{I^0K} C_{j^0K}^* \mu_{0I}{}^0 \cdot M_{j^00}{}^0 \tag{36}$$

$$D_{0K} = \sum_{I^0,J^0} C_{I^0K} C_{j^0K}^* (\mu_{0I^0}{}^0 \cdot \mu_{0J^0}^*) \tag{37}$$

The simplifications thus leave only transition moments from the ground state to the various singly excited states. Although the ground state plays no role in the secular equation, it is retained in Eq. (33) and thus in C to allow for the retention of transition moments that include it. Bayley *et al.* [19] point out that Eqs. (36) and (37) have a simple interpretation. The undeleted transition moments are the first rows of the μ^0 and M^0 matrices. The protein's transition moments are thus simple linear combinations of the moments of its constituent groups. The columns of C provide the coefficients for the linear transformation.

The physical meaning is perhaps best brought out by a comparison of Eqs. (36) and (37) with (7) and (9a). In Eq. (7) optical activity arises if an electron, upon promotion to an excited state, undergoes a charge displacement that has both linear and circular components whose vector directions are the same. For the $0 \rightarrow a$ transitions neither μ_{0a} nor m_{0a} is zero, nor are they perpendicular to one another. For the symmetric chromophores of proteins, however, these conditions are not met. By virtue of their symmetry, either one of the transition moments μ_{0a} or m_{0a} is zero or the moments are at right angles to one another.

Equations (36) and (37), however, expand the possibilities for optical activity to include the interaction of each group transition moment with all others in the protein. This is true from Eq. (36) for both μ_{0a}, which can yield rotatory strength in combination with any M_{0a}, and vice versa. It arises from delocalization of the excited state wave functions in principle over the entire macromolecule. The extent to which any individual group excited state participates in the delocalized protein state is given by its coefficient C_{p^0}. While, in principle, none of these are zero except the first row and column of Eq. (33) as noted, in practice, many will be negligibly small, especially those connecting spatially distant parts of the protein. The interaction potential falls off sharply with distance. This "blending" of locally excited states is called configuration interaction.

In our discussion thus far we have emphasized the fact that the individual group transition moments are observable quantities. While they can be calculated from theory alone [Eqs. (22) and (24)], it is preferable to use experimentally determined values when they are available. The experimental basis for measuring electric transition moments was discussed above. Magnetic moments, however, present a problem. Although observable in principle, there is as yet no experimental procedure for obtaining them on molecules of biological interest. Fortunately a procedure exists for obtaining some of them from electric transition moments.

Transitions can aquire magnetic moment in two ways. They may be intrinsically magnetic, and in Eq. (9) for a single transition on an optically active chromophoric group that is the case. The disulfide is of this type. Alternatively, an asymmetric coupling of two or more linear charge displacements can impart a net circularity to the combined motions of both (Fig. 5). In order to distinguish these, Eq. (23) for the one electron magnetic moment operator is rewritten, replacing $\mathbf{r}$ by $\boldsymbol{\rho}_{ai} + \mathbf{R}_{ai}$. The $\mathbf{R}_{ai}$ is a vector from an arbitrary molecular origin to a local origin in group i containing electron a. $\boldsymbol{\rho}_{ai}$ is the position vector of the electron relative to this local origin. If spin is neglected, Eq. (23) becomes

$$\mathbf{M} = \frac{e}{2m_ec} \sum_a \boldsymbol{\rho}_{ai} \times \mathbf{P}_{ai} + \frac{e}{2m_ec} \sum \mathbf{R}_{ai} \times \mathbf{P}_{ai} \tag{38}$$

$$= \mathbf{m}_i + \frac{e}{2m_ec} \mathbf{R}_i \times \sum_a \mathbf{P}_{ai} \tag{39}$$

The first term here is the intrinsic magnetic moment local to group i. By a judicious choice of local origin, it can usually be made to vanish. Transitions for which this is not possible are said to be intrinsically magnetic, the $n \rightarrow \pi^*$ transition of the peptide group being an example. The notation $\mathbf{R}_{ai}$ is used in Eq. (38) to indicate that the local origin need not be at the same point for all transitions of a single group. It is chosen to minimize the appropriate $\mathbf{m}_{ai}$. These subscripts are henceforward dropped, since one's computer program selects appropriately.

For the Kth transition of the macromolecule, the magnetic transition moment, $\mathbf{M}_{K0}$, is now

$$\mathbf{M}_{K0} = \mathbf{m}_{K0} + \frac{e}{2m_ec} (\mathbf{R} \times \mathbf{P})_{K0} \tag{40}$$

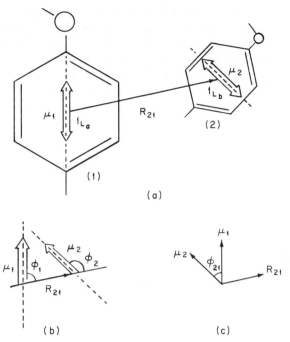

Fig. 5. μ-μ coupling of ^1L_a transition on one tyrosyl with ^1L_b on another. (a) Tyrosyl 2 lies into the plane of the paper behind tyrosyl 1. The vector $\mathbf{R}_{21}$ between the groups is given in Eq. (43b) as $(\mathbf{R}_{J^0} - \mathbf{R}_{I^0})$. The "handedness" of the transitions is indicated in (b) and (c). (From Sears and Beychok.[6])

But since

$$\mathbf{P}_{K0} = \frac{2\pi m_e c i}{\lambda_{K0} e}\, \mu_{K0} \tag{41}$$

$$\mathbf{M}_{K0} = \mathbf{m}_{K0} + \frac{i\pi}{\lambda_{K0}}\,\mathbf{R} \times \mu_{K0} \tag{42}$$

When the elements of $\mathbf{M}_{K0}$ from Eq. (42) are combined with the elements of μ_{0K} from Eq. (34), the scalar products of Eq. (36) may have two possible forms.

$$R_{0K} = \mathrm{Im} \sum_{I^0,J^0} C_{I^0 K} C_{J^0 K} \mu_{0I^0} \cdot \mathbf{m}_{J^0 0} \tag{43a}$$

$$= \mathrm{Im} \sum_{I^0 < J^0} \frac{i\pi}{\lambda_{K0}} C_{I^0 K} C_{J^0 K} (\mathbf{R}_{J^0} - \mathbf{R}_{I^0}) \cdot (\mu_{0J^0} \times \mu_{0I^0}) \tag{43b}$$

(The vector difference in Eq. (43b) arises from a reversal of order of cross product terms in combining symmetrically related matrix elements containing the same transition moments.)

Most of the terms involving interactions of aromatics in the near-uv are of the form in Eq. (43b). This is the familiar coupled oscillator mechanism of Kuhn, Kirkwood, and Moffitt.[27] In it circularity of charge displacement arises from the coupling of linear displacements as in Fig. 5. It should be noted that the quantities present in Eq. (43b) depend on the electric transition moments and transition wavelengths, which are experimentally determined, on molecular geometry, which we specify, and on the nature of the interaction potentials from which C is obtained. These we also specify. This simplification, however, is obtained at a price. The Heisenberg transformation which converts $\mathbf{p}$ to $\boldsymbol{\mu}$ (Eq. 41) is strictly valid only if applied to exact wave functions. But even the best wave functions presently attainable are in some degree approximate. The use of Eq. (41) then leads to rotational strengths whose sum over the entire spectrum will not be zero, which is a requirement of optical activity theory. The deviations are not generally large, and alternative procedures lead to worse problems. The reader is referred to Bayley et al.[19] for a discussion of how to deal with excessive departure from the sum rule.

Terms of the type in Eq. (43a) involve intrinsic magnetic moments, and these cannot be evaluated experimentally. They arise for the $n \rightarrow \pi^*$ transition of the peptide, for the disulfide, and, possibly, for prosthetic groups having nonbonding chromophonic electrons. They need be evaluated only once for each kind of chromophoric group, for

$$\mathbf{m} = \mathbf{m}^0 C \qquad (44)$$

(cf. Eq. 35). Given the wave functions, their computation is mathematically straightforward and is not further discussed here.

The interaction potentials V are needed to evaluate the off-diagonal elements of the secular equation. The interactions are between the excited state electronic charge distributions of each group with the ground state field of the rest of the groups. The charge distributions are represented as charge "monopoles,"[28] points of charge disposed about the atoms of the group to reproduce experimentally measured group properties, such as the electric transition moment, permanent dipole moment, etc. The interaction energy is simply a sum of Coulombic terms for all possible pairs of monopoles.

The formalism includes all of the specialized mechanisms by which optical activity is known to arise. The coupled oscillator mechanism[27] has already been mentioned. Its essence is that the net coupled charge displacements of two electrically allowed transitions has a circular component that neither has by itself. Since the transitions occur on different

[27] J. G. Kirkwood, J. Chem. Phys. **5**, 479 (1937); W. Moffitt, ibid. **25**, 467 (1956); W. Kuhn, in "Stereochemie" (K. Freudenberg, ed.), p. 317. Deuticke, Leipzig, 1933.
[28] R. G. Parr, J. Chem. Phys. **20**, 1499 (1952).

chromophores, the interaction is of the (1–2) type in the secular determinant, Eq. (33). The calculations on ribonuclease and insulin described below are coupled oscillator calculations. Also of the (1–2) type are interactions of the form of Eq. (43a). Here the electric moment of one group and the magnetic moment of another combine to produce rotatory power. This is the "μ–m" mechanism.[29] Finally, in the (1–1) term of Eq. (33), the asymmetric ground state field of the rest of the molecule breaks down the local group symmetry. As a result, electric and magnetic moments that would ordinarily be perpendicular to one another or would be zero ("forbidden") lose their symmetry derived properties. Rotatory power then results. This is the one-electron mechanism.[30] A good discussion of the history and fundamental bases of these classic papers is available.[6]

The Presentation of Results: Comparison of Theory with Experiment

Given the present state of the art, any interpretive calculation, to be convincing, must reproduce experimental data within some acceptable error. At risk of belaboring the obvious, we emphasize that the data here are the dichroism and absorbance as a function of wavelength. A convincing comparison requires, therefore, that calculated spectra be produced. Although other forms of presentation are often useful and sometimes essential, the reliability of the work cannot be judged adequately by a general readership without a direct comparison of spectra. In this section, then, we treat the generation and use of calculated spectra. The treatment owes much to Madison and Schellman.[31] An illustrative selection of other presentations, particularly with respect to conformational variation, is also described.

For each isolated group transition that is input to the calculation, a single macromolecular transition is produced. Each chromophore in the molecule usually gives rise to at least two transitions, and since far-uv transitions of both peptides and side chains interact with the near-uv transitions of interest, some, at least, of the far-uv data must usually be included in the calculation. The number of macromolecular transitions may thus be quite large.

Because their CD and absorption bands overlap strongly, individual protein transitions can seldom be distinguished experimentally, even with the best curve resolution methods available. As a result, a comparison at the level of single transitions is rarely possible. One can, however, re-

[29] J. A. Schellman, *Acc. Chem. Res.* **1**, 144 (1968).
[30] E. U. Condon, W. Altar, and H. Eyring, *J. Chem. Phys.* **5**, 753 (1937).
[31] V. Madison and J. A. Schellman, *Biopolymers* **11**, 1041 (1972).

verse the procedure described above for curve resolution, summing the contributions of individual bands as in Eq. (11) and Eq. (12) to obtain CD and absorption spectra, respectively. The starting point is the set of wavelengths λ_{0a}, dipole strengths D_{0a}, and rotational strengths R_{0a} produced by the optical activity calculation. If Gaussian functions are used to represent individual bands, one obtains the single band dichroism needed in Eq. (11) $[\theta_{0a}(\lambda)]$ from Eq. (13). To do this, in turn, one needs the ellipticity at the extremum $[\theta_{0a}]$, λ_a, the wavelength λ_a, and the half-width Δ_a. The wavelength obtained in the optical activity calculation may be used, or, alternatively, wavelengths of resolved transitions of model compounds in suitable solvents may be used [hydrophobic solvents to mimic buried residues and polar solvents for groups that are exposed at the protein surface (see Horwitz et al.[16])]. The ellipticity at the band extremum is obtained from R_{0a} and Eq. (15), but to use Eq. (15) [and Eq. (13)] one must choose the half-width. This is the one disadvantage to the use of calculated spectral curves, for they are rather sensitive to the value chosen for the Δ_a. The agreement between theory and experiment can be improved greatly by making adjustable parameters of them.[32] Some variability is probably justified on the grounds that band shape will be affected by the local environment. At the present level of refinement of the calculations, however, it is better to set the half-widths to values obtained for model compounds.[16,33] The more chemically sensible constraints one imposes, the more convincing a successful match between theory and experiment becomes.

It might be argued that given the novelty of whole-protein calculations, our lack of experience with them makes us suspect that a good correlation of experimental with calculated spectra is fortuitous. The cancellation of oppositely signed inaccuracies may yield a misleading impression of the agreement to be expected from the calculations. The emphasis on producing calculated spectra, subject to reasonable choices of half-widths and wavelengths, in no way reduces the need for a critical discussion of accuracy in all parts of the work. Strickland's careful analysis (see below) of the near-uv CD of ribonuclease is a good example.[33]

The ultimate aim in these calculations is to use them as a tool in understanding the interactions of proteins with one another (note discussion of insulin below), with substrates, with inhibitors, and with other cellular components. The variation of calculated optical activity with conformation must, therefore, be explored. Conformational energy calculations[34] become important here, for although one may start with a more or less

[32] R. W. Woody, J. Chem. Phys. 49, 4797 (1968).
[33] E. H. Strickland, Biochemistry 11, 3465 (1972).
[34] D. A. Brant, Annu. Rev. Biophys. Bioeng. 1, 369 (1972).

well-defined set of crystallographically determined coordinates, small changes in dihedral angles (bond rotations) may, for example, bring atoms closer together than the sum of their van der Waals radii. Such test conformations may be rejected immediately on those grounds and the optical results used to differentiate among a few chemically reasonable conformations. Work from Hooker's laboratory, described below, takes this approach.

Conformational energy calculations are especially useful in studies of model compounds in solution.[35,36] The single amino acids and other small compounds, often have degrees of conformational freedom not found in folded proteins. Two problems are raised in consequence. The first is that even with only two or three degrees of conformational freedom, as is common in model compounds, the number of conformations whose optical activity is computed is large. Calculated spectra for all of them cannot be presented intelligibly. Hooker has developed a useful presentation[35,36] for such cases. Two-dimensional maps showing the variation of two conformation angles are prepared. The sign, and sometimes magnitudes, of each CD band are shaded, each band differently. Thus the experimentally observed pattern of CD signs will have a unique and easily distinguishable zone on the map. The location of conformational energy minima are also indicated on the maps, giving an immediate, if qualitative, picture of the extent to which the optical activity calculated for the most stable conformation(s) agrees with experiment.

The second problem is more serious. When a substance having conformational freedom is dissolved, many of the molecules in solution will not have the most stable conformation. Other conformational states will be populated according to some distribution function (Boltzman's is usually used) of their energies. The rotatory properties of the solution are thus a weighted sum of the contributions from each conformer. The point is sometimes raised that when rotatory strengths are examined as a function of conformational angles, the optical results presented should include, in addition to values for particular conformations, data for a weighted sum of states.

Ultimately, as experience with interpretive spectroscopic calculations grows, and we become able to focus more on quantitative detail, this procedure may well become obligatory. At the present generally qualitative level of understanding, Boltzman distributions are rarely justified. The discussion of Grebow and Hooker[36] is pertinent here.

In the case of model compounds, it is probably more useful to use, wherever possible, substances of restricted conformational mobility.

[35] T. M. Hooker, Jr. and J. A. Schellman, *Biopolymers* 9, 1319 (1970).
[36] P. E. Grebow and T. M. Hooker, Jr., *Biopolymers* 14, 1863 (1975).

Constrained rings containing the chromophores of interest (provided their geometry is not excessively distorted) or to which chromophores are attached provide much information. Excellent use has been made of diketopiperazines (relatively rigid rings consisting of two peptide linkages) to which one or two chromophoric side chains are attached.[36-40]

Applications to Specific Chromophores

In this section we examine interpretive work on the side chains of tyrosine, tryptophan, and the disulfide. These are the major contributors to the near-uv CD of proteins. Phenylalanine and histidine are omitted partly because their contributions are generally much weaker and partly because the state of our understanding of them is less advanced. For each chromophore model compounds are examined first followed by protein studies where available. The examples chosen are illustrative. Comprehensive and critical literature reviews are found in Strickland[7] and Sears and Beychok,[6] and an extensive tabulation of near-uv CD data for numerous proteins is given by Strickland.[41]

Tyrosine Model Compounds

Hooker's laboratory[36-40] has presented a series of studies of small molecules containing both peptide chromophores and side chains that are optically active in the near-uv. The compounds were chosen to illustrate all spectroscopic interactions that are known to occur in proteins with the exception of the disulfide, which is a special case (see below). For each molecule experimentally measured absorption and CD spectra are presented, and both the conformational energy and rotatory properties are computed as a function of bond rotation angles.

Three of the compounds are especially interesting from an interpretive point of view: D-alanyl-*p*-hydroxy-D-phenylglycine diketopiperazine,[39] cyclo(L-alanyl-L-tyrosine), and cyclo(L-tyrosyl-L-tyrosine).[40] All three are cyclic dipeptides in which ring constraints limit severely the conformational freedom of the molecules. Instead of the usual ϕ, ψ angles, which would yield three significant degrees of freedom in a linear dipeptide, the cyclic form has but one, called β (see Fig. 6 of Hooker *et al.*,[38] but note the associated discussion of bond angle distortion), which governs devia-

[37] J. W. Snow and T. M. Hooker, Jr., *J. Am. Chem. Soc.* **96,** 7800 (1974).

[38] T. M. Hooker, Jr., P. M. Bayley, W. Radding, and J. A. Schellman, *Biopolymers* **13,** 549 (1974).

[39] J. W. Snow and T. M. Hooker, Jr., *J. Am. Chem. Soc.* **97,** 3506 (1975).

[40] J. W. Snow, T. M. Hooker, Jr., and J. A. Schellman, *Biopolymers* **16,** 121 (1977).

[41] E. H. Strickland, *Handb. Biochem. Mol. Biol., 3rd Ed.* **3,** 141 (1976).

tion of the ring from planarity. The alanyl side chain is. fixed in the staggered conformation because the barrier to its rotation is high. The phenoloic group has only one rotational degree of freedom in the hydroxyphenylglycine derivative (about the C_α-ring bond) and the tyrosyl side chains each have two. The bond connecting the aromatic ring to the rest of the molecule, moreover, need only be rotated between 0° and 180°. The twofold symmetry of the phenolic group ensures that the range from 180° to 360° is the same as 0° to 180°. Thus, between two (the hydroxyphenyl glycine derivative) and five (the dityrosyl derivative) degrees of rotational freedom about bonds specify all possible side chain–side chain and side chain–peptide interactions, and the range of possible values that the angles can take is sharply limited by steric and ring constraints and by symmetry. A full examination of the dependence of optical properties upon conformation is therefore possible.

The transitions included in the calculations were the ^1L_b, ^1L_a, ^1B_b, and ^1B_a of the phenolic group(s) and the $n–\pi^*$ and $\pi–\pi^*$ of the two peptides. The ^1B and the peptide transitions occur in the far-uv ($\lambda < 240$ nm) and will not concern us here. The ^1L_a transition generally occurs in the far-uv also but is red-shifted by a much as 20 nm into the near-uv upon deprotonation of the phenolic hydroxyl in alkali. The ^1L_b transition, of course, is the familiar 275–280 nm band which red-shifts upon titration to 295 nm.

The aim in these early studies was modest: to determine whether the formalism is capable of representing the rather complex pattern of signs and relative magnitudes of the readily measurable CD bands for the small number of conformations that are chemically reasonable. In this aim Hooker and his co-workers have succeeded, laying in the process much of the groundwork for subsequent application to proteins.

One result that is particularly worth noting is that the near-uv optical activity of tyrosine is due primarily to $\mu–\mu$ coupling. The one electron mechanism and $\mu–m$ coupling are both weaker and less sensitive to conformational variation. Since $\mu–\mu$ terms can be evaluated with greater accuracy than those due to the other mechanisms, this bodes well for whole-protein calculations.

The side chain bond about which tyrosine in proteins is most likely to be free to rotate is the $C^\beta–C^\gamma$ bond connecting the aromatic ring to the rest of the molecule. The ^1L_b transition is found to be exceedingly sensitive in both sign and magnitude to rotation about this bond (see Fig. 6 of Snow and Hooker[39]). The transition's direction of polarization is perpendicular to the rotation (see Fig. 1 above), and its interactions with other transitions would, accordingly be greatly affected. The ^1L_a band, being polarized in the direction of the rotation axis, is considerably less sensitive, although sign reversal can occur under some conditions (Fig. 7 of Snow and Hooker[39]).

Tyrosine in Proteins

The work of Strickland[33] on ribonuclease and of Strickland and Mercola[42] on insulin illustrate two important classes of problems with which one must be concerned. In the first the dichroism due to specific side chains is examined. The differing contributions of exposed and buried tyrosines is clearly distinguishable, and the effects on tyrosine-25 of converting ribonuclease A to ribonuclease S by subtilisin cleavage are also detailed. In the insulin study the effects upon the CD of subunit assembly are interpreted. These proteins are particularly useful in analyzing tyrosine spectra, for they contain no tryptophan.

Ribonuclease

The well-known tyrosine absorption spectrum in the vicinity of 280 nm contains a series of overlapping transitions that are poorly resolved at room temperature. These are due to symmetric vibrations of the phenolic ring.[16] Starting with the 0–0 transition there is an internal progression of bands to shorter wavelengths with a spacing of 800 cm^{-1} and a weaker progression with a 1250 cm^{-1} spacing. Harmonic combinations of these also appear, further complicating the spectra. Cooling to 77°K sharpens the bands to the point where the vibronic progressions can be located, a procedure that is useful in both model compound and whole protein studies.

Working with a variety of tyrosine model compounds, Horwitz et al.[16] and Strickland et al.[43] have measured the effect of solvent polarity on the absorption and CD at both 77°K and room temperature (298°K). The position of the 0–0 transition was found to vary with the solvent from 282 to 289 nm, but the vibrational spacing remained unchanged. Thus the entire 800 and 1250 cm^{-1} progressions and their harmonics are shifted as a unit (see Figs. 2, 3, and 6 of Horwitz et al.[16]). An increasingly nonaqueous environment corresponding to the burial of the side chain in a protein interior produces a red shift. The vibronic CD bands all have the same sign in a given solvent, and they occur at the same wavelengths as the corresponding absorption bands.

With this information, the resolved bands obtained with model compounds were used to fit the 77°K absorption spectrum of ribonuclease A. Three classes of tyrosine were identified by the positions of their 0–0 bands: 288.5, 286, and 283.5 nm. The first two have wavelengths and half-widths characteristic of nonaqueous solvents and the third of

[42] E. H. Strickland and D. Mercola, *Biochemistry* **15**, 3875 (1976).
[43] E. H. Strickland, M. Wilchek, J. Horwitz, and C. Billups, *J. Biol. Chem.* **246**, 572 (1972).

water–glycerol. Contributions to the total absorption intensity are estimated from the areas under the curves and are in the ratio $\frac{1}{6}:\frac{1}{3}:\frac{1}{2}$, respectively. For the six tyrosines in the protein the three classes thus contain one, two, and three residues. Based on comparison with model compound spectra of the wavelength of the 0–0 band and of the half-widths, the three residues at 283.5 nm correspond to the three exposed tyrosines of the protein.

In analyzing the CD of ribonuclease A at 77°K, they used the wavelengths and half-widths of the absorption spectrum, which were based, in turn, on the model compound studies (see discussion of curve resolution above). The ratio of CD intensity to absorption intensity was held constant for each class of residue (cf. Eq. 18). This last is a severe constraint, for within each of the three residue classes the intensities of all bands of the vibronic progressions were varied together by a constant factor in fitting the experimental spectrum.

Using this procedure, it is found that (a) the two tyrosines whose 0–0 transitions occur at 286 nm contribute negligibly to the dichroism, the intensity arising from the exposed residues (0–0 at 283 nm) and the one buried residue at 288.5 nm, and (b) the model compound data can only fit the ribonuclease spectrum if a band due to disulfides (see below) is included. The results of the fitting are shown in Fig. 6.

With this experimental background, Strickland[33] calculated the optical activity of ribonuclease S. The three-dimensional structure of ribonuclease A was not at a sufficiently high resolution to serve as the basis of calculations. The known high degree of structural similarity of the two forms of the protein,[44] however, allows most conclusions reached with one to be applied to the other.

A summary of some of the results is shown in Fig. 7.[33] The wavelengths and half-widths for the vibronic transitions found in the model compound measurements were used to determine the shapes and positions of calculated CD spectra. The intensities are from the calculated rotational strengths. The shape of the ribonuclease S vibrational fine structure at 77°K is reproduced, the difference in magnitude between the summed tyrosine contributions and the experimental curve being attributable to the four disulfides present in the molecule (see below).

The principal source of the tyrosine contribution is from nondegenerate dipole–dipole coupling of near-uv transitions on one group with far-uv transitions on others. Coupling with π–π^* transitions of nearby peptides also contributes significantly, but not as strongly as the aromatic–aromatic interactions. Not all the tyrosines contribute equally

[44] F. M. Richards and H. W. Wyckoff, *Enzymes* **4**, 647 (1971).

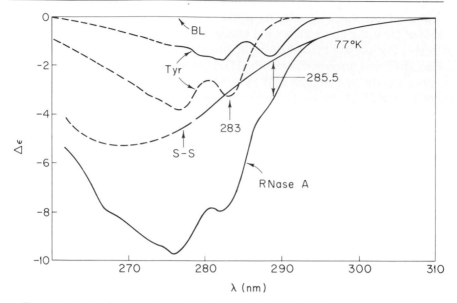

FIG. 6. Analysis of Ribonuclease A CD at 77°K. Buried (solid line) and exposed (dashed line) tyrosine contributions with wavelengths of their 0–0 bands indicated. SS is disulfide contribution (See text). BL is base line. The component spectra sum to yield the experimental spectrum which is indicated RNase A. (From Horwitz et al.[16])

to the latter, however. The ^1L_b rotatory strength of the Tyr-73–Tyr-115 pair, -7.3×10^{-40} cgs, exceeds the summed intensities of all other tyrosine pairs as well as the largest single value, $+1.0 \times 10^{-40}$ cgs for Tyr-25–Tyr-97. The 73–115 pair thus dominates the aromatic spectrum, indicating the crucial dependence of CD upon geometrical factors. The buried tyrosines, 25, 92, and 97, would be expected to add a shoulder at the wavelength of their 0–0 transitions, 286 nm, but this is obscured by the total exciton contribution, which at this wavelength is of opposite sign and only slightly lesser magnitude than the sum of the buried tyrosine terms.

It is interesting that tyrosine interactions with the $\pi-\pi^*$ transition of peptide bonds falls off rapidly with distance of the peptide from the ring. At separations greater than 10 Å no single interaction exceeds $\pm 0.3 \times 10^{-40}$ cgs in rotatory strength, and since the peptide bonds will be disposed more or less randomly about each ring, their total contributions to the near-uv ^1L_b tyrosine band will often cancel. Certainly beyond 10 Å their sum will rarely be significant. It may thus be possible to simplify future calculations, but this will have to be undertaken cautiously. More importantly, it provides a firm basis for using single residue CD contributions to probe local regions of the protein.

The potential use of single residues as local probes is illustrated by

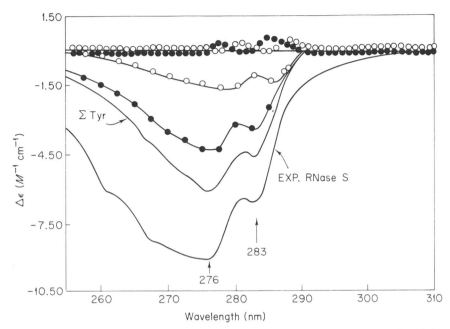

FIG. 7. Comparison between the tyrosyl CD bands calculated for ribonuclease S and the spectrum actually observed for ribonuclease S at 77°K (exp. RNase S). Σ Tyr, total for individual tyrosyl CD bands and their exciton components; ●—●, CD spectrum of Tyr-73, -115, and -76; ○—○, CD spectrum of Tyr-25, -92, and -97; ● ● ●, exciton CD component for Tyr-73–115 interaction; ○ ○ ○, exciton CD component for sum of Tyr 25–97 and Tyr 92–97 interactions. (From Strickland.[33])

Tyr-25. Of the six such side chains in the protein, Tyr-25 appears to be the only one that is significantly affected by the conversion of ribonuclease A to ribonuclease S.[44] The corresponding 0–0 absorption signal from this residue is blue-shifted 2.5 nm from 288.5 nm in ribonuclease A to 286 nm in the S form.[45] The CD shoulder at 289 nm in ribonuclease A is lost in S due in part to the wavelength shift and in part to a loss of dichroic intensity. Its rotational strength in ribonuclease S is calculated to arise primarily from interactions with nontyrosyl moieties, especially Phe-46 and His-48, a feature not shared by the other tyrosines. Some of the individual interactions with Tyr-25 are appreciable, but they are of opposite signs and lead, therefore, to a small net contribution to the CD of ribonuclease S. Because these interactions will vary differently in their intensity as the conformation changes (cf. ^1L_b and ^1L_a of tyrosyl itself discussed above), the overall contribution will be strongly sensitive to the local conformational change that subtilisin cleavage produces in this area.

[45] J. Horwitz and E. H. Strickland, *J. Biol. Chem.* **246**, 3749 (1971).

Effects of Subunit Assembly—Insulin

Conformational change is often involved as a catch-all explanation of otherwise inexplicable results. Subunit assembly is such an area, for one asks whether a change in a measured property is due solely to the assembly or to a conformational change attendant upon assembly or to both of these. Insulin is a good case study here, for it undergoes a complex assembly of monomers to dimers to hexamers.[46] Its tyrosines are amenable to study by CD, for, as mentioned above, it contains no tryptophan. Each of the assembly reactions increases the intensity of the dichroism, the overall enhancement on going from monomer to hexamer being about fourfold.[47] These considerations led Strickland and Mercola[42] to perform for insulin the kind of coupled oscillator calculations described above for ribonuclease.

Discussion of the detailed interactions is omitted here, as they are similar in kind to the ribonuclease results. The effect of assembly upon the CD spectrum is best stated by the original authors.

> Our calculations . . . show that intermolecular contacts between native insulin molecules are sufficient to explain the aggregation phenomenon without involving any mechanism of conformational change.[42]

Enhancement of CD intensity occurs in this case when aromatic residues in isolated species become closely juxtaposed upon interface formation. The resulting interactions lead to increased rotatory power. One pair of tyrosines produced most of the change in the CD seen upon dimerization: Tyr-B16 on molecule I and Tyr-B26 on molecule II. The primary contributor is the near-uv ^1L_b band of B26II by virtue of its coupling with the far-uv ^1L_a of B16I. Interaction of both of these tyrosine ^1L_b bands with peptide bonds on the other side of the interface also contribute significantly. The effect of hexamer formation is similar in kind but involves different residues.

The potential applications of this study to other assembly phenomena should be noted.

The Disulfide

Alone among the chromophores found in proteins the nonplanar disulfide is inherently asymmetric. In the manner of the α-carbon of an amino

---

[46] T. L. Blundell, G. Dodson, D. Hodgkin, and D. A. Mercola, *Adv. Protein Chem.* **26,** 279 (1972).

[47] M. J. Ettinger and S. N. Timasheff, *Biochemistry* **10,** 824 (1971); J. Goldman and F. H. Carpenter, *ibid.* **13,** 4566 (1974); C. Menendez and T. Herskovitts, *Arch. Biochem. Biophys.* **140,** 286 (1970); J. W. S. Morris, D. A. Mercola, and E. R. Arquilla, *Biochim. Biophys. Acta* **160,** 145 (1968); S. Wood, T. Blundell, A. Wollmer, N. Lazarus, and R. Neville, *Eur. J. Biochem.* **55,** 531 (1975).

FIG. 8. M (left-handed) and P (right-handed) disulfides. (Notation of Cahn *et al.*[48a])

acid, no rigid body rotation of one isomer can superimpose it on the other. The introduction of optical activity due to inherent asymmetry[48] makes the near-uv CD of disulfides more difficult to interpret than that of any other chromophore. There are several reasons for the difficulty.

1. The inherent optical activity is strongly dependent upon the geometry of the CSSC group.

2. Intensity questions are complicated by the presence in solutions of many model compounds of approximately, but not exactly, equimolar mixtures of oppositely handed disulfides. Their oppositely signed contributions to the CD yield weak net intensities.

3. The near-uv spectrum generally contains two closely spaced optically active transitions which, if of opposite signs, may also lead to weak net intensities.

4. Like any other chromophore, the disulfide transitions can be perturbed by their local environment. It is often difficult to disentangle vicinal effects upon disulfide transitions from geometric effects upon inherent optical activity.

The preferred geometry of unconstrained disulfides is illustrated in Fig. 8.[48a] It originates primarily in the mutual repulsion of nonbonding electrons on the sulfur atoms. These have approximate p-type orbital symmetry, and their repulsive interaction is least when the CSSC dihedral angle, ψ, is near $\pm 90°$.[49,50] (By convention $\psi = 0$ defines the planar form in which the carbons are cis with respect to the S–S bond. At $\psi = 180°$ the molecule is also planar, but the carbons are trans.)

Correlation of screw sense with the sign of the longest wavelength CD band was first made by Carmack and Neubert[51] in a study of

[48] S. Beychok, *Proc. Natl. Acad. Sci. U.S.A.* **53**, 999 (1965). S. Beychok, *Science* **154**, 1288 (1966).

[48a] R. S. Cahn, C. Ingold, and V. Prelog, *Angew. Chem. Int. Ed. Engl.* **5**, 385 (1966).

[49] W. G. Penney and G. B. B. M. Sutherland, *J. Chem. Phys.* **2**, 492 (1934).

[50] L. Pauling, *Proc. Natl. Acad. Sci. U.S.A.* **35**, 495 (1949).

[51] M. Carmack and L. A. Neubert, *J. Am. Chem. Soc.* **89**, 7134 (1967).

dithianes—six membered rings containing a disulfide constrained to a dihedral angle of 60°. A right-handed screw (P) was found to produce positive ellipticity at the extremum. They cautioned against applying their rule to open chain unconstrained disulfides without further work, however, because of the problems outlined above.

That their rule does apply to open chain compounds whose dihedral angles are near 90° is shown by work on cystine.[15b,52-54] Advantage was taken[52] of the fact that zwitterionic cystine, precipitated at neutral pH, crystallizes as a right screw,[55] while the dihydrochloride, prepared from strong acid, forms a left screw.[56] The chirality had been determined by X-ray diffraction. The CD was measured on solid samples, mulls and KBr pellets, to ensure that only one conformer was present in the samples.

The absolute intensities at the extrema of the long wave-length bands are in the range 6000–8000 deg cm²/dmole. It is of interest to note that a band at 250 nm with a half-width of 25 nm, which is typical of disulfides,[15b] and a peak ellipticity of 8104° has a rotatory strength of 0.1×10^{-38} cgs. This is squarely in the range to be expected for inherently dissymmetric chromophores.[57]

The dichroic intensities of unconstrained disulfides in solution are quite different from these values. Rarely do they exceed ±2100 deg cm²/dmole, and they are often less.[15b] L-Cystine in water at neutral pH, for example, peaks at 254–255 nm with an ellipticity of −2100 deg cm²/dmole, and in 1 M acid, the peak, now at 248 nm, measures −1900 deg cm²/dmole.[15b,48] These solutions are known to contain a mixture of P and M conformers. This was first shown[52] by the observation that the CD spectrum of L-cystine in 1 M acid does not depend on the screw sense in the crystals from which the solutions are prepared. Crystalline L-cystine dihydrochloride with M chirality, $[\theta]_{max} = -6000°$ to $-6500°$,[52-54] yields the same spectrum in solution as crystalline zwitterionic L-cystine, which has P chirality, $[\theta]_{max} = +7000°$,[53,54] the intensity differences being due at least partly to other differences in geometry. This can only happen if there is rotational equilibration about the S–S bond. The loss of dichroic intensity would then be due to cancellation of the oppositely signed P and M spectra. This conclusion is supported by the marked temperature dependence of the CD of L-cystine in solution.[58] Finally, Jung et al.[59] observe a

[52] P. C. Kahn and S. Beychok, J. Am. Chem. Soc. 90, 4168 (1968).
[53] A. Imawishi and T. Isemura, J. Biochem. (Tokyo) 65, 309 (1969).
[54] N. Ito and T. Takagi, Biochim. Biophys. Acta 221, 430 (1970).
[55] B. M. Oughton and P. M. Harrison, Acta Crystallogr. 12, 396 (1959).
[56] L. K. Steinrauf, J. Peterson, and L. H. Jensen, J. Am. Chem. Soc. 80, 3835 (1958).
[57] W. Moffitt and A. Moscowitz, J. Chem. Phys. 30, 648 (1959).
[58] T. Takagi and N. Ito, Biochim. Biophys. Acta 257, 1 (1972).
[59] G. Jung, P. Hartter, and H. Lachmann, Angew. Chem., Int. Ed. Engl. 14, 429 (1975); see also M. Ottnad, P. Harttner, and G. Jung, Eur. J. Biochem. 66, 115 (1976).

time dependence of the absorption and CD of *tert*-butyloxycarbonyl-L-cysteinylglycyl cysteine disulfide upon dissolving crystalline material. The spectra show isosbestic points, and a kinetic analysis indicates a two state transformation that proceeds in a "spectroscopically uniform manner."[59] The slowness of the transition they attribute to steric hindering of S–S bond rotation by bulky substituents.

Thus it seems inescapable that thermodynamically unconstrained disulfides equilibrate about their S–S bonds and yield solutions containing equal, or nearly equal, mixtures of P and M conformers. Samples known to contain one conformer produce absolute intensities of 6000 to 8000 deg cm²/dmole, while known mixtures of conformers rarely exceed 2000° and often have weaker dichroism.

Although it appears certain that solutions of L-cystine and related compounds contain a mobile equilibrium between conformers, the principal sources of their optical activity is in dispute.[60,61] Understanding the nature of the disagreement, however, requires that the nature of the near-uv electronic transitions be examined first. A detailed discussion is given in Kahn.[15b] It is abbreviated here.

As the dihedral angle of the disulfide is reduced from 90°, the longest wavelength transition red shifts from 255 nm at $\psi = 90°$ to 280 nm at $\psi = 60°$ and to 330 nm at $\psi \approx 27°$.[62] Bergson[63] developed a simple theory to account for this. He reasoned that the highest filled orbital would be an antibonding π-type combination of filled sulfur $3p$ atomic orbitals. Since the filled $3p$ orbitals repel one another, the antibonding nature [and thus the energy of the molecular orbital (m.o.)] would be least when the overlap is least. This is obtained when the $3p$ atomic orbitals are orthogonal (overlap = 0), which occurs, in turn, at a dihedral angle of 90°. Decreasing the dihedral angle from 90° causes them to overlap and leads to a rise in orbital energy which would be maximal when the $3p$'s are coplanar. This occurs at $\psi = 0°$. The lowest unfilled m.o. in Bergson's scheme is sulfur–sulfur σ antibonding. Because of its cylindrical symmetry about the S–S bond, its energy is, to a good approximation, expected to be independent of dihedral angle. The energy of the transition from highest filled to lowest unfilled m.o. thus decreases with decreasing dihedral angle, i.e., the wavelength increases.

The overlap of p-type atomic orbitals produces, in addition to the π^* orbital, a π bonding orbital. The latter is stabilized by increasing overlap, and its transition energy to σ^* therefore increases (its wavelength decreases) as $|\psi|$ decreases from 90°. At $|\psi| = 90°$ the π and π^* molecular or-

[60] J. Linderberg and J. Michl, *J. Am. Chem. Soc.* **92**, 2619 (1970).
[61] J. P. Casey and R. B. Martin, *J. Am. Chem. Soc.* **94**, 6141 (1972).
[62] J. A. Barltrop, P. M. Hayes, and M. Calvin, *J. Am. Chem. Soc.* **76**, 4348 (1954).
[63] G. Bergson, *Ark. Kemi* **12**, 233 (1958); **18**, 409 (1961).

bitals are degenerate and are best described as $3p$ non-bonding. The energies of the two longest wavelength transitions thus show opposite dihedral angle dependence. Because Bergson was concerned with the red shift of the longest wavelength band with decreasing dihedral angle, he did not discuss the second transition very much or the dihedral angle range from 90° to 180°. These, however, are crucial to an understanding of disulfide optical properties, and they are interpretable within the framework of his model.

Linderberg and Michl[60] and Woody[64] have calculated the optical properties of the disulfide, the first performing a CNDO-SCF computation on dihydrogen disulfide (H_2S_2), and the latter a direct application of Bergson's scheme to CD. Both obtained agreement with Carmack and Neubert's empirical rule for the longest wavelengths transition at $|\psi| < 90°$. They also found that the sign of the second band is opposite to that of the first and that they are of equal or nearly equal intensities.

In considering the range of dihedral angle between 90° and 180°, there is a change in the nature of the longest wavelength transition as a direct consequence of Bergson's model.[15b,60,64] It was mentioned above that at $\psi = 0°$ the energy difference between π and π^* is maximal. As ψ opens from 0°, π^* decreases in energy and π increases, until at 90° they are degenerate and nonbonding. As ψ continues to open toward 180°, they cross in energy smoothly, their separation now growing. The identity of the highest filled orbital and therefore of the longest wavelength transition has thereby changed, although it remains $\pi^* \rightarrow \sigma^*$ in character (see Kahn[15b] for a detailed discussion). This variation of wavelength with dihedral angle is shown in Fig. 9. The transition from each filled orbital to σ^*, however, is predicted to maintain the same sign throughout the range 0° < ψ < 180°, and their signs, as noted above, are opposite. As ψ opens past the point of orbital degeneracy, then, the sign of the longest wavelength CD band must change. This leads to a "quadrant rule" for the variation in sign of the long wavelength band with dihedral angle, a rule which was verified for $\psi \approx 120°$ by Ludescher and Schwyzer.[65]

Linderberg and Michl[60] argue that the equal intensity and opposite signs of the $\pi^* \rightarrow \sigma$ and $\pi \rightarrow \sigma^*$ bands leads at $|\psi| = 90°$ to their cancellation. Compounds having $|\psi| \cong 90°$ would therefore be expected to show little if any dichroism due to inherent asymmetry *even if only a single conformer is present*. The weak dichroism of L-cystine in solution is attributed to this and not to the presence in solution of unequal populations of oppositely handed disulfide molecules. In the absence of significant optical activity due to inherent asymmetry, these authors conclude that the near-uv CD of disulfides arises primarily from vicinal perturbations of the electronic transitions, as is the case for aromatics. In this conclusion they

[64] R. W. Woody, *Tetrahedron* **29**, 1273 (1973).
[65] U. Ludescher and R. Schwyzer, *Helv. Chim. Acta* **54**, 1637 (1971).

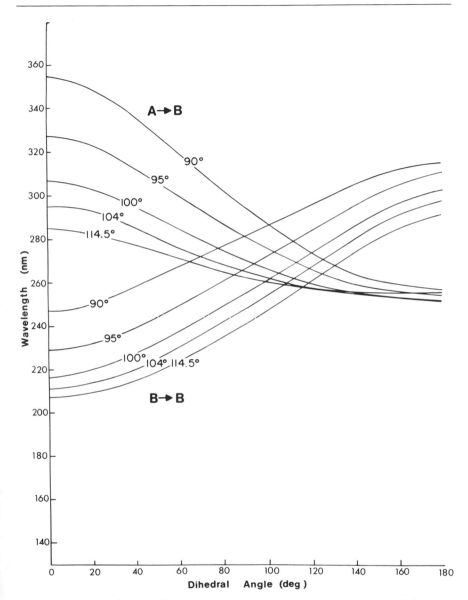

FIG. 9. Wavelength versus dihedral angle of $\pi \rightarrow \sigma^*$ and $\pi^* \rightarrow \sigma^*$ transitions of dicarbon disulfide for CSS bond angles as noted on the curves. (From Kahn.[15b])

are supported by Casey and Martin's interpretation of a combined and careful nmr and CD study of L-cystine and related compounds.[61] We disagree on two grounds.

1. The presence of a mixture of P and M conformers in L-cystine solutions and of a mobile equilibrium between them seems to be established.

Variation in CD attributed to direct vicinal effects on the disulfide's transitions could therefore arise instead from a displacement of the P ⇌ M equilibrium as suggested by the temperature dependence of the L-cystine CD intensity.[58]

2. The equal or nearly equal intensities of the $\pi \to \sigma^*$ and $\pi^* \to \sigma^*$ transitions predicted by Woody's application of Bergson's model[64] and by Linderberg and Michl[60] for H_2S_2 does not occur when the atoms bonded to the sulfurs are carbons rather than hydrogens.[15b] In the Bergsonian scheme, it should be noted, there are no atoms bonded to the sulfurs; it is treated as S_2, a homonuclear diatomic molecule. Thus the theoretical basis for weak inherent optical activity of a single conformer at ψ near 90° is eroded.

When carbon is bonded to sulfur rather than hydrogen, the dihedral angle at which the π and π^* orbitals are degenerate is shifted to significantly greater values, possibly as high as 107° (see Fig. 9 and Kahn[15b]). Single conformers at $|\psi| = 90° - 100°$, such as is often found in proteins, would then have a detectable difference in wavelength between their two lowest energy transitions, and their signs would be the same as is found for $|\psi| < 90°$. Casey and Martin also note this possibility.[61] The sizes of the quadrants in the angle convention diagram of Ludescher and Schwyzer[65] are thus unequal for organic disulfides. That diagram is redrawn for π, π^* degeneracy at 107° in Fig. 10. The cross-over value of 107° was chosen from Fig. 9 for a CSS bond angle of 104°, which is the value that is thought to obtain in solution.[66] It is presented here for illustrative purposes only, for it is certainly subject to quantitative refinement. It is interesting that for dihedral angles in the 90° to 100° range often found in proteins, the deviation from the 107° cross-over is 7°–17° and the single conformer ellipticity appears to be 6000–8000 deg cm²/dmole. A deviation of 13° beyond the cross-over point yields the 120° dihedral angle estimated for (2,7-cystine)-gramicidin S.[65] Given the symmetry properties of the sulfur $3p$ orbitals, one is led to anticipate similar intensities at similar deviations from the cross-over. Ludescher and Schwyzer report 6000 to 8000 deg cm²/dmole for their samples, depending upon the solvent.[65] If the true cross-over were at $|\psi| = 90°$, one would expect for gramicidin an intensity corresponding to a greater distance from degeneracy. The intensities at $|\psi| = 60°$, however, are 12,000 to 17,000 deg cm²/dmole.[51]

One consequence of the displacement of the dihedral angle of degeneracy to values greater than 90° is that Strickland's assignment of CD signs to disulfides 26–84 and 65–72 of ribonuclease S[33] probably ought to be reversed. The expectation of weak dichroism for several disulfides of insulin[37] is probably not correct either.

[66] O. Foss and O. Tjómsland, *Acta Chem. Scand.* **12,** 1810 (1958); C. Szantay, M. P. Kotick, E. Shefter, and T. J. Bardos, *J. Am. Chem. Soc.* **89,** 713 (1967).

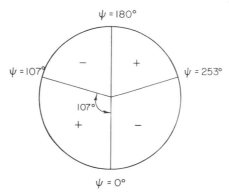

FIG. 10. "Quadrant" diagram for sign of longest wavelength disulfide band.   Redrawn from Ludescher and Schwyzer,[65] as described in text.

Based on the revised diagram in Fig. 10 and on Fig. 9, we can describe some rules for the interpretation of protein disulfide CD. Most of them concern situations in which the disulfide is not free to equilibrate its chirality about the S-S bond. Given the constraints of protein structure, this will include the majority.

1. If the dihedral angle is less than approximately 107°, positive ellipticity of the longest wavelength band is associated with P (right) chirality and negative ellipticity with M (left) chirality. This is the rule of Carmack and Neubert.[51]

2. If the dihedral angle is within approximately ±15° of the degeneracy point, the absolute ellipticity per disulfide is expected to fall in the range 6000 to 8000 deg cm²/dmole.

3. If the dihedral angle is in the vicinity of 60°, the intensity is expected to be higher, possibly as high as 12,000 to 17,000 deg cm²/dmole,[51] but see below also.

4. If the disulfide is free to equilibrate about the S–S bond, or if vicinal effects become very large, the intensity will fall in the range around 2000 deg cm²/dmole.

5. From Fig. 9 and from the Bergson model the wavelength positions and separation between the $\pi^* \to \sigma^*$ and $\pi \to \sigma^*$ transition are governed by the dihedral angle. The wavelengths are also sensitive to the CSS bond angle. Variation with respect to this angle is of the same order of magnitude as for dihedral angle, although the bond angle is less deformable. While the separation of $\pi \to \sigma^*$ from $\pi^* \to \sigma^*$ appears due to the dihedral angle, their mean position seems to be determined primarily by the bond angle.

6. The properties of the $\pi \to \sigma^*$ transition are expected on both empirical[51] and theoretical grounds[15b,64] to be more variable with vicinal conditions than the longest wavelength transition.

Disulfides in Proteins

For ribonuclease S three of the four disulfides are left-handed, and one is right-handed. If the CD contributions of the right screw cancels that from one left screw a net of two M forms remain. The total disulfide ellipticity per mole of ribonuclease S is estimated[7,16] to be $-16,500$ deg $cm^2/dmole$, which is $-8250°$ per SS.

For neurophysin II the one disulfide that is easily reduced has an apparent maximal ellipticity of approximately -8000 deg $cm^2/dmole$.[67] Its chirality is predicted to be left if its dihedral angle is less than approximately 107°.

The apparent extrema of this disulfide occur near 285 and 240 nm. In form and position it is rather similar to the constrained six-membered ring disulfides (dithianes) studied by Carmack and Neubert,[51] although the intensity of the 285 nm band is more similar to the value for cystine crystals.[52-54]

The wavelength shift in dithianes is usually attributed to their constrained dihedral angle (60°), but it is seen from Fig. 9 that distortion of the CSS bond angle also produces large shifts. The bond angle in dithianes is 100°,[66] and in unconstrained disulfides 104°.[66] The neurophysin II disulfide is unusually susceptible to reduction, and Menendez-Botet and Breslow[67] reason that it is sterically strained both because of its ease of reduction and because of the effect upon the CD of denaturants in the absence of reductants. It is hard to imagine dihedral angle distortion alone producing all these effects in neurophysin II. Some degree of bond angle deformation seems also to be necessary. A dihedral angle somewhat greater than 60° (70°?) and a bond angle near but somewhat greater than 100° would explain the data.

The two disulfides of human pituitary growth hormone yield intensities of -2100 and -3200 deg $cm^2/dmole$.[68] They are predicted either to have rotational freedom due to flexibility of the protein in their vicinity or, if constrained to be single conformers, to exist in highly unusual perturbing environments.

Tryptophan

Like tyrosine tryptophan has ^1L_b and ^1L_a bands polarized at right angles to one another. They are not oriented along the axes of the indole ring, however,[69,70] although the exact direction is not yet firmly established. Also in contrast to tyrosine, the ^1L_a transition does not occur in

---

[67] C. J. Menendez-Botet and E. Breslow, *Biochemistry* **14**, 3825 (1975); see inset, Fig. 2.
[68] T. Bewley, *Biochemistry* **16**, 209 (1977).
[69] W. J. Goux, T. R. Kadesch, and T. M. Hooker, Jr., *Biopolymers* **15**, 977 (1976).
[70] Y. Yamamoto and J. Tanaka, *Bull. Chem. Soc. Jpn.* **45**, 1362 (1972).

the far-uv. This complicates the interpretation of tryptophan spectra, for the ^1L_a transition is broad, intense, lacking in clear vibrational structure, and sensitive in its wavelength position to environmental conditions. As a result it often overlaps and obscures the ^1L_b transition, its long wavelength tail sometimes extending beyond the long wavelength end of the ^1L_b.

^1L_b is similar to the tyrosine ^1L_b, having a comparable dipole strength and vibrational fine structure. Strickland observed that although the overall intensity may be due primarily to the ^1L_a transition, tryptophan vibronic structure between 275 and 290 nm indicates the presence of ^1L_b intensity.[71] The absorption spectrum has been resolved into ^1L_a and ^1L_b components for solutions in organic solvents 1 which presumably mimic the hydrophobic interior of a protein (see Fig. 11 of Strickland[7]).

The ^1L_a and ^1L_b CD bands appear to vary independently of one another in sign and in intensity. For single indoles in uniform environments they have the same shapes as their absorption spectra. For such cases superposition of ^1L_a and ^1L_b type curves from model compounds can be used to resolve the spectra in a manner similar to that used for the different classes of tyrosine ^1L_b in ribonuclease. One problem with recognizing the CD contribution of tryptophan ^1L_a type spectra is their similarity to disulfides: a broad featureless band peaking at 260 to 270 nm. Some of these, at least, will be distinguishable by the long wavelength disulfide tail beyond 310 nm which tryptophan lacks.

Goux *et al.* have performed a matrix formalism calculation of the optical properties of yohimbinic acid.[69] Agreement with experiment for the ^1L_a and ^1L_b bands is reasonable, but further work, especially on the polarization directions, will be needed.

Concluding Remark

The reader may have noted that the whole protein calculations discussed here did not utilize the matrix formalism. Its application to proteins is new. Such calculations have been performed for lysozyme, ribonuclease, staphlococcal nuclease, and pancreatic trypsin inhibitor,[72] and their publication is expected shortly.

Note Added in Proof: The problem of origin dependence in the calculation of magnetic moments appears to have been solved within the framework of the matrix formalism in a way that does not lead to violation of the rotatory strength sum rule.[73]

[71] E. H. Strickland and C. Billups, *Biopolymers* **12,** 1989 (1973).
[72] T. Hooker, personal communication.
[73] T. M. Hooker, Jr. and W. J. Goux *In Excited States in Organic Chemistry and Biochemistry* (B. Pullman and N. Goldblum, eds.), p. 123, D. Reidel Publishing Co., Dordrecht, Holland, 1977.

Acknowledgments

Sherman Beychok was to have written this chapter, but because of the pressures of other work, he asked me to join him in it. I agreed, largely for the pleasure of working with him. The demands on his time and attention grew, however, and in the end he felt unable to do justice to this volume in the time available for its completion. Although the words here are therefore mine, his contribution is present throughout, for much of the chapter took shape in the course of our discussions. Being stubborn, I did not always follow good advice, and the resulting flaws are mine alone.

The sardonic skepticism of Phillip Pechukas served a useful purpose, and the assistance of Rhea McDonald in preparing the manuscript was essential.

The support of the New Jersey State Agricultural Experiment Station, The Rutgers University Research Council, and a Biomedical Resources Support Grant are also acknowledged.

[17] Time-Resolved Fluorescence Measurements

By Mugurel G. Badea and Ludwig Brand

Introduction

During the last decade, fluorescence techniques have been applied to numerous problems in biology and biochemistry. Both intrinsic and extrinsic fluorophores have been used to obtain information about proteins, nucleic acids, membranes and other biological materials. The increasing interest in the application of fluorescence methods in the life sciences has led to a continuing development of new instrumental techniques and procedures for data analysis. Nanosecond fluorometry represents an area of particularly intense activity. Ten years ago, the measurement of fluorescence decay times was still in its infancy and it was difficult, if not impossible, to measure and analyze multiexponential decay curves with any degree of confidence. Today, this situation has been completely altered and nanosecond decay data, nanosecond time-resolved emission spectra, and the decay of the emission anistropy can readily be obtained and analyzed. As a consequence, complex excited state interactions have become better understood and have been used to probe biological microenvironments.

It would be difficult to cover all the advances in experimental techniques that have been made in the last ten years and this will not be attempted. Numerous specialized topics such as differential fluorimetry, solute perturbation fluorescence, stopped flow fluorescence, fluorescence circular dichroism, have been or will be covered in other chapters in these volumes. Topics to be covered here include the following: advances in the instrumentation and procedures for data analysis required for nanosecond fluorometry; potentials of nanosecond time-dependent fluorescence emission spectroscopy, and nanosecond time-dependent emission anisotropy. In addition the treatment of decay data to obtain information regarding excited state interactions will be discussed.

The relations between fluorescence, excited state interactions, and probe environment are the themes to be emphasized. Reactions that occur in the excited state can either be inferred from or directly measured by the parameters characterizing the luminescence.[1]

There are a large number of excited state interactions that can occur prior to or can compete with fluorescence emission. These include vibrational relaxation of the photo-excited state, internal conversion and intersystem crossing, conformational change, hydrogen bonding, orientational relaxation in fluid media, proton transfer, electron ejection, exciplex and excimer formation, excited state charge transfer complexes, and nonradiative energy transfer which can be singlet-singlet or triplet-triplet. While this list is not complete, it does indicate the variety of excited state interactions that are known to exist and that have been studied in some detail. The extent to which these interactions take place will depend on the character of the fluorophore, on other molecules in the vicinity, and on general environmental factors, such as the solvent, the viscosity, and the temperature. It is desirable that the excited state interactions of a fluorescence probe first be characterized in detail in model solvent systems. In this way, when the fluorophore is used to probe a biological system, the information derived may be interpreted on a firm basis to give information about the microenvironment.

The shortest relaxation times measured for many of the excited state interactions indicated above are on the picosecond time scale. An excellent review of the published data on the picosecond time scale has been presented by Eisenthal.[1] In many cases of interest in biology, the environmental conditions are or can be made to be such that most excited state processes, including fluorescence, occur on the nanosecond time scale. The measured fluorescence decay time τ reflects the competition between fluorescence emission and other excited state reactions or quenching. If all other deexcitation channels are suppressed, then all the quanta absorbed will be emitted as fluorescence. Under these conditions the fluorescence decay time approaches the natural lifetime τ_0. The natural lifetime is related to the probability of the transition to the ground state and is often on the order of nanoseconds. The quantum yield approaches unity as the measured lifetime approaches the natural lifetime. These relations are summarized below.

$$q = \frac{\tau_m}{\tau_0} \qquad \tau_0 = \frac{1}{k_f} \qquad \frac{1}{k_f + k_{(other)}} \tag{1}$$

[1] K. B. Eisenthal, *Annu. Rev. Phys. Chem.* **28,** 207 (1977).

where τ_m and τ_0 are the measured and natural lifetimes respectively, q is the quantum yield, and k_f and $k_{(other)}$ are the rate constants for the fluorescence and nonfluorescence decay to the ground state, respectively.

The microenvironment which biomolecules and bioaggregates create around fluorophores can thus be probed through nanosecond fluorescence measurements. These measurements will provide information about processes that may be occurring concomitant with fluorescence. Changes in the rates of these processes can be correlated with changes in the biologically relevant parameters influencing them. The power of nanosecond fluorescence techniques lie in determining the *dynamics* of conformational changes of, or interaction between, biomacromolecules and biological macroassemblies. In some instances, nature has provided us with "built-in" fluorescence probes. Examples include the aromatic amino acids, especially tryptophan and tyrosine. Other examples include fluorescent coenzymes, such as pyridine nucleotides, pyridoxal phosphate, or flavins. In other cases, it is necessary to add a probe attached covalently or noncovalently to the system.

Our aim here is to provide an overview of the nanosecond fluorescence field at its present stage of development. As mentioned above, this section will not be comprehensive, but will rather be selective reflecting the experience (and bias!) of the authors. A comparative appraisal of the various instrumental techniques used in nanosecond fluorometry will be provided. Since suitable procedures for analysis of the data are as essential in nanosecond fluorometry as the instrumentation itself, the conceptual basis of the procedures for elimination of the convolution artifacts and thus obtaining the true fluorescence relaxation times will be described. The experimental procedures for performing time-resolved fluorescence measurements, i.e., time-resolved emission spectra (TRES) and decay of the emission anisotropy (DEA) will be explicitly dealt with. Finally, some selected examples of ways to handle complex decay data will be presented.

Fluorescence Decay Instrumentation

The most significant advances in fluorometry during the last decade are those that have been made in regard to the instrumentation required for fluorescence decay studies. Fluorescence lifetimes are usually of the order of nanoseconds, and, until recently, this short time scale made the measurements difficult. The development of the instrumental techniques required for decay measurements has proceeded in two directions.

The transient fluorescence properties of a molecular system can be studied by observing its response to either a continuous, high frequency modulated excitation or to a series of discrete, repetitive, short exciting pulses. In either case the desired information regarding fluorescence decay laws is extracted from the modification of the time profile of the exciting signal due to the lag introduced by the finite duration of the excited state.

Continuous Response Technique. In the literature, this method of determining the kinetics of the fluorescent decay is known as the "phase shift method." The kinetic information is usually obtained by measuring the phase shift (δ) between the excitation wave form and the corresponding fluorescence response wave form.[2] If a homogeneous population of *monoexponentially* decaying molecules is excited with sinusoidal modulated light with frequency f, then the decay constant τ is related to δ and f according to the formula

$$\tan \delta = 2\pi f \tau \tag{2}$$

In this case, the decay constant can also be obtained from the degree of modulation of the fluorescence relative to that of the exciting light given by

$$m = \text{relative modulation} = \frac{\text{modulation of fluorescence}}{\text{modulation of excitation}}$$

$$\frac{(F_{max} - F_{min})/(F_{max} + F_{min})}{(E_{max} - E_{min})/(E_{max} + E_{min})} = \frac{1}{(1 + 4\pi f \tau)^{1/2}} = \cos \delta \tag{3}$$

where F_{max} is maximal fluorescence and F_{min} is minimal fluorescence

The modulation frequency f is chosen to yield maximum measuring accuracy for both $\tan \delta$ and m. This happens when $f \sim 1/2\pi\tau$ which for decay constants in the range of 3–50 nsec corresponds to frequencies between 3 and 50 MHz.

Both Eqs. (2) and (3) are valid only for an emitting molecular system which decays monoexponentially, an assumption implicit in their derivation. A sufficient check for the validity of this assumption is the equality of the decay constants obtained from each equation, respectively. If the assumption is not valid, the decay constant measured by the degree of modulation will almost always be larger than that determined from the phase shift.[2]

In the multiexponential case the equations corresponding to Eqs. (2) and (3) are given by[3]

[2] R. D. Spencer and G. Weber, *Ann. N. Y. Acad. Sci.* **158**, 361 (1969).
[3] S. R. Schuldinger, D. Spencer, G. Weber, R. Weil, and H. R. Kaback, *J. Biol. Chem.* **250**, 8893 (1975).

$$\tan \bar{\phi} = \frac{\sum_i F_i \sin \phi_i \cos \phi_i}{\sum_i F_i \cos^2 \phi_1} \tag{2'}$$

$$\bar{M}^2 = \left(\sum_i F_i \cos^2 \phi_i\right) + \left(\sum_i F_i \sin \phi_i \cos \phi_i\right) \tag{3'}$$

where the ϕ and $\bar{M}$ are the experimentally observed phase lag and relative modulation and ϕ_i and F_i are the phase shift and fractional contribution of the ith component. To extract the individual parameters of an n-component system, n separate measurements of $\bar{\phi}$ and $\pm\bar{M}$ should be performed at n modulation frequencies.

Ideally, in order to ascertain the multicomponent nature of a given fluorescence decay the phase fluorimeter should possess the capability of a continuous variable frequency. Then the fluorescence intensity time profile could be obtained as the Fourier transform of the phase shift data taken at various frequencies.

While there are still technical difficulties associated with variable frequency measurements, good progress in this direction is being made.[4,5] Application of these techniques to the resolution of multi or nonexponential decay systems may be expected in the near future. At the time of writing, much of the work in this area has been performed on decay instruments based on pulse techniques (see below). Wherever the monoexponential character of the decay is firmly established the continuous response methods have excellent subnanosecond sensitivity and accuracy coupled with a relatively fast data acquisition. Several technical improvements[2,4,6] especially in the procedures used for phase detection have resulted in instruments capable of measuring differences of 0.1 nsec with an accuracy of 0.03 nsec between two decay constants each less than 1 nsec.[2] Suitable instrumentation for phase fluorometry is now available commercially.[7]

Discrete Response Technique. Significant advances have been made in the development of several types of nanosecond fluorometers based on pulse techniques such that this approach can be used to resolve complex fluorescence decay data. The pulse technique has also been used to obtain nanosecond time-resolved emission spectra and nanosecond time-resolved emission anisotropy data. Since pulse methods are used in our own laboratories their use will be emphasized here.

[4] M. Hauser and G. Heidt, *Rev. Sci. Instrum.* **46,** 470 (1975).
[5] F. E. Lytle, J. F. Eng, J. M. Harris, T. D. Harris, and R. E. Santini, *Anal. Chem.* **47,** 571 (1975).
[6] V. E.-P. Resewitz and E. Lippert, *Ber. Bunsenges. Phys. Chem.* **78,** 1227 (1974).
[7] SLM Instruments, Champaign, Illinois.

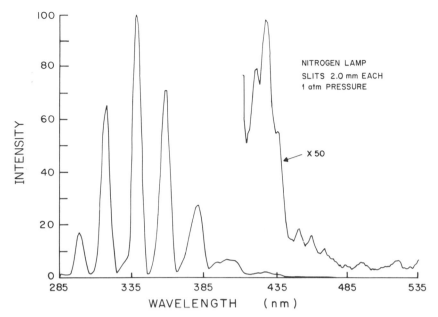

FIG. 1. Relative intensity distribution of a 1 atm nitrogen nanosecond flash lamp. The relative intensities from 400 to 530 nm are also shown amplified 50-fold. The emission was observed through a Bausch and Lomb No. 33-86-44 monochromator (33 Å/mm). The entrance and exit slits of the monochromator were set at 2 mm.

The pulsed excitation is provided by a nanosecond flash lamp, freerunning or gated at a particular frequency. The frequency is usually less than 50 kHz so that decays as long as 1000 nsec initiated by one flash are not perturbed by the next. It is our experience that gated lamps not only allow a convenient choice of a constant repetition rate but also exhibit improved lamp intensity per flash and pulse shape reproducibility. A thyratron gated nanosecond flash lamp with repetition rates up to 50 kHz is now available commercially from PRA[8] together with the required power supply. Based on a design due to Ware,[9] it allows the lamp gas to be changed with ease. Nitrogen gives intense light between 300 and 400 nm (Fig. 1) with lines at 226, 316, 337, 358 and 381 nm. Hydrogen and deuterium (Fig. 2) can be used to obtain excitation further in the ultraviolet. The timing characteristics of typical flash lamps now in use enable one to obtain flash profiles with a width at half-maximum about 2 nsec. Most lamps usually show a moderate tail as seen in Fig. 3. The electromagnetic

[8] PRA (Photochemical Research Associates Inc.), N6A 5B7. University of Western Ontario, London, Ontario, Canada.

[9] W. Ware, *in* "Creation and Detection of the Excited State" (A. A. Lamola, ed.), Vol. 1, Part A, p. 213. Dekker, New York, 1971.

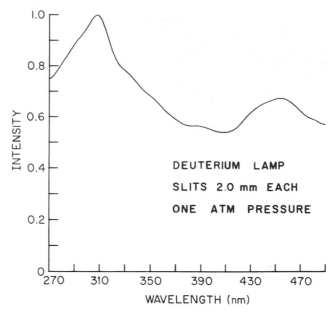

FIG. 2. Relative intensity distribution of a deuterium nanosecond flash lamp. The monochromator is as described in the legend to Fig. 1. The intensity scale is not related to that shown for Fig. 1.

radiation associated with the spark can be picked up by an antenna and used as a timing pulse to trigger the detection system. Alternatively, a suitable timing pulse can be provided by an electron multiplier phototube (an RCA IP28 is commonly used) which views the lamp directly.

Once the sample is excited a variety of methods are available for monitoring the time course of the fluorescence decay. From this point of view, the nanosecond pulse fluorometers are divided into two general types: (a) those that monitor many photons per pulse and (b) those that monitor only one photon per pulse (monophoton counting instruments). Both these types of instruments usually employ a combination of analog and digital techniques for data acquisition.

In the simplest case, the fluorescence decay of a sample is described by a single exponential:

$$I(t) = I_0 e^{-kt} = I_0 e^{-t/\tau} \tag{4}$$

where k represents the probability of a particular molecule emitting a fluorescence photon following excitation. This probability is assumed to be time independent. The exponential law is obtained by summing the photon contributions of an assembly of potentially fluorescent molecules at each infinitesimal time interval after synchronous excitation. In the

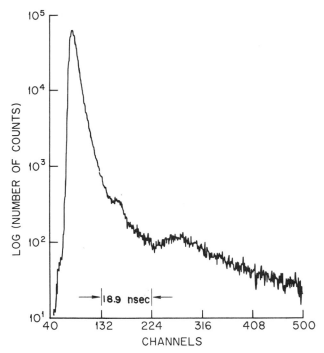

FIG. 3. Typical time–intensity profile of a 1 atm nitrogen flash lamp. There are 0.205 nsec/channel. The ordinate is a logarithmic scale in order to emphasize the tail.

limit of very small time intervals, $I(t)$ is thus proportional to the number of excited molecules making the transition to the ground state at time t after excitation.

At the present time, fast electronic circuitry capable of reproducing the decay of fluorescence light intensity on the nanosecond time scale is not widespread. The first type of existing instrumentation obviates this stringent requirement by prepositioning a constant, narrow time window (aperture) at variable delays after the triggering pulse and integrating the effect produced by the many photons received during the aperture opening. The delay is varied to cover the whole measurable decay curve, and the integration could be performed on various physical quantities proportional to the photon flux, i.e. photocurrent,[10] photovoltage,[11] or light emission from a slowly decaying phosphor detector screen.[12] The reader is referred to the original literature for details on these ingeneous techniques. The second type of instrumentation concentrates on the time axis

[10] R. G. Bennett, *Rev. Sci. Instrum.* **31,** 1275 (1960).
[11] O. J. Steingraber and I. B. Berlman, *Rev. Sci. Instrum.* **34,** 524 (1963).
[12] J. Yguerabide, *Rev. Sci. Instrum.* **36,** 1734 (1965).

of the decay curve rather than on the intensity. It accepts only one fluorescence photon per exciting flash and times its arrival relative to the latter. The histogram of a large number of such events (i.e., number of photons versus time of arrival) will represent, in the large limit, the actual decay curve. Only the first arriving photon is accepted such that it is essential that the conditions of collection be such that all possible times of arrival are equally represented.[9] In what follows, the operation and the conditions for proper use of the nanosecond instrumentation will be dealt with in some detail, stressing the respective advantages and disadvantages of the two types of instrumentation mentioned above.

MULTIPLE PHOTONS PER EXCITING PULSE. The block diagram of a typical instrument collecting multiple photons per excited pulse is shown in Fig. 4. It is easily assembled from commercially available units.[13] The heart of the instrument is a Princeton Applied Research Co., dual channel boxcar averager main frame model 162 fitted with two model 163 sampled integrators which incorporate two Tektronix S-2 sampling heads. One head is used to process successively the fluorescence and the excitation signals coming from a 56 TUVP Amperex photomultiplier which views either the fluorescent sample or the scatterer. The wavelength at which the emission is detected is selected by a monochromator and the excitation is performed through an interference filter. A beam splitter allows the simultaneous monitoring of the fluctuations in excitation by a 56 DUVP 03 photomultiplier whose output is fed into the second S-2 sampling head. By taking the instantaneous ratio of the fluorescence signal to the exciting light signal, the fluorescence decay data is obtained free of any fluctuations in the excitation intensity.[14]

The boxcar averaging is a process which involves measuring the amplitude of a particular point on a repetitive signal, integrating the result, and computing a representative average. Specifically, at any chosen time after a triggering signal, the nonadjustable sampling gate of 75 psec, built into the electronic circuitry of the S-2 sample head, opens and records the instantaneous peak amplitude of the repetitive signal. This process is performed a chosen number of times and the result averaged in a memory unit contained in the Model 163 sampled integrator. Then, controls from the Model 162 main frame allow a manual or an automatic new choice of the delay time (after the trigger) when the sampling gate opens and the process is repeated. The range to be scanned can be varied in steps from 100 nsec to 50 μsec, and the automatic scanning speed is selectable from 10 msec to 10^5 sec full range. The number of samples averaged at any particular time in the scanning range can be selected by a proper combination

[13] M. G. Badea and S. Georghiou, *Rev. Sci. Instrum.* **47,** 314 (1976).
[14] J. W. Longworth, *in* "Creation and Detection of the Excited State" (A. A. Lamola, ed.) Vol. 1, Part A, p. 343. Dekker, New York, 1971.

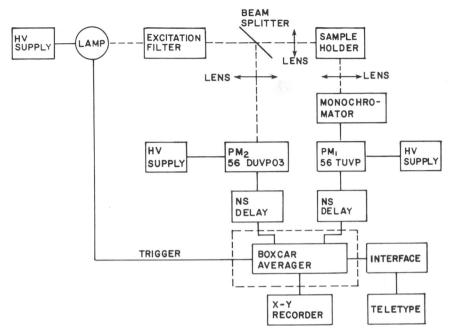

FIG. 4. Block diagram of a sampling fluorometer employing a boxcar averager. Details are given in the text.

of these two controls in relation to the repetition frequency of the signal to be measured.

Base line sampling, activated by a front panel switch, allows automatic correction for dc drifts. When the switch is activated, the averager samples and averages alternately the input signal and the base line for each gate opening. The output of the averager (for each gate opening) is then the difference between the two averaged samples. The time at which the base line is sampled, controlled by a front panel switch, is positioned well outside the time region of interest to avoid distortion of data. The output of the boxcar averager can be obtained either as a dc signal and/or in a digital format. It represents the time profiles of the excitation and the corresponding fluorescence decay over the previously selected scanning range. Subsequent data analysis can be performed on a suitable computer.

The principle of operation of this kind of instrumentation is identical with other pulse sampling scope instruments described in the literature by Steingraber and Berlman,[11] Hundly et al.,[15] and Hazan et al.[16] In the latter

[15] L. Hundley, T. Coburn, E. Gorwin, and L. Stryer, Rev. Sci. Instrum. **38**, 488 (1967).
[16] G. Hazan, A. Grinvald, M. Maytal, and I. Z. Steinberg, Rev. Sci. Instrum. **54**, 1602 (1974).

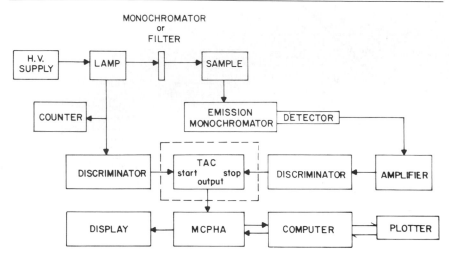

FIG. 5. Block diagram of a monophoton counting nanosecond decay fluorometer. Details are given in the text. MCPHA, multichannel pulse height analyzer.

a large number of sampling sweeps are averaged with the use of a minicomputer[15] or a multichannel analyzer[16] to eliminate the base line drifts and lamp intensity fluctuation. The P.A.R. boxcar averager can also be used in this mode of operation. Furthermore, its features of instantaneous ratio and base line sampling provide two additional capabilities for averaging the intensity fluctuations and the dc drifts. The performance characteristics of a particular fluorometer set-up employing a P.A.R. boxcar averager are described by Badea and Georghiou.[13]

ONE PHOTON PER EXCITING PULSE. The second method of detection used in pulse fluorometry is exemplified by instruments based on the monophoton counting procedure. This is an interval timing method. Instruments based on the monophoton counting method are available commercially[17,18] and numerous variations and improvements have been described in the literature.[19-22]

A schematic outline of a monophoton counting decay fluorometer is shown in Fig. 5. Light from the nanosecond flash lamp passes through an excitation filter or monochromator and is used to excite the sample. A po-

[17] Ortec, Inc., Oak Ridge, Tennessee.
[18] Photochemical Res. Assoc., Western Ontario, Canada.
[19] R. Schuyler and I. Isenberg, *Rev. Sci. Instrum.* **42**, 813 (1971).
[20] W. R. Ware, *in* "Fluorescence Techniques in Cell Biology" (A. A. Thaer and M. Sernetz, eds.), p. 15. Springer-Verlag, Berlin and New York, 1973.
[21] J. Yguerabide, *in* "Fluorescence Techniques in Cell Biology" (A. A. Thaer and M. Sernetz, eds.), p. 311. Springer-Verlag, Berlin and New York, 1973.
[22] B. Leskovar, C. C. Lo, P. Hartig, and K. Sauer, *Rev. Sci. Instrum.* **47**, 1113 (1976).

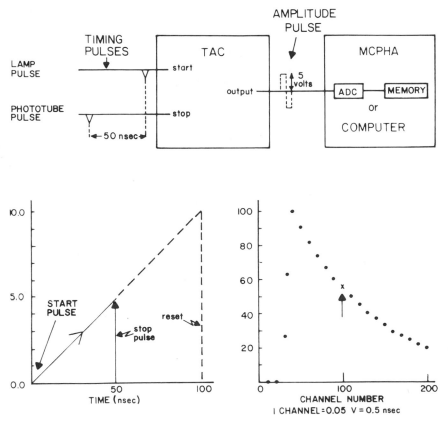

FIG. 6. Schematic illustration of the operation of a time to amplitude converter (TAC). A description is given in the text. ADC, analog to digital converter; MCPHA, Multichannel pulse height analyzer.

larizer may also be included in the light path. Fluorescence emission is usually observed at right angles to the excitation. The heart of a mono-photon counting instrument is the time to pulse height or time to ampli-tude converter (TAC). This is indicated in the dotted box in Fig. 5 and its function is shown in more detail in Fig. 6. This electronic device measures the time between the flash of the lamp and the arrival of a photon at the detector.

The start pulse for the TAC is obtained at the time of the lamp flash and initiates a voltage ramp linear with time on the nanosecond time scale. The start pulse has its origin either at an antenna placed near the spark gap or at a photomultiplier that directly views the lamp flash. The start pulses are shaped by a discriminator and counted before going to the TAC.

The voltage ramp is halted by a stop pulse. This signal has its origin at the photomultiplier which detects the photon emitted from the sample. The pulse is amplified and shaped by a discriminator before entering the TAC. This device now holds a signal whose amplitude is proportional to the time between the lamp flash and the photon emission. This information can be retained for several microseconds and then transferred through an analog to digital converter (ADC) to the memory of a multi-channel pulse height analyzer (MCPHA) or directly to a digital computer.

Typical operation of a TAC is illustrated in Fig. 6. The time between the lamp flash and a photomultiplier pulse is 50 nsec. This TAC gives a 10 V bipolar output pulse for 100 nsec. Thus in this case a 5 V pulse is passed through the ADC to the memory of the MPHA. A single count is added to the channel number proportional to the pulse amplitude (5 V in this case). The channel number is also proportional to the arrival time of the photon. In the present example, one channel equals 0.05 V which equals 0.5 nsec. The count is thus added to channel 100. The MPHA includes an oscillo-scope display unit which gives a continuous display of counts versus channel number. This represents the fluorescence decay curve or if a scattering solution is placed in the sample compartment, the lamp flash time profile.

This procedure thus involves recording of a time interval and conversion to an analog signal followed by conversion of the analog signal back to digital form. Since only the first of these events must take place on the nanosecond time scale less demands for "fast electronics" exist than is the case with other pulse methods.

As will be described below, fluorescence decay data obtained by pulse methods is usually distorted by convolution with a nondelta pulse lamp profile. For this reason, it is necessary to obtain a lamp flash profile along with each decay curve. This is usually done by measuring the scatter of the lamp flash with a colloidal material such as Ludox.[23] The desired impulse response $[F(t)]$ which describes the true kinetics of the fluorescence decay must then be extracted from the *experimental* convolved decay curve $R(t)$ (for *response* function) and $L(t)$ the experimental lamp flash profile.

If a timing drift occurs between the time that the lamp profile and the decay curve are obtained, adequate analysis of the data will be impossible. Hazan *et al.*[16] described an alternative procedure for overcoming this problem. Collection of data is alternated between a cuvette containing a scattering suspension and a cuvette containing the fluorescence sample. In this way $R(t)$ and $L(t)$ are collected during the same time period and errors due to drift tend to average out. Easter *et al.*[24] use a similar

[23] Collidal Silica, IBD-1019-69, trade name Ludox, Dupont, Wilmington, Delaware.
[24] J. H. Easter, R. P. DeToma, and L. Brand, *Biophys. J.* **16,** 571 (1976).

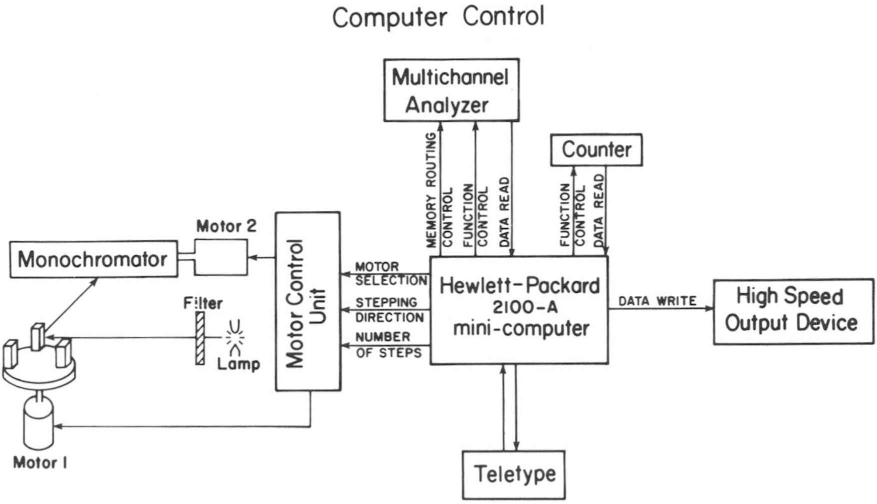

Computer Control

FIG. 7. Schematic diagram of the computer control over a monophoton counting decay fluorometer. The procedure used for the alternation of three samples is illustrated.

method which is shown in Fig. 7. Three cuvette holders are mounted on a turntable so that any one of three cuvettes can be positioned in the light path. This is accomplished by a stepper motor under control of a digital minicomputer. Light from a thyratron-gated flash lamp is focused on the sample with a lens and is passed through an interference filter to select the desired exciting wavelength. The cuvette turntable and the emission monochromator are driven by Slo-Syn stepper-motors models HS-50L which have 200 steps per 360° revolution. The motors are driven by Slo-Syn translators No. ST-1800BK. The translators are activated by a signal from a relay register in the minicomputer. Thus by means of the relay register output of the computer, a software program controls which motor is to be stepped (turntable or monochromator), the direction of the step, and the number of steps. Software control with the minicomputer thus enables the operator to alternate cuvettes with specified dwell times on each cuvette. Decay data can be collected at any desired emission wavelength.

Analysis of Fluorescence Decay

Procedures for data analysis are as important in nanosecond fluorometry as the instrumentation itself. This is particularly true in the case of pulse fluorometry. As was discussed above, the fluorescence is initiated by a short flash of light. The true fluorescence decay, $F(t)$ is the transient that follows excitation of the solution by a "delta" pulse i.e., an infinitely short pulse of light. $F(t)$, the true fluorescence decay, is also referred to as the impulse response. Since a delta pulse is infinitely narrow, excitation in this way would precisely define the time origin from which the decay is to be measured. In practice most light pulses used for excitation have a finite width which does not allow a clear definition of the true time zero. The experimentally obtained fluorescence decay will be distorted by convolution with the lamp flash profile. This artifact is described by the convolution integral

$$R(t) = \int_0^t L(t') F(t - t') \, dt' \tag{5}$$

where $F(t)$ is the true fluorescence decay, $L(t)$ is the time profile of the excitation light pulse, and $R(t)$ is the response that will actually be obtained from the instrument followed by the light pulse $L(t)$. In a more qualitative sense the convolution integral can be described as the summation of the true fluorescence decays that would be obtained following excitation by an infinitesimal pulse of width Δt, centered about t_i. t_i covers in discrete steps the range of $L(t)$ the excitation profile. Each of these narrow pulses at t_i induces a fluorescence response at time t given by

$$R_i(t) = L(t_i) \, \Delta t F(t - t_i) \tag{6}$$

The argument $(t - t_i)$ expresses the fact that the time course of fluorescence decay is measured relative to the inducing excitation which is centered at time t_i.

In the absence of nonlinear optical effects the superposition of the infinitesimal responses up to time t gives the total fluorescence response. Thus

$$R(t) = \sum_{t=0}^{t=t} L(t_i) \, F(t - t_i) \, \Delta t \tag{7}$$

which becomes the convolution integral in the limit $\Delta t \to 0$:

$$R(t) = \int_0^t L(t') F(t - t') \, dt' \tag{8}$$

Here t' is a dummy variable of integration. Making the change of variables $t' = t - u$, we have

$$R(t) = \int_0^t L(t - u) F(u) \, du \tag{9}$$

Thus a nanosecond pulse fluorometer gives $R(t)$ and $L(t)$, and our task is to extract $F(u) \equiv F(t)$ the true impulse fluorescence response which may then be compared to theoretical predictions regarding kinetic mechanisms. A large variety of numerical procedures have been devised and tested for carrying out this deconvolution procedure.[25-30]

It is convenient to divide these deconvolution procedures into two conceptual categories. First, those working in the real time domain, and, second, those working in a transformed domain. In the latter case, a linear operator is applied to the experimental data prior to performing the deconvolution.

Methods Working in the Real Time Domain. These are usually curve fitting techniques. They make use of a numerical convolution rather than a deconvolution procedure. A physically plausible analytical expression for the impulse response is assumed. In most cases a sum of exponential terms is assumed for $F(t)$:

$$F(t) = \sum_{i=1}^{n} \alpha_i e^{-t/\tau} \tag{10}$$

Thus the fluorescence decay is defined in terms of several decay constants, τ_i, and a set of amplitudes or preexponential terms α_i. In the fitting procedure suitable guesses are selected for the α_i's and τ_i's, and $F(t)$ is then numerically convoluted with the experimental $L(t)$. The theoretical $R(t)$ thus obtained is compared with the experimental $R(t)$. The free parameters in $F(t)$ are then adjusted until the best fit between the calculated and experimental $R(t)$ is obtained. Ware *et al.*[26] have described the use of a linear least squares procedure. In this method the decay times are set constant and the equations are solved for the best values of the preexponential terms. Grinvald and Steinberg[27,31] have used the nonlinear least squares method of Marquardt (see Bevington[32]) to solve for both the decay constants and amplitudes describing fluorescence decay data. This technique has also been found to be valuable in our own laboratory.[24]

The Marquardt algorithm of nonlinear least squares employs a dual search along the χ^2 hypersurface defined by

$$\chi^2 = \sum_{i=1}^{n} W_i[R(t)^{\text{expt}} - R(t)^{\text{calc}}]^2 \tag{11}$$

[25] I. Isenberg, R. D. Dyson, and R. Hanson, *Biophys. J.* **13,** 1090 (1973).
[26] W. Ware, L. J. Doemeny, and T. L. Nemzek, *J. Phys. Chem.* **77,** 2038 (1973).
[27] A. Grinvald and I. Z. Steinberg, *Anal. Biochem.* **59,** 583 (1974).
[28] A. Gafni, R. L. Modlin, and L. Brand, *Biophys. J.* **15,** 263 (1975).
[29] B. Valeur and J. Moírez, *J. Chim. Phys. Phys. Chim. Biol.* **70,** 500 (1973).
[30] J. N. Demas and A. W. Adamson, *J. Phys. Chem.* **75,** 2463 (1971).
[31] A. Grinvald, *Anal. Biochem.* **75,** 260 (1976).
[32] P. R. Bevington, "Data Reduction and Error Analysis for the Physical Sciences." McGraw-Hill, New York, 1969.

for a minimal value of χ^2. χ^2 is considered as a function of the parameter increments that are defined with respect to initial parameter guesses α_i^0 and τ_i^0, which must be specified at the start of each search. There are two searching paths, one for parameter guesses which establish χ^2 far removed from the minimum and the other for χ^2 values near the minimum. The two searching paths are apportioned automatically within the algorithm. ω_i is a statistical weighting factor defined by the principles of least squares and for the case of photon counting error may be approximated by

$$\omega_i = 1/R(t) \tag{12}$$

Grinvald and Steinberg[27] have emphasized the importance of using correct weighting factors in least squares analysis.

In our own laboratory, we have found it valuable to carry out the nonlinear least squares analyses with an interactive computer. Following Grinvald and Steinberg,[27] the "goodness of fit" is judged by (a) visual inspection of the superimposed $R(t)^{calc}$ and $R(t)^{expt}$. It is especially useful to inspect the rise and fall of the decay curves; (b) evaluation of χ^2; (c)

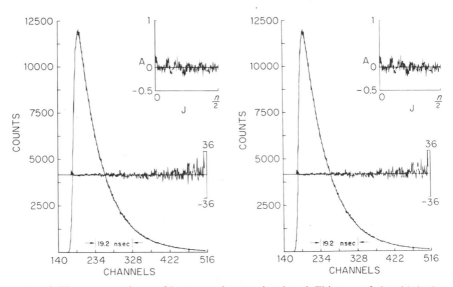

FIG. 8. Fluorescence decay of 9-cyanoanthracene in ethanol. This type of plot aids in determining the "goodness of fit" of fluorescence decay data. Shown are the experimental decay curve, the best theoretical data convolved with the lamp flash. The residuals between the theoretical and experimental decay curves are shown in the center and the autocorrelation of the residuals are shown in the inset at the upper right. The results of an analysis in terms of a single exponential decay law are shown on the left. A decay time of 11.8 nsec was obtained. An analysis in terms of a double exponential decay law is shown on the right.

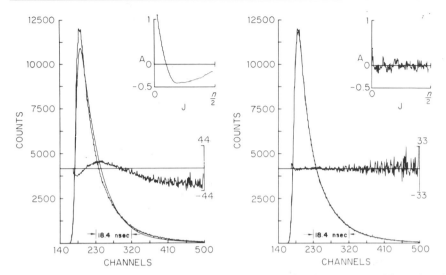

FIG. 9. Fluorescence decay of an equimolar mixture of anthracene ($\tau = 4.2$ nsec) and 9-cyanoanthracene ($\tau = 11.8$ nsec) in ethanol. The meaning of different portions of the figure is similar to that indicated in the legend to Fig. 8. The best fit to a single exponential decay law is shown on the left. The right shows the fit to a double exponential decay law. Decay times of 4.2 nsec and 11.8 nsec were obtained.

evaluation of the residual $[R^{\mathrm{calc}}(t) - R^{\mathrm{expt}}(t)]$; and (d) evaluation of the autocorrelation function of the residuals.

Typical results of an analysis of data showing both single and double exponential decay behavior are shown in Figs. 8 and 9. Figure 8 shows the experimental fluorescence decay, $R(t)^{\mathrm{expt}}$ of 9-cyanoanthracene dissolved in ethanol. The vertical axis indicates the number of counts accumulated and the X axis is in channels since time zero is ill-defined in convolved data. There are 0.2 nsec/channel. Figure 8 (left) the best fit $R(t)^{\mathrm{calc}}$ for a single exponential decay law superimposed on the experimental. The residuals between the theoretical and experimental data are shown in the center of the figure and the autocorrelation function of the residuals is shown in the inset at the upper right. The use of the autocorrelation function to betray a nonrandom distribution of the residuals was introduced for fluorescence decay data by Grinvald and Steinberg.[27] Figure 8 (right) shows the same data with an analysis in terms of a double exponential decay law. There is little or no improvement in any of the criteria for a good fit. This data is thus in accord with a single exponential decay law with a decay time of 11.8 nsec.

Figure 9 (left) shows the results of an analysis of the fluorescence decay of a mixture of 9-cyanoanthracene and anthracene in ethanol. In this case the best fit in terms of a single exponential decay law is clearly inadequate. The experimental and theoretical $R(t)$ show significant deviations. This systematic deviation is quite apparent in the residuals whose nonrandomness is indicated by the autocorrelation function. In contrast, Fig. 9 (right) indicates that a double exponential decay law is adequate to give a good fit to $R(t)$. Grinvald and Steinberg[27] have emphasized the use of standard compounds which show single exponential decay behavior to uncover systematic errors in nanosecond decay fluorometers. We also recommend their use in this way. The "real time" search methods are in actuality not procedures for deconvolution but rather iterative convolution methods.

Methods Making Use of Transformed Domains. In these numerical procedures the measured lamp flash $L(t)$ and the fluorescence response $R(t)$ are transformed to another domain by the application of a linear operator prior to further computations. The transformation can be carried out with operators that assume an infinite range for the time domain as is the case with the Laplace,[28,33] Fourier[34] and moment[35-39] operators. Alternatively, the transformation can assume a finite time range as is the case with the method of modulating functions[29] and the phase plane technique.[30]

There is an important difference between these two classes of transformations. Methods employing operators of the first class must contend with the difficulties imposed by the assumed infinite time range as compared with the finite time range over which experimental data is obtained. This problem does not exist with the second type of transformations.

Several of the transformation techniques aim at reducing the convolution integral to a system of simultaneous algebraic equations which are solved for the decay parameters appearing in the analytical expression for the fluorescence decay law.

As an example of the first class of transformations, the procedure for analysis of fluorescence decay curves by means of the *Laplace method*[28,32] will be described.

The convolution theorem states that if functions are related by a *convolution* product in the time domain, they will be related by a simple *algebraic* product in the Laplace domain. The basic computational procedure is as follows: The algorithm computes the Laplace transform of $L(t)$ and $R(t)$ and determines their ratio. This gives the Laplace transform of the

[33] W. P. Helman, *Int. J. Radiat. Phys. Chem.* **3**, 283 (1971).

[34] S. W. Provencher, *Biophys. J.* **16**, 27 (1976).

[35] I. Isenberg and R. D. Dyson, *Biophys. J.* **9**, 1337 (1969).

[36] I. Isenberg *in* "Biochemical Fluorescence Concepts" (R. Chen and H. Edelhoch, eds.), Vol. I, p. 43. Dekker, New York, 1975.

[37] I. Isenberg, *J. Chem. Phys.* **59**, 5696 (1973).

[38] I. Isenberg, *J. Chem. Phys.* **59**, 5708 (1973).

[39] Z. Bay, V. P. Henri and H. Kanner, *Phys. Rev.* **100**, 1197 (1955).

impulse response. Conversion from Laplace space to real time presents some numerical difficulties. Instead, the numerical Laplace of the impulse response is set equal to the Laplace transform of the analytical expression for the impulse response.

The Laplace transform $M(s)$ of a function $M(t)$ is defined:

$$M(s) = L[M(t)] = \int_0^\infty M(t)\, e^{-St}\, dt \qquad (S \geq 0) \qquad (13)$$

By applying the Laplace operator to the convolution integral, we obtain

$$L[R(t)] = L\left(\int_0^t L(t')\, F(t - t')\, dt'\right)$$
$$R(S) = L(S)\, F(S) \qquad (14)$$

Thus the true fluorescence decay in Laplace space is obtained by simply dividing the transforms of the response to that of the exciting pulse.

If the assumed decay law is a sum of exponential terms, it is easily shown that

$$L[F(t)] = L\left(\sum_{i=1}^n \alpha_i\, e^{\,t/\tau}\right) = \sum_{i=1}^n \frac{\alpha_i}{S + 1/\tau_i} \qquad (15)$$

S is the exponent in the Laplace transform defined by Eq. 13. Thus,

$$F(S) = \frac{L(S)}{R(S)} = \sum_{i=1}^n \frac{\alpha_i}{S + 1/\tau_i} \qquad (16)$$

A computer is used to evaluate $F(S)/L(S)$ for $2n$ values of S. A set of $2n$ simultaneous equations is obtained and solved for the n amplitudes and n decay constants. Procedures dealing with the finite limit of the data (the cutoff correction) and the choice of S value have been described[28] and will not be covered in detail here.

It must be emphasized that while the distortion in the fluorescence decay introduced by the finite width of the lamp flash is eliminated by the deconvolution procedure, if *actual* scattered light passes through the optics and contributes to the decay signal, it must be subtracted prior to analysis. As an alternative option a scatter term may be included in the Laplace algorithm. In the case of decay data which contains scattered light, the Laplace transform of the decay becomes

$$F(S) = L(S)\, R(S) + C\, L(S) \qquad (17)$$

with C being the relative contribution by scattered light. In this case, $2n + 1$ Laplace transforms must be computed, and $2n + 1$ simultaneous equations must be solved. The reliability of the numerical method decreases the more simultaneous equations there are. It is thus always desirable to measure the scatter contribution in an independent experiment and subtract it from $R(t)$ before analysis.

The *method of moments* belongs to the same class of techniques as the Laplace approach. It has been used for analyses of decay data for many years.[39,40] Isenberg has explored the application of this method to fluorescence decay studies.[34-37] He has investigated the statistics involved in these analyses and has shown that a contribution due to scattered light can be taken into account.

Since the method of moments was described in a previous article in this series,[41] it will not be described in detail here. Briefly, the approach is quite similar to that taken in the Laplace method.

The Kth moment transform of a function $F(u)$ is defined as

$$M^K[F(U)] = \int_0^\infty U^K F(U) \, dU \tag{18}$$

By applying the moment operator on both sides of the convolution integral the convolution product is transformed into a linear combination of moment transforms. Thus, we have

$$R(t) = \int_0^t L(t - t') F(t') \, dt' \tag{19}$$

$$M^K[R(t)] = M^K \left(\int_0^t L(t - t') F(t') \, dt' \right)$$

$$= \int_0^t M^K[L(t - t')] F(t') dt' \tag{20}$$

The last equality is possible due to the linear property of the moment operator. It can be shown that if the fluorescence decay [impulse response $= F(t)$] is represented as a sum of exponentials, the integral is equivalent to the following algebraic relation:

$$\frac{M^K[R(t)]}{1} = \sum_{s=1}^{K+1} G_s \frac{M^{(K+s-1)}[E(t)]}{(K + 1 - s)^1} \tag{21}$$

where G_s is combination of the assumed parameters for $F(t)$ given by

$$G_s = \sum_{n=1}^N \alpha_n \tau_n^{-1} \tag{22}$$

The linear set of algebraic equations obtained for different values of K can be solved for G_s's. A set of $2\,N\,G_s$'s completely characterizes a given

[40] Ph. Wahl and H. Lami, *Biochim. Biophys. Acta* **133**, 233 (1967).
[41] J. Yguerabide, Vol. 26, Part C, p. 498.

set of assumed decay parameters. In particular, the parameters could be extracted from the set $G_{d+1}, G_{d+2}, \ldots, G_{d+2N}$ where d, a positive integer, is called the order of moment index displacement.[36] Usually the set with $d = 0$ or $d = 1$ is used. The advantage of using the set with $d = 1$ is that it enables to correct for a variety of systematic instrumental errors like the time shifts and scattered light either completely or in a perturbation sense. It has been shown[42] that this moment index displacement of order one can completely correct for scattered light having a different wavelength than that of fluorescence, whereas it can correct only in a perturbation sense, i.e., to minimize an already small effect if the scattered light has the same wavelength as the fluorescence or if a time origin discrepancy exists between the measured curves $R(t)$ and $L(t)$ (zero time shift error). Moment index displacement will also reduce errors due to slow lamp drift during collection of an individual curve. The drawback of going to higher moments is that the collection time will have to be increased in order to calculate them with a satisfactory accuracy. This will increase errors associated with long-term time drifts.

As is the case with the Laplace method, the algebraic relations obtained by applying the moment operator to the convolution integral are theoretically valid only if the time range is infinite. As the real data is always collected over a finite range, a cut-off procedure has been devised which by iteration yields a self-consistent set of parameters. It has been found useful to exponentially depress the raw data prior to analysis in order to reduce the number of iterations and improve the convergence. This is always possible because the convolution integral is invariant to the multiplication by a time exponential.

The problems associated with the iterative convergence are not present when an operator which does not assume an infinite time range for data acquisition is used in the transformation. As an example of this second class operators the *method of modulating functions* will be described in some detail. This method first proposed by Loeb and Cahen[43] for the determination of unknown coefficients of linear differential equations has been introduced to the fluorescence field by Valeur and Moirez.[29] The authors observed that by assuming a sum of exponentials for the fluorescence decay the convolution integral can be put in the following differential form:

$$R + \alpha_1\dot{R} + \alpha_2\ddot{R} + \cdots \alpha_N R^{(N)} = \beta_1 L + \beta_2 \dot{L} + \cdots \beta_N L^{(N)} \qquad (23)$$

where (N) stands for the Nth derivative with respect to time, α_i is a set of constant coefficients dependent only on the lifetimes, and β_i is a similar set depending both on the lifetimes and amplitudes.

[42] E. Small and I. Isenberg, *Biopolymers* **15**, 1093 (1976).
[43] J. Loeb and G. Cahen, *Automatisme* **8**, 479 (1963); *IEEE Trans. Autom. Control* **ac-10**, 359 (1965).

The method aims at eliminating the first and all higher-order time derivatives of both $R(t)$ and $L(t)$ from Eq. (23) by multiplying it with a set of "modulating functions" followed by integration by parts. The "modulating functions" are the set of functions which together with their time derivatives to the Nth order are zero at the limits of the collection time range $(0, T)$. With $\rho(t)$ representing any such function Eq. (23) becomes

$$\alpha_1 \int_0^T R\dot{\rho}\, dt - \alpha_2 \int_0^T R\ddot{\rho}\, dt + \cdots \cdot (-1)^{N-1} \alpha_N \int_0^T R\rho^{(N)}\, dt$$

$$+ \beta_1 \int_0^T L\rho\, dt - \beta_2 \int_0^T L\dot{\rho}\, dt$$

$$+ \cdots \cdot (-1)^{N-1}\beta_N \int_0^T L\rho^{(N-1)}\, dt = - \int_0^T R\rho\, dt$$

Thus the time derivatives of $R(t)$ and $L(t)$ have been replaced by the time derivatives of an *analytical* function $\rho(t)$. By using $2N$ such "modulating function" a system of $2N$ linear equations is obtained from which α_i and β_i are obtained. From the set α_i the lifetimes are extracted and their values introduced in the set β_i to obtain the amplitudes. It is important to note that the integration has been performed only from 0 to T the range in which both $L(t)$ and $R(t)$ were numerically defined. There is, in principle, no error due to truncation.

The gist of the method consists of efficiently "modulating" the shape of the experimental curves $R(t)$ and $L(t)$. Therefore the choice of the "modulating functions" is somewhat dependent on the shape of the experimental curves. Valeur and Moirez used functions of the type $\rho(t) = t^n(T - t)^p$ for which the exponents n and p were arbitrary, but greater than $N - 1$. For further details regarding their choice and the use of the alternate method of truncated moments for the calculation of the amplitudes the reader is referred to the original paper.[29]

Comparison of Deconvolution Techniques. McKinnon et al.[44] have carried out a detailed evaluation of the various procedures for deconvolution of fluorescence decay data. They found that the iterative convolution procedure gave good recovery of decay parameters under a variety of conditions. The reader is referred to the original paper which includes an excellent discussion of the problems associated with the analysis of fluorescence decay data.

Time-Resolved Emission Spectra

Nanosecond time-resolved emission spectra (TRES) are fluorescence spectra obtained at discrete times during the fluorescence decay. For example with reference to the decay curve shown in Fig. 10, fluorescence emission spectra might be obtained during the time window indicated by

[44] A. E. McKinnon, A. G. Szabo and D. R. Miller, *J. Phys. Chem.* **81**, 1564 (1977).

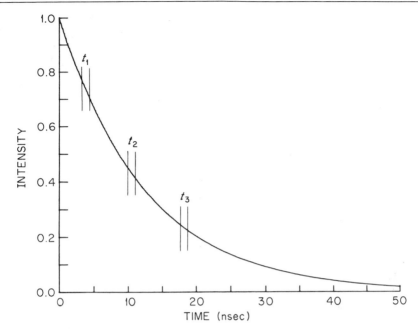

FIG. 10. Fluorescence decay curve with individual time windows.

t_1 or t_2 or t. These spectra will be defined in terms of both *spectral* and *time* resolution. The complete data matrix available may be described by a three dimensional surface $I(\lambda, t)$, representing the fluorescence intensity at all wavelengths and times during the fluorescence decay. Figure 11 shows an example of such a surface. It represents the decay of 2,6-toluidinonaphthalene (2,6-TNS) adsorbed to egg lecithin vesicles. In this case, there is a shift of the fluorescence emission to lower energies as a function of decay time. This phenomenon has been attributed to excited state solvation interactions.[45] Nanosecond time-dependent spectral shifts have been investigated by Ware,[46,47], Egawa *et al.*,[48] and others. Spectral shifts of this type have been observed with fluorophores bound to proteins[49,50] and adsorbed to phospholipid vesicles.[51–53] Apparent nanosec-

[45] J. H. Easter, R. P. DeToma, and L. Brand, *Biophys. J.* **16**, 571 (1976).
[46] W. R. Ware, P. Chow, and S. K. Lee, *Chem. Phys. Lett.* **2**, 356 (1968).
[47] W. R. Ware, S. K. Lee, G. J. Brant, and P. P. Chow, *J. Chem. Phys.* **54**, 4729 (1971).
[48] K. Egawa, N. Nakashima, N. Mataga, and C. Yamanaka, *Bull. Chem. Soc. Jpn.* **44**, 3287 (1971).
[49] A. Gafni, R. P. DeToma, R. E. Manrow, and L. Brand, *Biophys. J.* **17**, 155 (1977).
[50] L. Brand and J. R. Gohlke, *J. Biol. Chem.* **246**, 2317 (1971).
[51] R. P. DeToma, J. H. Easter, and L. Brand, *J. Am. Chem. Soc.* **98**, 5001 (1976).
[52] J. H. Easter, R. P. DeToma, and L. Brand, *Biochim. Biophys. Acta* **508**, 27 (1978).
[53] M. G. Badea, R. P. DeToma, and L. Brand, *in* "Biophysical Discussions: Fast Biochemical Reactions in Solutions, Membranes, and Cells" (V. A. Parsegian, ed.). The Rockefeller Univ. Press, Airlie House, Virginia, 1978.

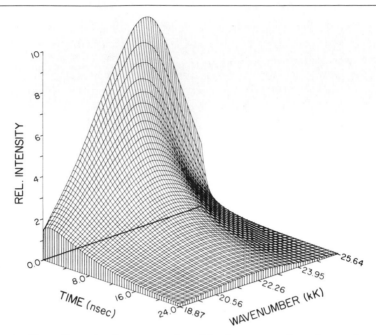

FIG. 11. Three-dimensional representation of the fluorescence intensity as a function of time and energy. This is for 2,6-p-toluidinonaphthalene (11 mM) adsorbed to egg L-α-lecithin vesicles (0.86 mM in lecithin) at 7°.

ond time-dependent spectral shifts may have their origin either in ground state heterogeneity or in excited state interactions. Both situations are of interest in biochemistry. While it is of interest to examine fluorescence decay curves as a function of emission wavelength, additional information can be obtained from the TRES.

Nanosecond pulse fluorometers can easily be used to generate time resolved emission spectra. Either of two instrumental approaches can be used. In the first, a "window" is set for a particular time after the lamp flash and with a desired time width. The light signal is scanned with an emission monochromator and time resolved spectra are directly produced. This can be done with either a sampling instrument or with a monophoton counting instrument. Thus, with reference to Figs. 5 and 6, a single channel analyzer (SCA) is inserted into the system between the TAC and the MCPHA. This electronic device (SCA) can be set to allow a pulse to pass only if it is within some specified voltage limits. Thus, in the example shown in Fig. 6 the SCA might be set to pass a pulse only if it has a voltage between 4.95 and 5.05 V. Since 0.05 V equals 0.5 nsec, the "time window" in this case would be 1 nsec. The MCPHA is used in a multichannel scaling (MCS) mode in this case. Channels are addressed sequentially in phase with the movement of the emission monochromator.

Each channel now represents a wavelength and the appropriate setting of the single analyzer enables one to obtain a TRES with the desired time and spectral resolution.

The major drawback to a procedure of this type is that the time resolved emission spectra obtained will be distorted by convolution errors.[45] In addition time-zero is ill defined. While there is no way of correcting for convolution artifacts when time resolved spectra are obtained as described above, an alternative procedure is available which will be described below.

The windowing technique with the use of a SCA as described above also has applications of value in steady-state fluorescence instrumentation. For example it can be used to reduce errors due to light scattering in fluorescence emission spectroscopy. Once again the MCPHA is used in the MCS mode. In our laboratory we adjust the scaling speed and the monochromator to give 1 nm/channel. The window on the SCA is now set to eliminate pulses arriving during the first 2nsec after the flash but to pass all later pulses. Thus photons originating in Rayleigh scattering are eliminated while delayed fluorescence photons are recorded. As an example, Fig. 12 shows the fluorescence emission spectrum of 9-

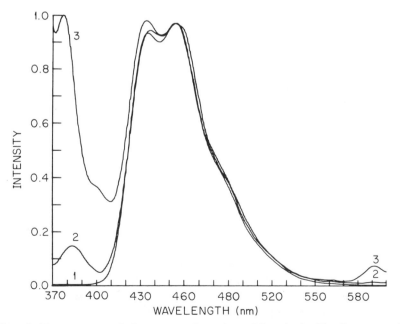

FIG. 12. Fluorescence emission spectra of 9-aminoacridine obtained by photon counting. (1) 9-Aminoacridine in water; (2) identical to (3), except the single channel analyzer is set so as to eliminate photons arriving within 3 nsec after the lamp flash; (3) 9-aminoacridine and Ludox to give rise to scattered light.

aminoacridine obtained under conditions of rather poor spectral resolution. Excitation was with a nitrogen flash lamp (see Fig. 1 for spectral emission) without the use of an excitation filter or monochromator. Ludox was included in the cuvette to enhance light scattering. Indeed a very significant scatter band is seen between 360 and 410 nm, and there is even a significant distortion at the main emission band of the fluorophore. The second-order scatter bands make a very significant contribution above 540 nm where the fluoresence intensity is low. When the time window is set to pass only photons arriving after 2–3 nsec, the scatter contribution is greatly reduced. The technique of time resolution can also be used to accentuate the contribution due to emission from one of a pair of fluorophores in a mixture. This assumes the two fluorophores have different decay times. This approach may also have value in Raman spectroscopy where errors due to contribution by fluorescence to the scatter may be significant.

It was indicated above that TRES obtained by the SCA method are distorted by convolution errors. As an alternative procedure for obtaining TRES, fluorescence decay curves are obtained at wavelength intervals over the spectral region of interest. This data is then deconvolved by the procedures already described.

Since a large number of decay curves must be obtained, these experiments are conveniently carried out under computer control as shown in Fig. 7. Lamp flash profiles and decay curves are collected making use of the alternation technique already described.

With reference to Fig. 7, the sequence of computer directives during the course of an experiment can be divided into three phases: collection, search, and output. In the collection phase the cuvette turntable is positioned with the fluoresence sample in the optical path, the monochromator is set to a previously specified wavelength, the MCPHA is started in the accumulation mode, and totalization of lamp flashes with the counter is initiated. After a designated decay dwell time (typically 5 to 15 min) data collection is stopped and the information in the MCPHA and the counter is transferred to computer memory. The cuvette turntable is then positioned so that a scattering is in the optical path, the emission monochromator is set to the exciting wavelength and data collection is initiated. After the specified collection time for the lamp profile, accumulation is stopped and the data is transferred to computer memory. In phase two of the computer operation the data representing the decay and the lamp flash are searched to determine whether specified peak and/or total counts have been obtained. If data collection is not complete, control is returned to the first phase of operation. When data collection is complete, the decay curve and the lamp profile are transferred to the output device (a disc, magnetic tape, or paper tape) together with the wavelengths, curve index number, peak counts, and counter monitor. The computer then initiates collection of the data for the next wavelength. The entire decay data matrix required for a TRES is now available.

As a first step each decay curve is deconvolved using the method of nonlinear least squares as an empirical deconvolution procedure. A sum of up to five exponential terms is usually sufficient to give a good empirical fit to any fluorescence decay curve. In this way reliable impulse response decay curves $F(t)$ are produced in spite of the fact that the α's and τ's obtained may or may not have direct physical meaning. It has been found with the use of computer simulation that nonlinear least squares with a multiexponential fitting function provides a powerful deconvolution method for fluorescence decay data.[45]

The impulse response curves $F(\lambda, t)$ are normalized to a steady-state emission spectrum obtained on the same instrument. The computer now contains the data matrix required to generate the time resolved spectra at any time t. The logical sequence of computer operations in constructing a TRES is shown in Fig. 13. Figure 13A indicates convolved decay curves obtained at three representative wavelengths. These curves have been collected to about the same number of counts at the peak. Figure 13B shows the same data deconvolved by the method of nonlinear least squares. Figure 13C shows that the three decay curves have been normalized to the corrected steady-state emission spectrum obtained using the same instrument. It is worth mentioning that in this example the initial rise seen in $F(t)$ at the longest wavelength provides evidence that an excited state reaction is taking place. Figure 13D shows TRES generated at the three times (0.5, 2, and 11 nsec) indicated by the arrows in Fig. 13C. A numerical derivative procedure can now be used to find the maximum of the decay at any time and the system can be characterized by a plot of $I_{max}(t)$. These curves can in turn be analyzed in terms of a multiexponential decay law.

Time-Resolved Emission Anisotropy

Fluorescence emitted by a sample is usually found to be polarized at least at early time during the fluorescence decay. This phenomenon can be understood in terms of the existence of fixed preferential directions **a** and **e** in the molecule along whose directions the transition moments for absorption and emission of light acquire their *maximum* value. Fluorescence polarization is observed even if the sample consists of a random distribution of fluorophores. This is because the processes of absorption and emission by individual molecules are always anisotropic. Thus the

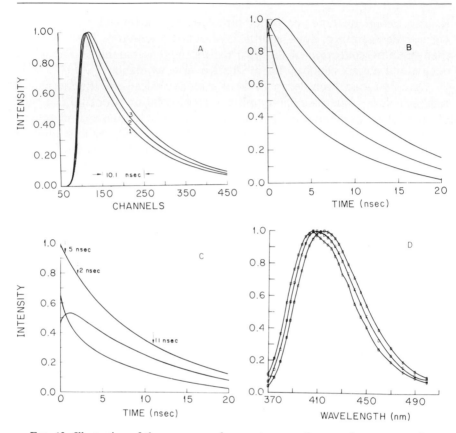

FIG. 13. Illustration of the sequence of computer operations used to generate deconvolved TRES. 2-Anilinonaphthalene (1 μM) in DML (dimyristogllecithin) vesicles. The lipid to dye ratio was about 700. The temperature was 37°. Excitation was at 315 nm and the emission wavelength was (1) 394, (2) 409 and (3) 445 nm. (A) shows the convolved decay curves collected to constant peak height. (B) shows the decay curves scaled according to the steady-state spectra. (C) shows the decay curves deconvolved to obtain the impulse response curves. (D) the intensity obtained from the impulse response at a particular time plotted against wavelength. The deconvolved TRES are normalized at the peak.

probability of absorption of polarized light having the electric vector **E** is proportional to the square of its component along the direction **a**. The emission from an individual fluorophore is always polarized with the electric vector in a plane defined by **e** and the direction of detection. Fluorescence polarization from a sample is observed even if the excitation is performed with unpolarized (natural) light. This is explained by the fact that light by its nature is anisotropic, since the possible orientations of the exciting electric vector **E** are always confined in a plane perpendicular to the direction of propagation. The simple model of linear dipole oscillators is based on the quantum mechanical picture of electric dipole transitions and is valid only in the limit of strict monochromaticity. It has been able to explain, however, at least in a phenomenological sense, all the known facts about polarization of broad band luminescence.

The interaction of light with a random array of anisotropic fluorophores results in the photoselection of an anisotropic ensemble of excited molecules from the original isotropic system. This photo-selected subensemble will have a nonrandom mutual orientation of their absorption and emission dipoles. The anisotropic distribution will be reflected macroscopically by a measurable polarization. Any process that reduces this nonrandomness will have a depolarizing effect.

Polarized light exhibits a variation in its intensity, when viewed through a polarizer whose transmission axis orientation in space is changed. Let us call $\mathbf{v}$ (vertical) the direction perpendicular to the plane of excitation detection and $\mathbf{H}$ (horizontal) *any* direction in this plane perpendicular to either the line of excitation or detection (as the case may be). The fluorescence polarization depends on the polarization of the exciting light and on the angle between direction of excitation and detection.

The parameters usually used to quantitate this dependence are the polarization (p) introduced by Perrin[54] and the emission anisotropy (r) introduced by Jablonski.[55] In time-dependent notation, they are defined by

$$p(t) = \frac{I_V(t) - I_H(t)}{I_V(t) + I_H(t)} = \frac{D(t)}{I_V(t) + I_H(t)} \tag{24}$$

and

$$r(t) = \frac{I_V(t) - I_H(t)}{I(t)} = \frac{D(t)}{S(t)} \tag{25}$$

where $I_V(t)$ and $I_H(t)$ are the fluorescence intensities detected through a polarizer whose transmission axis is aligned perpendicular and parallel to the excitation detection plane, respectively; $I(t)$ is a locally defined measure of the *total* emission intensity also called the sum function $S(t)$ and $D(t)$ is the difference function. Because it is defined in terms of the total emission, $r(t)$ has been found more convenient in theoretical calculations and is nowadays used preferentially. For later reference some of the expressions for $r(t)$ and the corresponding relationships for $p(t)$ are given below. For the usual 90° geometry of excitation detection we have the following.[56]

For linearly polarized excitation with $\mathbf{E} \parallel \mathbf{V}$

[54] F. Perrin, *J. Phys. Radium* [6] **7**, 390 (1926).
[55] A. Jablonski, *Acta Physiol. Pol.* **16**, 471 (1957).
[56] A. Jablonski, *Bull. Acad. Pol. Sci., Ser. Sci. Math., Astron. Phys.* **8**, 259 (1960).

$$r(t) = \frac{I_V(t) - I_H(t)}{I_V(t) + 2I_H(t)} \tag{26}$$

For natural light excitation, direction of $\mathbf{E}$ not defined

$$r_n(t) = \frac{I_V(t) - I_H(t)}{I_H(t) + 2I_V(t)} \tag{27}$$

$$r(t) = \frac{2p(t)}{3 - p(t)} \qquad p(t) = \frac{3r(t)}{2 + r(t)} \tag{28}$$

$$r_n(t) = \frac{r(t)}{2} \qquad p_n(t) = \frac{p(t)}{2 - p(t)} \tag{29}$$

$$r(t) = 2r_n(t) \qquad p(t) = \frac{2p_n(t)}{1 - p_n(t)} \tag{30}$$

$$I_V(t) = \tfrac{1}{3}S(t)[1 + 2r(t)] \quad I_H(t) = \tfrac{1}{3}S(t)[1 - r(t)] \tag{31}$$

For an isotropic sample the ranges of variations for r and p, respectively, are given by

$$-0.2 \le r \le +0.4 \qquad -0.33 \le p \le +0.5 \tag{32}$$

The total anisotropy due to i fluorophores characterized by r_i and contributing S_i to the total fluorescence is given by

$$r(t) = \frac{\sum_i r_i(t)S_i(t)}{\sum_i S_i(t)} \tag{33}$$

Steady state anisotropy is given by

$$\langle r \rangle = \frac{\displaystyle\int_0^\infty I_V(t)\,dt - \int_0^\infty I_H(t)}{\displaystyle\int_0^\infty I_V(t)\,dt + 2\int_0^\infty I_H(t)\,dt} \tag{34}$$

For an isotropic rotator whose fluorescence emission decays monoexponentially, i.e., $r(t) = \beta e^{-t/\phi}$ and $S(t) = \alpha e^{-t/\tau}$, we have

$$\begin{aligned} I_V(t) &= \tfrac{1}{3}\alpha(1 + 2e^{-t/\phi})e^{-t/\tau} \\ I_H(t) &= \tfrac{1}{3}\alpha(1 - e^{-t/\phi})e^{-t/\tau} \end{aligned} \tag{35}$$

In this case the expression for r becomes

$$\langle r \rangle = \beta\frac{\phi}{\tau + \phi} \qquad \frac{1}{\langle r \rangle} = \frac{1}{\beta}\left(1 + \frac{\tau}{\phi}\right) \tag{36}$$

This last expression is called the Perrin law. Perrin determined[54] that $\phi = \eta V / KT$, where η is the viscosity of the solvent, T its absolute temperature, V is the molecular volume, and K the Boltzman constant. Thus this law, valid only under the conditions stated above, becomes

$$\frac{1}{\langle r \rangle} = \frac{1}{r_0} \left(1 + \frac{KT\tau}{\eta V} \right) \tag{37}$$

where β, i.e., $r(t)$ at $t = 0$, has been called r_0.

The expressions for the denominator in Eqs. (26) and (27) are based on the symmetry induced in the photo-selected system by the polarization characteristics of the exciting light. Note that in Eqs. (26)–(35) the time dependence has been made explicit. At each particular time $I_V(t)$ and $I_H(t)$ will be, respectively, proportional to the number of emitted photons having these particular directions of polarization. As shown below, under analysis of time-dependent emission anisotropy, each of these two curves contains all the necessary information for determining the decay law of the emission anisotropy. The only prerequisite is knowledge of the decay of total emission $S(t)$ which, in principle, can be determined separately. However, in all but the simplest case of the isotropic rotator, the number of parameters to be determined is such as to make the cross-correlation of the analyses for either $I_V(t)$ and $I_H(t)$ or for any combination of them, e.g., $D(t)$ and $S(t)$ a practical necessity. Therefore, the anisotropy measurements must start with the collection of both curves.

The time-dependence that will be observed in $I_V(t)$ and $I_H(t)$ depends primarily on three processes.

a. The decay of the fluorescence emission.

b. Brownian rotation of the emission dipoles **e** randomizing the photo-selected subensemble; also intrinsic rotation of **e** within the molecule which could be due to excited state reactions.

c. Energy migration from the photo-selected subensemble to the nonselected molecules followed by emission from the latter; effectively this transfers the initial nonrandomness out of the photoselected subensemble and "dilutes" it in the whole.

Process (a) is characterized by the decay constants of the fluorophore. Its rate can be affected by the nature of the environment, i.e., solvent, temperature, viscosity, extent of solvation, and presence of foreign quenching substances. Brownian rotation (b) is mainly dependent on temperature, viscosity and the extent of solvation. Finally, the rate of energy migration is mainly controlled by the concentration of the fluorescent molecules and their relative orientations. While the rate of process (a) influences in an equal measure the rates of decay of both $I_V(t)$ and $I_H(t)$, the rates of processes (b) and (c) influence $I_V(t)$ and $I_H(t)$ in a differential manner and are primarily responsible for the time dependence of the emission anisotropy. If processes (b) and (c) are absent, the system will exhibit an emission anisotropy constant in time. This is obtained, for instance in dilute, highly viscous solutions or when the fluorophores have their motion restricted by bilayer membranes containing them.[57-60] The constancy in the last case is only relative and is due to the large discrepancy between the long time necessary for the whole membrane to rotate and the short nanosecond time window during which the fluorescence characteristics can be measured.

Time-dependent emission anisotropy measurements thus involve collection of *congruent* $I_V(t)$ and $I_H(t)$ fluorescence decay curves. These two curves are made congruent by contemporaneous collection or normalization.

a. Contemporaneous collection, i.e., during the same time interval, should ideally be used to eliminate any differential artifacts in the two decay curves due to long time drifts of the exciting flash and/or electronic circuitry. It can be performed in any of the following ways.

   i. Alternating the excitation polarizer orientation in front of a single fluorescent sample with emission detection through a fixed polarizer in front of a single photomultiplier[61] or the reverse optical situation.
   ii. Fixed vertical and horizontal excitation polarizers rigidly attached to two alternating identical fluorescent samples; emission detection through a fixed emission polarizer in front of a single photomultiplier.[57]
   iii. Fixed excitation polarizer in front of a single sample; emission detection through two fixed emission polarizers in front of two identical photomultipliers.

While procedure (iii) eliminates the necessity of mechanical alternation, it is our experience that matching identically the timing characteristics of two "factory identical" photomultipliers presents very difficult problems. In our laboratory procedure (ii) is used and some experimental details are presented below.

The alternation of the two identical fluorescent samples is performed as previously described (see section on instrumentation) by a computer-controlled turntable. Specifically, two identical cuvettes with double Polacoat filters rigidly attached are positioned alternately in the

[57] L. A. Chen, R. E. Dale, S. Roth, and L. Brand, *J. Biol. Chem.* **252**, 2163 (1977).
[58] K. Kinosita Jr., S. Mitaku, and A. Ikegami, *Biochim. Biophys. Acta* **393**, 10 (1975).
[59] R. A. Dale, L. A. Chen, and L. Brand, *J. Biol. Chem.* **252**, 7500 (1977).
[60] W. R. Veatch and L. Stryer, *J. Mol. Biol.* **117**, 1109 (1978).
[61] H. P. Tschanz and T. Binkert, *J. Phys. E.* **9**, 1131 (1976).

light path. The excitation light after passing through a quartz wedge scrambler plate is polarized vertically or horizontally by either one of the two Polacoat filters. The fluorescence is collected at 90° through a vertically oriented fixed double Polacoat filter. The emission wavelength can be selected by a computer-controlled monochromator. The third cuvette on the turntable contains the scattering Ludox solution. It is also provided with a rigidly attached excitation polarizer whose transmission axis is at 54.7° to the vertical (the so-called magic angle) in order to measure an excitation profile unbiased by polarization. Because the same photomultiplier views the scattered and the fluorescence light through the same fixed emission polarizer, the artifacts introduced by the detection train are, in principle, eliminated. The wavelength dependence of the photomultiplier transit time is dealt with as described under analysis.

(b) Normalization is necessary to eliminate any differential weighting in the two curves $I_V(t)$ and $I_H(t)$ of the artifacts associated with

   i. Intensity and frequency fluctuations of the exciting light
   ii. Anisotropy of the excitation train and/or detection train
   iii. Geometry of the excitation detection

The normalization process aims at equalizing the effect of these artifacts on the two decay curves. It consists of computing a total time-independent normalization factor by which one of the curves is scaled relative to the other. This factor is arrived at through a combination of hardware and software techniques.[62]

It is convenient to divide this total normalization factor into two partial ones to be determined in two separate steps: one to correct for the artifacts described under (i) and the other one to correct for those described under (ii) and (iii). First it should be noted that the artifacts (i) are practically eliminated when a contemporaneous collection procedure for the two curves $I_V(t)$ and $I_H(t)$ is used. In the case when a double check is desired or when such a capability is not available, one can monitor the number of TAC outputs n_V and n_H during the collection of the two curves. The partial normalization factor multiplying $I_H(t)$ is given by[63]:

$$N_1 = \frac{\displaystyle\int_0^T I_V(t)\,dt}{\displaystyle\int_0^T I_H(t)\,dt} \frac{n_H}{n_V} \tag{38}$$

Alternatively one may use a monitoring photomultiplier to view the exciting flash[13] or the fluorescence[61] directly and count the numbers n_V and n_H.

[62] C. E. Martin and D. C. Foyt, *Biochemistry* **17**, 3587 (1978).
[63] Ph. Wahl, *Biochim. Biophys. Acta* **175**, 55 (1969).

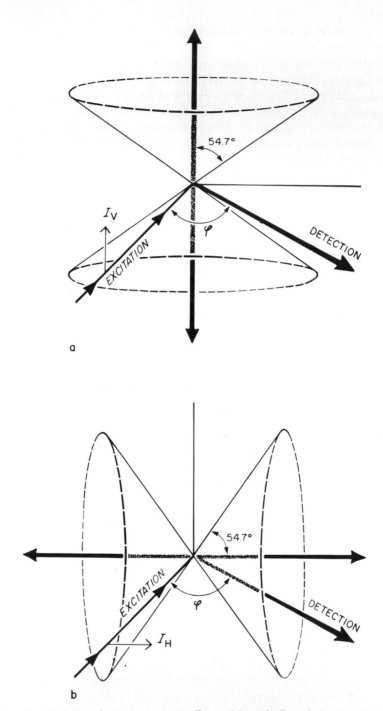

FIG. 14. "Magic" angle cones corresponding to (a) vertically polarized excitation, (b) horizontally polarized excitation; φ is the angle the detection direction makes with the excitation.

In the pulse sampling fluorometers this correction is automatically performed in the instantaneous ratio taking mode.[13]

The partial normalization factor for the artifacts under (ii) and (iii) can only be determined in a separate experiment during which the first factor is also computed. It is based on the observation that excitation with horizontally polarized light creates a symmetric axis in the photo-selected system (Fig. 14) coincident with the detection direction (in the usual 90° geometry). Thus, the intensity of the fluorescent light detected through a polarizer should be independent of the rotation of the latter in a plane perpendicular to the direction of detection. If this is not true, a normalization factor for $I_H(t)$, called G factor, is defined by

$$N_2 = G_{\text{for H excitation}} = \frac{\int_0^T I_V(t)\ dt}{\int_0^T I_H(t)\ dt} \neq 1 \tag{39}$$

Note that this normalization factor does not correct for the partial polarization that may be present in the lamp flash. For this reason a depolarizer, quarter wave scrambler plate is usually inserted in front of the excitation interference filter.

The new $I'_H(t)$ obtained by multiplying the experimentally determined $I_H(t)$ by $N = N_1 \times N_2$ is congruent with the $I_V(t)$. Thus we have

$$I'_H(t) = I_H(t)N_1N_2 \quad \text{and} \quad r(t) = \frac{I_V(t) - I'_H(t)}{I_V(t) + 2I'_H(t)} \tag{40}$$

Alternate Determinations of the Normalization Factor

As described above the normalization factor can be determined in two separate steps. The product of the partial factors gives the total one. Alternate determinations have been used[57,64] in which the total factor is obtained in one step. Their limitations will be pointed out below.

a. *Tail edge matching* is a procedure whereby the two curves $I_V(t)$ and $I_H(t)$ are scaled relative to one another until their tails coincide. It is based on the assumption that the randomization of the photoselected subensemble is practically achieved at a particular point in time. After that the decay of the emission alone is represented in both curves. This method could lead to substantial errors because the curves are matched in a region where the intensity is very low such that the random noise or any systematic errors could predominate. Additional error may be introduced

[64] G. R. Fleming, J. M. Morris, and G. W. Robinson, *Chem. Phys.* **17,** 91 (1976).

by starting the normalization too early in the decay when complete randomness is not yet achieved.

b. *Steady state normalization* is based on the separate determination of the steady state emission anisotropy. In this method the time integrated values of $I_V(t)$ and $I_H(t)$ are used to determine a *truncated* value of the steady state emission anisotropy with the normalization constant α as a variable parameter. By identifying this truncated value with the actual steady state emission anisotropy value measured separately with a continuous source of light, the value of the parameter is fixed. Thus

$$\langle r \rangle_{\text{truncated}} = \frac{\displaystyle\int_0^T I_V(t)\, dt - \alpha \int_0^T I_H(t)\, dt}{\displaystyle\int_0^T I_V(t)\, dt + 2\alpha \int_0^T I_H(t)\, dt} \tag{41}$$

and

$$\langle r \rangle = \frac{\displaystyle\int_0^\infty I_V(t)\, dt - \int_0^\infty I_H(t)\, dt}{\displaystyle\int_0^\infty I_V(t)\, dt + 2 \int_0^\infty I_H(t)\, dt} \tag{42}$$

If we identify $\langle r \rangle_{\text{truncated}}$ with $\langle r \rangle$ the value of the normalization constant, α, is given by

$$\alpha = \frac{1 - \langle r \rangle}{1 + 2\langle r \rangle} \frac{\displaystyle\int_0^T I_V(t)\, dt}{\displaystyle\int_0^T I_H(t)\, dt} \tag{43}$$

It is our experience that in order to be able to make this identification accurately, the decay of the two curves $I_V(t)$ and $I_H(t)$ should be followed and integrated over 2.5 to 3 decades. It is also preferable to measure $\langle r \rangle$ with the same instrument used for collecting the two curves. A continuous lamp may replace the flash lamp, and thus the artifacts associated with the geometry of collection are equally represented in $\langle r \rangle_{\text{truncated}}$ and $\langle r \rangle$.

Analysis of Time-Dependent Emission Anisotropy Measurements

In order to obtain valid results the convolution related artifacts should be eliminated from the two normalized decay curves $I_V(t)$ and $I_H(t)$. As described previously (see section on analysis of decay curves) this can be accomplished by a variety of procedures. Since either $I_V(t)$ or $I_H(t)$ contain all the desired information, they could be separately deconvolved and analyzed in terms of a multiexponential model. Correlation of these two analyses improves the accuracy.

Thus, in the simplest case of an isotropic rotator characterized by a single rotational relaxation time ϕ and emitting a monoexponentially decaying fluorescence with lifetime τ, theory predicts that

$$I_V(t) = \tfrac{1}{3}S(t)[1 + 2r(t)] = \tfrac{1}{3}\alpha e^{-t/\tau} + \tfrac{2}{3}\alpha\beta e^{-t(1/\tau + 1/\phi)} \qquad (44)$$

$$I_H(t) = \tfrac{1}{3}S(t)[1 - r(t)] = \tfrac{1}{3}\alpha e^{-t/\tau} - \tfrac{1}{3}\alpha\beta e^{-t(1/\tau + 1/\phi)} \qquad (45)$$

where $S = I_V + 2I_H$ is a measure of the total intensity emitted by the sample in all directions which in this case is given by $S(t) = \alpha e^{-t/\tau}$ and $r(t)$ is the decay of the emission anisotropy which for an isotropic rotator is given by $r(t) = \beta e^{-t/\phi}$. It is seen that separate deconvolution of $I_V(t)$ and $I_H(t)$ should yield the same two exponentials in both curves. The negative preexponential term in $I_H(t)$ should be two times larger and positive in $I_V(t)$. The other two corresponding preexponentials should be the same.

The correlation of the two separate analyses for $I_V(t)$ and $I_H(t)$ which, in principle, is redundant is found to be, in practice, indispensable for an adequate analysis. Moreover, the extension of this method to other analytical expressions for $r(t)$ (e.g., a two-exponential decay) although possible becomes considerably less reliable. Alternate methods of analysis become necessary. At the time of writing, the method of choice consists of analyzing the sum and difference curves[57,63] obtained from the normalized convoluted $I_V(t)$ and $I_H(t)$ curves and given by

$$S(t) = I_V(t) + 2I_H(t)$$
$$D(t) = I_V(t) - I_H(t)$$

and thus

$$r(t) = D(t)/S(t)$$

The curve $S(t)$ contains only terms describing the decay of the total emission independent of the model assumed for $r(t)$. It can also be obtained independently when the transmission axis of the excitation and emission polarizers are at an angle of 54.7°. The deconvolved impulse of $S(t)$ obtained as a sum of exponentials is multiplied by an assumed empirical form for $r(t)$ usually given by

$$r(t) = \sum \beta_i e^{-t/\phi} + c \qquad (46)$$

The significance of c can be understood through a limiting process: one of the empirical rotational relaxation time ϕ_i is large enough such that its exponential term is practically constant in the limited nanosecond time interval during which the collection of the emitted light is possible. The multiplication of $r(t)$ by the deconvolved $S(t)$ gives the new multiexponential model to be used in the deconvolution of $D(t)$. As a result the variable parameters of $r(t)$ β_i, ϕ_i, and c are fixed.

This method of analysis has been found to be more accurate than separate analysis of $I_V(t)$ or $I_H(t)$. The curve $S(t)$ is deconvolved with a multiexponential model containing only terms describing the decay of total emission, and $D(t)$ with a similar model which contains one term less than either polarized component [see Eqs. (24) and (25)].

From Nanosecond Fluorometry to Mechanisms

Magic Angle: Elimination of Polarization Bias. Before any attempt is made to extract information about the molecular characteristics of a fluorescent system, one should carefully consider the effects of the inherent photo-selection process. Fluorescence emission originates from a photo-selected ensemble of fluorophores which become excited following absorption. This set of molecules has an orientational distribution of emission dipoles different from that characterizing the bulk of molecules in the ground state. This is due to a greater probability of photon absorption as the angle between the molecular absorption dipole and the photon polarization vector approaches zero (probability of absorption is proportional to $\cos^2 \theta$). Thus, the excited molecules are selected from an isotropic random system by the anisotropic character of the excited light. It is important to note that photo-selection will occur in a three-dimensional system even if excitation is performed with natural light because the possible orientations of the exciting electric vector are confined to two dimensions.

The intensity and polarization characteristics emitted by such a nonrandom, photo-selected ensemble will be highly dependent upon the polarization of the exciting light, the wavelength-dependent orientation of molecular absorption and emission dipoles, and the spatial direction of fluorescence detection.

The dependence, however, is modulated by the relative magnitude of rotational relaxation and fluorescence decay times. If the former are much smaller than the latter, the photo-selection memory will be lost prior to light emission. As the ratio of the average rotational relocation time to the average fluorescence decay times increases the emission will reflect the photo-selection to an increasing extent. Thus any comparative series of fluorescence measurements (both steady state and/or transient) in which either this ratio or the parameters mentioned are altered will not be congruent.

A common example of such potentially erroneous comparative measurements is provided by the collisional quenching experiments in which the gradual addition of quencher affects the ratio by shortening the decay time of the excited state. Other examples have been discussed in the literature where it has been estimated that deviations up to 20% may be introduced into the experimental data.[65-67]

It is thus evident that in order to compare or to extract relevant information from fluorescence data it is necessary to eliminate this variability associated with the photo-selection process, namely, one should measure signals proportional to the total fluorescence which, by definition, is independent of the spatial directionality of fluorescence emission. It reflects only the number of the emitting fluorophores, their probability of emission being proportional to the product of these quantities.

By definition

$$I_{total} = Ix + Iy + Iz$$

where x, y, z are any three mutually orthozonal directions in space. It is desirable to measure a signal proportional to this sum along one single direction of detection. This direction is determined solely by the symmetry created in the emitting system through the photo-selection process. A linear polarized excitation will be absorbed preferentially by those molecules whose absorption dipole moments make smaller angles with the polarization direction. This being the only preferential direction in an isotropic system it becomes the symmetry axis. For instance, suppose that excitation is vertically polarized (let us call this axis x). Then all directions perpendicular to it will be equivalent (i.e., $Iy = Iz$) and I_{total} is given by

$$I_{total} = Ix + Iy + Iz = Ix + 2Iy$$

Thus, if one views the fluorescence signal at a particular angle to the symmetry axis which weights the perpendicular component twice as much as the one along the axis then a signal proportional to the total fluorescence will be detected. The weighting factors for the two components are the squared cosines of the respective angle they make the viewing (detection) direction. With θ the angle between the symmetry axis and the viewing direction, we have

$$I_{viewed} = Ix \cos^2 \theta + Iy \cos^2 (90° - \theta)$$

and I_{viewed} will be proportional to I_{total} if

$$\frac{\cos^2 (90° - \theta)}{\cos^2 \theta} = \tan^2 \theta = 2 \qquad \theta = 54.7°$$

The ensemble of viewing directions that satisfy this condition will form a cone as showed in Fig. 14 with half-angle 54.7° having the symmetry axis

[65] A. H. Kalantar, *J. Chem. Phys.* **48**, 4992 (1968).
[66] M. Shinitzky, *J. Chem. Phys.* **56**, 5979 (1972).
[67] K. D. Mielenz, E. D. Cehelnik, and R. L. McKenzie, *J. Chem. Phys.* **64**, 370 (1976).

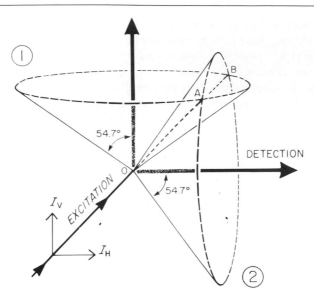

FIG. 15. "Magic" angle cones. Viewing the emission originating from O along the lines BO or AO eliminates the polarization related artifacts for any kind of excitation. See text for details.

parallel to the polarization direction of the exciting radiation and the apex in the center of the cuvette containing the fluorescent sample.

If the excitation is performed with natural light, i.e., all directions of polarization equally probable and confined to the plane perpendicular to the direction of excitation, then the transitive dipole moments of the pho-toselected molecules will be symmetrically distributed around this latter direction. The excitation direction becomes the symmetry axis of the photo-selected system. The "magic" angle 54.7° should be considered relative to it, and consequently it becomes the axis of the cone formed by the ensemble of viewing directions along which one measure a fluores-cence signal proportional to the total emission.

The general case in which one has a combination of the partially po-larized and unpolarized excitation could be approached in a similar manner. Any light signal can be thought of being composed of two po-larized beams, one vertical and the other horizontal coherently or inco-herently superimposed. The viewing directions along which the effect due to the vertically polarized beam is eliminated form the cone labeled 1 in Fig. 15; those corresponding to the horizontally polarized beam form the cone labeled 2 in Fig. 15. Along the lines of intersection of the two cones, OA and OB, one views a fluorescence signal free from any polarization re-lated artifacts.

A closer look at Fig. 15 reveals that the "right" (in the sense described above) directions of detection are quite difficult to adopt in the usual constraints of the 90° geometry built in many spectrophotometers.

One simple observation comes to the rescue. The transmission axis of the detection polarizer is equivalent to a viewing direction. Thus one only has to align the transmission axis of the detection polarizer parallel to the lateral surface of the proper cone determined by the excitation. An infinite number of combinations exist. In the confines of the 90° geometry three combinations are commonly used.

a. Vertically polarized excitation; detection with transmission axis of the polarizer at 54.7° to the vertical

b. Natural light excitation (scrambler plate); detection with transmission axis of the polarizer at 35.3° to the vertical

c. The optical reverse of the combinations (a) and (b)

Excited State Interactions: Limiting Cases. It is not our aim here to discuss in any detail the numerous excited state processes that are now well characterized both theoretically and experimentally.[68] Since the rationale behind experiments using nanosecond decay techniques is to obtain information regarding these interactions, it is worthwhile indicating some characteristics of the decay data that may be used to advantage.

A simple chromophore in dilute solution is expected to show single-exponential decay kinetics. This is probably more the exception than the rule. Double-exponential decay behavior is very commonly observed. This may have its origin in a mixture of two ground state species. This can sometimes be detected by a variation of the emission spectrum or the decay parameters with exciting wavelength. In some cases, the absorption and emission spectra of a chromophore may be quite similar but the fluorescence lifetimes may differ. This is certainly possible with chromophores associated with macromolecules and provides a powerful approach for investigating microheterogeneity in macromolecular systems.

Double-exponential decay kinetics may also have their origin in excited state reactions, even if only one chromophoric species exists in the ground state. A general scheme for a two-state system which provides for ground state equilibria and an excited state reaction which may be reversible or irreversible is shown below.

$$
\begin{array}{ccc}
A^* & \overset{k_{BA}}{\underset{k_{AB}[H^+]}{\rightleftharpoons}} & B^* \\
k_A \big\updownarrow & & \big\updownarrow k_B \\
A & \rightleftharpoons & B
\end{array}
$$

[68] J. B. Birks, ed., "Organic Photophysics," Vols. I and II. Wiley, New York, 1973.

This scheme is written for an excited state dissociation such as excited state deprotonation. In the reverse reaction a proton can add on, reforming A*. An example of a system showing this type of behavior is found with 2-naphthol. Depending on the pH, a significant ground state equilibrium may exist between A and B. Let us assume that the only significant ground state species is A. Excitation will lead to production of A*. In the case of 2-naphthol, the energetics of the excited state now favor formation of B*. (For 2-napththol $pK = 9.5$, while $pK* = 2.8$). At very high hydrogen ion concentration the reverse reaction will predominate and essentially no B* will be formed. Under these conditions the rate equation for A* is

$$-\frac{d(A)^*}{dt} = k_A(A^*)$$

where k_A includes terms both for radiation and quenching. At very acid pH the decay of 2-naphthol is well described by a monoexponential decay law.

At intermediate H^+ concentrations, the excited state reaction becomes kinetically significant and fluorescence due to A* *and* B* is observed, although only A exists to a significant extent in the ground state. At sufficiently low hydrogen ion concentrations (pH > 5 in the case of 2-naphthol) the back-reaction indicated by $k_{AB}[H^+]$ becomes negligible. Under these conditions the rate equation for A* is

$$dA^*/dt = (k_A + k_{BA})A^*$$

Once again the decay of A* is represented by a single exponential decay law. As is indicated below, the decay of B* is described by a double-exponential decay law under these conditions.

Thus the hallmark of a two-state reversible excited state reaction is that in general the decay of A* and B* are *both* characterized by a double-exponential decay law with the *same* decay constants for A* and B*.

The rate equations are

$$\frac{-d[A^*]}{dt} = (k_A + k_{BA})[A^*] - k_{AB}[H^+][B^*] \tag{47}$$

$$\frac{-d[B^*]}{dt} = (k_B + k_{AB}[H^+])[B^*] - k_{BA}[A^*] \tag{48}$$

It follows that the decay of A*, B* is

$$A^*(t) = \alpha_1 e^{-t/\tau_1} - \alpha_2 e^{t/\tau_2} \tag{49}$$

$$B^*(t) = \beta_1 e^{-t/\tau_1} - \beta_1 e^{-t/\tau_2} \tag{50}$$

Note that the identical decay constants appear in both cases and that the decay of B* has preexponential terms that are equal in magnitude but opposite in sign

$$\tau_1^{-1}, \tau_2^{-1} = \gamma_1, \gamma_2 = \tfrac{1}{2}[(X + Y) \mp \{(Y - X)^2 + 4k_{BA}k_{BA}k_{AB}[H^+]\}^{1/2}] \quad (51)$$

where $X = k_A + k_{BA}$, $Y = k_B + k_{AB}[H^+]$, and

$$\gamma_1 + \gamma_2 = (X + Y) = k_A + k_{BA} + k_B + k_{AB}[H^+]$$

Thus decay curves of either A* or B* can be obtained at various pH, and a plot of $(\gamma_1 + \gamma_2)$ versus pH can be used to obtain the reverse rate constant k_{AB}.

It can also be shown that

$$\gamma_1\gamma_2 = k_B X + k_A k_{AB}[H^+] \quad (52)$$

Thus

$$\gamma_1 + \gamma_2 = k_A + k_{BA}(1 - k_B/k_A) + \gamma_1\gamma_2/k_A \quad (53)$$

Thus in a graph suggested by DeToma[69], a plot of $(\gamma_1 + \gamma_2)$ versus $\gamma_1 \gamma_2$ yields a slope equal to $1/k_A = \tau_A$ and an intercept $= k_A + k_{BA}$ $(1 - k_B/k_A)$.

Once k_A is known, k_{BA} can be calculated from the decay of A* under irreversible conditions.

Quite analogous relationships obtain for a two-state excited state reaction involving complex formation in the excited state. Examples of reactions of this type include excimer formation, exciplex formation, or excited state proton addition reactions with heterocyclic molecules such as acridine. The appropriate scheme is indicated below.

$$M^* \underset{k_{ME}}{\overset{k_{EM}[Q]}{\rightleftharpoons}} E^*$$
$$k_M \downarrow \qquad \downarrow k_E$$
$$M \rightleftharpoons E$$

Once again, generally a double exponential decay will be observed at wavelengths where either species M or E emit. Equations (1)–(6) still apply with [Q] replacing [H⁺].

A plot of $\gamma_1\gamma_2$ versus $\gamma_1 + \gamma_2$ is linear with a slope $1/k_E = \tau_E$ and an intercept $= k_E + k_{ME} (1 - k_M/k_E)$.

The excited state solvation dynamics of 2-anilinonaphthalene (2 AN) show kinetic behavior that under suitable conditions, are in accord with this scheme.[69] Figure 16 shows the results obtained when increasing con-

[69] R. P. DeToma and L. Brand, Chem. Phys. Lett. 47, 231 (1977).

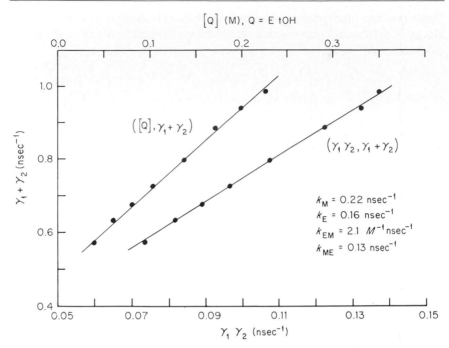

FIG. 16. Kinetic plots for a two-state reversible excited-state reaction. The data is for 2-anilinonaphthalene in cyclohexane with addition of ethanol in the range 0 to 0.35 M.

centrations of ethanol are added to 2 AN dissolved in cyclohexane. Both $\gamma_1 + \gamma_2$ versus $\gamma_1\gamma_2$ and $\gamma_1 + \gamma_2$ versus Q are straight lines as predicted and the rate constants indicated are obtained from the slopes and intercepts. It can be anticipated that a large number of excited state dissociation or association reactions will fit into excited state reversible or irreversible class. These reactions can be recognized by the fact that the reversible reactions will show double-exponential decay behavior with decay times independent of emission wavelength (since A* and B* or M* and E* show the same decay times) but dependent on the forward and reverse rates of the reaction. In addition the preexponential terms of B*, the product formed by dissociation will be equal in magnitude but opposite in sign. In the case of an irreversible excited state reaction the decay of the initially excited species (A* or M*) will be described by a single exponential.

There can be little doubt that as instrumentation for measurements in the nanosecond time domain become readily available, more complex excited state reaction schemes will be unraveled. The fluorophore, 2-naphthol-1-acetic acid, may be cited as an example of a system showing emission from *more* than two excited species.[70] Systems can be envisioned where a proton transfer reaction might be followed by a solvation reaction. Detailed lifetime measurements on energy transfer systems where the decay laws of donor and acceptor will depend on relative distances, orientations and nanosecond motions remain to be done. Ware and Nemzek[71] have investigated transient effects (time-dependent rate constants) in diffusion controlled reactions. These are likely to be important in biochemistry and will lead to complex decay laws. As the level of complexity increases it may not always be possible to determine values for all the rate constants. The decay measurements would be of value for obtaining evidence for or against a particular reaction mechanism.

Emission from two excited states represents a straightforward situation about which detailed information can be obtained. We now conclude this discussion by describing the other limit where emission appears to take place from a continuum of excited states. This situation appears to prevail for some excited state solvation reactions. Fluorescence decay data obtained with *N*-arylaminonaphthalene sulfonates adsorbed to liposomes,[24] adsorbed to some proteins[49] or dissolved in a viscous solvent[51] can be summarized in the following way.

The fluorescence emission spectra shift to lower energy with time (nano-seconds) with little or no change in band shape. The fluorescence intensity also decreases with time at all wavelengths. The combination of the band shift *and* the damping result in a complex nonexponential pattern of fluorescence decay at various emission wavelengths. While at first sight it might appear that only qualitative information can be obtained, this data can be treated according to a theory developed by Bakhshiev and his co-workers[72] to describe general excited state solvent relaxation. Their treatment represents the steady state fluorescence emission spectrum as a superposition of the elementary time-dependent spectra which evolve from some initial state that is continuously shifting in time to lower energy and is being damped at the same time. This phenomenon may be expressed in terms of the nonexponential decay law[51] indicated by Eq. (54).

$$I(\bar{\nu}, t) = i(t)\rho(\bar{\nu}, t) = i(t)\rho[\bar{\nu} - \bar{\nu}_{max}(t)] \tag{54}$$

$I(\bar{\nu}, t)$ is the data matrix for a time-resolved emission spectra obtained as described above. $I(\bar{\nu}, t) = F(\bar{\nu}, t)$ normalized to the corrected steady-state fluorescence spectra obtained with the same instrument.[49,51] This treatment allows the determination of an important new parameter defining the system, $i(t)$.

[70] A. Gafni, R. L. Modlin, and L. Brand, *J. Phys. Chem.* **80**, 888 (1976).

[71] W. R. Ware and T. L. Nemzek, *Chem. Phys. Lett.* **23**, 557 (1973).

[72] N. G. Bakhshiev, N. G. Mazurenko, and I. Y. Piterskaya, *Opt. Spectrosc. (USSR)* **21**, 307 (1966).

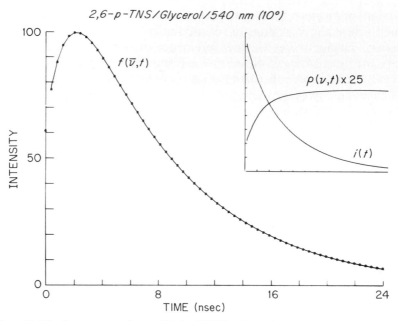

FIG. 17. The fluorescence decay of 2,6-p-TNS in glycerol at 540 nm at 10°. The inset at the upper right shows the breakdown of the fluorescence decay into the double-exponential damping function $i(t)$ and the wavelength-dependent shift function $\rho(\bar{\nu}, t)$.

This quantity is determined from the TRES data matrix (see Fig. 11) as follows: $\rho(\bar{\nu}, t)$ is first obtained as a function of t at each $\bar{\nu}$ by normalizing each "time slice" of the complete surface $I(\bar{\nu}, t)$ to constant emitted quanta and extracting the numeric values of $\rho(\bar{\nu}, t)$ at each $\bar{\nu}$. Once $\rho(\bar{\nu}, t)$ and $I(\bar{\nu}, t)$ are known, $i(t)$ can be determined. An example of this treatment is shown in Fig. 17 which shows data for 2,6-toluidinonaphthalene sulfonate (2,6-p-TNS) dissolved in glycerol. The deconvolved decay curve $F(\bar{\nu}, t)$ (designated as $f(\bar{\nu}, t)$) shows an initial rise indicating that a portion of the emission is due to species created in the excited state. The inset in the upper right shows $\rho(\bar{\nu}, t)$ at this wavelength and the derived damping term $i(t)$. $i(t)$ in this case can be described in terms of a double-exponential decay law. $\rho(\bar{\nu}, t)$ depends on $\bar{\nu}$ and is a complex function which depends on the kinetics of the spectral shift and the emission contour of the system. A nonlinear least squares algorithm was used to obtain $i(t)$ from $\rho(\bar{\nu}, t)$ and $F(\bar{\nu}, t)$ assuming that it could be expressed as a sum of exponential terms.

This treatment has been applied to several systems such as N-arylamino-naphthalene dyes adsorbed to a protein or liposomes. It should be pointed out that while the ability to analyze data in terms of this formulation does not *prove* any particular excited state molecular mechanism, it does require that a proposed mechanism be consistent with the phenomenological formulation. Of particular importance is the fact that the terms $i(t)$ and $\bar{\nu}_{max}(t)$ may be compared between different systems or as a function of intensive parameters such as temperature and pressure. Thus what might at first appear to be a nonanalyzable pattern of decay curves as a function of emission wavelength may in fact be categorized and subjected to further study.

Summary

Nanosecond fluorescence measurements can now be routinely performed and the data can be analyzed in terms of specified decay laws. It is anticipated that nanosecond fluorometry will continue to aid in the elucidation of a variety of excited state mechanistic pathways. This in turn will result in the more sophisticated use of fluorescence probes in biochemistry and cell biology.

At the beginning of this decade, Ware[9] indicated that the ideal decay fluorometer should permit measurements with samples where the absorbance quantum yield product is 10^{-10}, with high spectral resolution and picosecond time resolution over a wide spectral range. Many of the requirements which seemed so stringent then have now been met. It is likely that the next review on rapid decay techniques will be able to show that picosecond experiments can be carried out with the same degree of confidence that is now possible on the nanosecond time scale.

Acknowledgments

Portions of the work described in this report were supported by NIH grant No. GM 11632. One of us (L.B.) wishes to thank Drs. J. R. Gohlke, M. Loken, J. H. Easter, R. Dale, R. P. DeToma, L. Chen, A. Gafni and A. Grinvald for their collaboration in nanosecond fluorometry. This is Contribution No. 989 from the McCollum-Pratt Institute.

[18] Low Frequency Vibrations and the Dynamics of Proteins and Polypeptides

By WARNER L. PETICOLAS

I. Introduction

Proteins are large molecules which usually contain between 500 and 3000 atoms. As a result, they must have, by the laws of physics, between 1500 and 9000 (i.e., $3N$) normal vibrations. Six of these will be quasivibrations of zero frequency corresponding to three translations and three rotations. One of the interesting features of proteins is that they are usually linear peptide chains in which the same peptide group is repeated by means of rotation and translation along a single chain. In this respect, a protein may be regarded as a deformed one-dimensional crystal, with different masses at each site. Although most of these vibrations will occur at frequencies in the range of normal molecular group frequencies 200 to 3500 cm^{-1}), in this chapter we will consider only those vibrations which have a vibrational frequency of less than 200 cm^{-1}.

Vibrations with such low frequencies have many interesting and specific properties different from the group vibrations that normally are observed in infrared and Raman measurements. First of all, the concept of group frequencies, that is, of the vibrations of a particular chemical group of atoms in a molecule such as the peptide group or a CH_2 group, is not applicable for these low-frequency modes. These low-frequency modes must involve a collective motion of a large number of different atoms in the polymer or protein molecules. Small molecules simply do not have low-frequency modes. (However, in a crystal, small molecules can vibrate or twist relative to one another so as to produce low-frequency crystal modes.) The low-frequency modes in proteins arise because when many atoms are coupled together in a macromolecule they can move collectively like small molecules in a crystal so that the effective mass involved in the motion is quite large. Furthermore, the forces between the nonbonded atoms involve such interactions as hydrogen bonding, torsional, or van der Vaal's forces. Thus the forces which hold the distant molecules in a specific conformation are weak relative to chemical bonds so that the resulting vibration frequency is low because it corresponds to a large total mass moving under the action of weak but specific forces.

Another important feature of low-frequency vibrations of large molecules is that their excited vibration states are thermally populated at room temperature. From the laws of statistical mechanics, we know that the fraction of the molecules which will be in excited vibrational states is proportional to the exp $(-hc\lambda^{-1}/kT)$ where h is Planck's constant, λ^{-1} is the frequency of the vibration in cm^{-1}, c is the velocity of light, k is the Boltzman constant, and T is the absolute temperature. We can convert the common spectroscopic frequency unit, wave numbers (cm^{-1}) into cycles per second by multiplying by the velocity of light, 3×10^{10} cm/sec. Thus, if we consider a characteristic group frequency of proteins, such as the amide I vibration, which occurs approximately at 1660 cm^{-1}, this vibration will have a frequency of 5×10^{13} cycles/sec. On the other hand, a low-frequency vibration of a protein, such as, for instance, 10 cm^{-1}, will have a frequency of 3×10^{11} cycles/sec. The population of the excited vibrational levels of two such frequencies is much different.

A simple way to estimate the relative population of the excited vibrational state for molecular vibrations is to realize that the thermal energy kT at room temperature, equals approximately 200 cm^{-1}. Thus, the quantity which appears in the Boltzman factor given above is equal to the frequency of the vibration in wave numbers divided by 200. To compare the relative population distribution of two different vibrations of a protein, for instance the amide I at 1660 cm^{-1} and a low-frequency mode at 10 cm^{-1}, we must use the exact formula for P_n, the population of the nth level of a molecular vibration:

$$P_n = \exp\left(\frac{-nh\lambda^{-1}}{kT}\right)\bigg/ \sum_{n=0}^{\infty} \exp\left(\frac{-nh\lambda^{-1}}{kT}\right)$$

Setting $kT = 200$ cm^{-1}, we get $P_0 = 1$, $P_1 = 10^{-4}$ for the amide I vibration and $P_0 = 0.05$, $P_1 = 0.05$, $P_2 = 0.05$ for the 10 cm^{-1} vibration, so that while the amide I vibration is in the ground state, the 10 cm^{-1} vibration is populated appreciably through the first 20 vibrational quantum levels.

The next important feature of low-frequency vibrations which should be considered is the fact that the amplitudes of these vibrations at room temperature are much larger than those of higher frequency vibrations. This is because when the molecules are thermally excited to higher and higher levels, the vibrating units undergo larger and larger displacements.

On the other hand, if these large molecules are in solution, the viscosity of the solvent will dampen the low-frequency large-amplitude oscillations, and these can be transformed into a fluctuating stochastic motion if the friction constant for the motion is large relative to the product of the mass times the force constant. In general, a protein or polypeptide in solution may be considered to be a damped harmonic oscillator subjected to a random fluctuating force or torque $T(t)$ due to the Brownian motion in the solution, i.e.,

$$M\ddot{x} + \beta\dot{x} + fx = T(t) \tag{1}$$

where M is the effective mass, x a bulk displacement, β the viscous drag on the displacement, and f the Hooke's law force constant. As we will see, the relative value for these quantities will determine whether the system is vibrational or stochastic.

The calculation of the frequencies of vibrations of small molecules is a very precise science which involves the calculation of the potential energy between the various atoms in the molecules and the consideration of its structure in a very precise and mathematical way. This method was worked out by E. B. Wilson and his collaborators in the 1930's and 1940's, and these techniques are now well-known.[1] Recently, they have

[1] E. B. Wilson, Jr., J. C. Decius, and P. C. Cross, "Molecular Vibrations." McGraw-Hill, New York, 1955.

been applied to a variety of simple helical homopolypeptides.[2-7] When one wishes to discuss the vibrational properties at high frequencies of proteins, this is reasonably easy because the group frequencies of the various chemical groupings on the proteins retain largely their same force constants in the proteins as they do in small molecules. This is because the potential energy of interaction between the chemical groups (peptide, CH_2, etc.) is small in relation to the potential energy within the group which gives rise to the group frequency. Thus, the amide I frequency of the peptide group at 1660 cm^{-1} may vary from, for instance, 1640 to 1680 cm^{-1}, depending upon exactly the precise conformational environment of the peptide groups.[8-11] A protein will have a very broad band in this region, characteristic of each of these frequencies[9]; still, this band of frequencies centered around 1660 cm^{-1} will not be all that different from the frequency calculated for a simple model compound, such as, for example, in N-methyl acetamide. On the other hand, there is no really straightforward simple way of calculating a low-frequency vibration of a protein. The number of atoms is almost too large for ordinary GF-type calculation. Also, the proteins lack the symmetry of simple homopeptides; thus, a mathematical simplification which is allowed by virtue of the generation of a simple homopeptide structure by translation and rotation of the specific chemical repeating unit can no longer be applied for even the most simple proteins. There are, consequently, three approximate methods for calculating the low-frequency motion of proteins. These are (1) to calculate the frequencies for only certain structurally recognizable parts of the protein, such as the α-helix or β-structures and to simplify the actual molecular structure by a ball and spring model which preserves the essential topology of the structure but greatly simplifies the number of masses and force constants involved; (2) the globular mass approximation which considers the protein simply as an elastic ball immersed in water which would have normal coordinates of a solid vibrating sphere immersed in a viscous liquid; (3) the double-lobe model which recognizes that many important enzymes have deep clefts which separate the proteins into two specific lobes and then consider the motion of these lobes

[2] K. Itoh and T. Shimanouchi, *Biopolymers* **9**, 383 (1970).

[3] K. Itoh and T. Shimanouchi, *Biopolymers* **9**, 1413 (1970).

[4] K. Itoh and T. Shimanouchi, *Biopolymers* **10**, 1419 (1971).

[5] E. W. Small, B. Fanconi, and W. L. Peticolas, *J. Chem. Phys.* **52**, 4369 (1970).

[6] B. Fanconi, E. W. Small, and W. L. Peticolas, *Biopolymers* **10**, 1277 (1971).

[7] B. Fanconi and W. L. Peticolas, *Biopolymers* **10**, 2223 (1971).

[8] T. J. Yu, J. L. Lippert, and W. L. Peticolas, *Biopolymers* **12**, 2161 (1973).

[9] J. L. Lippert, D. Tyminski, and P. J. Desmeules, *J. Am. Chem. Soc.* **98**, 7075 (1976).

[10] N. T. Yu, *Crit. Rev. Biochem.* **4**, 229 (1977).

[11] J. L. Koenig and B. G. Frushour, *Adv. Infrared Raman Spectrosc.* **1**, 35 (1975).

relative to each other. In addition to those low-frequency motions of the whole protein molecules, torsional motions of the side chains give rise to low-frequency vibrations of a different type. These torsional vibrations will be discussed at the end of this chapter.

Another problem which must be considered is by what physical means can such low-frequency modes in proteins be observed and characterized. As we shall see, this is a difficult and as yet unfinished experimental problem, but several methods have been attempted or proposed including Raman scattering, Brillouin inelastic light scattering, thermal X-ray scattering, and neutron scattering. As we shall see, because of damping effects, it may be difficult to observe these vibrational frequencies in protein molecules in aqueous solutions.

Finally, the major question of interest to enzymologists, is what role does such low-frequency motion play in their enzymatic activity. Unfortunately, this is difficult to say with certainty. However, many enzymes contain active sites which are buried in deep clefts within the molecule. The opening and closing of these clefts may play a role in the action of the enzyme on the substrate. It is hoped that this chapter, in bringing together the fragmentary work in this field, will provide a further impetus to search for the link which certainly must exist between the molecular dynamics of an enzyme and its biological activity.

II. Low-Frequency Vibrations in Proteins and Polypeptides Involving Overall Deformations

A. *Polypeptides and Fibrous Proteins as Chain Molecules*

The infrared and Raman spectra of synthetic and biological polymers have been studied extensively over the past twenty years. Many different sets of normal coordinate treatments have been made, and the chemical structure and conformation have been revealed for many polymers by this technique. It will be worthwhile to consider the vibrations of chain molecules as an introduction to the more complex vibrations of proteins.

Vibrational waves in a periodic one-dimensional lattice, such as an ordered linear or helical polymer, are periodic both in time and in space. Thus, they have both a frequency ν and a wavelength λ. These vibrational waves are quantized and are often called phonons. Figure 1 is a diagram of a simple hypothetical linear polymer. In the top drawing all the little oscillators that make up the one-dimensional ordered array are shown in their equilibrium position. One can imagine them all oscillating in phase so that each oscillator reaches its maximum amplitude at the same time as its neighbor. If the polymer is infinite in extent, the wavelength of this vi-

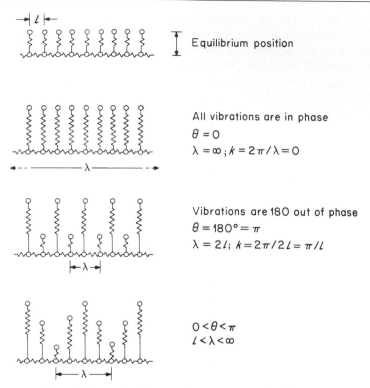

FIG. 1. Schematic diagram of a transverse optical phonon.

bration is also infinite in extent, the wavelength of this vibration is also infinite, and the phase angle between adjacent neighbors is zero, as is the wave vector $Q = 2\pi/\lambda$. This type of vibration at zero phase angle is shown in the second drawing. At the other extreme, one may imagine each oscillator reaching its maximum amplitude at the same time that its neighbor reaches its minimum amplitude, and vice versa, so that each oscillator is exactly 180° or π radians out of phase with its neighbor (third drawing). Between these two extremes, there can be a general sine wave motion such that each unit is out of phase with its neighbor by an amount θ, which is illustrated in the bottom drawing of Fig. 1. There are many optical modes such as, for example, the amide I (C=O stretch) in a helical polypeptide.

Another important type of motion in chain molecules are the longitudinal acoustical modes (LAM). This involves an overall stretching of the chain or helix and is shown diagramatically in Fig. 2. The longest wavelength mode has a wavelength of twice the length of the chain, while the shortest wavelength mode is twice the length L of the longitudinal spacing

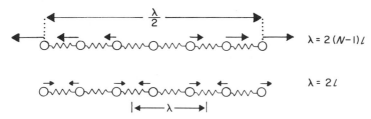

FIG. 2. Schematic diagram of a longitudinal acoustical phonon in a chain molecule.

between two chemical repeat groups. It can be shown that if a polymer chain has N chemical repeat units (i.e., N is the degree of polymerization) and each repeat unit has m atoms, then the number of nonzero vibrational frequencies for the chain is $3mN - 6$. When the polymer is in an ordered one-dimensional array, as in a helix, then the $3mN$ frequencies are distributed in the following way (including the motions of zero frequence). There are $3m$ dispersion curves of frequency versus phase angle that lie in a rather limited frequency range and do not overlap any other dispersion curve. On each of these dispersion curves lie N frequencies, which form a smooth curve from $\theta = 0°$ to $\theta = 180°$. A schematic example of this is shown in Fig. 3. Because these frequencies lie on a smooth dispersion curve, it is not necessary to calculate the whole $3mN$ different frequencies but merely to obtain the shape of the dispersion curve by calculation of a few points. The dispersion curves can be divided into two classes: optical and acoustical. In optical phonon motion, the atoms within the chemical

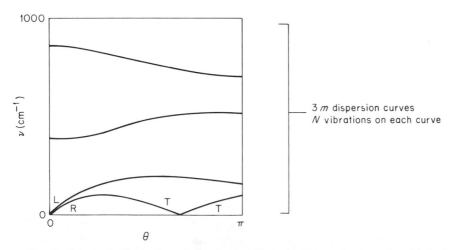

FIG. 3. Schematic illustration of the phonon dispersion curves for an ordered helical polymer showing the distribution of the $3mN$ vibrational frequencies where m is the number of atoms per repeat group and N is the degree of polymerization.

repeat unit vibrate against one another, whereas in acoustical phonon motion, the unit cells as a whole are displaced. As the wavelength of the acoustical phonons increases, one observes breathing and bending motions of the whole polymer. It is of interest that it is possible to calculate these acoustical motions of the chain itself by fitting the optical motions (N—H stretch, etc.) to the Raman and ir measurements of the optical modes.

The acoustical phonon curves at the bottom of Fig. 3 give the lowest frequency vibrations of the helix. As $\theta \to 0°$ the LAM curve, marked L in Fig. 3, becomes the simple accordianlike motion described at the top of Fig. 2. This LAM motion contains one point (called a node) at which there is no displacement. If there is no node, all the atoms move in the same direction and this motion becomes simple longitudinal translation. Similarly, the bending and twisting motions of the chain (labeled R and T in Fig. 3) become in the limit of long wavelength essentially motions of very low frequency. Less ordered structures, such as proteins, will have vibrations similar to the ones described here, but whose frequencies are much more difficult to calculate exactly. But proteins do have low-frequency large-amplitude vibrations with nodes in the interior of the molecule. For example, in the hinge-bending mode of lysozyme to be discussed in Section II,C, the node is at the hinge between the two lobes which are displaced.

One other characteristic of phonon dispersion curves is important. The occurrence in the frequency versus phase angle curve of a flat portion at a given frequency indicates a large number of vibrations with the same frequency. This means that the number of vibrations per unit frequency (usually called the density of states) is large at this point. This is important because neutrons are scattered more or less equally by all the vibrations, and hence, a large number of vibrations of the same frequency results in a maximum in the corresponding neutron-scattering peaks.

Figure 4 shows the phonon dispersion curves calculated by Small et al.[5] for the helical polypeptide polyglycine II. The flat portions of the frequency versus θ curves indicate a large number of vibrations all at the same frequency which give rise to maxima in the incoherent inelastic neutron-scattering curves obtained by Gupta et al.[12] The circles are measured Raman frequencies.[5]

Figure 5 shows similar calculations by Fanconi et al.[6,7] of α-helical poly-L-alanine (α-PLA) and Fig. 6 shows the low-frequency modes of α-PLA recalculated with various values of the H bond stretching force constant (from 0–0.30 mdynes/Å) including the values of 0.15 mdynes/Å, suggested by Itoh and Shimanouchi[2-4] who also calculated these curves,

[12] V. D. Gupta, S. Trevino, and H. Boutin, *J. Chem. Phys.* **48**, 3008 (1968).

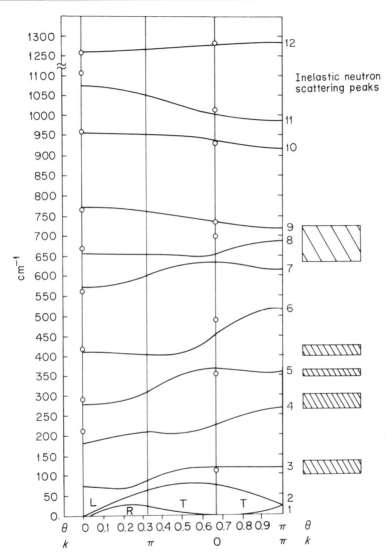

FIG. 4. The phonon dispersion curves for polyglycine II, taken from Yu.[10] The circles are measured Raman frequencies.

but obtained somewhat lower frequencies for the acoustical modes. Figure 7 shows these same low-frequency dispersion curves calculated by Itoh and Shimanouchi.[2-4] From these three sets of calculations one can obtain the corresponding frequency of the LAM accordion mode for the α-helix of poly-L-alanine as a function of its length. These calculated values are given in Table I.

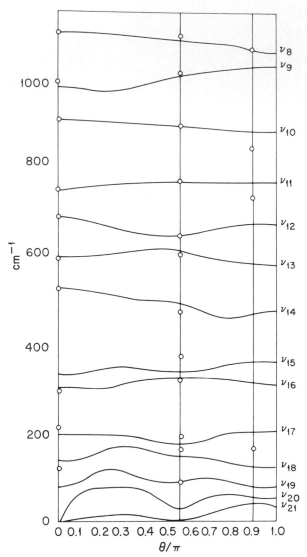

FIG. 5. The phonon dispersion curves for α-helical poly-L-alanine taken from Fanconi *et al.*[6] with the $C=O \cdot \cdot H-N$ hydrogen bonding force constant taken as 0.30 mdynes/Å. The circles are measured Raman frequencies.

   In Table I, the first column is taken from the exact GF matrix calculation of Itoh and Shimanouchi[2] with the $C=O \cdot \cdot H-N$ hydrogen bond
force constant taken as 0.15 mdynes/Å. Column 3 gives the values from
the exact GF calculations of Ref. 6 with the H-bond force constant equal

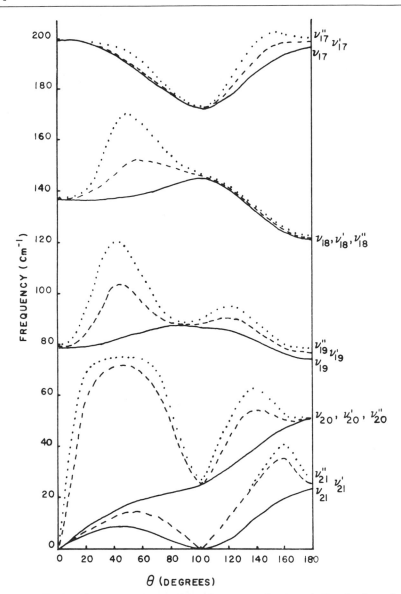

FIG. 6. The low-frequency phonon dispersion curves for the α-helix of poly-L-alanine taken from Fanconi and Peticolas[7] with the C=O · · H—N hydrogen bond force constant taken as 0.0(——); 0.15 mdynes/Å(- - -) and 0.3 mdynes/Å(. . . .).

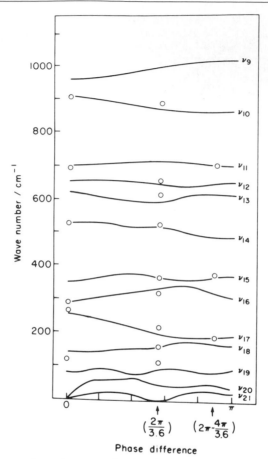

FIG. 7. The phonon dispersion curves for the α-helix of poly-L-alanine taken from Ito and Shimanouchi[2] with the C=O $\cdot\cdot$ H—N hydrogen bond force constant equal to 0.15 mdynes/Å. The circles are measured ir frequencies.

to $+0.15$ and 0.30 mdynes/Å, respectively. Notice that both the calculations of Fanconi et al.[6] are higher in frequency than those obtained in Itoh and Shimanouchi.[2] The last column shows a set of calculations by the present author of the LAM using the very simple formula:

$$\nu = 2\nu_0 \sin (\pi/2T) \tag{2}$$

where $\nu_0 = 42.5$ cm^{-1} in order to fit the Itoh and Shimanouchi data. The agreement between the frequencies in the first column and the last column of Table I is remarkable and illustrates the reliability of the simple model of a helix.[6,7,13,14] In Eq. (2) T the number of turns of the α-helix is related

[13] W. L. Peticolas and M. W. Dowley, Nature (London) 212, 400 (1966).
[14] W. L. Peticolas, Biopolymers, in press (1979).

TABLE I

CALCULATED FREQUENCIES FOR THE LONGITUDINAL MODES
OF THE α-HELIX OF POLY-L-ALANINE

Number of peptide groups (N)	Itoh and Shimanouchi[a]	Fanconi et al.[b]		Peticolas[c]
		1	2	
20	23	—	40	23
30	15	28	28	16
40	12	21	24	12
50	10	17	20	9
100	5	8.5	11	5

[a] From Itoh and Shimanouchi.[2] Calculated with H bond force constant equal to 0.15 mdynes/Å.

[b] From Fanconi et al.[6] Column 1 data calculated with H bond force constant equal to 0.15 mdynes/Å. Column 2 data calculated with H bond force constant equal to 0.30 mdynes/Å.

[c] From Peticolas.[14] Calculated from $\nu = (2 \times 42.5 \text{ cm}^{-1})$ per $(3.6\pi/2N)$.

to N the number of peptide groups by the relation $T = N/3.6$. Equation (2) also fits the data in the second column for very low frequencies, with $\nu_0 = 76 \text{ cm}^{-1}$. Although we have emphasized the low-frequency motions of the α-helix, more work has actually been done on all-*trans* hydrocarbon chains $(CH_2—CH_2)_n$ than any other kind of polymer. This work, with emphasis on the low-frequency modes has been recently reviewed by Shimanouchi[15] and Vergoten et al.[16]

The same accordion-like longitudinal acoustical mode (LAM) also occurs in all ordered hydrocarbon chains, such as the chains of fatty acids found in membranes bilayers.[16,17] The recognition that the concept of longitudinal acoustical modes could be applied to linear hydrocarbon chains was first made many years ago by Mizushima and Shimanouchi.[18] Schaufele and Shimanouchi subsequently measured the frequencies, by Raman scattering, of the LAM vibrations of hydrocarbon chains for chain-lengths varying between 8 and 94 carbon atoms.[19] Recently, these frequencies have also been calculated by the present author[14] using Eq. (2). What has been found is that it is possible to calculate very exactly the frequencies of these accordian-like modes for any type of helical polymer, using M_T the mass per turn and f_T the force constant between turns. The mass of a turn is given by the product of the number of residues per turn

[15] T. Shimanouchi, in "Structural Studies of Macromolecules by Spectroscopic Methods" (K. J. Ivin, ed.), p. 59. Wiley, New York, 1975.

[16] G. Vergoten, G. Fleury, and Y. Moschetto, Adv. Infrared Raman Spectrosc. 4, 195 (1978).

[17] J. L. Lippert and W. L. Peticolas, Biochim. Biophys. Acta 282, 8 (1972).

[18] S. Mizushima and T. Shimanouchi, J. Am. Chem. Soc. 71, 1320 (1949).

[19] R. F. Schaufele and T. Shimanouchi, J. Chem. Phys. 47, 3605 (1967).

TABLE II

FREQUENCIES OF THE LONGITUDINAL ACOUSTICAL
MODE OF HYDROCARBON CHAINS

Number of C atoms (m)	Observed frequency[a] (cm^{-1})	Calculated[b] (cm^{-1})	Calculated[c] (cm^{-1})
8	283	287	300
9	249	256	267
10	231	232	240
12	194	194	200
14	168	167	171
16	150	146	150
18	133	130	133
20	120	118	120
22	112	107	109
24	103	98	100
28	85	84	86
32	76	74	75
36	67	65	67
44	56	54	55
94	26	25	26

[a] From Schaufele and Shimanouchi.[19]
[b] From Peticolas.[14]
[c] From Mizushima and Shimanouchi.[15,18]

times the mass per residue. This calculation has been made both for linear hydrocarbon chains similar to those found in phospholipid membrane bilayers and much longer chains as well. Table II shows the frequencies of these longitudinal acoustical accordion-like vibrations of hydrocarbon chains which have so far been observed by Raman spectroscopy as well as those calculated from Eq. (2). Table III shows the simple parameters used in obtaining the low-frequency modes of hydrocarbon and α-helical peptide chains.

TABLE III

PARAMETERS FOR CALCULATION OF ν_0 FOR HELICAL CHAINS
(HYDROCARBONS AND α-HELICAL POLYPEPTIDES)[a]

Chain type	Mass per turn (MW)	Effective interturn force constant for LAM (mdynes/Å)
Hydrocarbon	28.0	2.33
α-Helical poly-L-alanine	255.6[b]	0.273

[a] From Peticolas.[14]
[b] For irregular helices in proteins, use actual molar mass per turn.

TABLE IV
THE ROOT-MEAN-SQUARE DISPLACEMENT OF THE OVERALL END-TO-END
LENGTH OF ALL-*trans*-HYDROCARBONS AND α-HELICAL
POLYPEPTIDE CHAINS AT ROOM TEMPERATURE[a]

| *trans*-Hydrocarbons | | α-Helical poly-L-alanine | |
No. of carbon atoms	$(l^2)^{1/2}$	No. of peptide groups	$(l^2)^{1/2}$
10	0.139013	10.8	0.208734
20	0.284714	18	0.380738
30	0.430363	36	0.807247
40	0.575999	54	1.23291
50	0.72163	72	1.65837
60	0.867259	90	2.08376
		108	2.50911

[a] From Peticolas.[14]

In addition to calculating the modes from the simple chain model, it has also been possible to estimate the root-mean-square amplitude of the overall end-to-end distance of the helical chain displaced from its equilibrium value and the function of the temperature.[14] Table IV shows the root-mean-square displacement of the end-to-end chain length from the equilibrium value as a function of the length of the number of residues in the chain for normal room temperature. As we may see, Table IV shows that the end-to-end root-mean-square displacement can vary from a few tenths of an angstrom for hydrocarbon chains of the length commonly found in membranes to several angstroms for long α-helical chains. In these calculations damping has not been considered but may play a part in the oscillation of the longer chains with their lower frequencies and larger amplitudes. The point we may gather from these calculations is that, for materials such as proteins, it is not unexpected to find in the crystal form, or in a film where there is little viscous damping, large amplitude vibrations at low frequencies. In addition to the overall longitudinal accordion-like vibrations and twisting and bending vibrations for the α-helix have also been calculated by Itoh and Shimanouchi, and their values are given in Shimanouchi.[15] It is found that these frequencies are much lower; these types of vibrations will be even more highly damped than longitudinal vibrations.

Very recently, the LAM vibrations of collagen and muscle fibers have been measured using the technique of laser Brillouin scattering.[20] In this technique the light is scattered coherently from the thermally excited

[20] T. Harney, D. James, A. Miller, and J. W. White, *Nature (London)* **267,** 285 (1977).

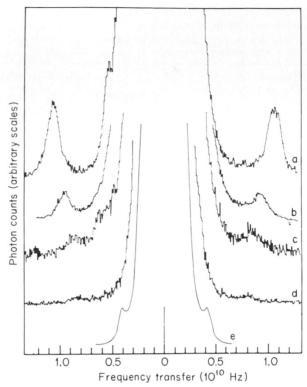

FIG. 8. Intensity of scattered (Brillouin) light versus frequency shift. Collagen fibers at relative humidities of (a) 0%, (b) 30%, (c) 84%, (d) 100%, (e) pure water.

longitudinal acoustical modes of the protein fibers and shifted in frequency from the laser light by the amount equal to the frequency of the LAM at its corresponding wavelength (see Fig. 2). The Bragg law holds for reflections from the parallel planes of compression in the protein fiber so that

$$\lambda_{\text{laser}} = 2\lambda_{\text{LAM}} \sin{(\theta/2)} \tag{3}$$

where θ is the scattering angle, $(\theta/2)$ is the Bragg angle, and the LAM wavelength λ_{LAM} is the distance between the Bragg planes which coherently reflect the laser light (see Fig. 2). Figure 8 shows a plot of the experimental observations of scattered intensity versus frequency, and Table V gives the comparison of the experimental data with the theoretical values of Fanconi and Peticolas[7] for poly-L-alanine α-helix as a function of the N—H $\cdots$ O—C bond force constant which was varied between 0.0 and 0.3 mdynes/Å (see Fig. 6). Q in column 2 is $2\pi/\lambda_{\text{LAM}}$, ν (GHz) is the observed or calculated LAM frequency in units of 10^9 Hz.

TABLE V

ACOUSTIC PROPERTIES OF COLLAGEN AND MUSCLE AT 25 °C[a]

Substance	Q (Å^{-1})	ν (GHz)	Sound velocity (cm/sec)	Density (g/cm³)	Elastic modulus (dynes/cm²)	H bond force constant (mdynes/Å)
Poly-L-alanine (theory)	0.18	1,800	6.4×10^5	1.51	6.18×10^{11}	0.3
	0.23	1,800	4.8×10^5	1.51	3.48×10^{11}	0.15
	0.23	258	6.9×10^4	1.51	7.2×10^9	0.0
Muscle 12% relative humidity	1.727×10^{-3}	9.69	3.53×10^5	1.40	1.74×10^{11}	0.1
Collagen at 0% relative humidity	1.727×10^{-3}	10.65	3.88×10^5	1.43	2.15×10^{11}	0.11
Collagen at 12% relative humidity	1.727×10^{-3}	10.29	3.74×10^5	1.4	1.96×10^{11}	0.11
Collagen at 30% relative humidity	1.727×10^{-3}	9.00	3.27×10^5	1.38	1.47×10^{11}	0.09
Collagen in 0.15 M NaCl solution (assumed equivalent to 85%)	1.727×10^{-3}	7.20	2.62×10^5	1.31	9.0×10^{10}	0.08
Collagen (air dried)						
(a) Macroscopic stress–strain measurements					$\sim 10^9$	
(b) Using the 660-Å reflection to measure strain					$\sim 10^9$	
(c) Using the 2.8-Å reflection to measure strain					$\sim 2 \times 10^9$	
Collagen (dry)						
X-Ray measurements					1.22×10^{10}	

[a] From Harney et al.[20]

These important measurements show that frequencies as low as 10^{10} Hz (0.3 cm^{-1}) can exist undamped in hydrated protein fibers. However, it must be noted that such long fibers are capable of sustaining the LAM wavelengths of the order of 3000 Å which is about the smallest λ_{LAM} one can see with visible laser light at 90° scattering angle. Enzymes are generally much smaller in radius than this and so cannot be studied in solution by this technique.

Recently Suzaki and Gō have calculated both the longitudinal and torsional fluctuations in the helices of polyglycine and poly-L-alanine.[20a] They use an infinite helix and calculate the tensile fluctuations (l^2) as a function of helical segment length within the infinite helix. For short α-helices (10–18 residues) their values are almost identical to those given in Table IV. However, for longer helices their values of (l^2) are considerably smaller than those in Table IV. This difference at long helical lengths may be due to the difference between the consideration of a length of a helical segment within a helix and a finite helix of the same length.

B. Globular Proteins Considered as Elastic Spheres

A globular protein or enzyme which is ridged and isolated possesses a number of modes of vibrations which may be considered analogous to those of a vibrating elastic sphere. This correspondence was pointed out originally by deGennes and Papoular[21] and more recently by Suezaki and Gō.[22] The latter article made an assignment to an observed low-frequency Raman band of globular proteins by Brown, Erfurth, Small, and Peticolas.[23] It will be necessary to give some consideration to the ideas of these authors in assigning the Raman spectra of low-frequency vibrations of proteins.

DeGennes and Papoular[21] point out that a ridged globular protein can have three types of overall vibrational motion: (1) a radial pulsation in which the sphere increases and decreases periodically in diameter (referred to by Brown et al.[23] as a breathing mode); (2) a shearing mode involving a counterrotation of the inside and outside of the sphere about a specific axis; and (3) a torsional or twisting mode involving the twisting of the top half of the sphere relative to the second half. Since these latter two

[20a] Y. Suezaki and N. Go, *Biopolymers* **15**, 2137 (1976).

[21] P. G. deGennes and M. Papoular, (1969) "Polarisation, Matière et Rayonnement. Volume in Honor of Alfred Kastler," p. 243. Presses Univ. Fr., Paris.

[22] Y. Suezaki and N. Gō, *Int. J. Pept. Protein Res.* **7**, 333 (1975).

[23] K. G. Brown, S. C. Erfurth, E. W. Small, and W. L. Peticolas, *Proc. Natl. Acad. Sci. U.S.A.* **69**, 1467 (1972).

motions are inactive in the Raman effect and have never been observed by other methods; we will not discuss them here.

The frequency for the radial pulsation is given by

$$\nu = \frac{1}{2R} \left(\frac{E + 2\mu}{\rho} \right)^{1/2} \tag{4}$$

where ρ is the density of the globular protein, E is the modulus for longitudinal stretch, μ is the shear modulus, and R is the radius of the protein. Since $E \gg \mu$, we need only to consider E. Estimates of E based on the LAM frequencies for the α-helix give $E = 2 \times 10^{11}$ dynes/cm^2 (see Table V). Globular proteins, although "cross-linked" by H bonds, will have a lower modulus—a reasonable estimate is 10^{11} dynes/cm^2. Assuming a radius of 20 Å and a density of 1 g/cm^3, the frequency of the vibration has been calculated by Suezaki and Gō[22] to be 26 cm^{-1}. DeGennes and Papoular[21] state that since these radially pulsating modes radiate their power acoustically in the solvent they will be substantially dampened in aqueous solution; however, they point out that these modes, when not over-damped, should be strongly Raman-active. The shear and torsion modes, however, will not be Raman-active. as will be discussed below, Mc-Cammon, Gelin, Karplus, and Wolynes[24] also point out that the hinge-bending mode in lysozyme will be over-damped in aqueous solution. Thus it is reasonable to look for low-frequency (25 cm^{-1}) vibrations in globular proteins which might be present in films or crystals but almost certainly not in aqueous solutions where the protein vibrations may be entirely damped by interaction with the viscous liquid.

Strong Raman-active modes have been found at 30 cm^{-1} in films and crystals of α-chymotypsin[23] and at 25 and 75 cm^{-1} crystals of lysozyme.[25] These Raman spectra are shown in Figs. 9 and 10. Brown et al.[23] assigned the 30 cm^{-1} mode in α-chymotypsin to an internal breathing mode of this globular protein. On the other hand, Genzel et al.[25] assigned a 25 cm^{-1} mode in crystalline lysozyme to an intermolecular crystalline mode. This conclusion was based on the fact that the 25 cm^{-1} that they had observed in crystalline lysozyme was not found to be present in solution as shown in Fig. 10. However, these pulsation modes of proteins might very well be damped out in solution but might be present in crystal or films. It is evident that the assignment of such low-frequency modes is a hazardous game. However, it seems to this author that the assignment of the 25 cm^{-1} mode observed in the Raman spectrum of crystal lysozyme is more likely

[24] J. A. McCammon, B. R. Gelin, M. Karplus, and P. G. Wolynes, Nature (London) 262, 325 (1976).
[25] L. Genzel, F. Keilmann, T. P. Martin, G. Winterling, and Y. Yacoby, Biopolymers 15, 219 (1976).

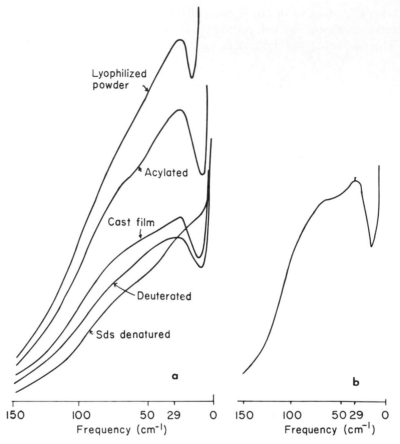

FIG. 9. The low-frequency mode in α-chymotrypsin at about 30 cm^{-1}. This vibration is only found in crystals and films and disappears upon denaturation with sodium dodecyl sulfate. (b) shows single crystal. From Brown et al.[23]

an internal mode of the lysozyme which is damped in solution than an intermolecular crystalline mode because of the following reasons. It is very common to find in crystals of organic molecules of about a molecular weight of 300, crystal modes with frequencies in the 20–50 cm^{-1} region. Consequently, if the 25 cm^{-1} mode is a crystal mode, it must follow that the intermolecular force constant between two protein molecules in a protein crystal must be as much stronger than the force between two hydrogen-bonded molecules in a molecular crystal, as the ratio of their masses, since the frequency ($\sim$25 cm^{-1}) is the same for both. Since the mass of the protein is in the order of 100 times that of a small organic molecule, the force constant between the molecules in the protein crystal

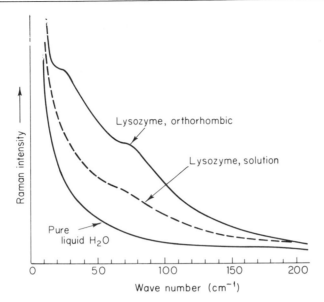

FIG. 10. Low-frequency modes of lysozyme both in the crystal and in solution. These low-frequency modes are obviously damped considerably upon going from the crystal to the solution. From Genzel et al.[25]

must be 100 times greater than the forces between small hydrogen-bonded molecules. It strikes this author as unlikely that the forces between protein molecules are sufficiently large to give a protein–protein vibration as large as 25 cm^{-1}. A more likely suggestion for the disappearance of these low wave number bands in going from the crystalline protein to solution, is that this band simply becomes so highly damped in aqueous solution that it is no longer visible by the Raman technique. This is in agreement both with the calculations of deGennes and Popoular[21] as well as those of McCammon et al.[24] However, neither possibility can be ruled out at the present time.

C. The Hinge-Bending Mode in Enzymes

Many enzymes contain active sites which are located in relatively deep clefts within the molecule. Certainly, the opening and closing of these clefts may generally be involved in the substrate protein interaction. Dickerson and Geis,[26] for example, point out that ribonuclease A, ribonuclease S, papain, and lysozyme all contain clefts in which their active sites are located. Figure 11 shows an idealized molecule containing a cleft such

[26] R. E. Dickerson and I. Geis, (1969) "The Structure and Action of Proteins." Benjamin, Menlo Park, California.

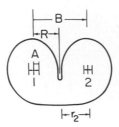

Fig. 11. Idealized diagram of enzyme with cleft containing active site. From Morgan and Peticolas.[27]

as this, taken from the paper of Morgan and Peticolas[27] who suggested an X-ray method for observing the motion of the sites of the cleft relative to each other. Another even more simple method of observing such motion would be to take the Debye–Waller factors of X-ray vector distances corresponding to that as shown by B in Fig. 11 and to measure the Debye–Waller factor as a function of the temperature from very low temperatures in the crystal to room temperature.

Recently, a very ambitious attempt has been made to calculate the potential energy well for this hinge-bending mode of the two globular lobes of lysozyme.[24] It is of interest that atomic displacements of up to 0.75 Å have been found in X-ray comparisons of the structure of the free enzyme and the enzyme inhibitor complex.[28] This indicates that in the enzyme–inhibitor complex, the two lobes have come closer together. However, because the protein surface moves appreciably during such a vibration, damping effects resulting from the viscous drag of the solvent were included in the calculations. The authors assumed a model of the hinge-bending mode in which two essentially ridged lobes (lobe 1, residues 5–36 and 98–129; and lobe 2, residues 40–94) are assumed to be connected by a hinged region consisting of the amino terminal end (residues 1–4) and two short polypeptide chain segments (residues 37–39, 95–97) running from one lobe to the other. A more accurate view of this cleft can be obtained from the diagram in Dickerson and Geis.[26] McCammon et al.[24] point out that the bending motion of lysozyme in the presence of a solvent must be determined by the Langevin equation given in Eq. (1), for a damped harmonic oscillator model. This model will be discussed in Section III in its general form.

The authors[24] have calculated the force constant f using an empirical energy function that expresses the total conformational energy as a func-

[27] R. S. Morgan and W. L. Peticolas, *Int. J. Pept. Protein Res.* **7**, 361 (1975).
[28] T. Imoto, L. N. Johnson, A. C. T. North, and J. A. Rupley, *In* "The Enzymes" (P.D. Boyer, ed.), Vol. 7. Academic Press, New York, 1972.

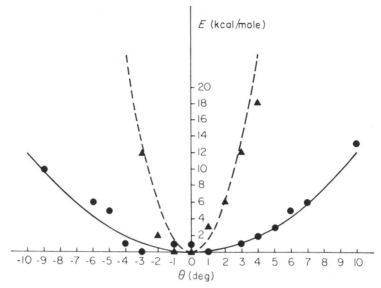

Fig. 12. The calculated potential energy minima produced by opening ($\theta > 0$) or closing ($\theta < 0$) the lysozyme cleft. Calculated values are for the rigid bending potential (▲) and adiabatic potential (●). The force constant for the harmonic motion is the second derivative of this curve at $\theta = 0$. From McCammon *et al.*[24]

tion of the Cartesian coordinates of the heavy atoms. That includes terms associated with bond length, bond angles, dihedral angles, hydrogen bonds, and nonbonded interactions. To obtain an equilibrium starting structure, the real space of the coordinates of lysozyme were greatly refined. The potential function for bending was then determined by rotating residues 39–96 rigidly about the chosen axis and computing the conformation energy of the distorted protein as a function of the bending angle. To obtain a more realistic potential function, a similar calculation was made but for each bending angle there was permitted an adiabatic relaxation of the protein. The bending angle was kept constant during the relaxation by imposing constraints on the residues at a distance from the hinge in the interior of each lobe. The results of the two calculations are shown in Fig. 12. A parabolic potential energy well is fit to the points and yields a force constant of 4.1×10^{14} erg rad^{-2} mole^{-1} (rigid) and 3.3×10^{13} erg rad^{-2} mole^{-1} (adiabatic). If hydrodynamic effects were neglected, the motion reduces to that of a simple harmonic oscillator and the resulting frequencies are 15 and 4.2 cm^{-1} for the rigid and adiabatic calculations, respectively. On the other hand, the authors point out that since β, the coefficient of the frictional drag on the lobes is sufficiently large relative to the product of mass times harmonic force constant, the motion in solution must be overdamped, i.e., rather than vibrating periodically with a calculated frequency, the two lobes move relative to each other in a stochastic diffusive manner in accord with Smoluchowski's diffusion equation. Thus, again we see that although the displacements are the same for either the stochastic or harmonic motion, the time dependence of the two motions is drastically different.

McCammon et al.[24] point out that it is possible that the cleft motion involving the active site cleft has a significant effect on the rate constants for substrate binding to lysozyme. The time required for a molecule the size of glucose to diffuse freely 5 Å is about 5×10^{-10} sec at 300° K and so is substantially greater than the damped relaxation time which is about 10^{-11} sec. Since the opening or closing fluctuations were on the order of about 1 Å on the outer part of the cleft region, the coupling between substrate binding and cleft vibrational mode requires a more detailed study, taking into account the enzyme–substrate interaction; for example, the motion could be significantly perturbed by the presence of a substrate. The authors point out that the Raman band associated with the lysozyme bending mode will be damped out in solution so the binding of an inhibitor could lead to underdamped motional and observable Raman bands. They do not discuss the possibility of this vibration being visible from a crystal and that is something which must certainly be considered.

III. Stochastic Fluctuations Giving Rise to Overall Deformations

A. Theory of a Diffusing, Vibrating Protein Molecule in Viscous Medium

In Section II of this chapter we have treated the protein from the point of view of a molecular solid with $3N$ vibrational modes. Several recent reviews have discussed the more stochastic nature of protein internal motions in a very qualitative sense.[29-31] One reviewer has even suggested that in solution a protein is "a kicking, screaming stochastic molecule."[31] In the view of the present author, this description must be highly qualified. Although the low-frequency vibrational motions in proteins certainly may be damped out so that their time dependence is no longer a simple sinusoidal variation of displacement of the normal modes with time, as we will show below, the actual atomic displacements which do occur are along exactly the same trajectories in space as for the vibrational motion. The atomic displacements of the damped modes occur randomly in time

[29] A. Cooper, Proc. Natl. Acad. Sci. U.S.A. **73**, 2740 (1976).
[30] G. Careri, P. Fasella, and E. Gratton, (1975) Crit. Rev. Biochem. **3**, 141 (1975).
[31] G. Weber, (1975) Adv. Protein Chem. **29**, 1.

but not in space as the spatial motion is restrained by the potential energy wells of the atoms in the protein, the position of whose minima are determined by X-ray diffraction.

Thus, in the view of this author, a good model for a protein is that of a series of masses connected by Hookian springs varying in strength from the very strong carbon–carbon or carbon–nitrogen bonds to the very weak hydrogen and hydrophobic (van der Waals) bonds as well as the weak forces which resist the torsional rotations of the dihedral angles. An enzyme is a finely tuned dynamic entity which would vibrate with all of its $3N - 6$ normal modes just as a crystal does if it were not for the dissipative effects of the medium—both interior and exterior in which the atoms of the protein find themselves. Consequently, only those vibrations in which the force constants are sufficiently large to overcome the viscous damping move in a harmonic sinusoidal oscillation; the low-frequency modes are constrainted to move along their normal coordinates, but the damped motion occurs as a stochastic random walk. This can be easily shown in a few simple mathematical steps in which the Wilson GF normal-coordinate treatment can be incorporated in the well-known Langevin equation for Brownian motion of a particle in a harmonic potential. This serves as an introduction to the more complete dynamic calculations discussed in Section III,B.

Let us consider a protein of N atoms. In the absence of damping (possibly even in a protein crystal) we can solve for the $3N$ normal modes of the protein in the following way: let $x_1, x_2, \ldots, x_{3N}$ be the $3N$ Cartesian displacement coordinate of the N atoms from their equilibrium position: x_1, x_2, x_3 are the x, y, and z displacement of atoms 1, etc. If the $x_j{}^{\text{th}}$ displacement is made it may exert a force $f_{ij}x_j$ along the $x_i{}^{\text{th}}$ displacement. Thus, in Hooke's law of harmonic approximation, Newton's second law for the atoms in the protein becomes

$$M_i\ddot{x}_i = - \sum_{j=1}^{3N} f_{ij}x_j \tag{5}$$

Transforming to mass-reduced coordinates as is customary[1]

$$q_i = M_i^{1/2}x_i \tag{6}$$

gives the equation:

$$\ddot{q}_i = - \sum_{j=1}^{3N} \frac{f_{ij}}{(M_iM_j)^{1/2}} q_j. \tag{7}$$

This equation in matrix notation becomes

$$\ddot{\mathbf{q}} = - \mathbf{F}^q\mathbf{q} \tag{8}$$

where

$$(F^q)ij = f_{ij}/(M_i M_j)^{1/2}. \tag{9}$$

At this point let us calculate f_{ij} in terms of the internal coordinates customarily used in molecular vibrational problems. The relation between the mass-reduced Cartesian coordinates q and the Wilson internal coordinates is well known.[1]

$$\mathbf{R} = \mathbf{D}\mathbf{q} \tag{10}$$

The matrix D contains the geometry of the problem and from it one can formulate the well-known Wilson G-matrix ($G = DD^T$) which will not be needed in this different but equivalent treatment. However, for well-established theoretical reasons[1] the internal potential energy of the protein, which will govern both its vibrational and its internal stochastic displacements must be set up in terms of the internal coordinates, the bond stretches, angle bends, torsions of the dihedral angles about the C_α—C and C_α—N bonds, etc.,

$$2V = \mathbf{R}^T\mathbf{F}^R\mathbf{R} = \sum_{i,j=1}^{3N-6} F_{ij}{}^R R_i R_j. \tag{11}$$

Using Eq. (10), we obtain

$$2V = \mathbf{q}^T\mathbf{D}^T\mathbf{F}^R\mathbf{D}\mathbf{q} = \mathbf{q}^T\mathbf{F}^q\mathbf{q}^T \tag{12}$$

so that we have now a simple formula for obtaining F^q used in Eq. (8), in terms of the realistic internal-coordinate force field,

$$\mathbf{F}^q = \mathbf{D}^T\mathbf{F}^R\mathbf{D} \tag{13}$$

To solve Eq. (8) we must solve the eigenvalue equation of the dynamical matrix F^q

$$\mathbf{F}^q\mathbf{A} = \mathbf{A}\Lambda \tag{14}$$

where Λ equals diag $\{\lambda_1, \lambda_2, \ldots, \lambda_{3N}\}$, the λ_i's being the $3N$ eigenvalues of F^q, six of which will be zero and correspond to the translations along and rotation of the whole protein about the three Cartesian axes. Since F^q is symmetric (or for a linear chain polypeptide, Hermitian) it follows that A will be orthogonal (or unitary); consequently, $A^T = A^{-1}$, and

$$\mathbf{A}^T\mathbf{F}^q\mathbf{A} = \Lambda \tag{15}$$

To solve Eq. (8) it will be necessary to transform to normal coordinates Q using the eigenvector matrix A or its transform A^T.

$$\mathbf{Q} = \mathbf{A}^T\mathbf{q} \tag{16}$$

Operating on both sides of Eq. (8) with A^T gives

$$\ddot{\mathbf{Q}} = \mathbf{A}^T \ddot{\mathbf{q}} = -\mathbf{A}^T \mathbf{F}^q \mathbf{A} \mathbf{A}^T \mathbf{q} = -\Lambda \mathbf{Q} \tag{17}$$

Since Λ is diagonal, we may write for the ith normal mode,

$$\ddot{Q}_i = -\lambda_i Q_i = -\omega_i^2 Q_i. \tag{18}$$

Thus, each of the $3N$ normal modes behaves like a harmonic oscillator with circular frequency ω_i, except for the six equations with $\lambda_i = 0$ which contain a Q_i given by Eq. (16) corresponding to either a rotational or a translational motion of the protein with no internal motion.

Now let us introduce $\beta_{ji} \dot{Q}_i$, the frictional force on the ith normal coordinate upon displacement of the jth normal coordinate. This is the damping force present in solution which must be added to Eq. (18) to obtain in matrix form

$$\ddot{\mathbf{Q}} + \beta \dot{\mathbf{Q}} + \Lambda \mathbf{Q} = T(t) \tag{19}$$

where $T(t)$ is the random force exerted by Brownian motion on the normal modes which are damped by the frictional forces of the medium. If, as is probable, β is diagonal, then each mode and its damping may be considered independently.

$$\ddot{Q}_i + \beta_i \dot{Q}_i + \omega_i^2 Q_i = T(t) \tag{20}$$

This equation is a generalization of Eq. (187) in the well-known review by Chandrasekhar[32] on Brownian motion. For the six rotational and translational motions $\lambda_i = 0$, since there is no constant external force on the protein and for these motions Eq. (20) simply changes to the well-known equations for rotational and translational diffusion. For high-frequency modes ($\omega_i^2 \gg \beta$) this equation is that of a simple harmonic oscillation. These terms give rise to the normal group vibration, the amide I, amide II, etc., which are readily observed in protein solutions by the Raman effect. For low-frequency modes such as bending or twisting the amplitudes will be damped so that all terms in Eq (20) may be of importance. Hence these modes cannot be considered a highly thermally excited quantum oscillator, but rather these atomic displacements will be diffusional or stochastic in nature but still directed along the corresponding normal mode displacement Q_i, as given in Eq. (16) where the linear transformation is obtained from the force constant matrix, $\mathbf{F}^R$, as shown in Eq. (14). Thus, the protein will not kick and scream, flopping about in any old manner, but will go along certain trajectories of the normal modes in a diffusive way.

At what frequency do the vibrations begin to be damped out? It seems likely to occur around 20 cm^{-1} (10^{11} Hz) in globular proteins in solution,

[32] S. Chandrasekhar, Rev. Mod. Phys. 15, 1 (1943).

that is, no frequencies in proteins or protein crystals have been found below 50 cm^{-1} which do not disappear when put into solution (see Fig. 10). Thus, we would imagine that in the crystal the protein behaves in a much more vibrational manner in its low-frequency motions and in solution, in a more stochastic manner. On the other hand, fibrous proteins, such as collagen and muscle, show frequencies 0.1 cm^{-1} (10^{10} Hz) (see Fig. 8).

One other point of interest is the fact that we have neglected anharmonicity, which certainly may be important in the very low-frequency motions. Because of the anharmonic nature of potential wells, the simple quadratic expansion [Eq. (11)] may not be sufficient and higher terms must be added. However, Raman scattering from protein crystals does show protein vibrations at about 25–30 cm^{-1}. Such vibration as pointed out in the beginning of this chapter must be thermally populated to their first 20 quantum levels. If, in fact, there were a great deal of anharmonicity for these vibrations, the effect of the anharmonicity would be to change the energy spacing between subsequent quantum levels with the result that the room temperature observation of such low-frequency vibrations by Raman spectroscopy would be impossible. The appearance of even a broad band at 25 and 30 cm^{-1} is highly indicative of the essentially harmonic nature of the potential which produced these vibrations. Thus, their disappearance in solution is probably due to damping by the solvent and not due to effects of anharmonicity. Second, the calculations of Karplus et al.[24] shown in Fig. 12 show beautiful parabolic harmonic potentials even for so complex a motion as the hinge-bending mode in lysozyme. The principles above, illustrated theoretically, have been applied by these same authors, in their detailed study of bovine pancreatic trypsin inhibitor.

B. Computer-Simulated Dynamics of Bovine Pancreatic Trypsin Inhibitor

Recently, an ambitious calculation of the dynamics of a folded globular protein, bovine pancreatic trypsin inhibitor (PTI), has been made and a calculation of the time averaged structure for this molecule has been obtained.[33] The approach used by these authors is of the molecular dynamics type, in which the practical equations of motion for all of the atoms in an assembly are solved simultaneously for a suitable time period, and information is abstracted by analyzing the resulting atomic displacements. The full interatomic potential can be used to obtain the forces on the atoms so that the method is applicable even if the system is highly anhar-

[33] J. A. McCammon, B. R. Gelin, and M. Karplus, (1977) *Nature (London)* **267**: 585.

monic. It is of interest that these authors found that the time-averaged structure which they obtained in the dynamic simulations is near the X-ray structure but is not identical to it. The root-mean-square deviation of the α-carbon atoms taken over all 58 amino acid residues was found to be a remarkably large 1.2 Å, and that for all the atoms, 1.7 Å. The largest deviations come from the two ends of the molecule in an internal loop, residues 25 and 28, which connects two β-sheets. Differences in the side chains were also found. The more ordered parts of the protein, that is, the two parts of the β-sheet (residues 18–24, 29–35) and the α-helix (residues 48–56) have significant deviations except for the atoms near 56, which, of course, lie near the carboxl terminus of the molecule. Their paper gives the peptide backbone (α-carbon atoms) and disulfide bonds of PTI, with the X-ray structure and the time evolved structure after 3.2 psec of dynamical simulation. Although the motion is by and large stochastic and diffusional, certain components of the relaxation correlation function showed dominant frequency components. For example, as might be expected, in the N_{22}—C_{22} bond, the main peaks were found at 800, 1010, 1110, 1170, and 1430 cm^{-1}. These frequencies correspond to the calculated normal modes that involve the largest displacement of the C—N bond of the isolated peptide fragment in this region, i.e., these are the local group frequencies discussed in Section II. Thus, we see that for high frequencies, the motion of the atoms in the protein molecule is largely harmonic.

Even for some low-frequency modes, harmonic behavior is often found. For example, backbone dihedral angles (ψ, ϕ) were found to exhibit considerable damping in general but the variations of ϕ_{22} for the twenty-second residue preserved a long-term correlation in a low frequency oscillation on the order of 2 psec which would correspond to a 66 cm^{-1} vibration. On the other hand, the χ_{22} value for the same residue produced a nearly monotonic decay with a relaxation time of 2 psec. This variation in χ_{22} is due to the torsional rotation of the tyrosine ring. It appears to be clear from the decay behavior of the correlation function that the fluctuations of the ψ, ϕ, and χ dihedral angles are dominated by the interaction of the protein matrix. Thus, the authors conclude that torsional motions which involved substantial displacement of large groups inside a protein, for example, aromatic side chains, will have collected diffusion-like behavior, while those subject to smaller steric interactions will retain more of a local vibrational character in agreement with the discussion given above.

One motion of considerable interest was the concerted atomic motion which was observed in the relatively flexible loop region 25–29 which seem to oscillate as a whole with the period of 6 psec, which would correspond to 200 cm^{-1}. Some damping of this motion is to be expected in solution due to the solvent accessibility of the loop. The general conclusion of these authors is that the interior of the protein is fluid-like overall, and the local atomic motions have a diffusional character.

IV. Torsional Side Chain Motions as a Source of Low-Frequency Motions in Proteins

Each of the amino acid residues found in a protein, except glycine, possesses either an aromatic or an aliphatic side chain. These side chains are capable of making vibrational or hindered rotational motion about their carbon–carbon bonds (C_α—C_β), (C_β—C_∂), etc. This rotation was one time thought to be potential free but for many years now it has been known that the potential barriers have to be surmounted in turning from one conformation to the other in any such molecule. For example, even in the simplest such molecule, ethane, the rotation of one methyl group with respect to another about the C—C bond is hindered by a barrier of 2.928 kcal/mole. To date, relatively little has been done on experimental measurement of the torsional side chain motions in proteins. For the purposes of this chapter, we will only discuss the rotation of the methyl group, which is much studied by vibrational spectroscopic techniques such as infrared and Raman, and inelastic neutron scattering,[34] and the rotation of the phenylalanine and tyrosine groups which occurs at a much slower rate and which has been observed in the nmr.[35]

All of the aliphatic side chains which possess one or two terminal methyl groups should possess a characteristic vibrational frequency in the 200–400 cm^{-1} region due to the torsional oscillation. Molecules, such as alanine and isoleucine, will have one methyl-hindered rotation, whereas valine and leucine belong to a group of molecules called "two-top rotor" because they have two methyl groups connected to a single carbon.[34] Model compounds with similar methyl structures have been much studied and the method for the determination of the torsional barriers is now well known.[34] However, only one study apparently has been made of this type of motion in a polypeptide.[36,37] Figure 13A and B shows the inelastic, incoherent neutron scattering from α-helical and β-sheet poly-L-alanine. The principal peak in the spectrum is the peak at 230 cm^{-1} which has been

[34] J. R. Durig, S. M. Craven, and W. C. Harris, (1972) *in* "Vibrational Spectra and Structure," (J. R. Durig, ed.), Vol. 1, p. 74. Dekker, New York.

[35] G. H. Snyder, R. Rowen, S. Karplus, and B. D. Sykes, *Biochemistry* **14,** 3765 (1975).

[36] W. Drexel and W. L. Peticolas, *Biopolymers* **14,** 715 (1975).

[37] W. L. Peticolas, *Brookhaven Symp. Biol.* **27,** VI-27 (1975).

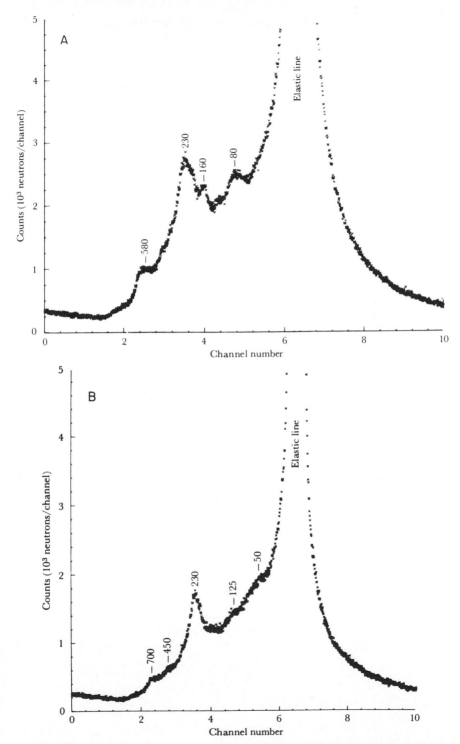

FIG. 13. (A) Inelastic neutron-scattering spectrum of α-helical poly-L-alanine. (B) Inelastic neutron-scattering spectrum of β-sheet poly-L-alanine. From Drexel and Peticolas.[36]

assigned to the torsional rotation of the methyl side chain of alanine. Torsional motions may be very active in neutron scattering because the displacement of the protons on the rotating group is large and the intensity of the scattered neutron flux at the frequency shift of the vibration or other motion increases with increasing amplitude of displacement of the hydrogen atoms involved in the motion.[37] This suggests that a possible experiment for examining the low-frequency torsional motions of the protein side chains would be to examine a fully deuterated protein crystallized in D_2O matrix, so that the only protons remaining would be the side chain protons. These motions could be studied both by inelastic scattering in the 5–300 cm^{-1} region, or by observing the broadening of the elastic scattering peak by lower frequency motions and rotations.

Recently, the side chain torsional potentials of the motion of amino acids in proteins has been the subject of a theoretical study by Gelin and Karplus.[38] In this study, the conformational potentials of the side chains in bovine pancreatic trypsin inhibitor were studied using an empirical potential energy function for each of the internal coordinates and nonbonded interactions. The calculated minimum energy positions were found to be in excellent agreement with the X-ray structure of the side chains in the core or at the surface of the protein. However, the angles for the side chains that are directed out into the solvent did not agree with the calculated values because the contribution from the solvent was not included. It is found that the minimum energy positions in the protein for each of the amino acids is close to that found in the free amino acids. To estimate the effective barrier for rotation of the aromatic rings (tyrosine and phenylalanine), calculations were done in which the protein was permitted to relax as a function of the ring orientation. The resulting barriers were found to be much lower than the rigid rotation barriers which are used to evaluate rotation rates. Using the calculated barrier results, it was possible to estimate the rate constants for the rotation of the tyrosine ring about the C_α—C_β bond by making a simple assumption that the rotation is a uni-molecular process that can be treated by transition state theory. It was found that with this model, all of the aromatic residues except Tyr-35 are expected to be freely rotating on a time scale of about 200 sec^{-1}. This turns out to be in good agreement with the nuclear magnetic resident measurements of Snyder et al.[35] Certainly these low-frequency vibrations and rotations will have to be considered in any thermodynamic analysis of protein behavior.

[38] B. R. Gelin and M. Karplus, *Proc. Natl. Acad. Sci. U.S.A.* **72**, 2002 (1975).

V. Thermodynamic Consequences of Low-Frequency Motions in Proteins

Recently, Sturtevant[39] has shown that there are six possible sources of large heat capacity changes as frequently observed for processes involving proteins. Among these he lists conformational, hydrophobic, and vibrational effects as being of the greatest importance. He also gives a method for estimating the magnitudes of the hydrophobic and vibrational contributions. As he points out, and as has been discussed in detail above, a protein has many soft internal degrees of freedom characterized by force constants weak enough to be significantly effected by chemical changes, such as unfolding or ligand binding. He concludes that it is necessary to include changes in the numbers of easily excitable, that is, thermally excitable, internal vibrational modes as a possible source of heat capacity changes. For example, there is evidence that a considerable fraction of a specific heat of solid proteins (usually close to 0.3 cal/°K/g) is due to internal modes having fundamental frequencies less than 500 cm^{-1} and therefore readily perturbed by chemical reactions. As he points out, a typical protein contains about 16 atoms per 100 daltons, so there are roughly 503 internal modes ($3N$) per 100 daltons. Because the heat capacity of solid proteins is approximately 30 cal/°K per 100 daltons at 25°, the average internal mode has a fundamental frequency corresponding to approximately 800 cm^{-1}, or $4kT$, at 25°. It may be noted that the heat capacity due to a vibrational mode having a fundamental frequency corresponding to $2.5kT$ to $4.5kT$ is changed by 0.12–0.13 cal/K/mole by a 10% change in frequency.

One of the many examples utilizing these concepts which is discussed by Sturtevant is the binding of NAD$^+$ to 2-glyceraldehyde-3-phosphate dehydrogenase (GPDH). This system is characterized by a large decrease in the apparent heat capacity with the necessarily resulting enthalpy–entropy compensation giving an almost constant free energy change. If one attributes the heat capacity decrease to the calculated hydrophobic effects, one is faced with the difficulty of understanding the large decrease in entropy at 25° and 40°. This difficulty is removed if the data are analyzed with the assumption that, as far as entropy is concerned, hydrophobic and vibrational effects are of roughly equal importance. He has suggested that the large volume decrease found in this system can be ascribed to the conversion of a few soft internal modes to stiffer modes as reflected in ΔC_p due to vibrations with the results of tightening of the structure of the protein. An interesting experiment suggested by these

[39] J. M. Sturtevant, *Proc. Natl. Acad. Sci. U.S.A.* **74**, 2236 (1977).

considerations is to measure the Raman spectrum at low frequencies of GPDH crystals with and without NAD^+ binding to see if low-frequency modes show a sharpening. Certainly this paper should be studied by anyone interested in the effects of low-frequency vibrations on the thermodynamics of protein reactions.

Cooper[29] has recently discussed the seeming conflict between the static crystalline model of proteins derived from X-ray crystallography and the fluctuating model of Weber[31] and others.[30] He uses the fact that proteins are small thermodynamic entities and consequently subject to fractionally large fluctuations in volume and energy. This chapter, plus the two review papers,[30,31] contain many references to work on stochastic properties of proteins.

In this chapter, we have emphasized the harmonic approximation to vibrations and fluctuations because it is easily visualized and simple mathematically. However, if a protein possesses several conformational potential energy minima lying close to one another they could give rise to a type of fluctuation not considered in this treatment. For a treatment of the large-amplitude conformational fluctuations, one should consult the recent paper by Veda and Gō.[40]

Acknowledgment

This work was generously supported by grants from the National Institutes of Health No. GM 15547 and the National Science Foundation No. 61329709.

[40] Y. Veda and N. Gō, *Int. J. Pept. Protein Res.* **8**, 551 (1976).

[19] Carbon-13 Nuclear Magnetic Resonance: New Techniques

By ADAM ALLERHAND

I. Introductory Remarks

The first carbon-13 Fourier transform nuclear magnetic resonance (nmr) spectrum of a protein was reported in 1970.[1] Since then, steadily improving instrumentation[2,3] has permitted the observation of *individual-carbon* resonances in natural-abundance ^{13}C nmr spectra of small pro-

[1] A. Allerhand, D. W. Cochran, and D. Doddrell, *Proc. Natl. Acad. Sci. U.S.A.* **67**, 1093 (1970).
[2] A. Allerhand, R. F. Childers, and E. Oldfield, *Biochemistry* **12**, 1335 (1973).
[3] A. Allerhand, R. F. Childers, and E. Oldfield, *J. Magn. Reson.* **11**, 272 (1973).

teins.[2,4-7] I shall use the term "single-carbon resonance" to denote a signal which arises from just *one* carbon of a protein. I shall use the term "individual-carbon resonance" to denote a signal that is either a single-carbon resonance or the sum of single-carbon resonances of two or more *equivalent* carbons (such as the two δ-carbons of a tyrosine residue which is undergoing fast internal rotation about the $C^\beta - C^\gamma$ bond, or two equivalent carbons from the two α-subunits of hemoglobin). Some confusion has been introduced into the literature by the use of the term "single-carbon resonance" to designate a signal that arises from numerous structurally similar carbon atoms, such as the signal from the γ-carbons of *all* tryptophan residues of a *random-coil* protein. The significant role that nmr can have for studies of *folded* proteins in solution depends largely on the ability to resolve signals from individual atomic sites. The γ-carbons of the six tryptophan residues of unfolded lysozyme yield essentially one signal in the ^{13}C nmr spectrum.[2] In contrast, the native protein yields fully resolved single-carbon resonances for C^γ of Trp-62, Trp-63, Trp-108, and Trp-123, and a two-carbon resonance for C^γ of Trp-28 and Trp-111.[2,8] This chapter deals with the observation of numerous single-carbon resonances in natural-abundance ^{13}C nmr spectra of small native proteins in solution, and with techniques and procedures which facilitate the use of the observed resonances for probing the environments of individual amino acid residues of native proteins in solution. The coverage is restricted mainly to the use of nonprotonated aromatic carbon resonances, as developed in the author's laboratory. Other important ways of using ^{13}C nmr spectroscopy for studying proteins are given brief coverage.

II. General Features of ^{13}C nmr Spectra of Proteins

A. Introduction

Carbon-13 nmr spectra of large molecules are normally recorded under conditions of broad band proton-decoupling, in order to remove the complicating effects of $^{13}C-^{1}H$ scalar coupling.[9-11] The natural abun-

[4] E. Oldfield and A. Allerhand, *Proc. Natl. Acad. Sci. U.S.A.* **70**, 3531 (1973).

[5] E. Oldfield, R. S. Norton, and A. Allerhand, *J. Biol. Chem.* **250**, 6368 (1975).

[6] E. Oldfield, R. S. Norton, and A. Allerhand, *J. Biol. Chem.* **250**, 6381 (1975).

[7] E. Oldfield and A. Allerhand, *J. Biol. Chem.* **250**, 6403 (1975).

[8] A. Allerhand, R. S. Norton, and R. F. Childers, *J. Biol. Chem.* **252**, 1786 (1977).

[9] G. C. Levy and G. L. Nelson, "Carbon-13 Nuclear Magnetic Resonance for Organic Chemists." Wiley (Interscience), New York, 1972.

[10] E. Breitmaier and W. Voelter, "^{13}C NMR Spectroscopy." Verlag Chemie, Weinheim, 1974.

[11] F. W. Wehrli and T. Wirthlin, "Interpretation of Carbon-13 NMR Spectra." Heyden, London, 1976.

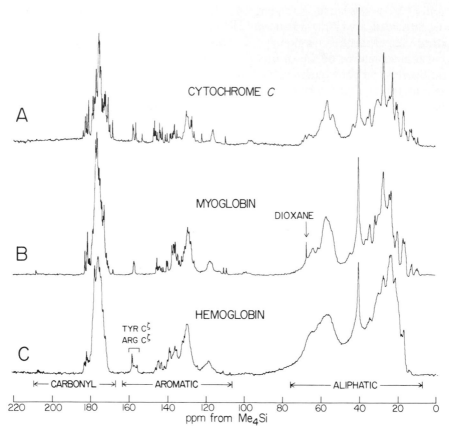

Fig. 1. Proton-decoupled natural-abundance ^{13}C nmr spectra (at 14.2 kG, with a 20-mm probe and 1.1-sec recycle time) of some diamagnetic heme proteins in H_2O (0.1 M NaCl, 0.05 M phosphate buffer). (A) 12 mM horse heart ferrocytochrome c at pH 6.7, 40°, after 5 hr signal accumulation. (B) 10 mM horse carbon monoxide myoglobin at pH 6.7, 36°, after 10 hr signal accumulation. (C) Human adult carbon monoxide hemoglobin (3.3 mM in tetramer) at pH 7.0, 34°, after 10 hr spectral accumulation. Taken from Oldfield et al.[5]

dance of the ^{13}C isotope is only 1.1%, which implies that $^{13}C-^{13}C$ scalar coupling has a negligible effect on ^{13}C nmr spectra of molecules of natural isotopic composition. Therefore, under conditions of proton-decoupling, each magnetically nonequivalent carbon of an amino acid or protein should yield one resonance (the resonance of a nucleotide carbon can be split by $^{13}C-^{31}P$ scalar coupling). In contrast, proton resonances are normally subject to splitting effects from $^1H-^1H$ scalar coupling.[12] The resolution in proton-decoupled natural-abundance ^{13}C nmr spectra of proteins

[12] E. D. Becker, "High Resolution NMR." Academic Press, New York, 1969.

is normally considerably greater than in proton nmr spectra, as a result of the large range of ^{13}C chemical shifts and the absence of $^{13}C-^{13}C$ coupling. However, natural-abundance ^{13}C nmr spectroscopy suffers from the problem of much lower sensitivity than proton nmr, while proton nmr is already a technique of extremely low sensitivity when compared with other common spectroscopic techniques (which operate at higher frequencies), such as electron spin resonance, infrared spectroscopy, and electronic spectroscopy. The Fourier transform technique and other developments have only partially overcome the limitation imposed by the poor sensitivity of ^{13}C nmr (see below). As of today, direct observations of single-carbon resonances of proteins are limited to small proteins available in large amounts.

Unless otherwise indicated, all spectra presented here are proton-decoupled natural-abundance ^{13}C nmr spectra obtained at 15.18 MHz (14.2 kG), on a spectrometer equipped with a 20-mm probe.[3] Spectra of horse heart ferrocytochrome c, horse carbon monoxide myoglobin, and human adult carbon monoxide hemoglobin are shown in Fig. 1.[5] Each spectrum can be divided into the region of carbonyl resonances, the region of aromatic carbon resonances (which also contains the resonances of C^ζ of arginine residues), and the region of aliphatic carbon resonances (see Fig. 1). Each region has some unique properties. An understanding of these properties requires some familiarity with nuclear spin relaxation.

B. Carbon-13 Relaxation

We need to be concerned with three parameters of ^{13}C relaxation: spin-lattice relaxation, spin-spin relaxation, and the nuclear Overhauser effect. Spin-lattice relaxation and the nuclear Overhauser effect influence the intensities of ^{13}C resonances and, therefore, affect the signal-to-noise ratio. Spin-spin relaxation is directly related to linewidths and, therefore, influences resolution. All three parameters can be used to extract information about molecular motions, but spin-lattice relaxation is most frequently used for this purpose.[9,11,13,14]

Spin-Lattice Relaxation. When a collection of molecules is placed in a magnetic field, the degeneracy of the nuclear spin energy states is removed. Consider spin-$\frac{1}{2}$ nuclei (^1H, ^{13}C, ^{15}N, ^{31}P, etc). Before the sample is placed in a magnetic field, the two degenerate spin states obviously have equal populations. Upon placement in a magnetic field, this equality of populations is no longer the equilibrium condition. The system will try

[13] J. R. Lyerla and D. M. Grant, *Phys. Chem., Ser. One* **4**, 155 (1972).
[14] A. Allerhand, D. Doddrell, and R. Komoroski, *J. Chem. Phys.* **55**, 189 (1971).

to establish the equilibrium ratio of populations by means of the process called spin-lattice relaxation. Most often, the equilibrium is reached exponentially. The time constant for this process is designated $1/T_1$, where T_1 is called the spin-lattice relaxation time. The transitions that result in spin-lattice relaxation are caused by fluctuating magnetic fields "within the sample." Molecular motions produce fluctuating magnetic fields by various mechanisms.[11,13,15] These fluctuating fields have a component at the resonance frequency, which causes transitions between the spin states. The magnitude of the component at the resonance frequency determines the value of the spin-lattice relaxation rate $(1/T_1)$.

Calculations for ^{13}C spin-lattice relaxation must take into account various relaxation mechanisms.[11,13,15] However, when dealing with proteins we need to consider only the following mechanisms.

a. Dipole-dipole relaxation[13,14] of $^{13}C-^1H$ results from the fluctuating magnetic fields produced by the rotational motions of C–H vectors. The strength of the dipole–dipole interaction between two nuclear spins (A and B) is a function of the angle between the external magnetic field and the $A-B$ vector. Overall rotational motion of the molecule and internal motions of the $A-B$ vector produce fluctuations of the dipole–dipole interaction. For the simple case of a C–H group which is part of a rigid molecule undergoing isotropic rotational reorientation, the $^{13}C-^1H$ dipolar spin-lattice relaxation time (T_1^D) is readily calculated.[13] A series of plots of T_1^D versus the correlation time (τ_R) for isotropic molecular rotation is shown in Fig. 2. Each curve was calculated for a different value of the applied magnetic field strength.[14,16] It is extremely important to keep in mind that the dipole-dipole relaxation rate $(1/T_1^D)$ is proportional to the inverse sixth power of the distance (r) between the interacting nuclei.[13,14] Hydrogens directly bonded to a carbon $(r \approx 1.1 \text{ Å})$ produce much shorter T_1^D values (Fig. 2) than nonbonded hydrogens $(r \geqslant 2 \text{ Å})$.[5] When dealing with a proton-bearing carbon of a large molecule, $^{13}C-^1H$ relaxation is overwhelmingly dominant $(T_1 = T_1^D)$,[5,14,17] if there are no paramagnetic centers very close to the carbon (see below). However, it is important to consider contributions to $1/T_1$ from relaxation mechanisms other than the $^{13}C-^1H$ dipolar one when dealing with *nonprotonated* carbons, because of the relatively long T_1^D values of such carbons.[5,17]

b. Chemical shift anisotropy (CSA) relaxation results from the fluctuating magnetic fields produced by the rotational motions of anisotropic

[15] A. Abragam, "The Principles of Nuclear Magnetism." Oxford Univ. Press, London and New York, 1961.

[16] D. J. Wilbur, R. S. Norton, A. O. Clouse, R. Addleman, and A. Allerhand, *J. Am. Chem. Soc.* **98**, 8250 (1976).

[17] R. S. Norton, A. O. Clouse, R. Addleman, and A. Allerhand, *J. Am. Chem. Soc.* **99**, 79 (1977).

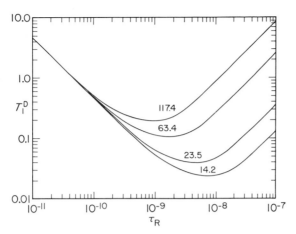

FIG. 2. Log-log plots of T_1^D *versus* τ_R (both in seconds) for a ^{13}C spin relaxing by a dipolar interaction with a single proton 1.09 Å away, in the case of isotropic rotational reorientation and under conditions of proton-decoupling. Plots are given for various magnetic field strengths, indicated in kilogauss. Taken from Wilbur *et al.*[16]

shielding tensors.[11,15] The magnetic field strength of the nmr apparatus is not "seen" in its entirety by the nuclei in atoms and molecules because of shielding by the electrons.[12,15] This effect gives rise to chemical shifts. In general, the chemical shift is anisotropic. For example, the ^{13}C nucleus of a carbonyl group has a very different chemical shift when the C=O bond is parallel to the magnetic field than when it is perpendicular to the field. Chemical shift anisotropy is relatively large for carbonyl and aromatic carbons, and relatively small for saturated carbons (see Norton *et al.*,[17] and references cited therein). For molecules in solution, the observed chemical shift is a rotational average (one-third of the trace of the shielding tensor).[15] However, rotational motion of the anisotropic shielding tensor produces a fluctuating magnetic field which contributes to relaxation. The contribution from CSA relaxation ($1/T_1^{CSA}$) to the value of $1/T_1$ is important only for resonances of *nonprotonated unsaturated* carbons at high magnetic field strengths (see Section III,B).

c. Paramagnetic (electron-carbon dipole-dipole) relaxation is important for carbons which are very near a center of unpaired electron density (see Sections IV,A and IV,B).

d. Dipole-dipole relaxation of ^{13}C–^{14}N is significant only when dealing with a nonprotonated carbon which is directly bonded to nitrogen *and* has fewer than two hydrogens two bonds away.[5]

Spin-Spin Relaxation. As mentioned above, spin-lattice relaxation refers to transitions (i.e., energy changes) caused by fluctuating local magnetic fields in the environment (the "lattice"). Spin-spin relaxation

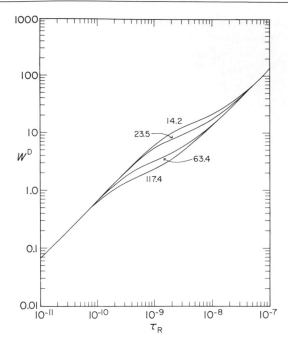

FIG. 3. Log-log plots of W^D (in hertz) versus τ_R (in seconds) for various magnetic field strengths (in kilogauss). See caption of Fig. 2 for other conditions, Taken from Wilbur *et al.*[16]

refers to the "spread" in resonance frequencies (i.e., line broadening) which results from fluctuating local magnetic fields within the sample.[15] The spin-spin relaxation time (T_2), in seconds, is related to the "natural" linewidth, (W), in hertz, by the expression

$$W = 1/\pi T_2 \qquad (1)$$

The various relaxation mechanisms which contribute to $1/T_1$ (see above) also contribute to $1/T_2$ and W. However, the mathematical expressions for $1/T_2$ are different from the equations for $1/T_1$ (compare Figs. 2 and 3). As in the case of T_1, the value of T_2 depends not only on the correlation time for molecular motion but also on the magnetic field strength (see Fig. 3). Also, $T_2 \leq T_1$.[15] In the "extreme narrowing limit", defined by Eq. (2), $1/T_2^D$ (the $^{13}C-^1H$ dipolar contribution to $1/T_2$) becomes equal to $1/T_1^D$, but $1/T_2^{CSA}$ (the CSA contribution to $1/T_2$) does not become equal to $1/T_1^{CSA}$.[15]

$$\tau_R \omega \ll 1 \qquad (2)$$

Here τ_R is the rotational correlation time (in seconds) and ω is the resonance frequency (in radians/second). Equation (2) is normally satisfied for small molecules, and even most large organic molecules in solution, but not for proteins.[5]

When $^{13}C-^1H$ dipolar relaxation is dominant, "natural" linewidths (W) of ^{13}C resonances are proportional to the inverse sixth power of the pertinent C–H distances.[13,14] Therefore, nonprotonated carbons should have narrower resonances than protonated carbons.[2,4-7,14] However, experimental linewidths contain contributions not only from the natural linewidth (W) but also from instrumental broadening (W*). When dealing with small molecules (and typical high-resolution nmr instruments) we often have $W^* \geqslant W$, so that the *observed* differences between the linewidths of protonated and nonprotonated carbons are often small or even undetectable. However, many ^{13}C resonances of proteins have $W \gg W^*$, so that nonprotonated carbons may yield much narrower observed resonances than protonated carbons.[2,4,5]

Nuclear Overhauser Effect. Carbon-13 nmr spectra of large molecules are nearly always recorded with simultaneous strong irradiation (saturation) of the ^1H resonances (proton decoupling). Proton decoupling produces an improvement in the signal-to-noise ratio of each ^{13}C resonance, in two ways. First, the splitting effects of $^{13}C-^1H$ scalar coupling are removed. Second, if ^{13}C relaxation is predominantly $^{13}C-^1H$ dipolar, then proton irradiation causes a favorable redistribution of the populations of the ^{13}C energy levels.[13] This phenomenon is called the nuclear Overhauser effect (NOE). In this chapter, I define the NOE of a ^{13}C resonance as the ratio of intensities with and without proton decoupling. The value of the NOE is a function of the rotational correlation time and of the relative contributions of $^{13}C-^1H$ dipolar and other relaxation mechanisms to $1/T_1$.[10,13]

$$NOE = 1 + \eta_0 T_1/T_1^D \qquad (3)$$

Here $1/T_1$ is the total spin-lattice relaxation rate, $1/T_1^D$ is the $^{13}C-^1H$ dipolar contribution to $1/T_1$, and η_0 is a function of the rotational correlation time. For a fixed value of η_0, the maximum NOE of $1 + \eta_0$ is achieved when relaxation is purely $^{13}C-^1H$ dipolar ($T_1 = T_1^D$). The maximum value of η_0 (1.988) is achieved in the extreme narrowing limit, given by Eq. (2).[10,13] Therefore, the absolute maximum for the NOE, achieved when $^{13}C-^1H$ relaxation is dominant *and* the extreme narrowing limit applies is 2.988. The effect of the rotational correlation time on the NOE is illustrated in Fig. 4, for the simple case of purely $^{13}C-^1H$ dipolar relaxation and isotropic rotational motion.

Different carbons within a molecule may have different NOE values, either because of differences in T_1/T_1^D ratios or because of differences in "effective" rotational correlation times (and therefore different η_0 values). Various carbons within the same molecule may have different effec-

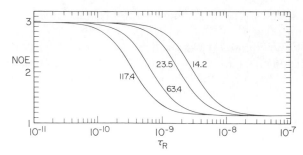

Fig. 4. Semilog plots of NOE versus τ_R (in seconds) for various magnetic field strengths (in kilogauss). See caption of Fig. 2 for other conditions. Here we define the NOE as the ratio of the intensities of the ^{13}C resonance in the presence and in the absence of proton-decoupling. Taken from Wilbur et al.[16]

tive rotational correlation times when overall molecular rotation is anisotropic, or when internal motions are present.[14]

Clearly, a knowledge of the NOE is necessary before one can use relative intensities in proton-decoupled ^{13}C nmr spectra for determining how many carbons contribute to each resonance (the "carbon count"). A common misconception is that carbons with long T_1 values (such as non-protonated carbons) normally have lower NOE values than carbons with short T_1 values (such as protonated carbons). Equation (3) indicates that the NOE is determined by the ratio T_1/T_1^D and not by the value of T_1. It is true, however, that when T_1^D is long, relaxation mechanisms other than the ^{13}C-^1H dipolar one have a greater opportunity for contributing to $1/T_1$ than when T_1^D is short. An NOE of about 3.0 is normally observed for all protonated carbons and most nonprotonated carbons of large organic molecules (such as sucrose, cholesterol, and nucleotides).[14,18] When dealing with proteins, Eq. (2) no longer applies, so that $\eta_0 < 1.988$, and NOE values much lower than 2.988 (Fig. 4) are observed for many carbons, even when $T_1 \approx T_1^D$. Details are given below.

C. Aliphatic Carbons

When we consider the region of aliphatic carbons (about 10 to 75 ppm downfield from Me_4Si) in each of the spectra of Fig. 1 we find that (a) broad bands and narrow resonances are discernible in each spectrum, (b) on the whole, there is less resolution in the spectrum of hemoglobin (Fig. 1C) than in those of myoglobin (Fig. 1B) and cytochrome c (Fig. 1A). Both observations are readily explained. Under the conditions of Fig. 1,

[18] R. S. Norton and A. Allerhand, J. Am. Chem. Soc. 98, 1007 (1976).

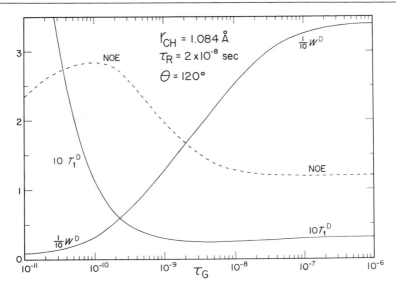

FIG. 5. Semilog plot of $T_1{}^D$ (in seconds), W^D (in hertz), and the NOE as a function of τ_G for a ^{13}C spin relaxing by a dipolar interaction with one proton 1.084 Å away, in the case of isotropic rotation of the molecule as a whole (τ_R = 20 ns), and one degree of internal rotation (correlation time τ_G). Equations (37)–(43) of Ref. 20 were used, with a magnetic field strength of 14.2 kG and an angle (θ) between the C–H vector and the axis of internal rotation of 120°.

the correlation times (τ_R) for overall rotation are about 20 nsec for cytochrome c and myoglobin and about 50 nsec for hemoglobin.[5] With the use of Fig. 3, we find that a methine carbon *which is not involved in fast internal motions* should have a linewidth of about 30 Hz (cytochrome c and myoglobin) or 70 Hz (hemoglobin) at 15.2 MHz (14.2 kG). These values should be multiplied by 2 and 3, for methylene and methyl carbons, respectively.[14] In Fig. 1, the poor resolution in the region which contains mainly α-carbon resonances (about 50 to 75 ppm) is consistent with the predicted linewidths. However, in the region which contains aliphatic side-chain carbons (about 10 to 50 ppm), there are many resonances which are narrower than expected on the basis of Fig. 3. Since all aliphatic carbons of a protein are protonated ones, we conclude that fast internal motions influence the effective rotational correlation time of *some* aliphatic side-chain carbons.[19,20]

Figure 5 shows an example of a computation of the effect of internal rotation on W, T_1, and the NOE, for the simple case of a C–H group un-

[19] A. Allerhand, D. Doddrell, V. Glushko, D. W. Cochran, E. Wenkert, P. J. Lawson, and F. R. N. Gurd, *J. Am. Chem. Soc.* **93**, 544 (1971).

[20] D. Doddrell, V. Glushko, and A. Allerhand, *J. Chem. Phys.* **56**, 3683 (1972).

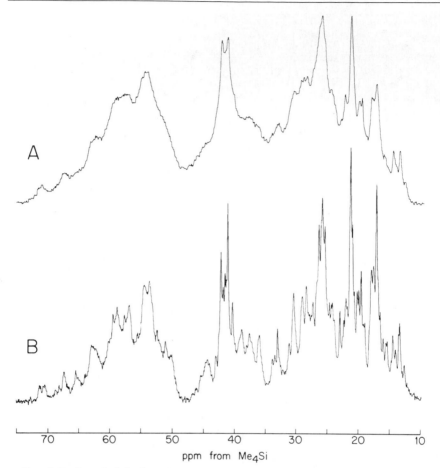

FIG. 6. Region of aliphatic carbons in the proton-decoupled natural-abundance ^{13}C nmr spectra of ~15 mM hen egg white lysozyme in H_2O, 0.1 M NaCl. (A) At 14.2 kG with a 20-mm probe, pH 4.0, 36°, after 20 hr accumulation time with a recycle time of 2.2 sec. (B) At 63.4 kG with a 10-mm probe, pH 3.1, 30°, after 27 hr accumulation time with a recycle time of 3.0 sec. Taken from Norton et al.[17]

dergoing one degree of internal rotation (with a correlation time τ_G), when overall molecular rotation is isotropic (with $\tau_R = 20$ nsec).[5,20] Note that internal rotation has a significant influence on the linewidth only when $\tau_G \lesssim \tau_R$.[5,14,20]

The decrease in resolution when going from cytochrome c and myoglobin to hemoglobin (Fig. 1) probably results not only from increases in natural linewidths but also from the presence of a greater number of nonequivalent carbons. Actually, we do not expect to find a significant number of resolved single-carbon resonances in the aliphatic carbon region of natural-abundance ^{13}C NMR spectra (at 14.2 kG) of proteins even as small as cytochrome c for two reasons. First, a relatively large number of carbons contribute to the aliphatic region. Second, chemical shift nonequivalence caused by folding of the protein (see below) is probably relatively small for side-chain carbons which are capable of internal rotation. The use of high magnetic field strengths should improve significantly the chances of observing single-carbon resonances in the aliphatic region of the spectrum.[17] In Fig. 6 we compare the aliphatic regions of ^{13}C nmr spectra of hen egg white lysozyme at 14.2 kG and 63.4 kG (67.9 MHz).[17] As expected, the resolution (reciprocal of the linewidth in parts per million) is considerably greater at 63.4 kG (Fig. 6B) than at 14.2 kG (Fig. 6A). However, even at 63.4 kG, few, if any, of the aliphatic carbons of hen egg white lysozyme yield resolved single-carbon resonances.[17] Bovine pancreatic trypsin inhibitor, with a sequence of only 58 amino acid residues, has yielded numerous single-carbon resonances in the aliphatic region of the natural-abundance ^{13}C nmr spectrum at 84.5 kG.[21] The use of even higher magnetic field strengths (such as 150 kG) may provide enough resolution for observing single aliphatic carbons of larger proteins. This is a nontrivial statement, because an increase in magnetic field strength does not necessarily provide increased resolution. It is necessary to consider chemical shift anisotropy (CSA) relaxation as a possibly significant broadening mechanism at high magnetic field strengths (see below).[17,22] It turns out, however, that CSA relaxation should be negligible for aliphatic carbons even at 200 kG.[17] A very different picture emerges for carbonyl and nonprotonated aromatic carbons (see below).

As an example of the potential of very high magnetic field strengths for studies of aliphatic carbons, consider the downfield portion of the aliphatic region (about 50 to 75 ppm in Fig. 6). This region contains the resonances of C^{α} of all residues except glycines, and those of C^{β} of threonine and serine residues.[23] From chemical shift considerations,[23,24] it appears that the two peaks at the downfield edge of Fig. 6B (at 70.5 and 71.3 ppm) arise from C^{β} of threonine residues (probably from 4 of the 7 threonine residues of lysozyme). It is obviously unrealistic to use the spectrum at 14.2 kG (Fig. 6A) to study individual threonine residues of lysozyme. There is more hope at 63.4 kG (Fig. 6B), but magnetic field strengths above 100 kG (not yet available for high resolution ^{13}C nmr) will probably

[21] R. Richarz and K. Wüthrich, *FEBS Lett.* **79**, 64 (1977).

[22] G. C. Levy and U. Edlund, *J. Am. Chem. Soc.* **97**, 5031 (1975).

[23] V. Glushko, P. J. Lawson, and F. R. N. Gurd, *J. Biol. Chem.* **247**, 3176 (1972).

[24] P. Keim, R. A. Vigna, J. S. Morrow, R. C. Marshall, and F. R. N. Gurd, *J. Biol. Chem.* **248**, 7811 (1973).

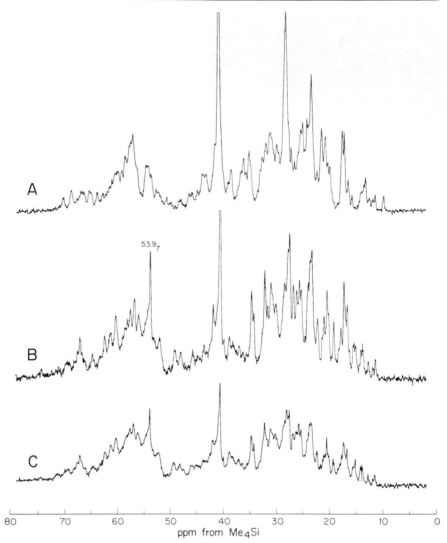

FIG. 7. Region of aliphatic carbons in natural-abundance ^{13}C nmr spectra at 63.4 kG (10-mm probe, recycle time of 2.0 sec) of ferrocytochromes c in H_2O, 0.1 M NaCl, 0.05 M phosphate buffer, 25°. (A) 13 mM protein from horse heart, pH 7.0, after 18 hr signal accumulation with full proton decoupling. (B) 15 mM protein from *Candida krusei,* pH 6.8, after 9 hr signal accumulation with full proton decoupling. (C) Same sample as in spectrum B, but after 18 hr signal accumulation with gated proton decoupling. Taken from Wilbur and Allerhand.[25]

be necessary for clear observations of C^β of individual threonine residues. It is unlikely that many individual C^α resonances of enzymes (of natural isotopic composition) will be observable in the near future.

A special case arises when an ϵ-N-trimethyllysine residue is present. The trimethylammonium group of such a residue gives rise to a resonance near the upfield edge of the α-carbon region (at about 54 ppm).[25] Because of internal motions in an ϵ-N-trimethyllysine side chain, the trimethylammonium carbon resonance has a smaller linewidth and greater NOE than the resonances of α-carbons, and, as a result, is easy to identify.[25] In Fig. 7 we show the aliphatic region in spectra (at 63.4 kG) of horse heart ferrocytochrome c (Fig. 7A) and *Candida krusei* ferrocytochrome c (Fig. 7B). The yeast protein has one ϵ-N-trimethyllysine residue, while the horse protein has none.[26] The sharp peak at 54 ppm in Fig. 7B arises from the methyl groups of the ϵ-N-trimethyllysine residue.[25] This resonance has an NOE of 2.8 ± 0.4, while the α-carbon envelope has an NOE of 1.3 ± 0.2. These differences cause changes in relative intensities when going from the fully proton-decoupled spectrum (Fig. 7B) to a spectrum recorded under conditions of gated decoupling (Fig. 7C). The gated-decoupling method used for the spectrum of Fig. 7C retains $^{13}C-{}^{1}H$ decoupling (elimination of splittings caused by $^{13}C-{}^{1}H$ scalar coupling) but eliminates the NOE effect.[27]

D. Carbonyl Carbons

Ordinarily, all carbonyl carbons of a protein are nonprotonated. Therefore, they should all yield relatively narrow resonances, even in the absence of line narrowing caused by internal motions.[5,17] The reported quantitative discussion for linewidths of nonprotonated aromatic carbon resonances,[5,17] summarized in the next section, is equally applicable to carbonyl resonances. In spite of the relatively small linewidth of each carbonyl resonance, the carbonyl regions of Fig. 1 (14.2 kG) yield few individual-carbon resonances. Even at 63 kG, few carbonyl carbons of proteins have yielded resolved individual-carbon resonances: Resolved resonances have been observed for some carboxylate groups (at the downfield edge of the carbonyl region in the spectrum of hen egg white lysozyme at 63.4 kG [28]) and for N-terminal glycine residues (at the upfield

[25] D. J. Wilbur and A. Allerhand, *FEBS Lett.* **74,** 272 (1977).

[26] M. O. Dayhoff, ed., "Atlas of Protein Sequence and Structure," Vol. 5. Natl. Biomed. Res. Found., Silver Spring, Maryland, 1972.

[27] R. Freeman, H. D. W. Hill, and R. Kaptein, *J. Magn. Reson.* **7,** 327 (1972).

[28] H. Shindo and J. S. Cohen, *Proc. Natl. Acad. Sci. U.S.A.* **73,** 1979 (1976).

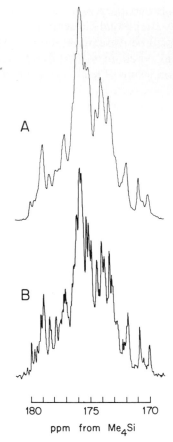

FIG. 8. Carbonyl region is proton-decoupled natural-abundance ^{13}C nmr spectra of ~15 mM hen egg white lysozyme in H_2O, pH 3.1, 0.1 M NaCl. (A) At 14.2 kG with a 20-mm probe, at 42°, after 10 hr accumulation time with a recycle time of 1.1 sec. (B) At 63.4 kG with a 10-mm probe, at 30°, after 27 hr accumulation time with a recycle time of 3.0 sec. Taken from Norton et al.[17]

edge of the carbonyl region in spectra of horse myoglobin[5,29] and human fetal hemoglobin[7] at 14.2 kG).

In estimating the expected improvement in resolution when going from low to high magnetic field strengths, we must consider the importance of chemical shift anisotropy (CSA) as a relaxation mechanism for carbonyl resonances at high magnetic field strengths.[17] Above about 40 kG, an increase in magnetic field strength is expected to cause a *decrease* in the resolution of carbonyl resonances.[17] Some details are given below in the discussion of nonprotonated aromatic carbons. Figure 8 shows the

[29] D. J. Wilbur and A. Allerhand, *FEBS Lett.* **79**, 144 (1977).

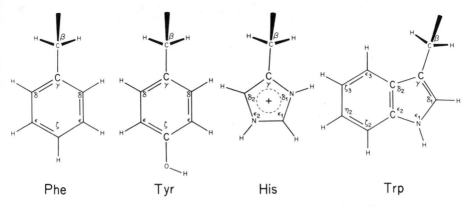

Phe　　　　Tyr　　　　His　　　　Trp

Fɪɢ. 9. Structures of side chains of aromatic amino acid residues.

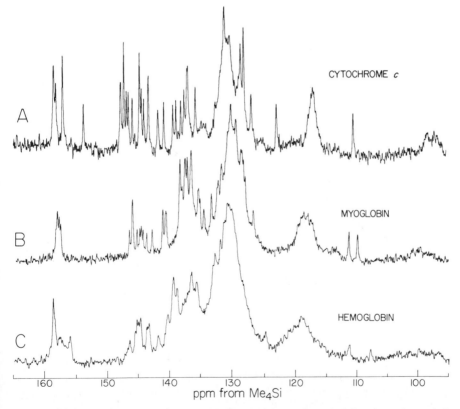

CYTOCHROME *c*

A

MYOGLOBIN

B

HEMOGLOBIN

C

160　　150　　140　　130　　120　　110　　100
ppm from Me₄Si

Fɪɢ. 10. Region of aromatic carbons and C^ζ of arginine residues in fully proton-decoupled natural-abundance ^{13}C nmr spectra at 14.2 kG of some diamagnetic heme proteins. (A) Horse heart ferrocytochrome *c*, from Fig. 1A. (B) Horse carbon monoxide myoglobin, from Fig. 1B. (C) Human adult carbon monoxide hemoglobin, as in Fig. 1C. Taken from Oldfield et al.[5]

carbonyl region in spectra of hen egg white lysozyme at 14.2 and 63.4 kG.[17]

E. Aromatic Carbons

The relatively few aromatic carbons of a protein yield resonances which cover a large range of chemical shifts (about 100 to 160 ppm downfield from Me$_4$Si).[2,4–7] The resonances of C$^\zeta$ of arginine residues appear at the downfield edge of the aromatic region (at about 158 ppm).[6] For convenience, these resonances are often included in discussions of the aromatic carbon resonances.[6]

The nature of the aromatic region of the ^{13}C nmr spectrum of a protein is greatly influenced by the fact that there are two types of aromatic carbons, methine and nonprotonated ones (Fig. 9). Figure 10 shows the aromatic regions of the spectra of Fig. 1. Note the presence of broad features and narrow resonances in each spectrum. Some of the narrow peaks are located on top or on the sides of the broad features. On the basis of calculated linewidths (see below), we can confidently assume that all of the nonprotonated aromatic carbons give rise to narrow resonances. However, our calculations cannot conclusively rule out the possibility that some narrow resonances arise from methine carbons: If an aromatic amino acid side chain is undergoing fast internal rotation (relative to the rate of overall molecular rotation), then a methine aromatic carbon *may* yield a narrow resonance (see Fig. 5). Fortunately, there is an experimental procedure for distinguishing resonaces of nonprotonated carbons from those of methine carbons, namely, "inefficient" proton decoupling. Inefficient proton decoupling produces a residual broadening of a ^{13}C resonance which is proportional to the square of the pertinent ^{13}C–^{1}H scalar coupling constant.[30] One-bond ^{13}C–^{1}H scalar coupling constants are larger than 100 Hz, while long-range ^{13}C–^{1}H scalar coupling constants are typically 10 Hz or less.[31] Therefore, it is possible to set up "inefficient" proton-decoupling conditions which will produce practically complete proton decoupling of nonprotonated carbon resonances, but a "smearing out" of methine carbon resonances. One convenient "inefficient" decoupling procedure is noise-modulated off-resonance proton-decoupling,[2,5,32] in which the center frequency of the noise-modulated ^{1}H irradiation is shifted enough off-resonance to cause incomplete decoupling of methine carbon resonances, but essentially full decoupling of

[30] R. R. Ernst, *J. Chem. Phys.* **45**, 3845 (1966).

[31] J. B. Stothers, "Carbon-13 NMR Spectroscopy." Academic Press, New York, 1972.

[32] E. Wenkert, A. O. Clouse, D. W. Cochran, and D. Doddrell, *J. Am. Chem. Soc.* **91**, 6879 (1969).

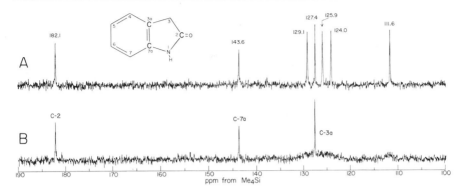

FIG. 11. Unsaturated carbon region in natural-abundance ^{13}C nmr spectra (at 14.2 kG with a 20-mm probe) of 56 mM of oxindole in H_2O (0.1 M NaCl, pH 3.1, 44°), recorded with a recycle time of 60 sec and 512 accumulations per spectrum. (A) Fully proton-decoupled. *Numbers* above the peaks are chemical shifts. (B) Noise-modulated off-resonance proton-decoupled. Taken from Oldfield et al.[6] Some details about ^1H irradiation are given in Oldfield et al.[5]

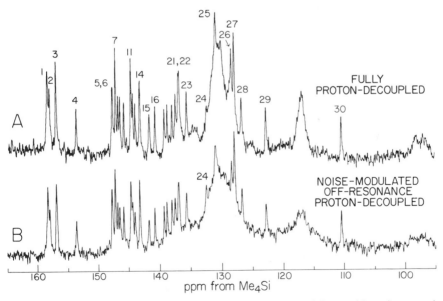

FIG. 12. Resonances of aromatic carbons and C^ζ of arginine residues in natural-abundance ^{13}C nmr spectra (at 14.2 kG, with a 20-mm probe, 1.1 sec recycle time, and 5 hr signal accumulation per spectrum) of 12 mM horse heart ferrocytochrome c in H_2O, 0.05 M phosphate buffer, pH 6.7, 40°. Peak numbers are those of Refs. 4 and 6. (A) Recorded with full proton decoupling (same spectrum as in Fig. 10A). (B) Recorded under conditions of noise-modulated off-resonance proton-decoupling (taken from Fig. 3B of Oldfield and Allerhand[4].

nonprotonated carbon resonances. As a simple example, consider the aromatic carbon resonances in the fully proton-decoupled ^{13}C NMR spectrum of oxindole (Fig. 11A). Because oxindole is a small molecule, instrumental broadening contributes significantly to all linewidths in Fig. 11A (see Section II,B). Therefore, no dramatic difference between the linewidths of methine and nonprotonated carbons is observed in Fig. 11A. However, under conditions of noise-modulated off-resonance proton-decoupling (Fig. 11B), the nonprotonated carbon resonances are just as narrow as in the fully proton-decoupled spectrum, but the resonances of the methine carbons are "smeared out." Figure 12A shows the aromatic region in the fully proton-decoupled spectrum of horse heart ferrocytochrome c (the same spectrum as in Fig. 10A). Figure 12B is the corresponding spectrum recorded with noise-modulated off-resonance proton-decoupling. All the narrow resonances of Fig. 12A remain narrow in Fig. 12B, and must, therefore, arise from nonprotonated carbons. The broad bands of Fig. 12A become even broader in Fig. 12B. They arise from methine carbons. The methine aromatic carbon resonances of cytochrome c are not as dramatically broadened by the use of inefficient decoupling as in the case of oxindole (Fig. 11), because they are very broad even in the fully proton-decoupled spectrum (Fig. 12A): The broadening effect of inefficient proton-decoupling becomes small when the linewidth of a ^{13}C resonance (in a fully proton-decoupled spectrum) approaches the value of the ^{13}C–^{1}H scalar coupling constant.

Noise-modulated off-resonance proton-decoupling experiments have shown that all the narrow resonances in the aromatic regions of the spectra of the proteins which have been examined in my laboratory arise from nonprotonated carbons. In principle, aromatic residues with sufficiently fast internal rotation (see Fig. 5) may yield narrow methine aromatic carbon resonances. Internal rotation of the phenyl rings of phenylalanine and tyrosine residues has been detected in some proteins.[33–41] However, it appears that, in general, internal rotation of aromatic amino acid side chains of folded proteins is not fast enough (see Fig. 5) to cause

[33] I. D. Campbell, C. M. Dobson, and R. J. P. Williams, *Proc. R. Soc. London, Ser. B* **189**, 503 (1975).

[34] G. H. Snyder, R. Rowan, III, S. Karplus, and B. D. Sykes, *Biochemistry* **14**, 3765 (1975).

[35] K. Wüthrich and G. Wagner, *FEBS Lett.* **50**, 265 (1975).

[36] G. Wagner, A. DeMarco, and K. Wüthrich, *Biophys. Struct. Mech.* **2**, 139 (1976).

[37] I. D. Campbell, C. M. Dobson, G. R. Moore, S. J. Perkins, and R. J. P. Williams, *FEBS Lett.* **70**, 96 (1976).

[38] S. J. Opella, D. J. Nelson, and O. Jardetzky, *J. Am. Chem. Soc.* **96**, 7157 (1974).

[39] D. J. Nelson, S. J. Opella, and O. Jardetzky, *Biochemistry* **15**, 5552 (1976).

[40] A. Cave, C. M. Dobson, J. Parello, and R. J. P. Williams, *FEBS Lett.* **65**, 190 (1976).

[41] K. Dill and A. Allerhand, *J. Am. Chem. Soc.* **99**, 4508 (1977).

significant narrowing of the methine aromatic carbon resonances.[5,41,42]
However, we must consider the possibility of aromatic residues with fast

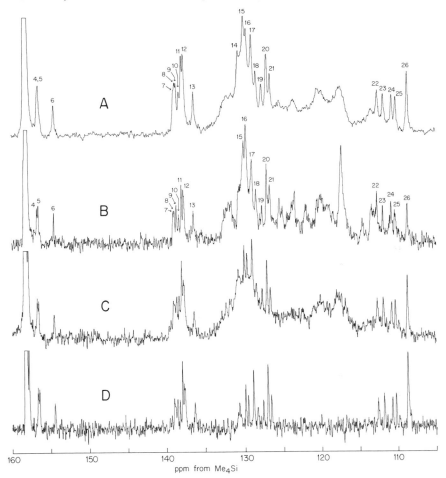

FIG. 13. Region of aromatic carbons in natural-abundance ¹³C nmr spectra of ~15 mM
hen egg white lysozyme in H$_2$O, 0.1 M HaCl, pH 3.1. Nonprotonated aromatic carbon resonances are numbered as in Allerhand *et al.*[5,6,8] The truncated band at about 158 ppm arises
from C$^\zeta$ of the 11 arginine residues (peaks 1–3 of Allerhand *et al.*[5,6,8]). (A) At 14.2 kG and 42°,
with a 20-mm probe, full proton decoupling, 1.1 sec recycle time, and 10 hr signal accumulation. (B) At 63.4 kG and 30°, with a 10-mm probe, full proton-decoupling, 3.0 sec recycle
time, and 27 hr signal accumulation. (C) As spectrum B, but with noise-modulated off-
resonance proton-decoupling, a recycle time of 5 sec, and 57 hr signal accumulation. (D)
Convolution-difference spectrum derived from the same raw data as spectrum C, with 2.8
Hz and 56 Hz digital broadening (see Section III,C). Taken from Norton *et al.*[17]

[42] D. T. Browne, G. L. Kenyon, E. L. Packer, D. M. Wilson, and H. Sternlicht, *Biochem.
Biophys. Res. Commun.* **50,** 42 (1973).

internal rotation (relative to the rate of overall rotation of the protein) in future studies of additional proteins. In any case, noise-modulated off-resonance proton-decoupling is a simple method for unambiguously distinguishing methine from nonprotonated aromatic carbon resonances. We shall see below that failure to apply this method to spectra at high magnetic field strengths (such as 63.4 kG) may result in erroneous conclusions even in the absence of internal rotations of the aromatic side chains.

The aromatic region of a fully proton-decoupled spectrum (at 14.2 kG) of hen egg white lysozyme is shown in Fig. 13A. The narrow resonances, labeled 4–26, arise from the 28 nonprotonated aromatic carbons.[2,5,6,8] The numbering system is that of Allerhand et al.[5,6,8] At 14.2 kG (Fig. 13A), the narrow nonprotonated carbon resonances are easily distinguished from the broad methine carbon bands, even when full proton-decoupling is used. However, at 63.4 kG (Fig. 13B), the difference between the line-widths of the two types of resonances diminishes[17] (details are given in Section III,B), and noise-modulated off-resonance proton-decoupling (Fig. 13C) becomes more necessary for distinguishing the two types of resonances (see Section III,C). The broad methine carbon bands of Fig. 13C can be eliminated by mathematical manipulation of the spectrum in the small computer which is normally an integral part of a Fourier transform nmr instrument. One simple procedure is the convolution-difference method,[43,44] described in Section III,C. The convolution-difference spectrum shown in Fig. 13D was obtained from the same raw data as the spectrum of Fig. 13C.

F. Effect of Protein Folding on ^{13}C Chemical Shifts

Figure 14B is the aromatic region of a convolution-difference ^{13}C nmr spectrum (at 14.2 kG) of native hen egg white lysozyme. The 28 nonprotonated aromatic carbons of this protein give rise to 18 single-carbon resonances (peaks 4–10, 13–16, 18, 19, 21, and 22–25) and 5 two-carbon resonances (peaks 11, 12, 17, 20, and 26).[5,8] Figure 14A is the aromatic region of the convolution-difference spectrum of lysozyme denatured with guanidinium chloride. In contrast to the native protein, each type of residue now yields chemical shifts essentially independent of its position in the sequence.[2] Figure 14 clearly indicates that the folding of lysozyme into its native conformation produces large changes in the chemical shifts of most nonprotonated aromatic carbons, and that these changes are quite dif-

[43] I. D. Campbell, C. M. Dobson, R. J. P. Williams, and A. V. Xavier, Ann. N. Y. Acad. Sci. **222,** 163 (1973).
[44] I. D. Campbell, C. M. Dobson, R. J. P. Williams, and A. V. Xavier, J. Magn. Reson. **11,** 172 (1973).

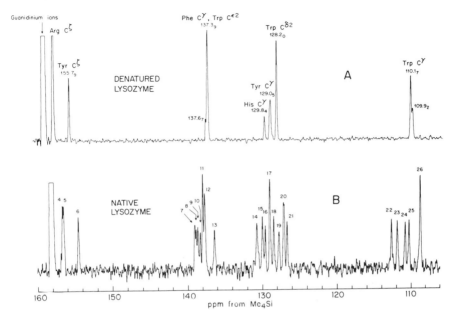

FIG. 14. Natural-abundance ^{13}C nmr spectra (at 14.2 kG, with a 20-mm probe) of 9 mM hen egg white lysozyme in H_2O, 0.1 M NaCl. Spectra were recorded under conditions of noise-modulated off-resonance proton decoupling. The convolution-difference method was applied with 0.9 Hz and 10 Hz digital broadening (see Section III,C). (A) Denatured protein, 6 M guanidinium chloride, pH 3.0, 50°, after 40 hr signal accumulation with a recycle time of 2.2 sec. Assignments are those of Allerhand et al.[2] Numbers above peaks are chemical shifts. This spectrum is taken from Fig. 3A of Norton and Allerhand.[82] (B) Native protein, pH 3.4, 38°, after 10 hr signal accumulation with a recycle time of 1.1 sec. Peak numbers are those of Fig. 13 and Allerhand et al.[5,6,8] This spectrum is taken from Fig. 1 of Norton and Allerhand.[92]

ferent for residues (of the same type) at different positions in the sequence. This fortunate phenomenon, which permits the observation of numerous resolved individual nonprotonated aromatic carbon resonances, is quite general for native globular proteins.[2,4,6,7,45–47]

In contrast to the situation with nonprotonated aromatic carbons, there are few reported observations of resolved individual-carbon resonances of methine aromatic carbons,[41] carbonyl carbons,[6,7,28,29] and aliphatic carbons.[21] On the basis of the few published results, it appears that protein folding causes substantial chemical shift nonequivalence for many aliphatic, methine aromatic, and carbonyl carbons. In particular, the use

[45] D. J. Wilbur and A. Allerhand, J. Biol. Chem. 251, 5187 (1976).

[46] K. Ugurbil, R. S. Norton, A. Allerhand, and R. Bersohn, Biochemistry 16, 886 (1977).

[47] D. J. Wilbur and A. Allerhand, J. Biol. Chem. 252, 4968 (1977).

of currently available high magnetic field strengths (such as 63.4 kG) may permit the observation of numerous individual methine aromatic carbons of small proteins.[41] Additional details are given in Sections III,C and V,B.

G. Fourier Transform nmr, Carbon Count, and Sensitivity

There are examples in the recent literature of overenthusiastic use of ^{13}C nmr spectra of proteins. In this section, I present some thoughts on important precautions (as I see them) which are required in order to minimize the chances of incorrect interpretations. First we shall consider a few pertinent features of ^{13}C Fourier transform nmr spectroscopy of proteins.

Fourier Transform nmr Spectroscopy. In Fourier transform nmr, a short burst of radio frequency excitation (typically about 20 μsec duration), at a frequency near the resonance frequencies of interest, is applied to the sample, and then the response of the spin system is observed after the excitation is turned off.[48,49] If the *field strength* of the radio frequency excitation (expressed in frequency units) is considerably greater than the difference between the excitation frequency and the resonance frequencies of all ^{13}C nuclei of interest (see Fig. 1), then each resonance will be fully excited even though the excitation is not strictly "on-resonance." Thus, Fourier transform nmr, unlike continuous wave nmr,[12] can easily provide simultaneous excitation of all resonances of interest.[48,49] For simplicity, I shall assume that the radio frequency excitation is so strong that resonances far from the excitation frequency are not significantly attenuated. This condition may not be satisfied on some Fourier transform nmr spectrometers (see Shaw[49] for information about how to handle such a limitation).

A carbon-13 nmr spectrum of a protein is normally obtained in a Fourier transform nmr instrument, with the use of many repetitive scans which are added together in the memory of a digital computer. The signal-to-noise ratio is proportional to the square root of the number of scans. The interval between successive scans is called the *recycle time*. The recycle time and the duration of the radio frequency excitation pulse affect the intensity of the signal.[48,49] The so-called 90° radio frequency pulse yields the maximum signal in a single scan.[48,49] All the "normal" spectra presented in this report were obtained with the use of 90° radio frequency pulse excitation (see Ref. 49 for circumstances which may warrant the use of shorter pulses). However, when measuring T_1 values it is common practice to apply first a 180° radio frequency pulse, which inverts

[48] R. R. Ernst and W. A. Anderson, *Rev. Sci. Instrum.* **37,** 93 (1966).
[49] D. Shaw, "Fourier Transform NMR Spectroscopy." Elsevier, Amsterdam, 1976.

the populations of the two spin energy levels.[49] The sequence of a 180° pulse, a waiting interval, and then a 90° pulse yields a so-called partially relaxed Fourier transform (PRFT) nmr spectrum (see Section IV,A).

Carbon Count. It is not always safe to assume that the integrated intensity of each observed peak is proportional to the number of carbons that contribute to that peak. Differences in NOE and T_1 values may produce differences in the intensities of single-carbon resonances.[50]

Carbon-13 nmr spectra of proteins are nearly always recorded under conditions of proton-decoupling, which causes NOE's. Different carbons may have different NOE values [see Eq. (3)]. The NOE can be measured by means of gated proton decoupling[27,51] (see Fig. 7B and C) and other methods.[5] One needs to be particularly concerned with differences in NOE values when dealing with the aliphatic carbon region (see Fig. 7B and C), because *some* aliphatic carbons are parts of side chains which have fast internal rotations (see Fig. 5). In contrast, there is experimental evidence that the nonprotonated aromatic carbon resonances of a native protein all have about the same NOE value, in the range 1.0 to 1.5 (the actual value depends on molecular weight and magnetic field strength).[5,17] It is likely that all methine aromatic carbon resonances of a native protein also have about the same NOE value (see below).

If two carbons have different T_1 values, and if the recycle time is not much greater than the longest T_1 value, then the two carbons may yield resonances with very different intensities.[49] Therefore, it is essential to know, at least approximately, the T_1 values of the carbons of interest. Additional details are given in Section III,C.

Sensitivity. Most of the Fourier transform nmr spectrometers which are in use today do not have sufficient sensitivity for detection of single-carbon resonances of a protein (with the use of a "reasonable" signal accumulation time per spectrum, such as 10 hr or less). Even when the best spectrometers are available, there may not be enough protein for observing single-carbon resonances (Table I). Sadly, there has been a tendency to overinterpret some published spectra (no literature will be cited!). Even when the signal-to-noise ratio is not sufficient for observing single-carbon resonances, "good looking" [13]C nmr spectra of proteins can be obtained. There has been an irresistible urge in some laboratories to believe that the *weakest* observed peaks are some exciting single-carbon resonances, when in fact they may be two-carbon, three-carbon, or even eleven-carbon resonances. Furthermore, some "peaks" in published spectra have similar intensities to noise spikes. The assignments of "peaks" and the structural conclusions based on such data should be

[50] A. Allerhand, *Pure Appl. Chem.* **41**, 247 (1975).
[51] S. J. Opella, D. J. Nelson, and O. Jardetzky, *J. Chem. Phys.* **64**, 2533 (1976).

TABLE I

REQUIREMENTS FOR OBSERVATION OF SINGLE-CARBON RESONANCES
OF A PROTEIN OF MOLECULAR WEIGHT 15,000

Magnetic field (kG)	Tube (mm)[a]	Sample size (ml)	Amount of protein (g)[b]	Time[c] (hr)
14.2[d]	20	11	1.7	4
23.5[e]	12	3	0.5	40[f]
63.4[g]	10	2	0.3	10
63.4[h]	15	4	0.6	4[i]

[a] Outside diameter of sample tube. Typical wall thicknesses are 0.5–1.0 mm.

[b] For a 10 mM protein solution.

[c] Signal accumulation time per spectrum required to get a signal-to-noise ratio of about 7 for single nonprotonated aromatic carbon resonances (with quadrature detection or crystal filter[3]). Here the signal-to-noise ratio is defined as 2.8 times the peak height divided by the peak-to-peak noise.

[d] Home-built instrument (see Allerhand et al.[3]).

[e] Varian XL-100 Fourier transform nmr instrument.

[f] Estimate based on spectra reported by Cozzone et al.[51a]

[g] Home-built instrument equipped with a Bruker high-resolution superconducting magnet (see Allerhand et al.[16,17,41]).

[h] Home-built instrument mentioned in footnote g, but equipped with a Bruker 15-mm probe.

[i] Based on unpublished spectra from the author's laboratory.

treated with caution. I urge readers to check the signal-to-noise ratios in published ^{13}C nmr spectra of proteins before reading the conclusions based on those spectra.

Misinterpretation is also possible if the spectra have been recorded under conditions which yield different intensities for different types of carbons (and these differences are not taken into account). For example, the recycle time may be too short for a valid direct comparison of intensities of methine and nonprotonated aromatic carbon resonances (see Section III).

In Figure 14B, *all* 28 nonprotonated aromatic carbons of hen egg white lysozyme yield observable resonances.[5,6,8] I call this a *complete carbon count* for the nonprotonated aromatic carbon resonances of lysozyme. When a complete carbon count can be made for a class of resonances, the possibility of a gross misinterpretation is much smaller, in my opinion, than when such a count cannot be made (as a result of poor sensitivity, poor resolution, uncertainties in T_1 and NOE values, or other problems). We shall see in the next sections that, with the use of good instrumentation and proper techniques, a complete carbon count is quite feasible when dealing with nonprotonated aromatic carbon resonances of small proteins. Some examples of proper and improper techniques (as I see them) are presented in the next section.

At this point, it seems appropriate to emphasize that a small minority of proteins can be realistically studied by natural abundance ^{13}C nmr. If the molecular weight is greater than about 30,000, not many resolved single-carbon resonances should be expected. Even when dealing with small proteins, one must have a large amount of the protein. Also, 2 to 20 hr of signal accumulation per spectrum will be needed normally. The amount of protein and the signal accumulation time per spectrum will depend on the type of available spectrometer. Details are presented in Table I.[3,16,17,41,51a] Because of wide variations in the sensitivities of individual nmr instruments, the numbers in Table I may not correspond to the performance of spectrometers in some laboratories. Also, the requirements listed in Table I refer to nonprotonated aromatic carbons. Other carbons, such as methine aromatic carbons, can be studied with the use of a smaller amount of protein (or less signal accumulation time per spectrum) than is indicated in Table I.

III. Optimum Conditions for Observing Aromatic Carbon Resonances

A. Introduction

Because the sensitivity of natural-abundance ^{13}C nmr spectroscopy is extremely low, the difference between success and failure to observe single-carbon resonances of a protein may depend very strongly on attention to detail. The first obvious detail is to determine if a particular nmr spectrometer has enough sensitivity for the task. Even with all parameters optimized, some types of ^{13}C Fourier transform nmr instruments do not yield single-carbon resonances of proteins in a reasonable amount of signal accumulation time (Table I). In particular, the combination of a small sample tube (diameter ≤ 15 mm) and a low magnetic field strength (such as 14 or 23 kG) will probably lead to difficulties. Instrumental characteristics vary widely and will not be discussed in this section. Instead, I will discuss parameters which are intrinsic characteristics of the sample, and how these parameters affect crucial decisions in the practice of ^{13}C nmr spectroscopy of proteins. The discussion will be restricted to aromatic carbons, with emphasis on nonprotonated aromatic carbons.

B. Relaxation Behavior of Aromatic Carbon Resonances

When dealing with most organic molecules ($\tau_R \leq 10^{-9}$ sec), the linewidths, spin-lattice relaxation times, and nuclear Overhauser enhancements (NOE) of ^{13}C resonances are generally independent of magnetic

[51a] P. J. Cozzone, S. J. Opella, O. Jardetzky, J. Berthou, and P. Jollès, *Proc. Natl. Acad. Sci. U.S.A.* **72**, 2095 (1975).

field strength. Notable exceptions are the resonances of nonprotonated *unsaturated* carbons (see below).[22] When dealing with folded proteins ($\tau_R \geq 10^{-8}$ sec), the linewidths and spin-lattice relaxation times of most (if not all) ^{13}C resonances are expected to depend on the magnetic field strength.[17,20] As a result, the choice of magnetic field strength (if such a choice is available!) is critical. It is not necessarily true that going to higher and higher magnetic field strengths will result in better spectra (see below). Also, spectrometer conditions (recycle time, etc.) which are suitable for one magnetic field strength may not be appropriate for a different magnetic field strength.

In general, only $^{13}C-^{1}H$ dipolar interactions and chemical shift anisotropy (CSA) are important relaxation mechanisms for ^{13}C resonances of proteins.[5,17] However, when dealing with a nonprotonated carbon which is directly bonded to nitrogen, $^{13}C-^{14}N$ dipolar relaxation contributes slightly.[5,18] Also, if a carbon is near a paramagnetic center (such as the iron of ferrimyoglobin), then paramagnetic relaxation may be the dominant effect.[52] Unless otherwise stated, the theoretical results presented below do not include the effects of paramagnetic centers. Also, I shall assume that internal rotations, if present, are too slow to affect the values of W, T_1, and NOE. For *methine* aromatic carbons, it is safe to assume that $^{13}C-^{1}H$ dipolar relaxation is overwhelmingly dominant even at magnetic field strengths as high as 150 kG.[17] For nonprotonated aromatic carbons, we must consider both $^{13}C-^{1}H$ dipolar and CSA relaxation.[17] The conclusions presented below also use the fact that the rotational correlation times (τ_R) of proteins are normally ≥ 10 nsec.

Linewidths. Figure 3 can be used to predict the linewidth of a methine aromatic carbon. For small proteins (τ_R in the range 10 to 20 nsec), the linewidth should decrease with increasing magnetic field strength. Therefore, when dealing with methine aromatic carbons of small proteins, the ratio of the resolutions at high and low magnetic field strengths should be *greater* than the ratio of the two field strengths.

Figure 15 shows the calculated values of W^D and W^{CSA} (the contributions to the linewidth form $^{13}C-^{1}H$ dipolar and CSA relaxation, respectively) for a nonprotonated aromatic carbon.[17] Note that the values of W^{CSA} (solid lines in Fig. 15) also apply to methine aromatic carbons.[17] The values of W^D (dashed lines in Fig. 15) were computed for a nonprotonated carbon which has three hydrogen two bonds removed (Fig. 16).[5] These values are good estimates for C^γ of histidine (in the $N^{\epsilon 2}$-H imidazole form) and tryptophan residues, and for C^ζ of tyrosine residues (Fig. 16).[5] For other nonprotonated aromatic carbons, the values of W^D in Fig. 15 must

[52] R. A. Dwek, "Nuclear Magnetic Resonance in Biochemistry." Oxford Univ. Press, London and New York, 1973.

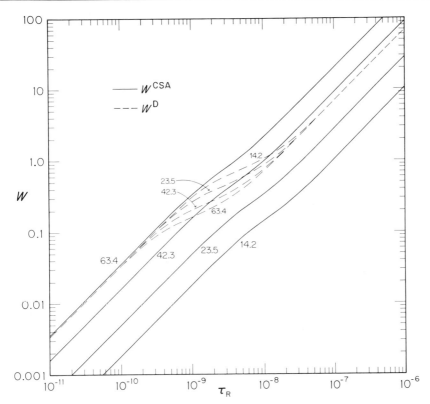

FIG. 15. Log-log plots of W^{CSA} and W^D (solid and dashed lines, respectively, both in hertz) versus τ_R (in seconds), in the case of isotropic rotational reorientation. Plots are given for four magnetic field strengths indicated in kilogauss. W^D was computed for a nonprotonated carbon with 3 hydrogens two bonds away ($r = 2.16$ Å). W^{CSA} was computed for a chemical shift anisotropy of 200 ppm (axially symmetric shielding tensor). Taken from Norton et al.[17]

be multiplied by $n/3$, where n is the number of hydrogens two bonds away.[5] Also, if D_2O and not H_2O is the solvent, the loss of exchangeable O—H and N—H hydrogens must be taken into account.[5] Furthermore, if a nonprotonated carbon has only one hydrogen two bonds away, interactions with hydrogens more than two bonds away and with directly bonded nitrogens become significant.[5] All of the above complications play a more significant *practical* role when dealing with T_1 values than with linewidths (see below).[5]

A comparison of Figs. 3 and 15 indicates that for a methine aromatic carbon the contribution of W^{CSA} is negligible. However, for a nonprotonated aromatic carbon, $W^{CSA} \gg W^D$ at magnetic field strengths much

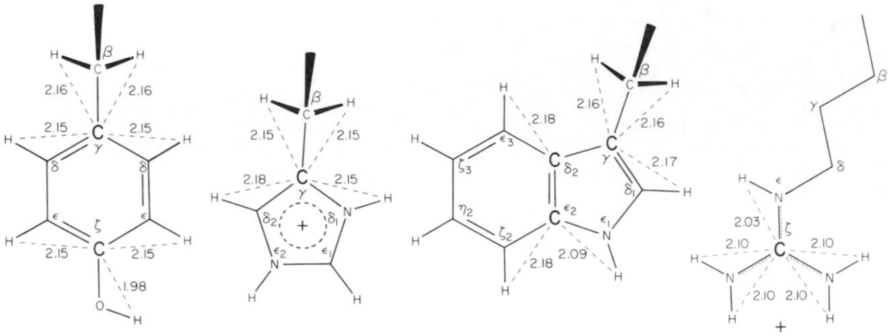

Tyr   His   Trp   Arg

FIG. 16. Estimated distances of nonprotonated side-chain carbons to hydrogens two bonds away, computed with the use of known bond lengths and angles (not involving hydrogen) in some crystalline amino acids and small peptides, and known C–H, O–H, and N–H bond lengths in small molecules. The calculated two-bond C–H distances for C^γ of phenylalanine are about the same as those shown for tyrosine. Taken from Oldfield et al.[5]

greater than 40 kG (Fig. 15).[17] The value of W^{CSA} increases approximately as the square of the magnetic field strength [see Eq. (2) of Norton et al.[17]]. Therefore, an increase in the magnetic field strength much above 40 kG will *decrease* the resolution of the nonprotonated aromatic carbon resonances.[17]

The above theoretical results indicate that the difference between the linewidths of methine and nonprotonated aromatic carbon resonances should decrease with increasing magnetic field strength. This prediction is verified experimentally in a comparison of the aromatic regions in fully proton-decoupled ^{13}C nmr spectra of lysozyme at 14.2 (Fig. 13A) and 63.4 kG (Fig. 13B). A consideration of linewidths indicates the desirability of different strategies at different field strengths. Additional differences in strategies at low and high magnetic field strengths are dictated by the magnetic field dependence of spin-lattice relaxation times. Therefore, I shall consider T_1 (and NOE) values before strategies are discussed.

Spin-Lattice Relaxation Times. I define $1/T_1^D$ and $1/T_1^{CSA}$ as the contributions from $^{13}C-^1H$ dipolar and CSA relaxation, respectively, to $1/T_1$ (see Section II,B). For a methine aromatic carbon, the contribution of $1/T_1^{CSA}$ is negligible at all magnetic field strengths available today for high resolution nmr,[17] so that the T_1^D values given in Fig. 2 can be used to estimate T_1. It follows from Fig. 2 that the T_1 value of a methine aromatic carbon of any protein ($\tau_R \geq 10$ nsec) will increase with increasing magnetic field strength. For example, in the case of hen egg white lysozyme ($\tau_R \approx 10$ nsec at 30°–40°) the T_1 value of a methine aromatic carbon should be about 0.03 sec at 14 kG and about 0.25 sec at 63 kG. The reported measured value at 63.4 kG (and 32°) is 0.20 ± 0.05 sec.[41] Note that Fig. 2 predicts longer T_1 values for proteins of higher molecular weight.

A more complex situation exists for nonprotonated aromatic carbons. The dashed lines of Fig. 17 are log-log plots of T_1^D versus τ_R for a nonprotonated carbon which has three hydrogens two bonds away.[17] These values are directly applicable to some nonprotonated aromatic carbons, but they must be multiplied by a number in the range 0.7–3 for other nonprotonated aromatic carbons (see Fig. 16).[5] The solid lines of Fig. 17 are calculated values of T_1^{CSA}. The value of T_1 is given by

$$1/T_1 = 1/T_1^D + 1/T_1^{CSA} \tag{4}$$

It follows from Fig. 17 that $1/T_1^{CSA} \ll 1/T_1^D$ at 14.2 and 23.5 kG, and that $1/T_1^D \ll 1/T_1^{CSA}$ much above 40 kG. Experimental T_1 values of nonprotonated aromatic carbons of proteins at 14.2 kG [5] and 63.4 kG [17] are in fairly good agreement with calculated values (Table II). Clearly, $T_1 \approx T_1^D$ at 14.2 kG, but $T_1 \approx T_1^{CSA}$ at 63.4 kG. As a consequence, at 14.2 kG there are significant differences in the T_1 values of the various types of nonprotonated aromatic carbons (Table II). These differences can be utilized for

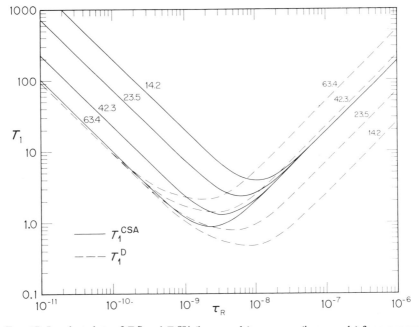

FIG. 17. Log-log plots of T_1^D and T_1^{CSA} (in seconds) versus τ_R (in seconds) for a nonprotonated aromatic carbon resonance at four magnetic field strengths (indicated in kilogauss). The meaning of solid and dashed lines is given in the legend of Fig. 15. Taken from Norton et al.[17]

TABLE II
CALCULATED AND EXPERIMENTAL T_1 VALUES (IN SECONDS) OF
NONPROTONATED AROMATIC CARBONS OF HEN EGG WHITE LYSOZYME AT 42°)[a]

| Carbon[b] | 14.2 kG | | 63.4 kG | | |
	Calc[c] $(T_1{}^D)$	Expt[d]	Calc[e] $(T_1{}^D)$	Calc[f] (T_1)	Expt[g]
Tyr C$^\zeta$ (H$_2$O)	0.49	0.5	6.8	1.8	1.4 ± 0.6
Tyr C$^\zeta$ (D$_2$O)	0.87	0.6	h	~2[i]	h
Phe C$^\gamma$	0.44	h	5.1	1.7	h
Trp C$^{\epsilon_2}$ (H$_2$O)	0.62	0.5	7.9	1.9	1.9 ± 0.8
Trp C$^{\epsilon_2}$ (D$_2$O)	1.03	0.7	h	~2[i]	h
His C$^\gamma$ (H$_2$O)[j]	0.39	0.3	4.4	1.6	2.5 ± 1.3
His C$^\gamma$ (D$_2$O)[j]	0.51	0.4	g	~2[i]	h
Tyr C$^\gamma$	0.44	0.4	5.1	1.7	1.2 ± 0.5
Trp C$^{\delta_2}$	1.20	0.8	20.5	2.2	1.8 ± 1.2
Trp C$^\gamma$	0.56	0.5	6.8	1.8	1.7 ± 0.7

[a] All T_1 values are in seconds. The T_1 values at 14.2 kG are from Table II of Oldfield et al.,[5] and those at 63.4 kG are from Table I of Norton et al.[17] (except as indicated in footnote i).

[b] Solvents are given in parenthesis for carbons which have theoretical T_1 values that change significantly when going from H$_2$O to D$_2$O solution.

[c] Dipolar interactions with ^{1}H nuclei two and three bonds removed and with directly bonded ^{14}N nuclei were considered. Calculated T_1 values for proteins in D$_2$O were obtained under the assumption that all O—H hydrogens of tyrosine residues and all N—H hydrogens of histidine and tryptophan side-chains are replaced by deuterium. The value of τ_R was set at 20 nsec.[5]

[d] Experimental value at about 43° (14 mM protein, 0.1 M NaCl, pH meter reading 3.1). Estimated accuracy is ±25%. See Ref. 5 for other details.

[e] Only dipolar interactions with ^{1}H nuclei two bonds removed and with directly bonded ^{14}N nuclei were considered. The value of τ_R was set at 13 nsec.[16,17]

[f] Values of T_1 were computed by assuming that $1/T_1 = 1/T_1{}^D + 1/T_1{}^{CSA}$. The values of $T_1{}^D$ were computed as described in footnote e. The values of $T_1{}^{CSA}$ were computed with the use of a 200 ppm chemical shift anisotropy (axially symmetric shielding tensor) for all aromatic carbons,[17] and with $\tau_R = 13$ nsec.[16,17]

[g] Experimental value at about 30° (16 mM protein in H$_2$O, 0.1 M NaCl, pH 3.1). See Norton et al.[17] for other details.

[h] Not reported.

[i] Value not given in Norton et al.[17] Estimated T_1 of about 2 sec is based on the assumption that $1/T_1 \approx 1/T_1{}^{CSA}$.

[j] Histidine residue in the imidazolium form (lysozyme at pH 3.1).

assigning the resonances (see below).[6] In contrast, at 63.4 kG all nonprotonated aromatic carbons have about the same T_1 value (Table II).[17]

Nuclear Overhauser Enhancement. In this chapter, I define the NOE as the *ratio* of the integrated intensity of a ^{13}C resonance under conditions of proton decoupling and the integrated intensity of the same resonance when no proton decoupling is applied. If $T_1 \approx T_1^D$, then Fig. 4 can be used to estimate the NOE of methine *and* nonprotonated aromatic carbon resonances, because the NOE does not depend on the actual value of T_1, but on the ratio T_1/T_1^D [see Eq. (3)]. Therefore, in the absence of fast internal rotations [which would affect the value of η_0 in Eq. (3)] the resonances of methine and nonprotonated aromatic carbon resonances of a native protein should all have the same NOE, *if* $T_1 \approx T_1^D$ (at low magnetic field strengths). This prediction has been verified experimentally for nonprotonated aromatic carbons.[5] For example, the measured NOE of nonprotonated aromatic carbon resonances of horse heart ferrocytochrome *c* at 14.2 kG is 1.3 ± 0.2, in agreement with the theoretical value of 1.2 ($\tau_R = 17$ nsec).[5] At high magnetic field strengths (such as 63 kG), the NOE of all nonprotonated aromatic carbon resonances of a native protein should be about 1.0 (because $T_1 \ll T_1^D$), while the NOE of the methine aromatic carbon resonances should be 1.15, the *minimum* NOE when $T_1 = T_1^D$ (see Fig. 4).[20] In summary, we expect that the NOE of the methine and nonprotonated aromatic carbon resonances of a native protein ($\tau_R \geq 10$ nsec) will never exceed a value of 1.3, unless fast internal rotation is present (see Fig. 5). Therefore, the *carbon count* should not be affected seriously by differences in NOE values. Note that an NOE of 3.0 is common for ^{13}C resonances of smaller molecules (see Fig. 4).[14] This implies that the minimum molar concentration required for observing single aromatic carbon resonances of a protein is 2 to 3 times *greater* than might be expected on the basis of sensitivity tests made with solutions of an organic molecule such as sucrose or cholesterol.

C. Techniques and Strategies

Choice of Recycle Time. When 90° radio frequency pulse excitation is used (see Section II,G), a recycle time about equal to T_1 will yield the best signal-to-noise ratio *for a fixed signal accumulation time*.[48,49] However, if the carbons of interest do not all have about the same T_1, and if one wishes to use peak intensities for a carbon count, then a recycle time about twice the longest T_1 may be desirable, even though such a recycle time may require a painfully long accumulation time per spectrum.

When dealing with nonprotonated aromatic carbons of a small protein (such as hen egg white lysozyme) at 14.2 kG, a recycle time of about 1 sec is optimum for most applications (see T_1 values in Table II). However, this recycle time will cause a slight selective attenuation of the resonances of C^{δ_2} of tryptophan residues (see Table II). Whenever practical, a recycle time of 2 sec is preferable (at 14.2 kG). In principle, a recycle time of 0.1 sec (or even less) would be satisfactory for methine aromatic carbon reso-

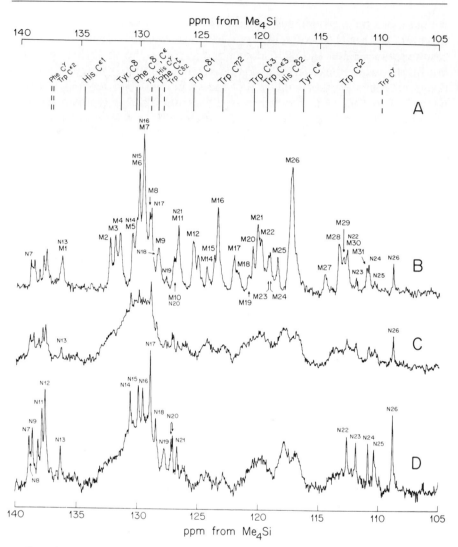

FIG. 18. (A) Chemical shifts of methine carbons (solid lines) and nonprotonated carbons (dashed lines) of aromatic amino acid side chains of small peptides. The values for tyrosine (in the phenolic form) and histidine (in the imidazolium form) are those reported for these residues in angiotensin II.[53] The resonance of C^ζ of tyrosine (at 155.6 ppm[53]) is not shown. The values for phenylalanine and tryptophan are those for these residues in lysine-vasopressin,[54] and luteneizing hormone-releasing hormone,[55] respectively. The assignments for $C^{\eta 2}$ and $C^{\zeta 3}$ of tryptophan are those of Bradbury and Norton.[56] (B) Region of aromatic carbons (except C^ζ of tyrosine residues) in the natural-abundance ^{13}C Fourier transform nmr spectrum of hen egg white lysozyme (15 mM protein in H_2O, pH 2.9, 32°) at 63.4 kG, with a 10-mm probe, under conditions of full proton decoupling, after 7 hr signal accumulation with a recycle time of 0.4 sec. The letters M and N designate methine and nonprotonated carbon

nances of a small protein, but in practice I do not expect much use of low magnetic field strengths for studying such resonances, because of poor resolution (see Section II,E).

At 63.4 kG, a recycle time of about 2 sec should be satisfactory for nonprotonated aromatic carbon resonances of a small protein (see Table II). A much shorter recycle time (0.3 to 0.5 sec) is adequate for the methine aromatic carbon resonances. In fact, if one wishes to study the methine aromatic carbons, a recycle time of much more than 0.5 sec may cause difficulties, because of interference from strong nonprotonated aromatic carbon resonances. For example, the fully proton-decoupled spectrum of lysozyme (at 63.4 kG) shown in Fig. 13B was recorded with a recycle time of 3 sec (peak numbers indicate nonprotonated aromatic carbons). A portion of a spectrum recorded with a recycle time of 0.4 sec is shown in Fig. 18B, where the letters M and N designate methine and nonprotonated carbon resonances, respectively (the numbers after the N correspond to the peak numbers of Fig. 13B). In Fig. 18B, but not in Fig. 13B, the nonprotonated carbon resonances are greatly attenuated relative to the methine carbon resonances.[41]

The recycle times mentioned above are not necessarily the best for proteins much larger than lysozyme. One must take into account the change in T_1 values caused by an increase in the rotational correlation time (see Figs. 2 and 17).

Choice of Proton-Decoupling Conditions. Figure 18A shows the chemical shifts of methine carbons (solid lines) and nonprotonated carbons (dashed lines) of aromatic amino acid side-chains of small peptides.[53–56] Only the resonance of C^ζ of tyrosine (at 155.6 ppm[53]) is not shown. Clearly, various types of resonances are in close proximity. With the additional chemical shift nonequivalence (of different positions in the amino acid sequence) caused by protein folding, we can expect rather complex patterns in the aromatic regions of the spectra of even small pro-

[53] R. Deslauriers, A. C. M. Paiva, K. Schaumburg, and I. C. P. Smith, *Biochemistry* **14,** 878 (1975).

[54] R. Walter, K. U. M. Prasad, R. Deslauriers, and I. C. P. Smith, *Proc. Natl. Acad. Sci. U.S.A.* **70,** 2086 (1973).

[55] R. Deslauriers, G. C. Levy, W. H. McGregor, D. Sarantakis, and I. C. P. Smith, *Biochemistry* **14,** 4335 (1975).

[56] J. H. Bradbury and R. S. Norton, *Biochim. Biophys. Acta* **328,** 10 (1973).

resonances, respectively. The methine carbon resonances are numbered consecutively from left to right. The numbering system for the nonprotonated carbon resonances is that of Figs. 13 and 14B and Allerhand *et al.,*[5,6,8] (C) As spectrum B, except recorded under conditions of noise-modulated off-resonance proton-decoupling. (D) As spectrum C, but with a recycle time of 2 sec and 18 hr accumulation time. Taken from Dill and Allerhand.[41]

teins (see Fig. 18B). Therefore, it is desirable (if not essential) to use conditions which permit the observation of the nonprotonated carbon resonances without interference from the protonated carbon resonances, and vice versa. This can be accomplished by taking advantage of the differences in properties of the two types of resonances (see above), as follows.

Consider, for example, hen egg white lysozyme at 63.4 kG. Full proton decoupling and a recycle time of 0.4 sec are suitable conditions for studying the methine aromatic carbon resonances. Under these conditions, *all* resonances are quite narrow, but the nonprotonated carbon resonances are relatively weak (Fig. 18B). Noise-modulated off-resonance proton-decoupling and a recycle time of 2 sec are good conditions for observing the nonprotonated aromatic carbon resonances (Figs. 13C and 18D). Full proton-decoupling with a 2 sec recycle time is a combination of conditions which yields, in principle, the greatest number of "useful" resonances, but in practice the spectrum is unnecessarily complex (Fig. 13B). At high magnetic field strengths, noise-modulated off-resonance proton-decoupling is essential, in my opinion, when studying nonprotonated aromatic carbon resonances, because with full proton-decoupling there is a definite risk of confusing methine and nonprotonated aromatic carbon resonances (see Fig. 13B). At low magnetic field strengths (such as 14 or 23 kG), the nonprotonated carbon resonances are much narrower than those of the methine carbons, even in fully proton-decoupled spectra (see Figs. 12A and 13A). Nevertheless, it is still desirable to use noise-modulated off-resonance proton-decoupling (see Fig. 12).

Convolution-Difference Spectra. Multiplication of a digitally accumulated time-domain signal by e^{-t/τ_1} produces a "digital broadening" of $1/\pi\tau_1$ Hz after Fourier transformation.[49] In order to improve the signal-to-noise ratio without considerable loss of resolution, the digital broadening is normally chosen to be about equal to (or slightly smaller than) the linewidths of the resonances of interest.[49] For example, the spectrum of Fig. 19A was obtained when the accumulated time-domain data (recorded under conditions of noise-modulated off-resonance proton-decoupling, and stored on a disk) from horse heart cyanoferricytochrome *c* was processed with 0.93 Hz digital broadening. When the same time-domain data was processed with 9.3 Hz digital broadening, the spectrum of Fig. 19B resulted. In this spectrum, the nonprotonated aromatic carbon resonances are severely broadened, while the methine carbon bands are not significantly broader than in Fig. 19A. When the spectrum of Fig. 19B (multiplied by 0.9) was digitally subtracted from Fig. 19A, the result was the *convolution-difference* spectrum of Fig. 19C, in which the broad methine carbon bands have been essentially eliminated. The final form of

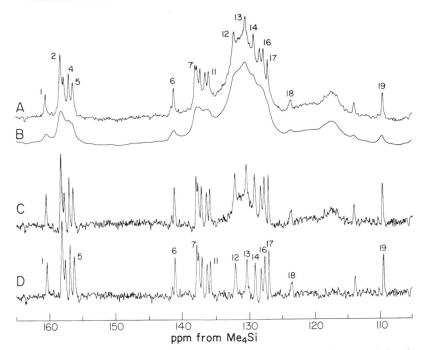

FIG. 19. Region of aromatic carbons and C$^\zeta$ of arginine residues in the natural-abundance ^{13}C nmr spectrum (at 14.2 kG, with a 20-mm probe) of 19 mM horse heart cyanoferricytochrome c in H$_2$O (0.1 M NaCl, 0.05 M phosphate buffer, pH 6.9, 36°), recorded under conditions of noise-modulated off-resonance proton decoupling with a recycle time of 1.1 sec and 10 hr signal accumulation. Resonances of nonprotonated aromatic carbons and C$^\zeta$ of the 2 arginine residues are numbered as in Ref. 6. The peak at about 114 ppm arises from excess free HCN which is in fast-exchange with about 0.5% free CN$^-$. All four spectra were obtained from the same raw time-domain data (stored on a disk) as follows: (A) With 0.93 Hz digital broadening. (B) With 9.3 Hz digital broadening. (C) Convolution-difference spectrum, obtained by digital subtraction of spectrum B (multiplied by 0.9) from spectrum A. (D) Spectrum C, after a digital base line adjustment.

the convolution-difference spectrum (Fig. 19D) was obtained by applying a digital base line adjustment to Fig. 19C.

For horse heart cyanoferricytochrome c, application of the convolution-difference method, although convenient, is not essential, because the spectrum is very simple to begin with (Fig. 19A).[6] A more dramatic demonstration of the use of convolution-difference spectra for studying nonprotonated aromatic carbon resonances is shown in Fig. 20.

Note that the effectiveness of the convolution-difference method for removing the resonances of methine aromatic carbons hinges on a large difference between the linewidths of nonprotonated and methine aromatic

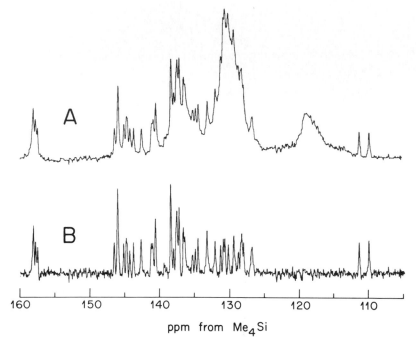

FIG. 20. Region of aromatic carbons and C^ζ of arginine residues in the fully proton-decoupled natural-abundance ^{13}C nmr spectrum (at 14.2 kG, with a 20-mm probe) of 9 mM horse carbon monoxide myoglobin (0.1 M NaCl, 0.05 M phosphate buffer, pH 6.4, 36°), after 19 hr spectral accumulation with a recycle time of 2.1 sec. (A) With 0.93 Hz digital broadening. (B) Convolution-difference spectrum of the same time-domain data as in spectrum A, processed with 0.93 Hz and 9.3 Hz digital broadening, as in Fig. 19D. Taken from Fig. 4A of Oldfield $et\ al.$[6]

carbon resonances. At low magnetic field strengths (such as 14 or 23 kG), the convolution-difference method will work even if the spectrum has been recorded with full proton-decoupling (see Fig. 20). However, at high magnetic field strengths (such as 63 kG), it is essential to use noise-modulated off-resonance proton-decoupling (or some other suitable "inefficient-decoupling" scheme) if one wishes to remove the methine aromatic carbon resonances by means of the convolution-difference method (see Fig. 13).

IV. Assignments of Nonprotonated Aromatic Carbon Resonances

A. Assignments to Types of Carbons

There are seven types of nonprotonated aromatic carbons of amino acid residues: C^γ of phenylalanines, C^γ and C^ζ of tyrosines, C^γ of histidines, and C^γ, C^{δ_2}, and C^{ϵ_2} of tryptophans (Fig. 9). When dealing with diamagnetic heme proteins, the 16 nonprotonated aromatic carbons of the heme also yield narrow resonances in the aromatic region of the spectrum.[4,6] It is often necessary to include the resonances of C^ζ of arginine residues in the discussion, because they may overlap with resonances of C^ζ of tyrosine residues.[6,45] Before one tries to assign the resonances to specific residues in the amino acid sequence, it is normally desirable to determine which of the various types of nonprotonated aromatic carbons gives rise to each resonance. This section describes several methods which have been used for this purpose.

Chemical Shift Considerations. Figure 21 shows the chemical shift ranges for the various types of nonprotonated aromatic carbons of diamagnetic proteins, based on assignments for several proteins.[6] Also shown in Fig. 21 are the relatively invariant positions (*thick vertical lines*) of the corresponding resonances in denatured proteins and small peptides.[6]

The resonances of C^γ of tryptophan residues do not overlap with any other type of *nonprotonated* carbon resonances (Fig. 21), and, therefore, they can be readily identified. However, these resonances may overlap with those of C^{ζ_2} of tryptophan residues (see Figs. 18A and 18B). For this reason, I do not recommend full proton-decoupling when studying C^γ resonances of tryptophan residues *at high magnetic field strengths*. This is just a more specific form of a general recommendation given in Section III,C.

The range of resonances of C^ζ of tyrosine residues overlaps only with that of C^ζ of arginine residues (Fig. 21). These two types of resonances

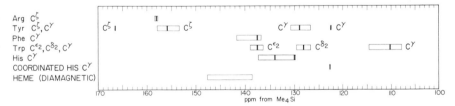

FIG. 21. Observed ranges of chemical shifts (boxes) of nonprotonated aromatic carbons and C^ζ of arginine residues of some native diamagnetic proteins in H_2O. The thick lines are the relatively invariant chemical shifts of these carbons in a denatured protein and in small peptides, except for the chemical shift of C^γ of coordinated histidine, which is the value observed for His-18 of horse heart ferrocytochrome c. The histidine line at 130 ppm refers to a histidine residue in the imidazolium form. The histidine line at 134 ppm refers to a histidine in the imidazole form, but predominantly in the N^{ϵ_2}-H tautomeric state. The tyrosine lines at 129 and 122.5 ppm refer to C^γ of the phenolic and phenolate forms, respectively. The tyrosine lines at 156 and 166.5 ppm refer to C^ζ of the phenolic and phenolate forms, respectively. Taken from Oldfield *et al.*[6]

can be distinguished by means of selective proton decoupling (see below).[6] The resonances of C^{ϵ_2} or tryptophan residues and those of C^γ of histidine (in the N^{ϵ_2}-H imidazole form) and phenylalanine residues have overlapping ranges of chemical shifts.[6,46,47] The resonances of C^{δ_2} of tryptophan residues and those of C^γ of histidine (in the imidazolium and N^{δ_1}-H imidazole forms) and tyrosine residues have chemical shift ranges that overlap.[6,46,47] The resonances of C^{ϵ_2} and C^{δ_2} of tryptophans can be identified by their relatively long T_1 values (protein in D_2O must be used for identifying the C^{ϵ_2} resonances), and those of C^γ of *titratable* histidine and tyrosine residues can be identified by the effect of pH on their chemical shifts (see below).[6]

The presence of a heme can have large effects on the ^{13}C nmr spectrum of a protein. The 16 nonprotonated aromatic carbons of the heme contribute narrow resonances if the heme is in a diamagnetic state, as in ferrocytochrome c (Fig. 12),[4,6] carbon monoxide myoglobin (Fig. 20),[6] and carbon monoxide hemoglobin.[7] Some of these resonances are readily identified on the basis of their chemical shifts, while others overlap with resonances of aromatic amino acid residues (see Fig. 21). The C^γ resonance of the histidine coordinated to the iron of a diamagnetic heme protein is shifted several parts per million upfield of the normal range of C^γ resonances of uncoordinated histidine residues (Fig. 21).[6] Part of this upfield shift arises from the strong ring current effect of the porphyrin ring.[52] Ring current effects may also be important for other amino acid carbons near the heme.

The chemical shift ranges of Fig. 21 may not apply to some carbons near the iron of a heme protein in a paramagnetic state (see discussion of effects of paramagnetic centers in this section).

Selective Proton Decoupling. If we apply single-frequency proton irradiation at the resonance frequency of H^δ of arginine residues (about 3.2 ppm downfield from the ^1H resonance of Me_4Si), of low enough power to produce negligible decoupling of aromatic protons of tyrosine residues (^1H chemical shifts in the range 6.8–7.2 ppm), then the ^{13}C nmr spectrum yields a sharp resonance for each C^ζ of arginine, but a broad, partly split band for each C^ζ of tyrosine.[6] The effect is illustrated in Fig. 22. It is desirable to use D_2O and not H_2O as the solvent when using this method.[6]

Partially Relaxed Fourier Transform (PRFT) Spectra. Whenever two classes of carbons have measurably different T_1 values, PRFT nmr spectra[57] (see below) can be used to distinguish their resonances.[58–60] At

[57] R. L. Vold, J. S. Waugh, M. P. Klein, and D. E. Phelps, *J. Chem. Phys.* **48,** 3831 (1968).
[58] A. Allerhand and D. Doddrell, *J. Am. Chem. Soc.* **93,** 2777 (1971).
[59] D. Doddrell and A. Allerhand, *Proc. Natl. Acad. Sci. U.S.A.* **68,** 1083 (1971).
[60] E. Oldfield and A. Allerhand, *J. Am. Chem. Soc.* **97,** 221 (1975).

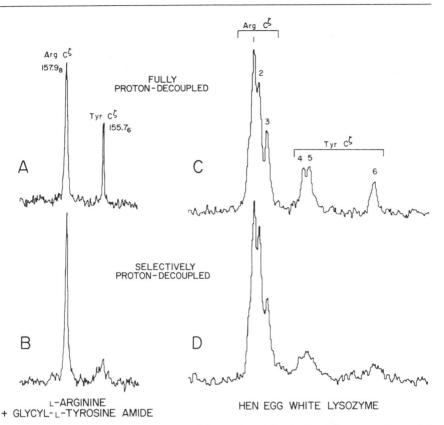

FIG. 22. (A) Resonances of C^ζ of arginine and tyrosine in the fully proton-decoupled natural-abundance ^{13}C nmr spectrum of a sample of 0.29 M glycyl-L-tyrosine amide hydrochloride and 1.13 M L-arginine hydrochloride in D_2O (pH meter reading 3.0) at 37°, recorded at 14.2 kG, with a 20-mm probe, a recycle time of 30 sec, and 8 min accumulation time. Numbers are chemical shifts in parts per million downfield from Me_4Si. Estimated accuracy is ±0.05 ppm. (B) Same as spectrum A, but selectively proton-decoupled spectrum, at 30°. Low-power single-frequency ^1H irradiation was set about 3.2 ppm downfield from Me_4Si (see Oldfield *et al.*[6]). (C) Fully proton-decoupled spectrum of 15 mM hen egg white lysozyme in D_2O, 0.1 M NaCl, pH meter reading 3.0, at 44°, recorded essentially as spectrum A, but with a recycle time of 2.2 sec and 10 hr accumulation time. Peak numbers are those of Oldfield *et al.*[6] Chemical shifts are given in Table IV (protein in D_2O). (D) Same sample and conditions as spectrum C, but selectively proton-decoupled (as spectrum B), at 31°. Taken from Oldfield *et al.*[6]

low magnetic field strengths, the $^{13}C-^1H$ dipolar mechanism dominates the relaxation of nonprotonated aromatic carbon resonances of proteins (Table II).[5] The most important contributions to the $^{13}C-^1H$ dipolar relaxation of a nonprotonated carbon are those from hydrogens two bonds away (Fig. 16).[5] The only nonprotonated aromatic carbons which have

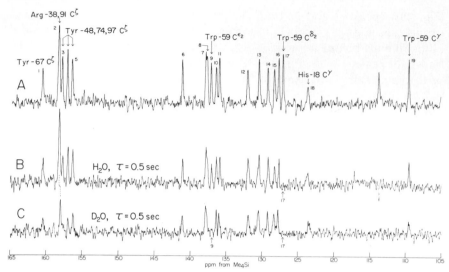

FIG. 23. Region of aromatic carbons and C^ζ of arginine residues in convolution-difference natural-abundance ^{13}C nmr spectra of 15 mM horse heart cyanoferricytochrome c, 0.1 M NaCl, 0.05 M phosphate buffer, at 36°. Each spectrum was recorded at 14.2 kG with a 20-mm probe, under conditions of noise-modulated off-resonance proton-decoupling, and with 32,768 accumulations. The convolution-difference procedure was carried out with 0.9 Hz and 9 Hz digital broadening (see Section III,C). Peak numbers are those of Oldfield et $al.$[6] and Fig. 19. The peak at about 114 ppm arises from excess free HCN which is in fast exchange with about 0.5% free CN^-. (A) Normal spectrum of protein in H_2O, pH 6.7. Recycle time was 2.1 sec (19 hr accumulation time). (B) PRFT spectrum of same sample as in A, with $\tau = 0.5$ sec (see Section IV,A). Recycle time (interval between successive 90° pulses) was 2.6 sec (24 hr accumulation time). (C) As spectrum B, but after deuterium exchange, in D_2O, pH meter reading 6.7. Taken from Oldfield et $al.$[6]

only one hydrogen two bonds away are $C^{\delta 2}$ of tryptophan residues (in H_2O and D_2O solution) and $C^{\epsilon 2}$ of tryptophan residues (when D_2O is the solvent and the N—H hydrogens have been exchanged by deuterium). As a result, at low magnetic field strengths the T_1 values of $C^{\delta 2}$ (protein in H_2O or D_2O) and $C^{\epsilon 2}$ (protein in D_2O) of tryptophan residues are significantly longer than those of C^γ of phenylalanine, histidine, and tyrosine residues (Table II).[5] Therefore, the PRFT method (see below) can be used to identify the resonances of $C^{\delta 2}$ and $C^{\epsilon 2}$ of tryptophan residues.[6,60]

Figure 23A shows the aromatic region of the *normal* convolution-difference ^{13}C nmr spectrum of horse heart cyanoferricytochrome c in H_2O. Here I define a *normal* spectrum as one obtained with the use of 90° radio frequency pulse excitation (see Section II,G), and with a recycle time long enough to yield negligible attenuation of all resonances. Each of the 18 nonprotonated aromatic carbons of amino acid residues of horse heart cyanoferricytochrome c gives rise to a single-carbon resonance in

Fig. 23A (peaks 1 and 3–19).[6] On the basis of chemical shift consider- ations (see Fig. 21), the $C^{\epsilon 2}$ resonance of the single tryptophan residue (Trp-59) must be one of peaks 6–11, and the $C^{\delta 2}$ resonance of Trp-59 must be one of peaks 12–18. Figure 23B is a PRFT spectrum (see Section II,G) of the same sample as used for Fig. 23A, recorded with an interval (τ) of 0.5 sec between each 180° radio frequency pulse and the following 90° pulse. In a PRFT spectrum, a resonance will appear negative (relative to its polarity in the *normal* spectrum) if $\tau < T_1 \ln 2$, nulled if $\tau = T_1 \ln 2$, and positive if $\tau > T_1 \ln 2$.[14,57] Peaks 12–16 and 18 are positive in Fig. 23B, while peak 17 is nulled. Therefore, peak 17 has a considerably longer T_1 value than peaks 12–16 and 18. We assign peak 17 to $C^{\delta 2}$ of Trp-59. Peaks 6–11 all have similar positive intensities in Fig. 23B. However, the PRFT spectrum of the protein in D_2O (Fig. 23C) clearly indicates that peak 9 arises from $C^{\epsilon 2}$ of Trp-59.[6] Note that the $C^{\epsilon 2}$ resonance of Trp-59 moves slightly upfield when going from H_2O to D_2O solution, while peaks 6–8, 10, and 11 do not move measurably (see Table IV of Oldfield *et al.*[6]). Deu- terium isotope effects on ^{13}C chemical shifts may complicate the task of establishing one-to-one connections between ^{13}C resonances of a protein in H_2O and in D_2O.[6] The excellent resolution in the spectra of horse heart cyanoferricytochrome c (Fig. 23) makes it possible to establish the con- nections by inspection.

Effect of pH. Studies of small peptides indicate that the resonances of C^ζ and C^γ of a tyrosine residue should move about 10 ppm downfield and about 6 ppm upfield, respectively, when going from the phenolic to the phenoxy form of the residue.[61] Studies of small peptides and other model compounds indicate that the C^γ resonance of a histidine residue in the imi- dazolium form (Fig. 24A), with a chemical shift of about 130 ppm, moves either upfield or downfield when the residue is converted into the imida- zole form.[62,63] The direction of the shift depends on which of the two ring nitrogens is deprotonated.[62,63] Most often, deprotonation yields mainly the $N^{\epsilon 2}$–H imidazole tautomer (Fig. 24B). Such deprotonation causes the C^γ resonance to move about 4–6 ppm downfield.[6,8,47,62,63] However, some histidine residues of some proteins yield mainly the $N^{\delta 1}$–H imidazole taut- omer (Fig. 24C).[46] In such cases, an upfield shift of about 2 ppm is ex- pected upon deprotonation.[62,63]

The behavior of a ^{13}C resonance of a titrating residue will be affected by the rate of exchange between the protonated and deprotonated states

[61] R. S. Norton and J. H. Bradbury, *J. Chem. Soc., Chem. Commun.* p. 870 (1974).
[62] W. F. Reynolds, I. R. Peat, M. H. Freedman, and J. R. Lyerla, Jr., *J. Am. Chem. Soc.* **95**, 328 (1973).
[63] R. Deslauriers, W. H. McGregor, D. Sarantakis, and I. C. P. Smith, *Biochemistry* **13**, 3443 (1974).

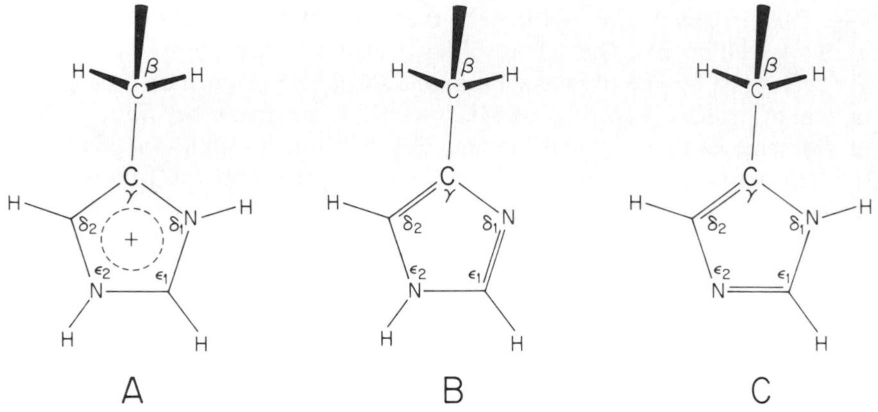

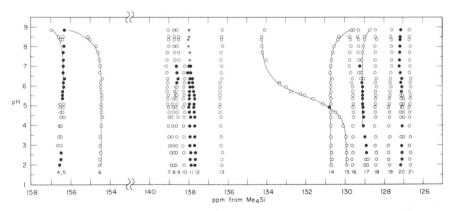

FIG. 24. Side chain of a histidine residue. (A) Imidazolium state. (B) Imidazole state, N^{ϵ_2}-H tautomeric form. (C) Imidazole state, N^{δ_1}-H tautomeric form.

(see Section IV,B, discussion on chemical exchange). If this rate is fast relative to the difference in the chemical shifts (in radians/second) of the two states, then a single sharp exchanged-averaged resonance is observed throughout the titration range, with a chemical shift which is the weighted average of the values for the protonated and deprotonated forms of the

FIG. 25. Effect of pH on the chemical shifts of the nonprotonated aromatic carbons (except C^{γ} of tryptophan residues) of hen egg white lysozyme in H_2O, 0.1 M NaCl, 38°. Peak numbers are those of Figs. 13 and 14B. Open circles, closed circles, and asterisks indicate peaks that arise from 1, 2, and 5 carbons, respectively. Protein concentration was 9.0 mM, with the following exceptions: 7.3 mM at pH 8.84, and 14.5 mM at pH values of 2.96, 3.97, 4.65, 4.90, and 5.58. Only the chemical shift of peak 15 is shown at pH 5.58 and pH 6.09. The solid lines are best-fit (single pK) titration curves. For each titrating tyrosine residue, the chemical shifts of C^{ζ} and C^{γ} of the phenoxy form of the residue were assumed to be 10 ppm downfield and 6 ppm upfield, respectively, from the corresponding values of the phenolic form of the residue. Taken from Allerhand et al.[8]

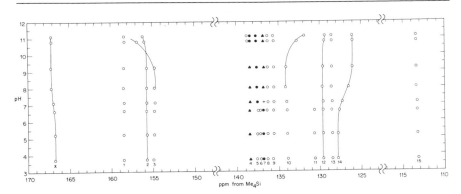

FIG. 26. Effect of pH on the chemical shifts of some nonprotonated carbons of reduced *P. aeruginosa* azurin in H_2O, 0.05 M ammonium acetate, 31°. Peak designations are those of Ugurbil *et al.*[46] and Table X. Open circles, closed circles, and triangles indicate peaks that arise from 1, 2, and 3 carbons, respectively. The symbol + indicates a peak with an intensity intermediate between 2 and 3 carbons (see text). The solid lines are best-fit titration curves, assuming a single pK in each case. The chemical shifts of C^ζ and C^γ of the phenoxy form of each tyrosine residue were constrained to values 10.4 ppm downfield and 6.2 ppm upfield, respectively, of the corresponding values of the phenolic form of the residue. Taken from Ugurbil *et al.*[46]

residue. This is the most common situation. However, in some cases the exchange between the two states of a titrating residue is slow relative to the difference in the chemical shifts. In such situations, distinct resonances of the protonated and deprotonated states will coexist in the titration range. For intermediate rates of exchange, two broad peaks or one broad feature may be observed in the titration range. Because of the limited sensitivity of natural-abundance ^{13}C nmr, fast exchange is best for detection of the resonances of a titrating residue at pH values near the pK.

The effect of pH on the chemical shifts of the nonprotonated aromatic carbon resonances of hen egg white lysozyme (Fig. 25) indicates that peak 15 arises from C^γ of the single histidine residue (His-15), that exchange between the imidazole and imidazolium states of His-15 is fast (relative to a difference in chemical shifts of about 410 rad/sec at 14.2 kG), and that the imidazole state is predominantly the common $N^{\epsilon 2}$–H tautomer (Fig. 24B). In contrast, the C^γ resonance of one of the two titrating histidine residues of azurin from *Pseudomonas aeruginosa* (peak 14 of Fig. 26) also exhibits fast exchange behavior, but the direction and magnitude of the titration shift indicate that the imidazole form of this residue is mainly the uncommon $N^{\delta 1}$–H tautomer (Fig. 24C).[46] The other titrating histidine of azurin exhibits slow or intermediate exchange.[46,64] The C^γ resonance of

[64] H. A. O. Hill, J. C. Leer, B. E. Smith, and C. B. Storm, *Biochem. Biophys. Res. Commun.* **70**, 331 (1976).

this histidine is a sharp single-carbon resonance at pH $\leqslant 6.5$ (peak 11 of Fig. 26) and at pH $\geqslant 8$ (one contributor to peak 7), but it is difficult to detect near neutral pH.[46] There is a 6 ppm downfield shift when going from peak 11 to peak 7. We conclude that the imidazole form of the titratable

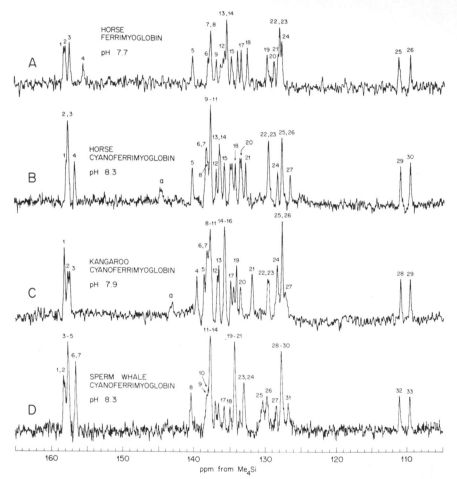

FIG. 27. Region of aromatic carbons (and C^{ζ} of arginine residues) in the convolution-difference natural-abundance ^{13}C nmr spectra of myoglobins (about 12 mM protein in H_2O, 0.1 M KCl) recorded at 14.2 kG, with a 20-mm probe, under conditions of noise-modulated off-resonance proton-decoupling, with 32,768 accumulations per spectrum, and a recycle time of 0.555 sec (for horse ferrimyoglobin and kangaroo cyanoferrimyoglobin) or 1.05 sec (for horse and sperm whale cyanoferrimyoglobins). Because of the complex pH dependence of many chemical shifts, the peak numbering system assigns one number to each carbon and not to eack peak, consecutively from left to right at pH ≈ 8. (A) Horse ferrimyoglobin, pH 7.7, 36°. (B) Horse cyanoferrimyoglobin, pH 8.3, 39°. (C) Kangaroo cyanoferrimyoglobin, pH 7.9, 37°. (D) Sperm whale cyanoferrimyoglobin, pH 8.3, 36°. Taken from Wilbur and Allerhand.[47]

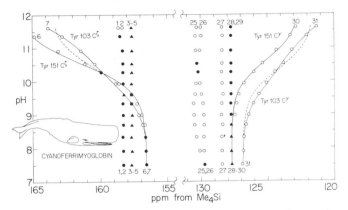

FIG. 28. Effect of pH on the chemical shifts of some nonprotonated aromatic carbons and C^ζ of arginine residues of sperm whale cyanoferrimyoglobin at 36°. Observed nonprotonated aromatic carbon resonances which are not shown (105–115 ppm and 130–145 ppm) have chemical shifts which are practically pH independent above pH 8. The peak numbering system is that of Fig. 27D (described in the legend of Fig. 27). Open circles, closed circles, triangles, and squares indicate peaks that arise from 1, 2, 3, and 4 carbons, respectively. Figure 27D gives typical sample and spectral conditions. The solid lines are best-fit theoretical titration curves, with one pK for Tyr-151 and two pK values for Tyr-103 (see Wilbur and Allerhand[45]). Dashed lines are best-fit single pK titration curves for Tyr-103. Below pH 10.5, the dashed curve for C^ζ of Tyr-103 coincides with the solid curve for C^ζ of Tyr-151. Taken from Wilbur and Allerhand.[45]

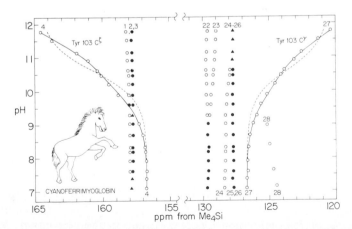

FIG. 29. Effect of pH on the chemical shifts of some nonprotonated aromatic carbons and C^ζ of arginine residues of horse cyanoferrimyoglobin at 38°. Peak numbers and typical sample and spectral conditions are given in Fig. 27B. The behavior of omitted resonances and the meaning of symbols and curves are given in the caption of Fig. 28. Taken from Wilbur and Allerhand.[45]

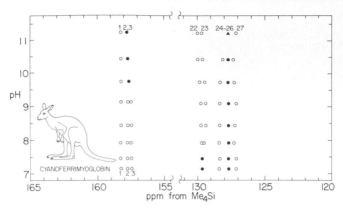

FIG. 30. Effect of pH on the chemical shifts of some nonprotonated aromatic carbons and C^ζ of arginine residues of red kangaroo cyanoferrimyoglobin at 36°. Peak numbers and typical sample and spectral conditions are given in Fig. 27C. The behavior of omitted resonances and the meaning of symbols are given in the caption of Fig. 28. Taken from Wilbur and Allerhand.[45]

histidine with slow exchange behavior exists mainly in the $N^{\epsilon 2}$–H tautomeric state (Fig. 24B).

Figure 27 shows the aromatic regions of convolution-difference ^{13}C nmr spectra of some myoglobins near pH 8.[47] The effect of high pH values (≥ 9) on the chemical shifts (Figs. 28–30) clearly identifies the resonances of C^ζ and C^γ of titratable tyrosine residues.[45] Two of the three tyrosines of sperm whale myoglobin (Fig. 28) and one of the two tyrosines of the horse protein (Fig. 29) are titratable residues.[45] The single tyrosine of kangaroo myoglobin (Tyr-146) does not exhibit titration behavior (Fig. 30).[45] The effect of low pH (≤ 7.5) on the resonances of horse ferrimyoglobin (Fig. 31) and horse cyanoferrimyoglobin (Fig. 32) identifies the C^γ resonances of titratable histidine residues.[47] Eight of the eleven histidine residues of horse cyanoferrimyoglobin exhibit titration behavior (Fig. 32), but only six titrating histidine residues are detected in the spectrum of the ferrimyoglobin (Fig. 31).[47] This difference does not necessarily reflect a change in the number of titrating histidines, because some nonprotonated aromatic side-chain carbon resonances of the ferrimyoglobin (spin-5/2) are broadened beyond detection, while *all* nonprotonated aromatic side-chain carbons of the cyanoferrimyoglobin yield detectable resonances (see below).

Effects of Paramagnetic Centers. Some proteins contain "built-in" paramagnetic metal ions. Some metal-free proteins may have a strong specific binding site for a paramagnetic metal ion or an organic free radical. Furthermore, in some cases it is easy to attach a free radical cova-

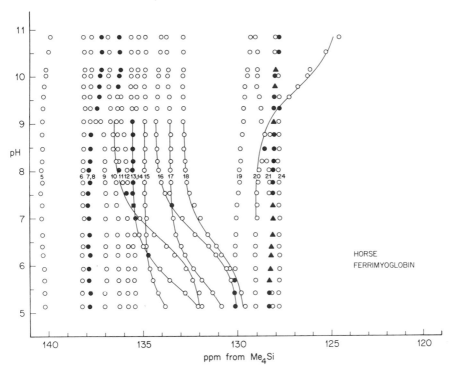

FIG. 31. Effect of pH on the chemical shifts of some nonprotonated aromatic carbons of horse ferrimyoglobin at 36°. Peak numbering system and typical sample and spectral conditions are given in Fig. 27A. Open circles, closed circles, triangles, and squares indicate peaks that arise from 1, 2, 3, and 4 carbons, respectively. The solid lines are best-fit (single pK) theoretical titration curves. Taken from Wilbur and Allerhand.[47]

lently at a specific point in the protein. A paramagnetic center can change the chemical shift of a ^{13}C resonance.[52] It may also provide an effective relaxation mechanism, thereby causing a line broadening and a decrease in the spin-lattice relaxation time.[52] At one extreme, there are some paramagnetic species (those with relatively short electron relaxation times) that produce large changes in chemical shifts but negligible effects on T_1 and on the linewidths. They are called *shift probes*.[52] At the other extreme are the paramagnetic centers (with relatively long electron relaxation times) that do not significantly affect chemical shifts but contribute greatly to the linewidths and $1/T_1$ values of resonances of nearby carbons. They are called *relaxation probes*.[52] The paramagnetic contribution to the linewidth and to $1/T_1$ of a ^{13}C resonance is proportional to r^{-6}, where r is the distance between the carbon and the paramagnetic center.[52] If the values of r for the various carbons are known (or correctly inferred from

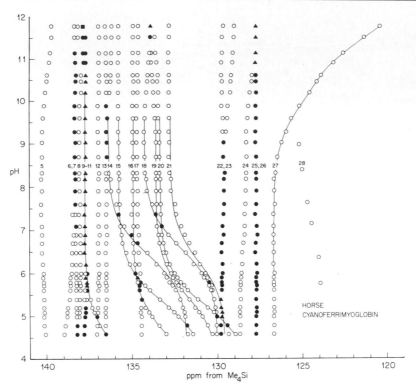

FIG. 32. Effect of pH on the chemical shifts of some nonprotonated aromatic carbons of horse cyanoferrimyoglobin at 38°. Peak numbers are those of Figs. 27B and 34. The meaning of symbols and curves is given in the legend of Fig. 31. Typical sample and spectral conditions are those of Fig. 34 (for peak 28) and Fig. 27B (for all other peaks). Taken from Wilbur and Allerhand.[47]

crystallographic data), then differences in the paramagnetic contributions to $1/T_1$ and/or W of the various resonances can be used for making assignments. It is possible to assign resonances to specific residues in the sequence by this procedure (see Section IV,B). In this section, I shall discuss the less specific task of identifying the resonances of carbons which are very near the iron atom of a *diamagnetic* heme protein, by comparing the spectrum with that of a paramagnetic form of the protein.

Consider the aromatic regions of the convolution-difference spectra of horse heart ferrocytochrome c (diamagnetic) and ferricytochrome c (spin-$\frac{1}{2}$ paramagnetic), shown in Figs. 33A and 33B, respectively.[6] All 34 nonprotonated aromatic carbons (including 16 from the heme) and the ζ-carbons of the 2 arginine residues of the ferrocytochrome c yield narrow resonances in Fig. 33A. However, 17 of the 36 carbons under consider-

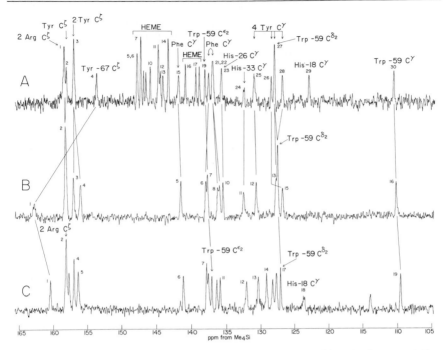

FIG. 33. Regions of aromatic carbons and C^ζ of arginine residues in the convolution-difference natural-abundance ^{13}C nmr spectra of horse heart cytochrome c in H_2O (0.1 M NaCl, 0.05 M phosphate buffer). Each spectrum was recorded at 14.2 kG, with a 20-mm probe, noise-modulated off-resonance proton-decoupling, and a recycle time of 1.1 sec. The convolution-difference procedure was carried out with 0.6 Hz and 9 Hz digital broadening (see Section III, C). (A) 11.5 mM ferrocytochrome c, pH 6.7, 40°, after 5 hr accumulation time. Peak numbers are those of Fig. 12 and Oldfield *et al.*[4,6] (B) 19.4 mM ferricytochrome c, pH 6.9, 36°, after 10 hr accumulation time. Peak numbers are those of Oldfield *et al.*[4,6] (C) 19.4 mM cyanoferricytochrome c, pH 6.9, 36°, after 10 hr accumulation time. Peak numbers are those of Figs. 19 and 23, and Oldfield *et al.*[6] The peak at about 114 ppm arises from excess free HCN which is in fast exchange with about 0.5% free CN^-. Taken from Oldfield *et al.*[6]

ation do not yield detectable resonances in the spectrum of the ferricytochrome c (Fig. 33B).[4,6] The 17 undetectable resonances are those of the 16 nonprotonated porphyrin carbons and of C^γ of the coordinated His-18 residue.[4,6] A simple inspection of the two spectra does not automatically reveal which resonances of ferrocytochrome c are missing from the spectrum of the ferric protein, because there are some chemical shift changes when going from the ferrous to the ferric state.[4,6] The one-to-one connections between peaks 2–16 of ferricytochrome c (Fig. 33B) and peaks of the ferrous protein (Fig. 33A) were made by examining spectra of mixtures of the two states.[4] Because of the fast electron transfer between the two states,[65] these spectra yield exchange-averaged resonances which correspond to peaks 2–16 (but not peak 1) of ferricytochrome c.[4] Additional details are given in Section IV,B (discussion of chemical exchange).

When the chemical exchange method does not yield one-to-one connections between resonances of a diamagnetic and a paramagnetic state of a heme protein, as in the case of peak 4 of ferrocytochrome c (Fig. 33A) and peak 1 of ferricytochrome c (Fig. 33B), it may be necessary to make direct assignments of the resonances of the paramagnetic protein. Note that the chemical shift information of Fig. 21 may not apply to some nonprotonated aromatic carbon resonances of a paramagnetic protein, if the paramagnetic center causes large changes in the chemical shifts of these resonances. Furthermore, we must consider the possibility that paramagnetically shifted carbonyl resonances will show up in the aromatic region of the spectrum (noise-modulated off-resonance proton-decoupling will eliminate paramagnetically shifted aliphatic carbon resonances). As a first approximation, the paramagnetic contribution to the chemical shift should be proportional to the reciprocal of the absolute temperature.[52] Extrapolation to $1/T = 0$ should yield an approximate chemical shift which lacks the paramagnetic contribution. For example, the chemical shift of peak 1 of horse heart ferricytochrome c (Fig. 33B) is 162.7 ppm at 36°, but about 150 ppm at $1/T = 0$.[6] The value of $1/T = 0$ is similar to the observed chemical shift of peak 4 of ferrocytochrome c (153.5 ppm at 36°).[6]

Published results for cytochrome c,[4,6] myoglobins,[6,47] and hemoglobins[7] indicate that the 16 nonprotonated porphyrin carbons of paramagnetic heme proteins normally do not yield *narrow* resonances. There are indications that some of these carbons may yield detectable resonances in spectra of cyanoferrimyoglobins (spin-$\frac{1}{2}$),[6,47] but not in spectra of ferrimyoglobins (spin-$\frac{5}{2}$).[47] Application of the convolution-difference method (as in Fig. 27) degrades the detectability of broad resonances. Figure 34 is a spectrum of horse cyanoferrimyoglobin which covers a larger range of chemical shifts than shown in Fig. 27, and without application of the convolution-difference method. Peaks a, b, and c in Fig. 34 may arise from nonprotonated heme carbons, but firm evidence is lacking.[47]

Consider now the effect of a spin-$\frac{1}{2}$ paramagnetic heme on the detectability of the resonances of nonprotonated side-chain carbons of aromatic amino acid residues. The γ-carbon of the histidine coordinated to the heme yields a relatively broad but readily detectable resonance in spectra of horse heart cyanoferricytochrome c (peak 18 of Fig. 33C),[6] horse cyanoferrimyoglobin (peak 28 of Fig. 34),[47] and kangaroo cyanoferrimyoglobin (peak 27 of Fig. 27C),[47] but it is not detected in spectra of horse heart ferri-

[65] R. K. Gupta, S. H. Koenig, and A. G. Redfield, *J. Magn. Reson.* **7,** 66 (1972).

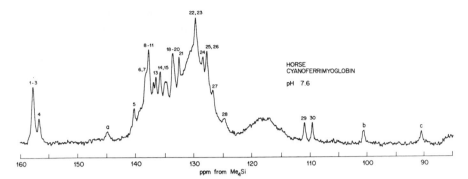

FIG. 34. Region of aromatic carbons (and C^ζ of arginine residues) in the natural-abundance ^{13}C nmr spectrum of 18 mM horse cyanoferrimyoglobin in H_2O (pH 7.6, 0.1 M KCl, 39°), recorded at 14.2 kG, with a 20-mm probe, noise-modulated off-resonance proton-decoupling, a recycle time of 0.555 sec, and 10 hr accumulation time. The convolution-difference method was not used, in order not to degrade the detectability of broad resonances (such as peak 28). Instead, a digital broadening of 0.9 Hz was applied. Peak numbers are those of Figs. 27B, 29, and 32. Taken from Wilbur and Allerhand.[47]

cytochrome $c^{4,6}$ and sperm whale cyanoferrimyoglobin.[47] The C^ζ resonance of Tyr-67 of horse heart ferricytochrome c (peak 1 of Fig. 33B) is relatively broad, but not necessarily as a result of paramagnetic broadening (see discussion of chemical exchange in Section IV,B). All other nonprotonated aromatic carbons of amino acid residues of all the spin-$\frac{1}{2}$ heme proteins mentioned above have yielded resonances which are not measurably broadened by the paramagnetic center.[4,6,47]

In spectra of horse ferrimyoglobin (spin-$\frac{5}{2}$), appreciable paramagnetic broadening extends to carbons at a greater distance from the iron than in spectra of the spin-$\frac{1}{2}$ heme proteins.[47] Crude calculated estimates of paramagnetic contributions to the linewidths of the resonances of the 30 nonprotonated side-chain carbons of aromatic and arginine residues of horse ferrimyoglobin have been reported.[47] The C^γ resonance of each of the following residues has an estimated paramagnetic broadening of more than 20 Hz (number in brackets is the estimated broadening, in hertz): His-93 [800], His-97 [160], His-64 [60], Phe-43 [30]. All other resonances under consideration have an estimated paramagnetic broadening of 5 Hz or less.[47] These results suggest that the C^γ resonances of His-93, His-97, His-64, and Phe-43 of the ferrimyoglobin should be difficult to detect. This prediction is consistent with the experimental observation that only 26 of the 30 nonprotonated side-chain carbons of aromatic and arginine residues of horse ferrimyoglobin yield detectable resonances (Fig. 27A).[47]

Deuterium Isotope Effect on Chemical Shifts. Most proton nmr studies of proteins have been done with D_2O as the solvent. In general, there is no need to employ this unnatural solvent for ^{13}C nmr studies of aqueous proteins. Most of the spectra presented in this report were obtained on proteins in H_2O. However, each of the following circumstances dictates the use of D_2O as the solvent: (a) simultaneous ^1H and ^{13}C nmr studies on the same protein solution; (b) the use of deuterium field-frequency stabilization on many nmr instruments, although H_2O with a small proportion of D_2O is normally adequate for this purpose; (c) identification of C^{ϵ_2} resonances of tryptophan residues by means of PRFT spectra; (d) identification of C^{ζ} resonances of tyrosine residues by means of selective proton decoupling (if some arginine residues have guanidinium N—H hydrogens which are not in fast exchange with solvent protons).[6]

Now some words of caution about possible problems when comparing the ^{13}C nmr spectrum of a protein in D_2O with the spectrum of the protein in H_2O. The resonances of carbons two bonds removed from a labile hydrogen (such as C^{ϵ_2} of tryptophan residues and C^{γ} of histidine residues in the imidazolium or N^{δ_1}–H imidazole form) may undergo slight upfield shifts ($\leqslant 0.2$ ppm) when the labile hydrogen is replaced by deuterium. If most carbons yield well separated resonances, as in the case of horse heart cyanoferricytochrome c (Fig. 23A), it is easy to make one-to-one connections (by inspection) between peaks in spectra of H_2O and D_2O solutions. In such a case, assignments of C^{ϵ_2} resonances of tryptophan residues obtained for the protein in D_2O (Fig. 23C) automatically yield the corresponding assignments for the protein in H_2O. However, it may be difficult to establish a one-to-one correspondence between peaks in spectra of H_2O and D_2O solutions if many peaks are crowded together, as in the region of resonances of C^{γ} of phenylalanines and C^{ϵ_2} of tryptophans in the spectrum of lysozyme (Fig. 35).[6,8] It may be tempting to assume that peaks 11 and 12 (each a two-carbon resonance) of lysozyme in H_2O (Fig. 35A) correspond to peaks 11 and 12, respectively, of the protein in D_2O (Fig. 35B). Actually, peak 12 of lysozyme in H_2O has a chemical shift (137.7_3 ppm) nearly identical to that of peak 11 of the protein in D_2O (137.7_9 ppm).[6] Peak 12 of lysozyme in D_2O is at 137.5_8 ppm.[6] Peaks 11 and 12 arise from C^{γ} of one phenylalanine and C^{ϵ_2} of three tryptophans.[6,8] The γ-carbon of the phenylalanine gives rise to one-half of peak 12 of the protein in H_2O (Fig. 35A), and to one-half of peak 11 of the protein in D_2O (see Allerhand et al.[8] for evidence).

Studies of model compounds give an indication of expected deuterium isotope effects on chemical shifts of nonprotonated side-chain carbons of aromatic and arginine residues of proteins (Table III): (a) The effect should be appreciable (about 0.1–0.2 ppm upfield shift when going from H_2O to D_2O) for resonances of C^{ζ} of arginine and tyrosine residues, C^{ϵ_2} of

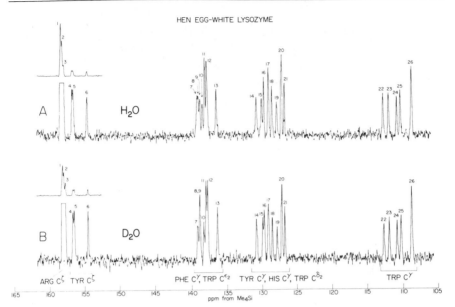

FIG. 35. Region of aromatic carbons and C^ζ of arginine residues in convolution-difference natural-abundance ^{13}C nmr spectra of hen egg white lysozyme. Each spectrum was recorded at 14.2 kG, with a 20-mm probe, 49,152 accumulations, and a recycle time of 2.2 sec. Peak numbers are those of Allerhand *et al.*,[5,6,8] and Figs. 13, 14B, and 25. The insets (peaks 1 to 6) are shown with one-eighth the vertical gain of the main spectra. (A) 14.6 mM protein in H_2O, pH 3.1, 0.1 M NaCl, 44°. (B) 13.8 mM protein in D_2O, pH meter reading 3.1, 0.1 M NaCl, 42°. Taken from Allerhand *et al.*[8]

tryptophan residues, and C^γ of histidine residues in the imidazolium and (presumably) the uncommon N^{δ_1}–H imidazole form. (b) The effect should be negligible for resonances of carbons which are not bonded to hydrogen-bearing oxygens or nitrogens (including C^γ of histidine residues in the common N^{ϵ_2}–H imidazole form).[6] On the whole, these expectations are confirmed in studies of proteins, but there are exceptions (see Tables I–IV of Oldfield *et al.*[6]). Some labile hydrogens of a folded protein in D_2O may not be exchanged appreciably with deuterium when the nmr spectrum is recorded. Also, protein folding may cause deuterium isotope effects beyond those predicted from studies of model compounds. For example, the chemical shift of C^γ of Trp-108 of hen egg white lysozyme exhibits a significant deuterium isotope effect (see peak 22 of Table IV; assignment is from Section IV,B). These problems, together with the small values of the deuterium isotope effects, prevent the use of these effects for making reliable assignments. This is unfortunate, because there is no reliable method now for distinguishing resonances of C^γ of nonti-

TABLE III
DEUTERIUM ISOTOPE EFFECTS ON SOME ^{13}C CHEMICAL
SHIFTS OF MODEL COMPOUNDS[a]

Compound	pH	Carbon	δ_H	$\delta_H - \delta_D$
L-Arginine	6.8	C^ζ	158.2_7	0.1_9
Gly-Tyr amide	6.8	Tyr C^ζ	155.8_9	0.1_3
		Tyr C^γ	129.4_2	0.0_0
Gly-Phe amide	6.7	Phe C^γ	137.7_2	-0.0_3
L-Tryptophan	4.0	$C^{\epsilon 2}$	137.7_5	0.1_3
		$C^{\delta 2}$	127.9_7	0.0_0
		C^γ	108.7_1	-0.0_6
Gly-His-Gly	3.0	His C^γ	129.3_8	0.0_9
	9.6	His C^γ	133.4_6	-0.0_3

[a] δ_H and δ_D are the chemical shifts (in parts per million downfield from Me$_4$Si) for samples in H_2O and D_2O, respectively. Estimated accuracy is ± 0.05 ppm. Other details are given in Table V of Oldfield et al.[6]

trating tyrosine residues from those of C^γ of nontitrating histidine residues in the imidazolium or $N^{\delta 1}$–H imidazole form. Table III suggests that the deuterium isotope effect may be used for this purpose. However, the resulting assignments should be considered as very tentative.

Note that no method (reliable or unreliable) has been reported for distinguishing resonances of C^γ of phenylalanine residues from those of C^γ of nontitrating histidine residues in the $N^{\epsilon 2}$–H imidazole state.

Summary of Assignments to Types of Carbons for Hen Egg-White Lysozyme. Chemical shift considerations, the effect of pH, and the PRFT method yield the assignments given in Table IV.[8]

B. Assignments to Specific Residues in the Amino Acid Sequence

Relaxation Probes. The broadening of a ^{13}C resonance by a paramagnetic species which behaves as a relaxation probe[52] is inversely proportional to r^6, where r is the distance from the relaxation probe to the pertinent carbon.[52] Consequently, if the values of r are known, it is possible to use differences in paramagnetic contributions to the linewidths for making specific assignments.[6,8,47] In principle, this technique is applicable to proteins which have "built-in" relaxation probes, such as some metalloproteins, and to proteins which have a binding site for a paramagnetic metal ion or an organic free radical. However, because of the strong dependence of the paramagnetic broadening on r, the spectrum of a sample in which all protein molecules are in the paramagnetic state may not yield many resonances with a detectable paramagnetic broadening: For carbons with large values of r, the paramagnetic broadening may be much

TABLE IV

DEUTERIUM ISOTOPE EFFECTS ON CHEMICAL SHIFTS, AND ASSIGNMENTS TO TYPES
OF CARBONS FOR THE RESONANCES OF NONPROTONATED AROMATIC CARBONS
(AND C^ζ OF ARGININES) OF HEN EGG WHITE LYSOZYME[a]

Assignment	Peak[b]	δ_H	$\delta_H - \delta_D$
11 Arg C^ζ { 1	1	158.1_9	$0.2_2{}^c$
	2	158.0_7	$0.2_4{}^c$
	3	157.8_6	$0.2_7{}^c$
Tyr C^ζ	4	156.6_1	0.0_9
Tyr C^ζ	5	156.4_9	0.1_5
Tyr C^ζ	6^d	154.5_4	0.1_2
Phe C^γ	7	139.0_1	-0.0_3
Trp $C^{\epsilon 2}$	8	138.8_6	0.1_6
Phe C^γ	9	138.7_0	0.0_0
Trp $C^{\epsilon 2}$	10	138.3_1	0.1_6
Phe C^γ + 3 Trp $C^{\epsilon 2}$	11^e	137.9_6	$0.1_7{}^f$
	12^e	137.7_3	$0.1_5{}^f$
Trp $C^{\epsilon 2}$	13	136.3_9	0.1_5
Tyr C^γ	14^d	130.6_9	0.0_0
His C^γ	15	129.9_8	0.1_4
Trp $C^{\delta 2}$	16	129.6_3	0.0_0
2 Tyr C^γ	17^e	129.0_2	-0.0_3
Trp $C^{\delta 2}$	18	128.5_0	0.0_6
Trp $C^{\delta 2}$	19	127.7_8	0.0_4
2 Trp $C^{\delta 2}$	20^e	127.1_3	0.0_3
Trp $C^{\delta 2}$	21	126.6_9	0.0_2
Trp C^γ	22	112.6_9	0.1_2
Trp C^γ	23	111.9_0	-0.0_3
Trp C^γ	24	110.8_4	0.0_7
Trp C^γ	25	110.3_2	0.0_9
2 Trp C^γ	26^e	108.7_6	0.0_3

[a] Taken from Table I of Oldfield et al.[6] δ_H and δ_D are the chemical shifts (in parts per million downfield from Me$_4$Si) for protein solutions in H_2O and D_2O, respectively (about 14 mM protein, pH meter reading 3.1, 0.1 M NaCl, about 43°).

[b] Peak numbering system of Fig. 35.

[c] Because of poor resolution, these values do not necessarily represent the isotope effects for C^ζ of all arginine residues.

[d] Peaks 6 and 14 arise from the same tyrosine residue (based on effect of pH).

[e] Two-carbon resonance.

[f] Peaks 11 and 12 (each a two-carbon resonance) arise from C^γ of one phenylalanine and $C^{\epsilon 2}$ of three tryptophans. The phenylalanine contributes to peak 12 of the protein in H_2O (137.7_3 ppm), but to peak 11 of the protein in D_2O (137.7_9 ppm). Therefore, the indicated deuterium isotope effects for peaks 11 and 12 probably correspond to $C^{\epsilon 2}$ of two tryptophans (one contributor to peak 11 and one to peak 12 of the proteins in H_2O and D_2O). The third tryptophan presumably contributes to peak 11 of the H_2O solution and to peak 12 of the D_2O solution, and has a deuterium isotope effect of about 0.3 ppm.

smaller than other contributions to the linewidths; the resonances of carbons with small values of r may be too broad for detection. Indeed, experimental results confirm that a detectable paramagnetic broadening occurs for a very limited range of values of r, if the protein is studied as a *pure* paramagnetic species.[46,47] For example, in the case of horse ferrimyoglobin the resonances of the heme carbons and of C^γ of His-93, His-97, His-64, and Phe-43 have not been detected; the resonance of C^ζ of Tyr-146 (peak 4 of Fig. 27A) has a paramagnetic broadening of about 5 Hz, and the resonances of all other nonprotonated aromatic carbons are not significantly broadened by the paramagnetic center.[47]

In many cases it is possible to observe the paramagnetic broadening of resonances which are too broad for detection in the spectrum of the *pure* paramagnetic protein: If there is fast interchange between the paramagnetic and a diamagnetic state of the protein, then the paramagnetic broadening can be "diluted" by observing spectra of mixtures of the two states.[4,6,8,46] The key to success is sufficiently fast exchange between the two states (see Section IV, B and Ugurbil *et al.*[46]). If the protein has a built-in (strongly bound) paramagnetic center, then the pertinent exchange is electron transfer between two redox states, as in the case of cytochrome c. If we are dealing with a relatively weak paramagnetic ligand, then the pertinent exchange is between the free protein and the protein–ligand complex, as in studies of Gd^{3+} binding to hen egg white lysozyme.

Consider the ^{13}C nmr spectra of the Cu(I) and Cu(II) states of azurin (from *Pseudomonas aeruginosa*), a small "blue copper" protein.[66] As expected, all 17 nonprotonated aromatic carbons, and also C^ζ of the single arginine residue, give rise to narrow resonances in the spectrum of the diamagnetic Cu(I) azurin (Fig. 36A).[46] However, only 11 of these 18 carbons yield detectable resonances in the spectrum of the Cu(II) protein (Fig. 36B).[46] Figures 37A and 37D show the same spectra as in Figs. 36A and 36B, respectively, but without application of the convolution-difference procedure (in order not to degrade the detectability of broad resonances). Figures 37B and 37C show spectra of samples with a 90:1 and a 23:1 ratio, respectively, of Cu(I) to Cu(II) protein. In the absence of exchange between the two redox states, these spectra would yield all the resonances observed in the spectrum of the Cu(I) protein (Fig. 37A). However, peak x (a carbonyl resonance), peak 6, and one contributor to peak 4 (a three-carbon resonance) of Cu(I) azurin are broadened beyond detection even in the presence of only 1% Cu(II) protein (Fig. 37B).[46] If the fast exchange condition applies (see Ugurbil *et al.*[46]), then the paramagnetic broadening of resonances in Fig. 37B should be about 1% of the broadening expected for resonances of pure Cu(II) azurin. Although the

[66] J. A. Fee, *Struct. Bonding (Berlin)* **23**, 1 (1975).

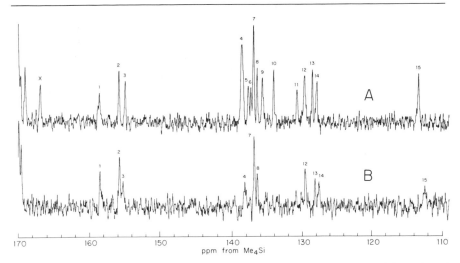

FIG. 36. Region of aromatic carbons (and C^ζ of arginine) and upfield edge of the carbonyl region in convolution-difference natural-abundance ^{13}C nmr spectra of reduced and oxidized *P. aeruginosa* azurin in H_2O, 0.05 *M* ammonium acetate, 31°. Each spectrum was recorded at 14.2 kG, with a 20-mm probe, noise-modulated off-resonance proton-decoupling, and a recycle time of 1.1 sec. Peaks are labeled as in Fig. 26 and Table X. (A) 7.4 m*M* reduced azurin, pH 5.2, after 65,536 accumulations. (B) 5.8 m*M* oxidized azurin, pH 5.3, after 131,072 accumulations. Taken from Ugurbil *et al.*[46]

resonances of 10 nonprotonated aromatic carbons are broadened beyond detection in the spectrum of pure Cu(II) azurin (Figs. 36B and 37D), only two of these resonances are severely broadened in the spectrum of a 90:1 mixture of Cu(I) and Cu(II) azurin (Fig. 37B).[46] Clearly, although many aromatic carbons are close to the copper ion of azurin, two of them are much closer than the rest.[46] The paramagnetic broadening of resonances in spectra of azurin has not yielded any assignments to specific residues in the amino acid sequence, because the values of r are not known. However, the spectra of Figs. 36 and 37 have been used to extract information about the number and types of aromatic amino acid residues near the copper ion.[46] Details are given in Section V,A.

Hen egg white lysozyme has a site, in the vicinity of the carboxylate groups of Glu-35 and Asp-52, that binds lanthanide ions.[43,67–69] Also, the nitroxide free radical 4-*N*-acetamido-2,2,6,6-tetramethylpiperidine-1-oxyl (Tempo-acetamide) binds to lysozyme at the amino-sugar binding sites A and C, and in a hydrophobic pocket near Trp-123.[70,71] We have used Gd^{3+}

[67] R. A. Dwek, R. E. Richards, K. G. Morallee, E. Nieboer, R. J. P. Williams, and A. V. Xavier, *Eur. J. Biochem.* **21**, 204 (1971).
[68] I. I. Secemski and G. E. Lienhard, *J. Biol. Chem.* **249**, 2932 (1974).
[69] K. Kurachi, L. C. Sieker, and L. H. Jensen, *J. Biol. Chem.* **250**, 7663 (1975).
[70] L. J. Berliner, *J. Mol. Biol.* **61**, 189 (1971).
[71] R. W. Wien, J. D. Morrisett, and H. M. McConnell, *Biochemistry* **11**, 3707 (1972).

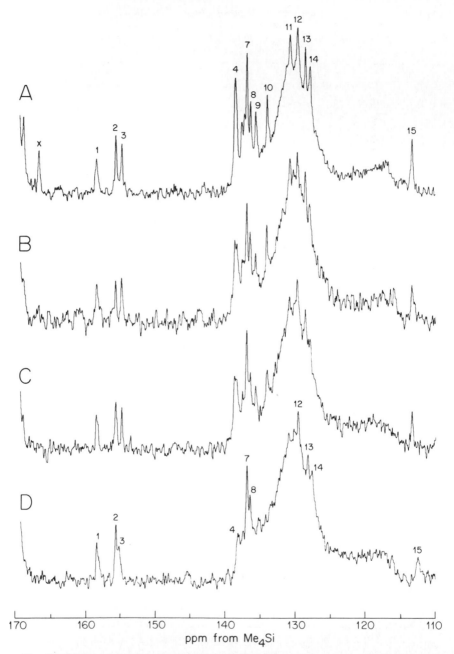

FIG. 37. Region of aromatic carbons (and C^ζ of arginine residues) and upfield edge of the carbonyl region in natural-abundance ^{13}C nmr spectra of *P. aeruginosa* azurin in H_2O, 0.05 *M* ammonium acetate, 31°. Each spectrum was recorded at 14.2 kG, with a 20-mm probe, noise-modulated off-resonance proton-decoupling, a recycle time of 1.1 sec, and 1.5 Hz digital broadening. The convolution-difference procedure was not applied. Peak designa-

TABLE V
VALUES OF THE SIXTH POWERS OF THE DISTANCES FROM THE NONPROTONATED
AROMATIC CARBONS OF LYSOZYME TO TWO REPORTED
LANTHANIDE ION BINDING SITES[a]

Carbon	r^6 (nm^6)	
	Tetragonal[b]	Triclinic[c]
Tyr-53 C$^\zeta$	1.7	4.3
Tyr-23 C$^\zeta$	39.6	23.3
Tyr-20 C$^\zeta$	108.6	84.2
Trp-108 C$^{\epsilon_2}$	2.3	1.1
Trp-111 C$^{\epsilon_2}$	6.8	2.7
Trp-63 C$^{\epsilon_2}$	8.2	12.3
Trp-62 C$^{\epsilon_2}$	9.5	20.4
Trp-28 C$^{\epsilon_2}$	14.8	7.1
Trp-123 C$^{\epsilon_2}$	53.4	19.4
Phe-34 C$^\gamma$	4.2	0.8
Phe-38 C$^\gamma$	20.3	7.6
Phe-3 C$^\gamma$	46.6	24.7
His-15 C$^\gamma$	136.0	95.9
Tyr-53 C$^\gamma$	1.0	2.0
Tyr-23 C$^\gamma$	48.1	26.7
Tyr-20 C$^\gamma$	96.7	66.9
Trp-108 C$^{\delta_2}$	1.2	0.5
Trp-63 C$^{\delta_2}$	4.5	7.1
Trp-62 C$^{\delta_2}$	5.3	12.2
Trp-111 C$^{\delta_2}$	8.1	3.0
Trp-28 C$^{\delta_2}$	23.3	11.3
Trp-123 C$^{\delta_2}$	50.6	18.0
Trp-108 C$^\gamma$	0.6	0.2
Trp-63 C$^\gamma$	2.7	4.6
Trp-62 C$^\gamma$	3.5	8.2
Trp-111 C$^\gamma$	6.3	2.0
Trp-28 C$^\gamma$	29.6	13.8
Trp-123 C$^\gamma$	57.2	21.0

[a] Taken from Table IV of Allerhand *et al.*[8] Both sets of values of r^6 were computed with the use of the carbon coordinates of tetragonal lysozyme, but with slightly different Gd^{3+} coordinates (see footnotes b and c).
[b] Computed with the use of the Gd^{3+} coordinates in tetragonal crystals of the Gd^{3+}–lysozyme complex.
[c] Computed with the use of the reported Gd^{3+} position in triclinic crystals of the Gd^{3+}–lysozyme complex.

tions are those of Figs. 26 and 36 and Table X. (A) 7.4 mM reduced azurin, pH 5.2, after 65,536 accumulations (same time-domain data as in Fig. 36A). (B) 90:1 ratio of reduced to oxidized azurin, 5.1 mM total protein concentration, pH 5.3, after 49,152 accumulations. (C) 23:1 ratio of reduced to oxidized azurin, 5.2 mM total protein concentration, pH 5.3, after 65,536 accumulations. (D) 5.8 mM oxidized azurin, pH 5.3, after 131,072 accumulations (same time-domain data as in Fig. 36B). Taken from Ugurbil *et al.*[46]

TABLE VI

Values of the Sixth Powers of the Distances from the Nonprotonated Aromatic Carbons of Lysozyme to the Centroid of Tempo-Acetamide in Each of Three Binding Sites[a]

Carbon	r^6 (nm^6)		
	Site A[b]	Site C[b]	Pocket[c]
Tyr-23 C$^\zeta$	d	5.03	d
Phe-34 C$^\gamma$	d	d	0.57
Phe-38 C$^\gamma$	d	d	1.95
Trp-28 C$^{\epsilon 2}$	d	6.35	d
Trp-62 C$^{\epsilon 2}$	0.05	0.43	d
Trp-63 C$^{\epsilon 2}$	0.46	0.56	d
Trp-108 C$^{\epsilon 2}$	4.86	0.21	d
Trp-111 C$^{\epsilon 2}$	d	1.39	d
Trp-123 C$^{\epsilon 2}$	d	d	0.01
Tyr-23 C$^\gamma$	d	9.59	d
Tyr-53 C$^\gamma$	d	7.38	d
Trp-62 C$^{\delta 2}$	0.07	0.19	d
Trp-63 C$^{\delta 2}$	0.46	0.25	d
Trp-108 C$^{\delta 2}$	5.82	0.17	d
Trp-111 C$^{\delta 2}$	d	2.35	d
Trp-123 C$^{\delta 2}$	d	d	0.01
Trp-62 C$^\gamma$	0.12	0.12	d
Trp-63 C$^\gamma$	0.89	0.25	d
Trp-108 C$^\gamma$	7.06	0.13	d
Trp-111 C$^\gamma$	d	2.70	9.37
Trp-123 C$^\gamma$	d	d	0.03

[a] Taken from Table V of Allerhand et al.[8] Carbons with values of $r^6 \geq 10$ nm^6 (to all three binding sites) are not listed.
[b] Site designation of Berliner[70] and Imoto et al.[72]
[c] Hydrophobic pocket (see Berliner[70]).
[d] Values of $r^6 \geq 10$ nm^6.

and Tempo-acetamide as relaxation probes for assigning some ^{13}C resonances of lysozyme.[8] We used crystallographic information to estimate the distances (r) from the nonprotonated aromatic carbons to the bound relaxation probes.[8] Values of r^6 are given in Tables V and VI.[8,70,72] In the case of the lanthanide–lysozyme complex, two sets of estimated values of r^6 are presented (Table V), based on two slightly different reported positions for Gd^{3+} bound to crystalline lysozyme. The binding of lanthanide ions and Tempo-acetamide to lysozyme occurs under conditions of fast exchange.[43,71] Therefore, the amount of paramagnetic broadening can be

[72] T. Imoto, L. N. Johnson, A. C. T. North, D. C. Phillips, and J. A. Rupley, in "The Enzymes" (P. D. Boyer, ed.), 3rd ed., Vol. 7, Chapter 21. Academic Press, New York, 1972.

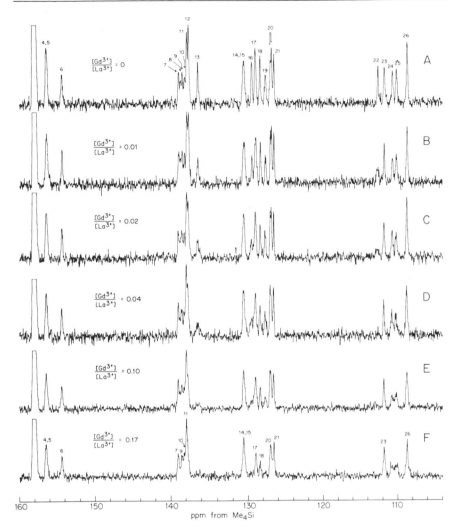

FIG. 38. Effect of various molar ratios of $LaCl_3$ to $GdCl_3$ (but 45 mM total lanthanide ion concentration) on the resonances of nonprotonated aromatic carbons of 15 mM hen egg white lysozyme in H_2O (pH 5.0, about 40°). Each spectrum was recorded at 14.2 kG, with a 20-mm probe, a recycle time of 2.2 sec, and 32,768 accumulations (spectra A–D) or 65,536 accumulations (spectra E and F). The convolution-difference method was applied. Peak numbers are those of Figs. 13, 14B, 25, and 35A. The molar ratios of $GdCl_3$ to $LaCl_3$ are indicated in each spectrum. Taken from Allerhand *et al.*[8]

controlled by changing the ratio of diamagnetic to paramagnetic protein. In the case of Tempo-acetamide binding, the diamagnetic protein is free lysozyme. For studies of Gd^{3+} binding, it is preferable to use the La^{3+}–lysozyme complex as the diamagnetic species which is in fast exchange with the Gd^{3+}–lysozyme complex.[43,44]

Figure 38A shows the aromatic region of the convolution-difference ^{13}C nmr spectrum of 15 mM lysozyme at pH 5 in the presence of a 45 mM concentration of La^{3+} ions. Under these conditions, only about 10 to 25% of the protein molecules are not binding lanthanide ions.[8] The spectrum does not differ significantly from one recorded at the same pH in the absence of La^{3+}.[6] We conclude that binding of lanthanide ions causes no major changes in the conformation of lysozyme. Figure 38 (spectra B–F) shows the effects of gradually replacing the La^{3+} ions by Gd^{3+}, without changing the total lanthanide ion concentration. We can use the observed differences in the extent of broadening of the various resonances, in conjunction with the r^6 values of Table V, to make specific assignments. However, we must take into account the uncertainty in the Gd^{3+} coordinates (Table V) and the possibility of secondary binding sites for lanthanide ions. Secondary binding does occur, but it does not significantly affect the linewidths of the resonances in Fig. 38.[73] Also, regardless of the exact Gd^{3+} coordinates (for the main binding site), binding of Gd^{3+} to the carboxylate groups of Glu-35 and Asp-52 places the metal ion closer to Trp-108 than to the other five tryptophan residues, closer to Tyr-53 than to the other two tyrosine residues, and closer to Phe-34 than to the other two phenylalanine residues (Table V). On this basis (together with the assignments to *types* of carbons given in Table IV), peaks 13, 16, and 22 are assigned to C$^{\epsilon 2}$, C$^{\delta 2}$, and C$^\gamma$, respectively, of Trp-108. One-half of peak 4–5 and one-half of peak 17 are assigned to C$^\zeta$ and C$^\gamma$, respectively, of Tyr-53. It is more difficult to identify the resonance of C$^\gamma$ of Phe-34, because the region of peaks 7–12 suffers from poor resolution (Fig. 38A). Nevertheless, it is clear that peaks 7 and 9, which arise from C$^\gamma$ of two phenylalanine residues (Table IV), are not significantly broadened even by high Gd^{3+} concentrations (Figs. 38E and 38F). Therefore, peaks 7 and 9 are assigned to Phe-3 and Phe-38 (not on a one-to-one-basis). By elimination (Table IV), C$^\gamma$ of Phe-34 contributes to peak 12. This assignment is consistent with the observation that low Gd^{3+} concentrations (Figs. 38C and 38D) seem to broaden one component of peak 12. Additional assignments based on the effects of Gd^{3+} are given in Allerhand et al.[8]

Figure 39 shows the effect of various concentrations of Tempo-acetamide on the resonances of nonprotonated aromatic carbons of lysozyme.[8] Even at high concentrations of Tempo-acetamide (Fig. 39D), the resonances assigned above to Trp-108 (peaks 13, 16, and 22) are not severely broadened. We conclude that site C is relatively free of Tempo-acetamide under the sample conditions of Fig. 39 (see Table VI).[8] Table VI then indicates that the resonances which are broadened beyond detection in Fig. 39D arise from Trp-62, Trp-63, and Trp-123.[8]

[73] K. Dill and Allerhand, *Biochemistry* 16, 5711 (1977).

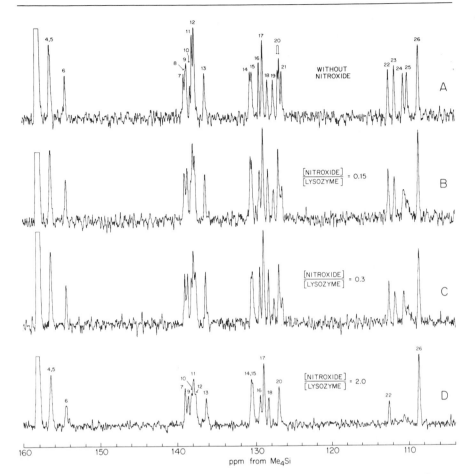

FIG. 39. Effect of the nitroxide free radical Tempo-acetamide on the resonances of non-protonated aromatic carbons of 13.3 mM hen egg white lysozyme in H_2O (pH 4.4, 38°). Each spectrum was recorded at 14.2 kG, with a 20-mm probe, a recycle time of 2.2 sec and 16,384 accumulations (spectra A–C) or 32,768 accumulations (spectrum D). The convolution-difference procedure was applied. Peak numbers are those of Figs. 13, 14B, 25, and 35A. (A) Without nitroxide. (B) With 2.0 mM nitroxide. (C) With 4.0 mM nitroxide. (D) With 26.6 mM nitroxide. Taken from Allerhand et al.[8]

The combination of results based on Gd^{3+} and Tempo-acetamide yields the assignments given in Table VII.[8]

Specific Chemical Modifications. A comparison of the ^{13}C nmr spectrum of an unmodified protein with that of a chemically modified version (at a specific residue) of the same protein may yield specific assignments. Possible difficulties are that (a) the ^{13}C nmr spectrum of the modified pro-

TABLE VII
Specific Assignments for Hen Egg White Lysozyme Based
on the Effects of Relaxation Probes[a]

Peak[b]	Assignment
4 + 5	$\begin{cases} \text{Tyr-53 C}^\zeta \\ \text{Tyr-20 or 23 C}^\zeta \end{cases}$
6	Tyr-23 or 20 C^ζ
7 + 9	Phe-3 + 38 C^γ
8 + 12[c]	Trp-62 + 63 $C^{\epsilon 2}$
10 + 11[c]	Trp-28 + 111 $C^{\epsilon 2}$
11[c]	Trp-123 $C^{\epsilon 2}$
12[c]	Phe-34 C^γ
13	Trp-108 $C^{\epsilon 2}$
14	Tyr-23 or 20 C^γ
15	His-15 C^γ
16	Trp-108 $C^{\delta 2}$
17[d]	$\begin{cases} \text{Tyr-53 C}^\gamma \\ \text{Tyr-20 or 23 C}^\gamma \end{cases}$
18	Trp-111 $C^{\delta 2}$
19 + 20[c]	Trp-62 + 63 $C^{\delta 2}$
20[c]	Trp-28 $C^{\delta 2}$
21	Trp-123 $C^{\delta 2}$
22	Trp-108 C^γ
23	Trp-123 C^γ
24 + 25	Trp-62 + 63 C^γ
26[d]	Trp-28 + 111 C^γ

[a] Taken from Table VI of Ref. 8. These assignments are based on the effects of Gd^{3+} and Tempo-acetamide, together with the assignments to types of carbons given in Table IV.

[b] Peak numbering system of Table IV and Figs. 13, 14B, 25, 35A, 38, and 39.

[c] One-half of this two-carbon resonance contributes here.

[d] Two-carbon resonance.

tein may indicate that the reaction is less specific than had been suggested in published reports and (b) the ^{13}C nmr spectrum may indicate that the chemical modification has caused extensive conformational reorganization. In any case, a comparison of the ^{13}C nmr spectrum of a chemically modified protein with the spectrum of the intact protein will yield information about the possible existence of the above two problems (see Section V,A).

We shall deal with only two types of chemical modifications, i.e., iodination of tyrosine residues and oxidation of tryptophan residues. Although a variety of oxidation products of tryptophan residues are possible,[74] here we shall encounter only oxindolealanine residues (Fig. 40C)

[74] A. N. Glazer, in "The Proteins" (H. Neurath and R. L. Hill, eds.), 3rd ed., Vol. 2, Chapter 1. Academic Press, New York, 1976.

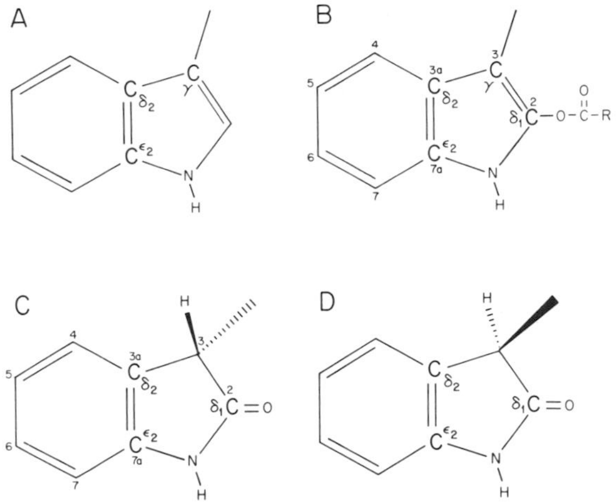

FIG. 40. Structures. (A) Side chain of a tryptophan residue. (B) Side chain of a δ_1-hydroxytryptophan residue which has the hydroxyl group esterified. (C) and (D) Side chains of the two diastereoisomers of an oxindolealanine residue. In each case, the $C^\beta H_2$ group is not shown.

and esters of δ_1-hydroxytryptophan (the enol form of oxindolealanine) residues (Fig. 40B).

Consider first the reaction of hen egg white lysozyme with an equimolar (or less than equimolar) amount of I_2. At pH $\geqslant 6$, the predominant reaction is iodination of Tyr-23 (conclusion based on results which did not involve ^{13}C nmr).[75,76] At pH 5.5, there is iodination of Tyr-23 (^{13}C nmr evidence[8,77]) and oxidation of Trp-108 (based on evidence which did not involve ^{13}C nmr).[78,79]

The ^{13}C nmr spectrum of a lysozyme sample which had been treated at pH 8.5 with an equimolar amount of I_2 clearly indicates that peak 14 of intact lysozyme (Figs. 14B and 35A) arises from C^γ of Tyr-23, but does not yield a firm assignment for C^ζ of Tyr-23.[8] However, the effect of high pH on the chemical shifts of lysozyme (Fig. 25) indicates that peak 6 and peak 14 arise from the same tyrosine residue.[8]

The spectra of lysozyme samples which had been treated with iodine (I_2/protein molar ratios of 0.5 and 1.0) at pH 5.5 yield assignments for C^{ϵ_2}, C^{δ_2}, and C^γ of Trp-108 which are in agreement with the assignments based on Gd^{3+} binding (Table VII).[8] Although the samples that were actually

[75] K. Hayashi, T. Shimoda, K. Yamada, A. Kumai, and M. Funatsu, *J. Biochem. (Tokyo)* **64**, 239 (1968).
[76] K. Hayashi, T. Shimoda, T. Imoto, and M. Funatsu, *J. Biochem. (Tokyo)* **64**, 365 (1968).
[77] R. S. Norton and A. Allerhand, *J. Biol. Chem.* **251**, 6522 (1976).
[78] T. Imoto, F. J. Hartdegen, and J. A. Rupley, *J. Mol. Biol.* **80**, 637 (1973).
[79] F. J. Hartdegen and J. A. Rupley, *J. Mol. Biol.* **80**, 649 (1973).

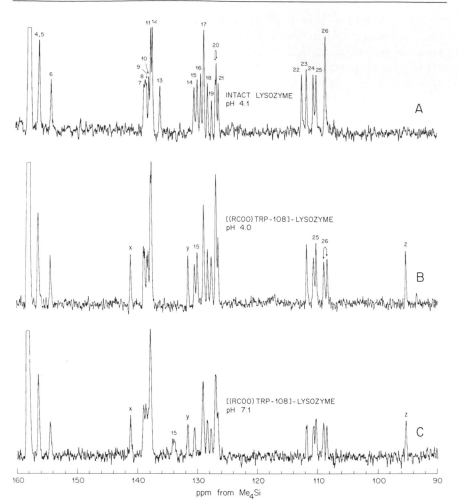

FIG. 41. Region of aromatic carbons (and C^ζ of arginine residues) in convolution-difference natural-abundance ^{13}C nmr spectra of intact hen egg white lysozyme and [(RCOO)Trp-108]lysozyme in H_2O (0.1 M NaCl). Each spectrum was recorded at 14.2 kG, with a 20-mm probe and a recycle time of 1.1 sec. (A) 11 mM intact lysozyme at pH 4.1 and 33°, after 65,536 scans. (B) 8 mM [(RCOO)Trp-108]lysozyme at pH 4.0 and 37°, after 96,566 scans. (C) 8 mM [(RCOO)Trp-108]lysozyme at pH 7.1 and 36°, after 65,536 scans. Taken from Dill and Allerhand.[73]

used for making the assignments were mixtures of lysozyme iodinated at Tyr-23 and lysozyme oxidized at Trp-108,[8] it is more informative to examine the spectrum of a pure fraction of lysozyme modified only at Trp-108.[73] The modified residue 108 is actually δ_1-hydroxytryptophan esterified to the side-chain carboxyl group of Glu-35 (Fig. 40B).[77,80,81] We shall call this modified residue (RCOO)Trp-108. Figure 41 shows the region of aromatic carbons in convolution-difference ^{13}C nmr spectra of intact hen egg white lysozyme at pH 4.1 (Fig. 41A), [(RCOO)Trp-108]lysozyme at pH 4.0 (Fig. 41B), and [(RCOO)Trp-108]lysozyme at pH 7.1 (Fig. 41C). Only Figs. 41A and 41B will be discussed at this point. Peaks x, y, and z of the modified protein (Fig. 41B) arise from C^{ϵ_2}, C^{δ_2}, and C^γ, respectively, of (RCOO)Trp-108.[77] The δ_1-carbon of this residue is also a nonprotonated one (Fig. 40B), but it has not been detected,[73,77] probably because it overlaps with the large peak of C^ζ of the eleven arginine residues (truncated peak at about 158 ppm in Fig. 41B).[77] Peaks 13, 16, and 22 of the intact protein (Fig. 41A) are "missing" from the spectrum of the modified protein (Fig. 41B) and are therefore assigned to Trp-108.[8]

The conversion of Trp-108 of lysozyme into (RCOO)Trp-108 causes minor changes in the ^{13}C chemical shifts of some unmodified aromatic residues (Fig. 41).[73] Most affected is peak 26, which arises from C^γ of Trp-28 and Trp-111 (Table VII). One component of peak 26 moves about 0.3 ppm downfield and the other moves about 0.3 ppm upfield.[73] However, these effects are not large enough to prevent the use of Fig. 41 for identifying the resonances of Trp-108 of the intact protein. A less simple situation is encountered when we try to use the spectrum of [oxindolealanine-62]lysozyme for identifying the resonances of Trp-62 of the intact protein.[6,8,82]

A widely quoted example of a specific and quantitative chemical modification of a protein is the oxidation of Trp-62 of hen egg white lysozyme when the protein is treated with an equimolar amount of N-bromosuccinimide (in 0.1 M acetate buffer at pH 4.5).[83] However, ^{13}C nmr spectra of the modified protein reveal several complications: (a) The composition of the modified protein is time-dependent.[82] Initially, there is an equimolar mixture of [oxindolealanine-62]lysozyme and [δ_1-acetoxytryptophan-62]lysozyme. The latter converts to [oxindolealanine-62]lysozyme with a half-life of about 2 days (at 25° and pH 4).[82] (b) [Oxindolealanine-62]lysozyme is a mixture of two diastereoisomeric species.[6,77,82] The conversion of lysozyme into [oxindolealanine-62]lysozyme generates a new asymmeteric center at C^γ of residue 62 (Figs. 40C and D). (c) Conversion of Trp-62 into oxindolealanine causes slight changes in the chemical shifts of Trp-63.[6,82]

[80] T. Imoto and J. A. Rupley, *J. Mol. Biol.* **80**, 657 (1973).

[81] C. R. Beddell, C. C. F. Blake and S. J. Oatley, *J. Mol. Biol.* **97**, 643 (1975).

[82] R. S. Norton and A. Allerhand, *Biochemistry* **15**, 3438 (1976).

[83] K. Hayashi, T. Imoto, G. Funatsu, and M. Funatsu, *J. Biochem. (Tokyo)* **58**, 227 (1965).

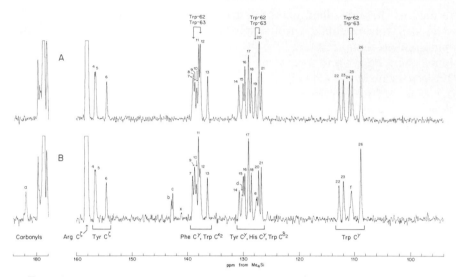

FIG. 42. Region of aromatic carbons and downfield edge of the carbonyl region in convolution-difference natural abundance ^{13}C nmr spectra of hen egg white lysozyme and [oxindolealanine-62]lysozyme. Each spectrum was recorded at 14.2 kG, with a 20-mm probe, noise-modulated off-resonance proton-decoupling, 65,536 accumulations, and a recycle time of 2.2 sec. (A) 14.6 mM lysozyme in H$_2$O, 0.1 M NaCl, pH 3.1, 44°. (B) 14.1 mM [oxindolealanine-62]lysozyme in H$_2$O, 0.1 M NaCl, pH 3.1, 42°. Taken from Norton and Allerhand.[82]

The spectral changes that occur upon conversion of lysozyme (Fig. 42A) into essentially pure [oxindolealanine-62]lysozyme (after complete conversion of δ_1-acetoxytryptophan-62 into oxindolealanine, Fig. 42B) yield assignments for Trp-62 and Trp-63 of the intact protein, but not on a one-to-one basis.[6,82] These assignments (shown in Fig. 42A) are in full agreement with those based on relaxation probes (Table VII).[8] One-to-one assignments for Trp-62 and Trp-63 of the intact protein can be extracted from the spectrum of [δ_1-acetoxytryptophan-62]lysozyme,[8,82] and from other evidence.[8]

Note that conversion of a tryptophan residue (Fig. 40A) into oxindolealanine (Fig. 40C) converts C$^\gamma$ into a *methine* aliphatic carbon, but generates a carbonyl carbon (C$^{\delta_1}$). The effect on the region of unsaturated carbon resonances is the disappearance of the C$^\gamma$ resonance, and the appearance of a carbonyl resonance (peak a of Fig. 42B).[6,82] Peaks b and c of Fig. 42B arise from C$^{\epsilon_2}$ of the two diastereoisomeric forms of [oxindolealanine-62]lysozyme.[82] Other details about Fig. 42B are given in Norton and Allerhand.[82]

Proteins from Different Species. Figure 43 shows the C$^\gamma$ resonances of tryptophan residues of carbon monoxide hemoglobins from various

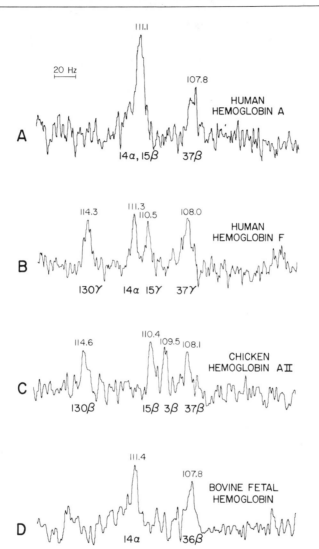

FIG. 43. Region of C^γ resonances of tryptophan residues in proton-decoupled natural-abundance ^{13}C nmr spectra at 14.2 kG (with a 20-mm probe and a recycle time of 1.1 sec) of some carbon monoxide hemoglobins in H_2O, 0.1 M NaCl, 0.05 M phosphate buffer, pH 7.0, about 35°. Chemical shifts are given in parts per million downfield from Me$_4$Si. (A) 3.3 mM (in tetramer) human adult hemoglobin, after 40 hr signal accumulation. (B) 2.3 mM (in tetramer) human fetal hemoglobin, after 77 hr signal accumulation. (C) 2.6 mM (in tetramer) chicken AII hemoglobin, after 40 hr signal accumulation. (D) 3.3 mM (in tetramer) bovine fetal hemoglobin, after 30 hr signal accumulation. Taken from Oldfield and Allerhand.[7]

species.[7] Human adult hemoglobin (A_0) has three nonequivalent trypto-phan residues: Trp-14α, Trp-15β, and Trp-37β.[26] The α-chains of human fetal hemoglobin (F_0) have the same amino acid sequence as the α-chains of hemoglobin A_0.[26] The sequence of the γ-chains of hemoglobin F_0 has 39 substitutions relative to the sequence of the β-chains of hemoglobin A_0.[26] However, the only substitution which involves a tryptophan residue is the replacement of Tyr-130β by Trp-130γ. Chicken AII hemoglobin has no tryptophan residues in the α-chains[26] and four tryptophan residues in the β-chains, three of which occupy the same positions as the three trypto-phan residues of the γ-chains of human hemoglobin F_0 (the fourth one is Trp-3β).[84] Bovine fetal hemoglobin has only two nonequivalent trypto-phan residues, Trp-14α and Trp-36β.[26] The latter is analogous to Trp-37β of the other hemoglobins.[26] The only peak present in all four spectra of Fig. 43 (at about 108 ppm) is assigned to Trp-37β, which is the only con-served tryptophan residue.[7] The peak at about 114 ppm, present only in the spectra of human F_0 and chicken AII hemoglobins, is assigned to Trp-130β.[7] The peak at about 111.3 ppm, missing from the spectrum of chicken AII hemoglobin, is assigned to Trp-14α. The peak at 109.5 ppm in the spectrum of chicken AII hemoglobin, not observed in any of the other spectra, is assigned to Trp-3β. By elimination, the peak at about 110.5 ppm in the spectra of human fetal and chicken AII hemoglobins (and one component of the two-carbon resonance at about 111 ppm in the spectrum of human adult hemoglobin) is assigned to Trp-15β.[7]

The change in the chemical shift of C^γ of Trp-15β when going from human fetal (and chicken AII) to human adult hemoglobin gives a hint of possible difficulties in the use of proteins from different species for as-signing ^{13}C resonances. This particular change does not create problems because all but one of the resonances of Fig. 43 exhibit chemical shifts which are remarkably invariant from species to species, and because there are only a few resonances over a large range of chemical shifts. In less favorable situations, the method of species comparisons may fail. For example, when going from horse heart cytochrome c to *Candida krusei* cytochrome c, Phe-46 "becomes" Tyr-52 and there are two new aromatic residues (Phe-4 and His-45). A comparison of the ^{13}C nmr spectra of the two proteins (at one pH value) did not yield any *firm* assignments.[6]

A rather favorable situation occurs when *titratable* residues are not in-variant in proteins from different species. Consider the tyrosine residues of myoglobins from red kangaroo, horse, and sperm whale. Red kangaroo myoglobin has only one tyrosine (Tyr-146), horse myoglobin has two (Tyr-146 and Tyr-103), and sperm whale myoglobin has three (Tyr-151,

[84] G. Matsuda, T. Maita, K. Mizuno, and H. Ota, *Nature (London), New Biol.* **244**, 244 (1973).

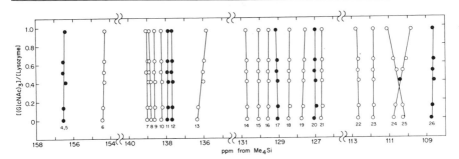

FIG. 44. Effect of (GlcNAc)$_3$ on the chemical shifts of the nonprotonated aromatic carbons of 14.6 mM hen eggs white lysozyme in H$_2$O, 0.1 M NaCl, pH 4.0, about 36°. The molar ratio [(GlcNAc)$_3$]/[Lysozyme] is essentially equal to the mole fraction of protein which has bound inhibitor.[72] Open and closed circles indicate one- and two-carbon resonances, respectively. Peak numbers are those of Figs. 13, 14B, 25, and 35A. The lines are linear least-square fits. Taken from Allerhand et al.[8]

Tyr-146, and Tyr-103).[26] The effect of pH on the nonprotonated aromatic carbon resonances of kangaroo cyanoferrimyoglobin (Fig. 30) indicates that Tyr-146 does not titrate.[45] Therefore, the effect of pH does not yield the assignments for C$^\zeta$ and C$^\gamma$ of Tyr-146. However, in the case of horse cyanoferrimyoglobin the resonances of C$^\zeta$ and C$^\gamma$ of one titrating tyrosine are observed (peaks 4 and 27, respectively of Fig. 29).[45] These resonances are assigned to Tyr-103.[45] Peaks 7 and 31 of sperm whale cyanoferrimyoglobin (Fig. 28) are practically superimposable with peaks 4 and 27, respectively, of the horse protein (Fig. 29) at all pH values. They are assigned to Tyr-103 of the sperm whale protein.[45] However, sperm whale cyanoferrimyoglobin exhibits an additional titrating tyrosine residue (peaks 6 and 30 of Fig. 28). Peaks 6 and 30 (Fig. 28) are assigned to C$^\zeta$ and C$^\gamma$, respectively, of Tyr-151.[45] A comparison of the spectra of the cyanoferrimyoglobins from kangaroo, horse, and sperm whale *at just one pH value* does not necessarily yield obvious assignments for Tyr-103 and Tyr-151, because of interference from resonances of nontitrating residues (see Fig. 27).

Chemical Shift Changes Caused by Ligands. In our hands, chemical shift changes caused by ligands (paramagnetic shift probes and diamagnetic species) have not played a useful role in assignments of ^{13}C resonances of proteins. Two problems must be overcome: (a) One-to-one connections between resonances of the free protein and resonances of the protein–ligand complex must be established. Such one-to-one connections are not always obvious by inspection (see below). (b) There is no guarantee that the chemical shifts of all carbons close to the ligand will be influenced by the ligand. Examples follow.

Problem (a) can be solved if mixtures of free protein and the

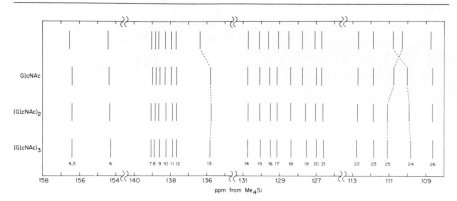

FIG. 45. Chemical shifts of the nonprotonated aromatic carbons of hen egg white lyso-zyme (in H_2O, 0.1 M NaCl, pH 4.0, about 36°) in the absence and presence of amino sugar inhibitors. Peak numbers are those of Figs. 13, 14B, 25, and 35A. (A) 9.0 mM protein, no inhibitor. (B) 12.5 mM protein, 0.76 M GlcNAc. (C) 14.1 mM protein, 20.4 mM (GlcNAc)$_2$. (D) 14.6 mM protein, 14.2 mM (GlcNAc)$_3$. The connections between peaks in spectrum A and those in spectra B, C, and D are based on concentration-dependence studies, as shown in Fig. 44 for (GlcNAc)$_3$. Taken from Allerhand *et al.*[8]

protein–ligand complex yield exchanged-averaged resonances. This is a common situation. For example, exchange-averaged ^{13}C resonances are encountered in spectra of mixtures of free hen egg white lysozyme and the complex of the protein with *N*-acetyl-D-glucosamine (GlcNAc).[8] The same situation arises when the inhibitor is the $\beta(1 \rightarrow 4)$-linked disac-charide or trisaccharide of GlcNAc.[8] Figure 44 shows the effect of various concentrations of (GlcNAc)$_3$ on the chemical shifts of the nonprotonated aromatic carbons of a fixed concentration of lysozyme. Because of the very high binding constant of (GlcNAc)$_3$ to lysozyme,[72] the molar ratio [(GlcNAc)$_3$]/[Lysozyme] in Fig. 44 is practically equal to the mole frac-tion of protein which has bound (GlcNAc)$_3$. Clearly, there is a cross-over of peaks 24 and 25 when going from free lysozyme to the complex (Fig. 44).[8]

Figure 45 illustrates problem (b). The inhibitors GlcNAc, (GlcNAc)$_2$, and (GlcNAc)$_3$ significantly affect the chemical shifts of C^γ of Trp-62 (peak 25), C^γ of Trp-63 (peak 24), and $C^{\epsilon 2}$ of Trp-108 (peak 13) of lyso-zyme. These effects are consistent with the presence of Trp-62, Trp-63 and Trp-108 at the active site of the enzyme.[72] However, the $C^{\epsilon 2}$ reso-nances of Trp-62 (one-half of peak 12) and Trp-63 (peak 8), and C^γ resonance of Trp-108 (peak 22), and the $C^{\delta 2}$ resonances of all three of these tryptophan residues (one-half of peak 20, peak 19, and peak 16, respectively) are not significantly affected by inhibitor binding. The

TABLE VIII

SPECIFIC ASSIGNMENTS FOR HEN EGG WHITE LYSOZYME[a]

Assignment	Peak[b]	Assignment	Peak[b]
Tyr-20 or 53 C^ζ	4	Trp-108 C^{δ_2}	16
Tyr-53 or 20 C^ζ	5	Tyr-20 C^γ ⎫	17
Tyr-23 C^ζ	6	Tyr-53 C^γ ⎭	
Phe-3 or 38 C^γ	7	Trp-111 C^{δ_2}	18
Trp-63 C^{ϵ_2}	8	Trp-63 C^{δ_2}	19
Phe-38 or 3 C^γ	9	Trp-28 C^{δ_2} ⎫	20
Trp-28 or 111 C^{ϵ_2}	10	Trp-62 C^{δ_2} ⎭	
Trp-111 or 28 C^{ϵ_2} ⎫	11	Trp-123 C^{δ_2}	21
Trp-123 C^{ϵ_2} ⎭		Trp-108 C^γ	22
Phe-34 C^γ ⎫	12	Trp-123 C^γ	23
Trp-62 C^{ϵ_2} ⎭		Trp-63 C^γ	24
Trp-108 C^{ϵ_2}	13	Trp-62 C^γ	25
Tyr-23 C^γ	14	Trp-28 C^γ ⎫	26
His-15 C^γ	15	Trp-111 C^γ ⎭	

[a] From Allerhand et al.[8]

[b] Peak designations are those (for the protein in H_2O) given in Tables IV and VII and in Figs. 13, 14B, 25, 35A, 38, 39, 41A, 42A, 44, 45, and 46A.

specific assignments mentioned here are taken from Table VIII (see below).

We explored the feasibility of using paramagnetic shift probes (such as Co^{2+}, Eu^{3+}, and Pr^{3+}) for assigning ^{13}C resonances of lysozyme.[8] The binding of these species to lysozyme (near Glu-35 and Asp-52) produces rather small changes in the chemical shifts of the aromatic carbons which are near the binding site.[8]

Chemical Exchange. I have already presented various examples of studies which involve spectra of mixtures of two states of a protein (two redox states, two ionization states, free protein and protein–ligand complex, or two protein–ligand complexes). We have also seen that if there is fast exchange between the two states, then spectra of mixtures of the two states will yield one-to-one connections between resonances of the two states. As a result, assignments for one state will automatically yield the corresponding assignments for the other one. As an illustration, consider the ferrous and ferric states of horse heart cytochrome *c*.

Ferrocytochrome *c* and ferricytochrome *c* undergo electron transfer in solution with a rate that depends on ionic strength.[65] Under typical sample conditions (0.2 ionic strength, about 40°), the rate constant (*k*) for electron transfer is about $10^4 \, M^{-1} \, sec^{-1}$.[65] The electron transfer produces a fluctuation in the chemical shifts of the ^{13}C resonances between the values of the two oxidation states. Equations (5) and (6) define slow and fast exchange, respectively.[12,52]

$$2\pi|\nu_R - \nu_0| \gg kC \tag{5}$$

$$2\pi|\nu_R - \nu_0| \ll kC \tag{6}$$

Here ν_R and ν_0 are the chemical shifts (in hertz) for a particular carbon of the reduced and oxidized protein, respectively, and C is the total molar concentration of the protein.

The 2 arginines, 4 tyrosines, 4 phenylalanines, 3 histidines, and single tryptophan (Trp-59) of horse heart cytochrome c contain 20 nonprotonated side chain carbons, and the heme contains 16 nonprotonated aromatic carbons. All 36 of these carbons yield narrow resonances in the spectrum of pure ferrocytochrome c (Fig. 33A).[4,6] Eighteen of these carbons (those which give rise to peaks 1–3, 15, 19–28, and 30 of Fig. 33A) yield narrow resonances in spectra of mixtures of the ferrous and ferric protein, with chemical shifts equal to $x_R\nu_R + x_0\nu_0$ (where x_R and x_0 are the mole fractions of ferrous and ferric protein, respectively).[4] This behavior is characteristic of fast exchange. A plot of chemical shifts as a function of x_0 (see Fig 4 of Oldfield and Allerhand[4]) yields the connections between peaks 1–3, 15, 19–28, and 30 of ferrocytochrome c and resonances of ferricytochrome c (vertical lines between Fig. 33A and Fig. 33B).

The resonances of C^γ of the coordinated His-18 and of the 16 nonprotonated aromatic carbons of the heme of horse heart cytochrome c are too broad for detection in the spectrum of the ferric protein (Fig. 33B).[4] Even if the fast exchange condition applies to any of these resonances, they are not expected to be narrow in spectra of mixtures of the two states, because the natural linewidth is approximately equal to $x_RW_R + x_0W_0$ (where W_R and W_0 are the natural linewidths for pure ferrous and pure ferric protein, respectively).[4]

The resonance of C^ζ of Tyr-67 (peak 4 of the ferrous protein, Fig. 33A) is a special case. It is broad when $x_R = 0.9$ and undetectable for smaller values of x_R, but then reappears when $x_R = 0$ (pure ferricytochrome c, peak 1 of Fig. 33B).[4] This behavior is probably a consequence of exchange broadening [Eq. (6) not valid], because C^ζ of Tyr-67 has $2\pi|\nu_R - \nu_0| = 873$ sec^{-1}, a value much larger than those of all other carbons which yield detectable resonances in the spectrum of ferricytochrome c.[6] Under our sample conditions,[4] $kC \approx 100$ sec^{-1}, so that Eq. (6) is clearly not valid for C^ζ of Tyr-67. Obviously, the connection between peak 4 of Fig. 33A and peak 1 of Fig. 33B is not based on spectra of mixtures of the two states.[6] Note that peak 1 of "pure" ferricytochrome c (Fig. 33B) is somewhat broader than other resonances, either as a result of paramagnetic broadening (C^ζ of Tyr-67 is relatively close to the heme[85]), or because the sample had a trace of ferrocytochrome c (exchange broadening).

C. Summary of Assignments for Various Proteins

Hen Egg White Lysozyme. Out of the 129 amino acid residues of this enzyme, 3 are tyrosines, 3 are phenylalanines, 6 are tryptophans, 1 is a histidine, and 11 are arginines.[26] The ζ-carbons of the arginines give rise to a partly split feature at about 158 ppm (peaks 1–3 of Fig. 35).[6] The 28 nonprotonated aromatic carbons yield peaks 4–26 (Fig. 35). The various methods discussed above have yielded the assignments of Table VIII (protein in H_2O). Details are given in Allerhand *et al.*[8] Some of the assignments were derived by two *independent* procedures. At this time, more assignments of ^{13}C resonances (to specific amino acid residues in the sequence) are available for lysozyme than for any other protein.

Horse Heart Cytochrome c. Out of the 104 amino acid residues of this protein, 2 are arginines, 4 are tyrosines, 4 are phenylalanines, 1 is a tryptophan, and 3 are histidines.[26] These residues contain a total of 20 nonprotonated side-chain carbons. In the case of the diamagnetic ferrocytochrome c, we must also consider the 16 nonprotonated aromatic carbons of the heme (Fig. 33A).[4,6] The known assignments[6] for ferrocytochrome c, ferricytochrome c, and cyanoferricytochrome c are indicated in Fig. 33.

Myoglobins. Out of the 153 amino acid residues of horse myoglobin, 2 are arginines, 2 are tyrosines, 7 are phenylalanines, 2 are tryptophans, and 11 are histidines.[26] These residues contain a total of 30 nonprotonated side-chain carbons. Because of the complexity of the spectra of myoglobins, the peak numbering system in Fig. 27 assigns one number to each carbon and not to eack peak, consecutively from left to right (at pH ≈ 8). At other pH values (Figs. 28–32) each resonance retains the number it was assigned in Fig. 27. Reported assignments for horse cyanoferrimyoglobin[45,47] are given in Table IX. Some assignments for other myoglobins are given in Wilbur and Allerhand.[45,47]

Azurin. Out of the 128 amino acid residues of *P. aeruginosa* azurin, 1 is an arginine (Arg-79), 2 are tyrosines, 6 are phenylalanines, 1 is a tryptophan (Trp-48) and 4 are histidines.[26] These residues contain a total of 18 nonprotonated side-chain carbons. Only the relatively simple specific assignments for C^ζ of Arg-79 and for $C^{\delta 2}$ and C^γ of Trp-48 have been made.[46] Other resonances have been assigned to "types" of carbons.[46] The known assignments are listed in Table X.

[85] N. Mandel, G. Mandel, B. L. Trus, J. Rosenberg, G. Carlson, and R. E. Dickerson, *J. Biol. Chem.* **252,** 4619 (1977).

TABLE IX

Assignments for Horse Cyanoferrimyoglobin[a]

Assignment	Peak[b]	Assignment	Peak[b]
Tyr-146 C$^\zeta$	1	Phe C$^\gamma$	17
2 Arg C$^\zeta$	2, 3	His C$^\gamma$ (6.6)	18
Tyr-103 C$^\zeta$	4	His-113 C$^\gamma$ (5.5)	19
4 Phe C$^\gamma$	5–8	His C$^\gamma$ (5.3)	20[c]
2 Trp C$^{\varepsilon_2}$	9, 10	His C$^\gamma$ (6.3)	21[d]
His C$^\gamma$ (≤4.4)	11[c]	Tyr-146 C$^\gamma$ ⎫	
2 Phe C$^\gamma$	12, 13	2 His C$^\gamma$ ⎬	22–24[e]
His C$^\gamma$ (6.4)	14[d]	2 Trp C$^{\delta_2}$	25, 26
His C$^\gamma$ (5.4)	15	Tyr-103 C$^\gamma$	27
His C$^\gamma$ (≤4.6)	16	His-93 C$^\gamma$	28[f]

[a] From Wilbur and Allerhand.[45,47] Number in parenthesis after a histidine C$^\gamma$ assignment is pK value from Wilbur and Allerhand.[47]

[b] Peak numbers of Figs. 27B, 29, and 32.

[c] Peaks 11 and 20 have been assigned to His-64 and His-97, but not on a one-to-one basis.

[d] His-116 gives rise to peak 14 or peak 21.

[e] Two nontitrating histidines contribute here.

[f] Tentatively assigned to the coordinated His-93 residue.

Hemoglobins. There are 287 nonequivalent amino acid residues in a molecule of human hemoglobin A$_0$. Out of these residues, 6 are arginines, 6 are tyrosines, 15 are phenylalanines, 3 are tryptophans, and 19 are histidines, with a total of 61 nonequivalent nonprotonated side-chain carbons.[26] Specific assignments for C$^\gamma$ of the tryptophan residues of some hemoglobins (Fig. 43) have been discussed above.[7] No firm specific assignments have been reported for other types of nonprotonated aromatic carbons of hemoglobins.[7]

V. Applications of Natural-Abundance ^{13}C nmr Spectra of Proteins

This section deals only with applications which involve the use of ^{13}C resonances of the protein itself, and only when the protein has natural isotopic composition. Section VI deals very briefly with the use of ^{13}C resonances of ^{13}C-labeled proteins and with studies which involve ^{13}C resonances of ligands.

A. Uses of Nonprotonated Aromatic Carbon Resonances

Conformation of Azurin in Solution. The environments of the aromatic residues of azurin from *Pseudomonas aeruginosa* were investigated by observing the nonprotonated aromatic carbon resonances of the Cu(I) and Cu(II) proteins (Figs. 26 and 36), and also mixtures of the two states (Fig. 37).[46]

TABLE X
ASSIGNMENTS AND CHEMICAL SHIFTS (AT pH 5.2) OF ^{13}C RESONANCES
OF *P. aeruginosa* AZURIN[a]

Assignment	Peak[b]	Chemical shift[c]	
		Cu(I)	Cu(II)
Carbonyl	x	166.7$_5$	
Arg-79 C$^\zeta$	1	158.5$_0$	158.4
Tyr C$^\zeta$	2[d]	155.7$_0$	155.7
Tyr C$^\zeta$	3[e]	154.8$_4$	155.2
6 Phe C$^\gamma$ ⎫	4[f,g]	138.5$_2$	138.1
2 His C$^\gamma$ ⎬	5	137.6$_2$	
Trp-48 C$^{\epsilon 2}$ ⎭	6	137.2$_4$	
	7[g,h]	136.8$_8$	136.9
	8[g]	136.3$_6$	136.4
	9	135.6$_6$	
Tyr C$^\gamma$	10[e]	133.9$_8$	
His C$^\gamma$	11[i]	130.6$_9$	
Tyr C$^\gamma$	12[d]	129.6$_1$	129.6
Trp-48 C$^{\delta 2}$	13	128.5$_0$	128.1
His C$^\gamma$	14	127.8$_3$	127.6
Trp-48 C$^\gamma$	15	113.2$_7$	112.5

[a] Taken from Table I of Ugurbil *et al.*[46]

[b] Peak designations are those of Figs. 26, 36, and 37.

[c] In parts per million downfield from Me$_4$Si. Values for reduced and oxidized azurin were obtained from the spectra of Fig. 36A (pH 5.2) and 36B (pH 5.3), respectively.

[d] Peaks 2 and 12 arise from the same tyrosine residue (see Fig. 26).

[e] Peaks 3 and 10 arise from the same tyrosine residue (see Fig. 26).

[f] Three-carbon resonance in the spectrum of reduced azurin, and single-carbon resonance in the spectrum of the oxidized protein.

[g] One-to-one connections between peaks 4, 7, and 8 of oxidized azurin and the corresponding peaks of the reduced protein are tentative (see Ugurbil *et al.*[46]).

[h] Two-carbon resonance at pH ≤6.5, but three-carbon resonance at pH ≥8 (peak 11 "moves" to this position at high pH; see footnote *i*).

[i] Not detected at pH ≥7. Assigned to the imidazolium state of a titratable histidine whose imidazole form contributes to peak 7.

(a) Histidines. All four histidine residues are strongly affected by protein folding: Two histidines do not exhibit titration behavior, the imidazole form of one titrating histidine is the "uncommon" N^{δ_1}–H imidazole tautomer, and the second titrating histidine exhibits slow proton exchange between the imidazole and imidazolium states.

(b) Residues coordinated to the copper: A comparison of the spectra of Cu(I) and Cu(II) azurin (Fig. 36) is not particularly informative about this crucial question, because the resonances of *many* aromatic residues are too broad for detection in the spectrum of Cu(II) azurin.[46] In other words, even uncoordinated residues which are close to the Cu(II) yield severely broadened resonances. However, spectra of mixtures of the two redox states (Fig. 37) indicate that two resonances (peak 6 and one contributor to peak 4 of the reduced protein) arise from carbons which are *much* closer to the copper (of the oxidized protein) than any of the other nonprotonated carbons which are close to the copper (see Section IV,B, discussion of relaxation probes). Therefore, it is reasonable (but not completely safe) to conclude that peak 6 and one contributor to peak 4 arise from residues coordinated to the copper. These peaks can only be assigned to C^γ of phenylalanine or nontitrating histidine residues and to $C^{\epsilon 2}$ of Trp-48 (Fig. 21). Since we reject the possibility of phenylalanine and tryptophan coordination to the copper,[46] we conclude that there are two coordinated histidines.

It will be possible to obtain much more detailed information about azurin when more specific assignments than those shown in Table X become available, and when better signal-to-noise ratios than those of Figs. 36 and 37 will permit quantitative measurements of paramagnetic broadening.[46]

Ionization Behavior of Tyrosine Residues. The effect of pH on resonances of C^ζ and C^γ of each titrating tyrosine residue yields the pK value for that residue. The existence of two usable signals for each tyrosine is particularly useful when anomalous titration behavior (not consistent with a single pK) is detected.[45] Carbon-13 nmr spectra of myoglobins indicate that Tyr-146 is a nontitrating residue (see Fig. 30), Tyr-151 of sperm whale myoglobin titrates with a single pK of 10.6 (see Fig. 28), and Tyr-103 of horse and sperm whale myoglobins has a titration behavior that cannot be fitted with a single pK value (see Figs. 28 and 29).[45]

Figures 25 and 26 yield only limited information about the ionization behavior of the tyrosine residues of hen egg-white lysozyme and Cu(I) azurin, respectively, because the titrations were not carried out to high enough pH for accurate determinations of pK values. Nevertheless, Fig. 25 indicates that Tyr-23 of lysozyme (peaks 6 and 14) has a lower pK than the other two tyrosine residues, and that a second tyrosine begins to titrate above pH 8.5.[8] A comparison of these results with various conflicting reports is given in Allerhand *et al.*[8] In the case of azurin (Fig. 26), one tyrosine (peaks 3 and 10) shows titration behavior above pH 9, while the other tyrosine (peaks 2 and 12) barely begins to titrate above pH 11.[46]

Ionization Behavior of Histidine Residues. Proton nmr spectroscopy has been the favorite method for measuring the ionization behavior of individual histidine residues of proteins (for a review, see Markley[86]), because the resonances of $H^{\epsilon 1}$ (and sometimes $H^{\delta 2}$) of histidine residues can often be observed as resolved single-hydrogen resonances. I believe that

[86] J. L. Markley, *Acc. Chem. Res.* **8**, 70 (1975).

^{13}C Fourier transform nmr is an attractive alternative (or complement) to proton nmr, especially when dealing with a protein which contains many histidine residues. However, in many cases there may be enough protein available for a proton nmr study, but not enough for ^{13}C nmr (see Table I).

It is noteworthy that the effect of pH on the chemical shift of the C^γ resonance of a histidine residue not only yields the pK value but also indicates the predominant tautomeric state of the imidazole form of the residue. We have seen that if the imidazolium form (Fig. 24A) of a histidine deprotonates at $N^{\delta 1}$ to yield the $N^{\epsilon 2}$–H imidazole tautomer (Fig. 24B), then the C^γ resonance should move about 6 ppm *downfield*. If deprotonation yields the $N^{\delta 1}$–H tautomer, the C^γ resonance should move about 2 ppm upfield.[46,62,63] In contrast, the proton resonance of H^ϵ moves upfield upon deprotonation either at $N^{\delta 1}$ or at $N^{\epsilon 2}$.[46]

Figure 25 indicates that the imidazole state of His-15 of hen egg white lysozyme exists mainly as the common $N^{\epsilon 2}$–H tautomer. Figure 25 yields a pK of 5.5 ± 0.1 (at 38°, 0.1 M NaCl), in reasonable agreement with reported values measured by other methods.[8]

Figure 26 indicates that only two of the four histidines of *P. aeruginosa* azurin titrate. Clearly, peak 14 arises from a titrating histidine (pK = 7.5 ± 0.2) whose imidazole form is mainly the uncommon $N^{\delta 1}$–H tautomer. Peak 11 (at low pH) and one contributor to peak 7 (at high pH) arise from a titrating histidine (p$K \approx 7$) whose imidazole form is mainly the common $N^{\epsilon 2}$–H tautomer. However, this histidine exhibits unusually slow proton exchange between the imidazole and imidazolium states.[46]

It is not surprising that horse myoglobin, with 11 histidine residues, exhibits a complex pattern of pH-dependent chemical shifts.[47] In the case of horse cyanoferrimyoglobin (Fig. 32), 8 of the 11 histidines exhibit titration behavior. The pK values and available assignments are listed in Table IX. Results (based on ^{13}C nmr) for horse ferrimyoglobin and kangaroo cyanoferrimyoglobin, and a comparison with published results based on proton NMR, can be found in Wilbur and Allerhand.[47]

Chemical Modifications of Proteins. Studies of the properties of a protein which has been chemically modified at a specific amino acid residue can be used for determining which portions in the intact protein are involved in its biological function and the maintenance of its native conformation.[74,87,88] A prerequisite for these studies is the determination of the position in the amino acid sequence of the chemically modified residue. It is also desirable to determine the structure of the modified residue. These tasks have typically involved procedures of chemical analysis which are

[87] G. E. Means and R. E. Feeney, "Chemical Modifications of Proteins." Holden-Day, San Francisco, California, 1971.
[88] J. R. Knowles, *Biochem., Ser. One* 1, 149 (1974).

lengthy and vulnerable to secondary reactions during the analysis. It has recently become apparent that, in favorable instances (examples are given below), ^{13}C nmr spectroscopy can be used for determining quickly and nondestructively which aromatic amino acid residues of a small protein have been chemically modified,[77,82] for establishing the structure of the modified residues,[77,82] and for studying various properties of the modified protein.[73]

The reaction of hen egg white lysozyme with a half-molar amount of I_2 at pH 5.5 yields a protein mixture that can be separated chromatographically into three major fractions.[78] One of these fractions (component C of Imoto et al.[78]) is lysozyme modified only at Trp-108.[78,79] The modified residue is δ_1-hydroxytryptophan (the enol form of oxindolealanine) esterified to Glu-35 (Fig. 40B).[77,80,81] Here I call the modified residue (RCOO)Trp-108. In Section IV,B (discussion of specific chemical modifications) I discussed the use of the ^{13}C nmr spectrum of component C (Fig. 41B) for identifying the resonances of Trp-108 of the intact protein. However, the spectrum of Fig. 41B also provides evidence for the structure of component C. Peaks 13, 16, and 22 of intact lysozyme (Fig. 41A), which can be assigned to Trp-108 without the use of Fig. 41B (see Table VII), are replaced by peaks x, y, and z in the spectrum of component C. The chemical shifts of most other resonances (of nonprotonated aromatic carbons) do not change when going from the intact protein to component C. A few chemical shifts do change (see peak 26), but only by 0.3 ppm or less.[73] Clearly, residue 108 of component C is the only *aromatic* residue which is chemically modified. The chemical shifts of peaks x, y, and z of component C indicate that the modified residue 108 is either δ_1-hydroxytryptophan or an ester of this species. For details, see Norton and Allerhand.[77]

It is not necessary to fractionate the product of the reaction of lysozyme with iodine in order to get information about the location and structure of the modified residues.[77] Figure 46 shows the region of aromatic carbons (and also the downfield edge of the carbonyl region) in convolution-difference spectra of intact lysozyme (Fig. 46A) and lysozyme samples treated at pH 5.5 with I_2/protein molar ratios of 0.5 (Fig. 46B) and 1.0 (Fig. 46C). The reaction with iodine causes a decrease in the intensities of the resonances of C^{ϵ_2}, C^{δ_2}, and C^γ of Trp-108 (peaks 13, 16, and 22, respectively) and C^γ of Tyr-23 (peak 14). Also, new resonances appear (peaks a, b, x, y, and z). Peaks a and b can be assigned to modified tyrosine by examining the spectrum of a lysozyme sample which had been treated with I_2 at pH 8.5 (see Fig. 3 of Norton and Allerhand[77]), because at high pH iodine reacts with Tyr-23 but not with Trp-108.[77] Also, the chemical shift of peak b is characteristic of an iodine-bearing aromatic carbon.[77] The chemical shifts of peaks x, y, and especially peak z yield information

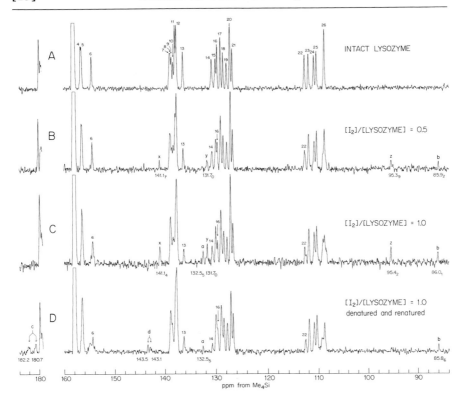

FIG. 46. Region of aromatic carbons and downfield edge of the carbonyl region in convolution-difference natural-abundance ^{13}C nmr spectra of intact hen egg white lysozyme, and lysozyme samples treated at pH 5.5 with iodine. Each spectrum was recorded at 14.2 kG, with a 20-mm probe, noise-modulated off-resonance proton-decoupling, and a recycle time of 2.2 sec. Small numbers are peak designations of Figs. 13, 14B, 25, and 35A. Peaks designated by letters arise from chemically modified residues (see text). The large numbers below the peaks designated by letters are chemical shifts. (A) Intact lysozyme: 14.6 mM protein in H$_2$O, 0.1 M NaCl, pH 3.1, 44°, 65,536 accumulations. (B) Lysozyme treated with an I$_2$/protein molar ratio of 0.5; 14.8 mM protein in H$_2$O, 0.1 M NaCl, pH 3.0, 33°, 49,152 accumulations. (C) Lysozyme treated with an I$_2$/protein molar ratio of 1.0; 14.2 mM protein in H$_2$O, 0.1 M NaCl, pH 3.0, 34°, 57,344 accumulations. (D) Lysozyme treated with an I$_2$/protein molar ratio of 1.0, then incubated for 24 hr in 6.3 M guanidinium chloride at pH 3.1 and 36°, and then renatured; 16.1 mM protein in H$_2$O, 0.1 M NaCl, pH 3.1, 33°, 98,304 accumulations. Taken from Norton and Allerhand.[77]

about the structure of the modified Trp-108 residue.[77] The intensities of peaks 13, 16, 22, x, y, and z in the spectra of the modified protein (Figs. 46B and 46C) give the mole fraction of protein modified at residue 108. The intensities of peaks 14, a, and b yield information about the extent of iodination of Tyr-23.[77]

The Glu-35 ester of δ_1-hydroxytryptophan-108 (Fig. 40B) can be hy-

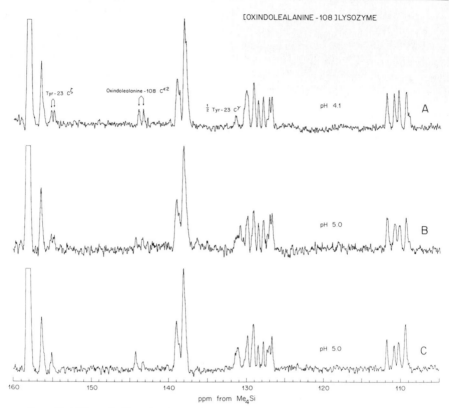

[OXINDOLEALANINE -108]LYSOZYME

FIG. 47. Effect of pH and sample history on the nonprotonated aromatic carbon resonances of 14 mM [oxindolealanine-108]lysozyme in H_2O at 39°. Each spectrum was recorded at 14.2 kG, with a 20-mm probe and a recycle time of 1.1 sec. The convolution-difference method was applied (see Section III, C). (A) At pH 4.1, with 65,536 scans, recorded immediately after removal of guanidinium chloride by dialysis at pH 3. (B) At pH 5.0, with 32,768 scans, recorded after the spectrum at pH 4.1 was obtained. (C) At pH 5.0, with 65,536 scans, recorded after the pH had been taken to 5.5 (for 10 hr) and 6.3 (for 30 hr). Taken from Dill and Allerhand.[73]

drolyzed irreversibly to oxindolealanine (Fig. 40, structures C and D) when [(RCOO)Trp-108]lysozyme is unfolded.[77,80] The ^{13}C nmr spectrum clearly demonstrates the conversion process.[77] The modified protein mixture, after denaturation and renaturation, yields the spectrum of Fig. 46D. In this spectrum, peaks a and b of the modified Tyr-23 residue are still present, but peaks x, y, and z of (RCOO)Trp-108 are missing. Instead, we have peaks c and d (each a doublet), characteristic of $C^{\delta 1}$ (a carbonyl) and $C^{\epsilon 2}$, respectively, of the two diastereoisomers of an oxindolealanine residue (Fig. 40, structures C and D).[77] These peaks are more clearly observed in a spectrum of *pure* [oxindolealanine-108]lysozyme (Fig. 47).[73]

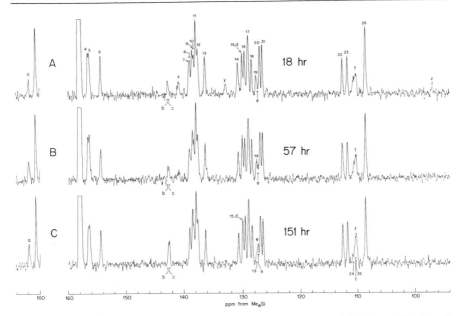

FIG. 48. Time dependence of the region of aromatic carbons (and the downfield edge of the carbonyl region) in convolution-difference natural-abundance ^{13}C nmr spectra of hen egg white lysozyme treated with an equimolar amount of N-bromosuccinimide in acetate buffer (see Norton and Allerhand[82]). Each spectrum is the sum of two spectra obtained from two different samples at similar times after transfer to 25°. In each case, spectra were recorded at 14.2 kG, with a 20-mm probe, noise-modulated off-resonance proton-decoupling, a recycle time of 2.2 sec, and 32,768 accumulations (20 hr total time) per spectrum. Peak designations are those of Fig. 42 and Ref. 81. Arrows labeled b, c, e, f, 19, 24, and 25 indicate chemical shifts of corresponding resonances of Fig. 42. Solution conditions for each sample were: 14.7 mM protein in H$_2$O, pH 3.9, 0.1 M NaCl, 25°. The resonance of C$^\gamma$ of His-15 (peak 15) shifts slightly downfield when going from pH 3.1 (Fig. 42) to pH 3.9, and as a result it coincides with peak d at pH 3.9. (A) 18 hr after transfer of each sample to 25°. (B) About 57 hr after transfer to 25°. (C) 151 hr after transfer to 25°. Taken from Norton and Allerhand.[82]

When the pH of folded [oxindolealanine-108]lysozyme is maintained at a value ≤5, the oxindolealanine-108 residue exists as an equimolar mixture of the two diastereoisomers (see Fig. 47, spectra A and B). However, when the pH is raised to a value ≥6, one diastereoisomer becomes predominant. When the pH is lowered again, the protein does not revert back to the equimolar mixture of diastereoisomers (see Fig. 47C).[73]

Application of ^{13}C nmr spectroscopy to the study of the reaction of equimolar amounts of N-bromosuccinimide and hen egg-white lysozyme yielded the following previously unknown facts.[82] (a) When the reaction is carried out under the conditions of Hayashi *et al.* (acetate buffer),[83] the initial ^{13}C nmr spectrum (Fig. 48A) indicates the presence of about equal amounts of oxindolealanine-62 (peak a and peak b–c) and δ_1-acetoxytryptophan-62 (peaks x, y, and z). The involvement of an acetoxy group was actually established with the use of [1-^{13}C]acetate buffer, which yielded a strong acetyl carbonyl resonance (with a time-dependent intensity) in the spectrum of the modified protein.[82] (b) The δ_1-acetoxytryptophan-62 residue hydrolyzes to oxindolealanine with a half-life of about 2 days (see Figs. 48B and 48C). (c) When the reaction is carried out in the absence of acetate buffer, the *initial* ^{13}C nmr spectrum shows essentially complete conversion of Trp-62 into oxindolealanine. (d) Both diastereoisomers of oxindolealanine-62 are formed (see Fig. 42B).

Carbon-13 nmr spectra have been used recently to study the effects of chemical modifications at Trp-108 of lysozyme on some properties of the protein.[73] The following findings were made: (a) The strong binding of Gd^{3+} in the vicinity of Glu-35 and Asp-52 of lysozyme is weakened (by a factor ≥ 20) when the protein is converted into [(RCOO)Trp-108]lysozyme. (b) The modified protein has a weak binding site for lanthanide ions near Asp-101. (c) Unlike intact lysozyme, [(RCOO)Trp-108]lysozyme does not show evidence of self-association at neutral pH (see below). (d) The more stable of the two diastereoisomers of [oxindolealanine-108]lysozyme (see Fig. 47C) binds Gd^{3+} in the vicinity of Glu-35 and Asp-52 with a binding constant similar to that of the intact protein.

Self-Association of Lysozyme. Self-association of lysozyme has been studied by various methods.[72] A head-to-tail model which involves the active site has been proposed.[89,90] Thermodynamic measurements on [oxindolealanine-62]lysozyme and [(RCOO)Trp-108]lysozyme suggest that the intermolecular contact is asymmetric (head to tail), and that Trp-62 and Glu-35 are among the residues in the contact.[91] Carbon-13 nmr spectra provide direct evidence for the participation of Trp-62 in the aggregation process.[92]

Figure 49 shows the effect of pH on the chemical shifts of the C^γ resonances of the six tryptophan residues of 9 mM lysozyme (0.1 M NaCl, 38°). A few results are also shown for 14.5 mM protein in the pH-independent range below pH 5. Only Trp-108 and Trp-62 exhibit significant pH dependence, with apparent pK values of 5.9 ± 0.3 and 6.1 ± 0.3, respectively.[92] These apparent pK values reflect the ionization of Glu-35 (pK ≈ 6[72]). However, ionization of Glu-35 *directly* changes the chemical shift of the C^γ resonance of Trp-108, while for Trp-62 the effect is indirect,

[89] A. J. Sophianopoulos, *J. Biol. Chem.* **244**, 3188 (1969).
[90] J. F. Studebaker, B. D. Sykes, and R. Wien, *J. Am. Chem. Soc.* **93**, 4579 (1971).
[91] S. K. Banerjee, A. Pogolotti, Jr., and J. A. Rupley, *J. Biol. Chem.* **250**, 8260 (1975).
[92] R. S. Norton and A. Allerhand, *J. Biol. Chem.* **252**, 1795 (1977).

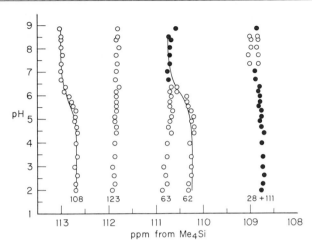

FIG. 49. Effect of pH on the chemical shifts of C^γ of tryptophan residues of hen egg white lysozyme in H_2O, 0.1 M NaCl, 38°. Protein concentration was 9.0 mM at all pH values except the following: At pH 2.96, 3.97, 4.65, and 4.90 the concentration was 14.5 mM; at pH 8.84 it was 7.3 mM. Numbers below points at pH 2 are assignments to specific residues (Table VIII). Open and solid circles indicate single-carbon and two-carbon resonances, respectively. Lines for Trp-62 and Trp-108 are best-fit theoretical "titration" curves, assuming a single pK in each case. For Trp-62, only the values for 9 mM protein were used for the best-fit calculation. Taken from Norton and Allerhand.[92]

i.e., ionization of Glu-35 promotes self-association of the protein, and the self-association process changes the chemical shift of C^γ of Trp-62.[92] The evidence for this statement comes from Fig. 50 (spectra B, C, and D), which shows the concentration dependence of the chemical shifts of the C^γ resonances of the tryptophan residues at pH 7.[92] When the concentration is lowered from 9 mM to 1 mM, the chemical shift of C^γ of Trp-108 remains invariant (and quite different from the value at low pH). However, the chemical shift of C^γ of Trp-62 is concentration dependent at pH 7. The value for 1 mM protein (Fig. 50D) is very similar to the corresponding chemical shift of 9 mM lysozyme at low pH (Fig. 50A).[92]

A comparison of Fig. 50E (9 mM lysozyme at pH 7, 18 mM LaCl$_3$) with the other spectra in Fig. 50 clearly indicates that the binding of La^{3+} ions in the vicinity of Glu-35 and Asp-52 disrupts the self-association of lysozyme.[92]

There is no change in the chemical shift of C^γ of Trp-62 of 8 mM [(RCOO)Trp-108]lysozyme when the pH is changed from 4.0 (Fig. 41B) to 7.1 (Fig. 41C), which strongly suggests that formation of the ester bond between Glu-35 and the modified Trp-108 residue greatly reduces (or entirely suppresses) self-association of the protein.[73]

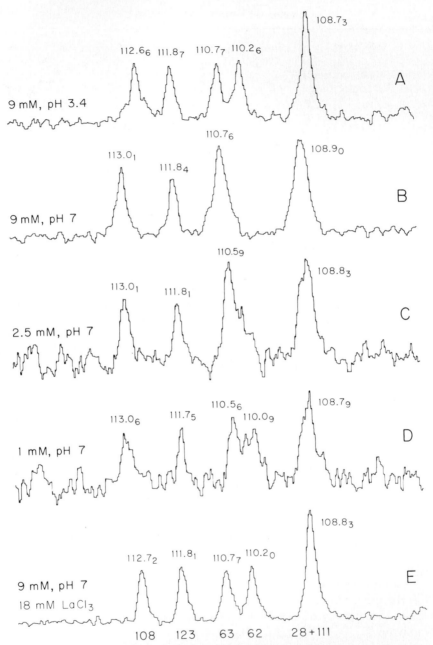

FIG. 50. Resonances of C^γ of tryptophan residues in the natural-abundance ^{13}C Fourier transform nmr spectra of hen egg white lysozyme in H_2O, 0.1 M NaCl, 38°. Spectra were recorded at 14.2 kG, with a 20-mm probe, noise-modulated off-resonance proton-decoupling, a recycle time of 1.1 sec, and a digital broadening of 0.9 Hz. The convolution-difference procedure was not applied. Numbers above peaks indicate chemical shifts. Numbers below the

B. Uses of Methine Aromatic Carbon Resonances

As far as I know, only one example of the use of methine carbon resonances of individual aromatic amino acid residues of a protein of natural isotopic composition has been reported (see Fig. 18B).[41] Each of peaks M2, M3, and M4 of Fig. 18B is a resonance which arises from the two δ-carbons of a tyrosine residue (Tyr-23, Tyr-53, and Tyr-20, respectively) of hen egg white lysozyme (at 63.4 kG, 32°).[41] Thus, the two δ-carbons of each tyrosine appear to be indistinguishable, as is also the case for the reported ^{1}H nmr signals of the corresponding hydrogens of lysozyme (at 54°).[33] These results have been interpreted as indicative of flipping (180° rotation about the C^β–C^γ bond) of each phenolic ring at rates exceeding the (unknown) difference in the chemical shifts of the two δ-carbons.[33,41] The linewidths of peaks M2, M3, and M4 (Fig. 18B), together with the assumption that the chemical shift difference between the two δ-carbons is at least 0.1 ppm (7 Hz), yielded a *lower limit* to the rate of flipping (at 32°) of 30 sec^{-1}.[41] An *upper limit* of 10^8 sec^{-1} was obtained from the T_1 and NOE values of peaks M2, M3, and M4.[41]

C. Uses of Carbonyl Resonances

Single-carbon resonances have been reported for some carboxylate groups of hen egg white lysozyme.[28] A few extremely tentative assignments were given.[28] Some pK values, probably of the C-terminal carboxyl and of some aspartic acid residues, were determined.[28]

Individual-carbon resonances have been reported for the carbonyl carbons of the N-terminal glycine residues of horse myoglobin,[6,29] kangaroo myoglobin,[29] and the γ-chains of human fetal hemoglobin.[7] The effect of pH on the chemical shift of the carbonyl carbon of Gly-1 of horse cyanoferrimyoglobin (the 9 spectra of Fig. 51 plus spectra at 20 additional pH values) yielded a pK of 7.81 ± 0.05 (protein in H_2O, 39°, 0.1 M KCl).[29]

D. Uses of Aliphatic Carbon Resonances

Very few resolved individual aliphatic carbon resonances of proteins of natural isotopic composition have been identified so far.[21,25] It is likely that the use of high-field nmr spectrometers will encourage the identifica-

---

peaks in spectrum E indicate assignments to specific residues (Table VIII). (A) 9.0 mM protein at pH 3.4, after 32,768 accumulations. (B) 9.0 mM protein at pH 7.0, after 98,304 accumulations. (C) 2.5 mM protein at pH 7.0, after 131,072 accumulations. (D) 1.0 mM protein at pH 7.0, after 294,912 accumulations. (E) 9.0 mM protein and 18 mM $LaCl_3$ at pH 7.0, after 81,920 accumulations. Taken from Norton and Allerhand.[92]

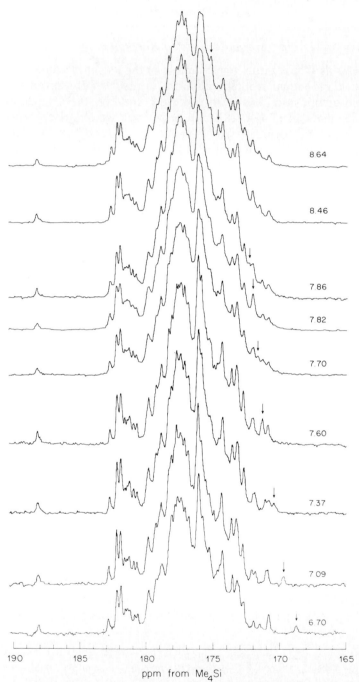

8.64

8.46

7.86

7.82

7.70

7.60

7.37

7.09

6.70

190 185 180 175 170 165

ppm from Me$_4$Si

FIG. 51. Effect of pH on the region of carbonyl resonances in proton-decoupled natural-abundance ^{13}C nmr spectra of horse cyanoferrimyoglobin (9–13 mM protein in H$_2$O, 0.1 M KCl, 39°). Each spectrum was obtained at 14.2 kG, with a 20-mm probe, 32,768 accu-

tion and use of individual aliphatic carbon resonances in the future. In the past, the aliphatic region in spectra of proteins has enjoyed some popularity, but mainly for measuring "overall" relaxation behavior of multiple-carbon bands, especially the α-carbon region.[16,19,23,93,94]

VI. Other Ways of Using ^{13}C nmr for Studying Enzymes

This chapter has dealt entirely with ^{13}C nmr spectra of proteins of natural isotopic composition. The emphasis has been on the observation and use of resonances of individual nonprotonated aromatic carbons. I have not discussed two important areas of ^{13}C nmr research which the reader may wish to explore, i.e., studies of incorporated ^{13}C-enriched amino acid residues into an otherwise intact (or nearly intact) protein[42,95–106] and studies of ^{13}C resonances of ligands or covalently attached reporter groups.[107–141] Some brief comments about these techniques are presented below.

[93] R. B. Visscher and F. R. N. Gurd, *J. Biol. Chem.* **250**, 2238 (1975).

[94] K. Wüthrich and R. Baumann, *Org. Magn. Reson.* **8**, 532 (1976).

[95] D. J. Saunders and R. E. Offord, *FEBS Lett.* **26**, 286 (1972).

[96] D. T. Browne, G. L. Kenyon, E. L. Packer, H. Sternlicht, and D. M. Wilson, *J. Am. Chem. Soc.* **95**, 1316 (1973).

[97] M. W. Hunkapiller, S. H. Smallcombe, D. R. Whitaker, and J. H. Richards, *J. Biol. Chem.* **248**, 8306 (1973).

[98] M. W. Hunkapiller, S. H. Smallcombe, D. R. Whitaker, and J. H. Richards, *Biochemistry,* **12**, 4732 (1973).

[99] E. L. Packer, H. Sternlicht, and J. C. Rabinowitz, *Proc. Natl. Acad. Sci. U.S.A.* **69**, 3278 (1972).

[100] E. L. Packer, H. Sternlicht, and J. C. Rabinowitz, *Ann. N. Y. Acad. Sci.* **222**, 824 (1973).

[101] M. W. Hunkapiller, S. H. Smallcombe, and J. H. Richards, *Org. Magn. Reson.* **7**, 262 (1975).

[102] W. C. Jones, Jr., T. M. Rothgeb, and F. R. N. Gurd, *J. Am. Chem. Soc.* **97**, 3875 (1975).

[103] W. C. Jones, Jr., T. M. Rothgeb, and F. R. N. Gurd, *J. Biol. Chem.* **251**, 7452 (1976).

[104] D. T. Browne, E. M. Earl, and J. D. Otvos, *Biochem. Biophys. Res. Commun.* **72**, 398 (1976).

[105] I. M. Chaiken, J. S. Cohen, and E. A. Sokoloski, *J. Am. Chem. Soc.* **96** 4703 (1974).

[106] I. M. Chaiken, *J. Biol. Chem.* **249**, 1247 (1974).

[107] A. M. Nigen, P. Keim, R. C. Marshall, J. S. Morrow, and F. R. N. Gurd, *J. Biol. Chem.* **247**, 4100 (1972).

[108] R. B. Moon and J. H. Richards, *J. Am. Chem. Soc.* **94**, 5093 (1972).

[109] P. J. Vergamini, N. A. Matwiyoff, R. C. Wohl, and T. Bradley, *Biochem. Biophys. Res. Commun.* **55**, 453 (1973).

mulations and a recycle time of 1.1 sec. In each spectrum, an arrow indicates the position of the carbonyl resonance of Gly-1. Numbers on the right-hand side are pH values. Spectra recorded at twenty additional pH values (see Fig. 2 of Wilbur and Allerhand[29]) are not shown. Taken from Wilbur and Allerhand.[29]

A. Incorporation of ^{13}C-Enriched Amino Acid Residues

Although the information presented in this report deals with proteins of natural isotopic composition, much of this information can be used directly when dealing with ^{13}C nmr studies of ^{13}C-enriched proteins. For example, the values of chemical shifts, linewidths, spin-lattice relaxation times, and nuclear Overhauser enhancements presented here should be useful to researchers who are contemplating studies of ^{13}C-enriched proteins. However, ^{13}C enrichment greatly reduces the sensitivity problems presented in Table I. Also, a protein which is ^{13}C enriched at *selected* carbons of *selected* amino acid residues should yield much simpler ^{13}C nmr spectra (with fewer assignment headaches) than a protein of natural isotopic composition.[97-106] Therefore, with selected ^{13}C enrichment, it becomes attractive to use ^{13}C nmr spectroscopy for studying larger proteins than those which are readily studied by natural-abundance ^{13}C nmr.[104] However, the ^{13}C nmr spectrum of a protein which has a high level ($\geqslant 30\%$) of *nonspecific* ^{13}C enrichment will be more complicated than the spectrum of a protein of natural isotopic composition, as a consequence of the splittings caused by $^{13}C-^{13}C$ scalar coupling.[142,143]

[110] N. A. Matwiyoff, P. J. Vergamini, T. E. Needham, C. T. Gregg, J. A. Volpe, and W. S. Caughey, *J. Am. Chem. Soc.* **95**, 4429 (1973).

[111] E. Antonini, M. Brunori, F. Conti, and G. Geraci, *FEBS Lett.* **34**, 69 (1973).

[112] C. H. Fung, A. S. Mildvan, A. Allerhand, R. Komoroski, and M. C. Scrutton, *Biochemistry* **12**, 620 (1973).

[113] A. M. Nigen, P. Keim, R. C. Marshall, J. S. Morrow, R. A. Vigna, and F. R. N. Gurd, *J. Biol. Chem.* **248**, 3724 (1973).

[114] P. Rodgers and G. C. K. Roberts, *FEBS Lett.* **36**, 330 (1973).

[115] J. Feeney, A. S. V. Burgen, and E. Grell, *Eur. J. Biochem.* **34**, 107 (1973).

[116] J. S. Morrow, P. Keim, R. B. Visscher, R. C. Marshall, and F. R. N. Gurd, *Proc. Natl. Acad. Sci. U.S.A.* **70**, 1414 (1973).

[117] C. F. Brewer, H. Sternlicht, D. M. Marcus, and A. P. Grollman, *Proc. Natl. Acad. Sci. U.S.A.* **70**, 1007 (1973).

[118] C. F. Brewer, H. Sternlicht, D. M. Marcus, and A. P. Grollman, *Biochemistry* **12**, 4448 (1973).

[119] S. H. Koenig, R. D. Brown, T. E. Needham, and N. A. Matwiyoff, *Biochem. Biophys. Res. Commun.* **53**, 624 (1973).

[120] J. S. Morrow, P. Keim, and F. R. N. Gurd, *J. Biol. Chem.* **249**, 7484 (1974).

[121] G. Robillard, E. Shaw, and R. G. Shulman, *Proc. Natl. Acad. Sci. U.S.A.* **71**, 2623 (1974).

[122] R. B. Moon and J. H. Richards, *Biochemistry,* **13**, 3437 (1974).

[123] R. B. Moon, M. J. Nelson, J. H. Richards, and D. F. Powars, *Physiol. Chem. Phys.* **6**, 31 (1974).

[124] D. C. Harris, G. A. Gray, and P. Aisen, *J. Biol. Chem.* **249**, 5261 (1974).

[125] H. M. Miziorko and A. S. Mildvan, *J. Biol. Chem.* **249**, 2743 (1974).

[126] B. Boettcher and M. Martinez-Carrion, *Biochem. Biophys. Res. Commun.* **64**, 28 (1975).

B. Carbon-13 Resonances of Ligands and Reporter Groups

Instead of observing the ^{13}C resonances of the protein itself, it can be quite profitable to examine the ^{13}C nmr spectrum of a ligand or covalently attached "reporter" group.[107-141] Obviously, it is attractive to use a ^{13}C-enriched ligand or "reporter" group.[107-129,131-141] In many cases, the ^{13}C-enriched species of interest is commercially available or readily synthesized.

[127] P. L. Yeagle, C. H. Lochmüller, and R. W. Henkens, *Proc. Natl. Acad. Sci. U.S.A.* **72,** 454 (1975).

[128] D. Mansuy, J.-Y. Lallemand, J.-C. Chottard, B. Cendrier, G. Gacon, and H. Wajcman, *Biochem. Biophys. Res. Commun.* **70,** 595 (1976).

[129] L. O. Morgan, R. T. Eakin, P. J. Vergamini, and N. A. Matwiyoff, *Biochemistry,* **15,** 2203 (1976).

[130] U. Kragh-Hansen and T. Riisom, *Eur. J. Biochem.* **70,** 15 (1976).

[131] M. F. Roberts, S. J. Opella, M. H. Schaffer, H. M. Phillips, and G. R. Stark, *J. Biol. Chem.* **251,** 5976 (1976).

[132] G. M. Giacometti, B. Giardina, M. Brunori, G. Giacometti, and G. Rigatti, *FEBS Lett.* **62,** 157 (1976).

[133] R. Banerjee, F. Stetzkowski, and J. M. Lhoste, *FEBS Lett.* **70,** 171 (1976).

[134] R. Banerjee and J.-M. Lhoste, *Eur. J. Biochem.* **67,** 349 (1976).

[135] J. S. Morrow, J. B. Matthew, R. J. Wittebort, and F. R. N. Gurd, *J. Biol. Chem.* **251,** 477 (1976).

[136] R. G. Khalifah, *Biochemistry* **16,** 2236 (1977).

[137] R. G. Khalifah, D. J. Strader, S. H. Bryant, and S. M. Gibson, *Biochemistry* **16,** 2241 (1977).

[138] P. J. Stein, S. P. Merrill, and R. W. Henkens, *J. Am. Chem. Soc.* **99,** 3194 (1977).

[139] C.-H. Niu, H. Shindo, J. S. Cohen, and M. Gross, *J. Am. Chem. Soc.* **99,** 3161 (1977).

[140] E. Stellwagen, L. M. Smith, R. Cass, R. Ledger, and H. Wilgus, *Biochemistry* **16,** 3672 (1977).

[141] R. B. Moon, K. Dill, and J. H. Richards, *Biochemistry* **16,** 221 (1977).

[142] J. A. Sogn, L. C. Craig, and W. A. Gibbons, *J. Am. Chem. Soc.* **96,** 4694 (1974).

[143] S. Tran-Dinh, S. Fermandjian, E. Sala, R. Mermet-Bouvier, and P. Fromageot, *J. Am. Chem. Soc.* **97,** 1267 (1975).

Author Index

Numbers in parentheses are footnote reference numbers and indicate that an author's work is referred to although his name is not cited in the text.

Subject Index

A

Acetic-water solution, density measurements on, 24

N-Acetyl-L-phenylalanine methyl ester, α-chymotrypsin reaction with, 325–326

Adsorption scanners, for ultracentrifuge, improvements in, 99–100

Aggregates, of enzymes, small-angle x-ray scattering studies on, 239–240

D-Alanyl-p-hydroxy-D-phenylglycine diketopiperazine, near uv circular dichroism of, 362–363

Albumin, erythrocyte binding of, calorimetry, 265

Alcohol dehydrogenase, reaction intermediates of, stopped-flow absorbance studies of, 319

Aldolase
microcalorimetric studies of binding by, 302
preferential solvent interaction of, 33
sedimentation coefficient of, 108

Aliphatic carbons, ^{13}C nmr in region of, 466–471

Alkaline phosphatase, spectrokinetic probe of, 323

Amino acids, aromatic, side chain structures of, 473

9-Aminoacridine, fluorescence emission spectra of, 403

Aminoacyl-tRNA, binding at recognition site, 250

Amphiphiles, protein interaction with, 60

Ampholines, tRNA binding of, 143–144, 147

Amplitude converter (TAC), 389

Amylase, effector changes on, x-ray, scattering studies on, 227

Analog to digital converter (ADC), in fluorescence measurements, 390

Analytical ultracentrifugation, protein-detergent interaction in, 61–62

2-Anilinonaphthalene, fluorescence measurements on, 406, 422

Anisotropy, calculation of, 348–349

Anthracene fluorescence decay of, 395

Antigen-antibody interactions, calorimetry, 268

Antithrombin III, heparin complex with, x-ray scattering studies on, 232

Anodes, for x-ray tubes, 183–184

Anton Paar precision density meter, 42

Apparent free energy, microcalorimetric determination of, 298–301

Argon ion laser, 3
polychromatic emission of, 7–8
for pulsed laser interferometry, 3–12

"Array detector method," in rapid scan spectrometry, 321

Arrhenius relationship, 323

N-Arylaminonaphthalene sulfonates, fluorescence decay data on, 423, 424

ASCII2 language, for computer used for dilution method, 95

Ascorbate, low-temperature dynamics of, 328

Asparaginase, sedimentation coefficient of, 108

Avidin, sedimentation coefficient of, 108

Azurin
^{13}C nmr spectroscopy of, 501, 514–515, 517
conformation studies, 534–536
summary, 533

B

Bacteriophage øCbK capsid, immune electron microscopy of, 257

Bacteriophage T4, immune electroscopy of proteins in, 250–251

Beer-Lambert law, 340
equation for, 70

Beneckea harveyi, growth of, calorimetric studies, 285

Benzene-cyclohexane solutions, density measurements on, 24

Bis-Tris, hemoglobin binding of, 87

Block camera, for small-angle x-ray scattering, 199–204
resolution by, 201–204

Bovine serum albumin

T